# WORKSHOP ON INSTABILITIES OF HIGH INTENSITY HADRON BEAMS IN RINGS

# Workshop on Instabilities of High Intensity Hadron Beams in Rings
## is supported by

# WORKSHOP ON INSTABILITIES OF HIGH INTENSITY HADRON BEAMS IN RINGS

Upton, New York   June/July 1999

EDITORS
T. Roser
S. Y. Zhang
*Brookhaven National Laboratory*

**American Institute of Physics**

AIP CONFERENCE PROCEEDINGS 496

Melville, New York

**Editors:**

T. Roser and S. Y. Zhang
Collider Accelerator Department
Building 911B
Brookhaven National Laboratory
Upton, NY 11973-5000
USA

E-mail:  roser@bnl.gov
         syzhang@bnl.gov

The articles on pp. 285–294 and 371–380 were authored by U. S. Government employees and are not covered by the below mentioned copyright.

Authorization to photocopy items for internal or personal use, beyond the free copying permitted under the 1978 U.S. Copyright Law (see statement below), is granted by the American Institute of Physics for users registered with the Copyright Clearance Center (CCC) Transactional Reporting Service, provided that the base fee of $15.00 per copy is paid directly to CCC, 222 Rosewood Drive, Danvers, MA 01923. For those organizations that have been granted a photocopy license by CCC, a separate system of payment has been arranged. The fee code for users of the Transactional Reporting Service is: 1-56396-910-6/99/$15.00.

© 1999 American Institute of Physics

Individual readers of this volume and nonprofit libraries, acting for them, are permitted to make fair use of the material in it, such as copying an article for use in teaching or research. Permission is granted to quote from this volume in scientific work with the customary acknowledgment of the source. To reprint a figure, table, or other excerpt requires the consent of one of the original authors and notification to AIP. Republication or systematic or multiple reproduction of any material in this volume is permitted only under license from AIP. Address inquiries to Office of Rights and Permissions, Suite 1NO1, 2 Huntington Quadrangle, Melville, N.Y. 11747-4502; phone: 516-576-2268; fax: 516-576-2450; e-mail: rights@aip.org.

L.C. Catalog Card No. 99-068238
ISBN 1-56396-910-6
ISSN 0094-243X
Printed in the United States of America

## CONTENTS

**Preface** .................................................................. ix
**List of Participants** ............................................... xi

### INVITED TALKS

**Overview of Beam Instabilities** (*Plenary*) ................................. 3
    A. Hofmann
**Aspects of Beam Stability at ISIS** ........................................ 17
    G. H. Rees
**High Intensity Beam Operation of the Brookhaven AGS** ...................... 22
    T. Roser
**Proton Driver Study at Fermilab** .......................................... 30
    W. Chou
**Collective Instabilities in the LHC: Electron Cloud
and Satellite Bunches** ..................................................... 40
    F. Ruggiero and X. Zhang
**Stability Issues of Low Energy Intense Beams** ............................. 49
    K. Y. Ng and A. V. Burov
**Impedance Issues in the CERN SPS** ........................................ 64
    T. Linnecar
**Analytic Methods for Impedance Calculations** ............................. 77
    R. L. Gluckstern and A. V. Fedotov
**Nonlinear Features of the Longitudinal Instability for
High-Current Machines** .................................................... 86
    I. Hofmann and O. Boine-Frankenheim
**Impedances and Wakes in High-Energy Proton Accelerators** (*Plenary*) ...... 99
    B. Zotter
**Collective Effects in the CERN-PS Beam for LHC** .......................... 116
    R. Cappi, R. Garoby, and E. Métral
**Impedance Budget and Beam Instabilities of the JHF 50-GeV
Proton Synchrotron** ....................................................... 131
    Y. Mori and M. Yoshii
**Instability Issues at the SNS Storage Ring** .............................. 136
    S. Y. Zhang
**Instability Issues for ESS Linac and Rings** .............................. 151
    G. H. Rees
**Longitudinal Impedance Tuner Using a New Material, FINEMET** .............. 169
    K. Koba
**Head-Tail Instability and Microwave Instability in the KEK-PS** ........... 182
    T. Toyama, D. Arakawa, S. Igarashi, J. Kishiro, K. Koba,
    E. Nakamura, K. Takayama, and M. Yoshii
**Intensity Dependent Effects in RHIC** ..................................... 197
    J. Wei

Short Bunch Production and Microwave Instability Near Transition ......... 213
    K. Y. Ng and J. Norem

## CONTRIBUTED PAPERS

Convergence of Basis Expansions ......................................... 231
    M. Blaskiewicz
Non-Linear Mode Coupling and Saw-Tooth Instability ..................... 240
    S. Heifets
Analysis of Coupled Bunch Instability Spectra ........................... 256
    E. Shaposhnikova
The Coupling Impedance of the RHIC Injection Kicker System ............. 266
    H. Hahn
Longitudinal Space Charge Impedance. ..................................... 276
    J. G. Wang
Impedance Considerations for the Intense Pulsed Neutron Source (IPNS)
Rapid Cycling Synchrotron (RCS) ........................................ 285
    J. C. Dooling, F. R. Brumwell, and G. E. McMichael
3D Multispecies Nonlinear Perturbative Particle Simulations
of Collective Instabilities in Intense Particle Beams ....................... 295
    H. Qin, R. C. Davidson, and W. W. Lee
A Simple Simulation of Electron-Proton Instability ........................ 305
    T. F. Wang
Multipacting on the Trailing Edge of Proton Beam Bunches
in the PSR and SNS..................................................... 315
    V. Danilov, A. Aleksandrov, J. Galambos, D. Jeon, J. Holmes,
    and D. Olsen
The Fast Loss Electron Proton Instability ................................ 321
    M. Blaskiewicz
Longitudinal Relaxation Oscillations Induced by HOM's .................. 331
    J. Sebek and C. Limborg
Surface Impedance and Synchronous Modes. ............................. 341
    G. V. Stupakov
Nonlinear Longitudinal Waves in High Energy Stored Beams .............. 351
    S. I. Tzenov.
Space Charge Impedance Calculations in Long-Wavelength
Approximation ......................................................... 361
    S. S. Kurennoy
Bunch Stabilization Using rf Phase Modulation in the
Intense Pulse Neutron Source (IPNS) Rapid Cycling
Synchrotron (RCS)..................................................... 371
    J. C. Dooling, F. R. Brumwell, and G. E. McMichael
Comment to the Kinematics of e-p Multipactoring. ....................... 381
    S. Heifets, G. Stupakov, and V. Danilov

## SUMMARIES OF WORKING GROUPS

**Impedance Group Summary** .................................................... 385
    M. Blaskiewicz, J. Dooling, M. Dyachkov, A. Fedotov,
    R. Gluckstern, H. Hahn, H. Huang, S. Kurennoy,
    T. Linnecar, E. Shaposhnikova, G. Stupakov, T. Toyama,
    J. G. Wang, W. T. Weng, S. Y. Zhang, and B. Zotter

**Summary of Instabilities and Damping Group** ........................... 391
    S. Koscielniak

**Short High-Intensity Bunches Group Summary** .......................... 407
    M. Brennan, K. Brown, R. Cappi, W. Chou, T. Linnecar,
    C. Prior, G. Rees, T. Roser, E. Shaposhnikova, M. Yoshi,
    and W. Van Asselt

**Author Index** .................................................................... 411

# PREFACE

Hadron beams with intensities exceeding present performance records are a central feature of many planned and proposed new facilities, such as SNS, ESS, JHF, NSP at JAERI, and the Muon-Collider proton driver, as well as of possible upgrades of existing facilities, such as the AGS as proton driver and CERN-PS as spallation driving facility. To examine the beam dynamics of these high intensity machines, the Workshop on Instabilities of High Intensity Hadron Beams in Rings has been held at Brookhaven National Laboratory from June 28 to July 1, 1999.

The workshop was devoted to:

1. Instability issues associated with high intensity beams.
2. Evaluate beam instabilities in the SNS storage ring, proton driver of a muon collider, and other hadron facilities.
3. Develop understanding of instability issues from comparison of theoretical models and machine measurements.

The workshop consisted of invited talks and three working groups on impedance issues, instability thresholds and damping, and the production of intense short bunches.

Although the main focus of the workshop was on high intensity hadron machines operating below or around transition energy, instability issues from other machines were also included.

Prior to the workshop, a set of topics was developed and circulated among the potential attendants of the workshop. During the workshop these topics were used as the guidance for the working group discussions. These workshop topics are:

1. Impedance sessions:
   - High intensity proton rings will require large apertures. Are impedance calculations reliable for large vacuum chambers, large steps, and large aperture kickers?
   - Is it beneficial to have the rf shielding or vacuum chamber follow the betatron envelope?
   - What is the impedance of ceramic chambers with or without metallic strips?
   - Is it beneficial to reduce the broad band impedance to a few Ohm for high intensity proton machines?
   - How do ferrite window frame, C frame, traveling wave, and stripline kickers compare in terms of impedance and engineering requirements?
   - Is it practical and/or useful to compensate the longitudinal space charge impedance?
   - What are the best methods to measure longitudinal and transverse impedances?
   - What are the key issues for the impedance budget for a) high power, b) high peak current (short bunch), c) very low loss hadron machines?

2. Instability sessions:
   - Stability against longitudinal microwave instability at high intensity: is a large momentum spread sufficient to stabilize the beam?
   - Transverse stability at high intensity: what is the effect of space charge?
   - What is an appropriate description of fast transverse instabilities for long bunches and large space charge tune shifts?
   - What is an appropriate description of fast transverse instabilities for short bunches and very large space charge tune shifts?
   - What is the effect of uneven longitudinal phase space distributions?
   - Do space charge stabilized "hot spots" exist and do they affect overall beam stability?
   - e-p instability: How to identify it and how to cure it.

3. Short bunches session:
   - For the muon collider proton driver, short bunches with 1 ns rms length and 5e13 protons at about 20 GeV are required. How can this be achieved and what techniques are most promising (operation near transition, bunch rotation, high rf voltage, ... )?
   - What are the requirements on impedance?
   - What are the relevant stability criteria and growth rates?
   - What can be learned from the experience with short electron bunches?

Very active discussions have produced ample results, which are included in the summaries for the working groups. As expected, some issues in the topics have been resolved, but most of the issues still lack clear cut solutions. The discussions will continue, beyond the workshop time period.

The Organizing Committee of the workshop was: J. Alonso (ORNL), C. Ankenbrandt (Fermilab), R. Cappi (CERN), A. Chao (SLAC), W. Chou (Fermilab), R. Macek (LANL), Y. Mori (KEK), G. Rees (RAL), T. Roser (Chair), F. Ruggiero (CERN), W.T. Weng (BNL), and S.Y. Zhang (Co-chair).

Mary Campbell served as the secretary of the workshop.

The active support for the workshop given by the SNS project and the Brookhaven National Laboratory are greatly appreciated.

T. Roser and S.Y. Zhang

# LIST OF PARTICIPANTS

Dan Abell
Brookhaven National Laboratory
Bldg. 911B
Upton, NY 11973-5000
dabell@bnl.gov

Mei Bai
Brookhaven National Laboratory
Bldg. 1005S
Upton, NY 11973-5000
mbai@bnl.gov

Mike Blaskiewicz
Brookhaven National Laboratory
Bldg. 911B
Upton, NY 11973-5000
mmb@bnl.gov

Mike Brennan
Brookhaven National Laboratory
Bldg. 911B
Upton, NY 11973-5000
brennan@bnl.gov

Roberto Cappi
CERN, PS Division
CH-1211 Geneva 23
Switzerland
roberto.cappi@cern.ch

Alex Chao
SLAC
P.O.Box 4349
Stanford, CA 94309
achao@slac.stanford.edu

Weiren Chou
Fermilab, MS 220
PO Box 500
Batavia, IL 60510
chou@fnal.gov

Mickael Craddock
TRIUMF, 4004 Wesbrook Mall
Vancouver BC
Canada V6T 2A3
craddock@triumf.ca

Slava Danilov
Oak Ridge National Laboratory
104 Union Valley Road
Oak Ridge, TN 37830
danilovs@ornl.gov

Ronald Davidson
PPPL, Plasma Physics Laboratory
Princeton University
PO Box 451, Princeton, NJ 08543
rdavidson@pppl.gov

Jeffrey Dooling
Argonne National Laboratory
360 IPNS, 9700 S. Cass Ave.
Argonne, IL 60439
jcdooling@anl.gov

Mikhail D'yachkov
TRIUMF, 4004 Wesbrook Mall
Vancouver BC
Canada V6T 2A3
dyachkov@triumf.ca

Alexey Fedotov
Brookhaven National Laboratory
Bldg. 911A
Upton, NY 11973-5000
fedotov@bnl.gov

Robert Gluckstern
Univ. of Maryland
801 Drottomer Place
Baltimore, MD 21210
rlg@physics.umd.edu

Harald Hahn
Brookhaven National Laboratory
Bldg. 1005S
Upton, NY 11973-5000
hahnh@bnl.gov

Robert Hardekopf
LANL, SNS Project Office
LANSCE-DO, MS-H824
Los Alamos, NM 87545
hardekopf@lanl.gov

Samuel Heifets
SLAC
P.O. BOX 4349
Stanford, CA 94309
heifets@slac.stanford.edu

Albert Hofmann
CERN, SL Division
CH-1211 Geneva 23
Switzerland
albert.hofmann@cern.ch

Ingo Hofmann
GSI, Planckstr. 1
64291 Darmstadt
Germany
i.hofmann@gsi.de

Haixin Huang
Brookhaven National Laboratory
Bldg. 911B
Upton, NY 11973-5000
huanghai@bnl.gov

Kiyomo Koba
KEK, 1-1 Oho
Tsukuba, Ibraaki 305-0801
Japan
koba@hatokyo9.tanashi.kek.jp

Shane Koscielniak
TRIUMF, 4004 Wesbrook Mall
Vancouver BC
Canada V6T 2A3
shane.koscielniak@triumf.ca

Sergey Kurennoy
LANL
MS H808
Los Alamos, NM 87545
kurennoy@lanl.gov

S.Y. Lee
Indiana University
2401 Milo B. Sampson Lane,
IUCF, Bloomington, IN 47408
shylee@indiana.edu

Y.Y. Lee
Brookhaven National Laboratory
Bldg. 911B
Upton, NY 11973-5000
yylee@bnl.gov

Andreas Lehrach
Brookhaven National Laboratory
Bldg. 911B
Upton, NY 11973-5000
lehrach@bnl.gov

Trevor Linnecar
CERN, SL Division
CH-1211 Geneva 23
Switzerland
trevor.linnecar@cern.ch

Robert Macek
LANL, LANSCE-DO
MS-H848
Los Alamos, NM87545
macek@lanl.gov

Elias Metral
CERN, PS Division
CH-1211 Geneva 23
Switzerland
elias.metral@cern.ch

K.Y. Ng
Fermilab, MS 220
PO Box 500
Batavia, IL 60510
ng@fnal.gov

James Niederer
Brookhaven National Laboratory
Bldg. 911B
Upton, NY 11973-5000
jn@bnl.gov

Robert Palmer
Brookhaven National Laboratory
Bldg. 901A
Upton, NY 11973-5000
palmer@bnl.gov

Yannis Papaphilippou
CERN, SL Division
CH-1211 Geneva 23
Switzerland
yannis@mail.cern.ch

Christopher Prior
Rutherford Appleton Laboraory
Chilton, Didcot, Oxfordshire
OX11 0QX, United Kingdom
c.r.prior@rl.ac.uk

Hong Qin
PPPL, Plasma Physics Laboratory
Princeton University, PO Box 451
Princeton, NJ 08543
hongqin@pppl.gov

Graheme H. Rees
Rutherford Appleton Laboraory
Chilton, Didcot, Oxfordshire
OX11 0QX, United Kingdom
g.h.rees@rl.ac.uk

Thomas Roser
Brookhaven National Laboratory
Bldg. 911B
Upton, NY 11973-5000
roser@bnl.gov

Alessandro Ruggiero
Brookhaven National Laboratory
Bldg. 911B
Upton, NY 11973-5000
agr@bnl.gov

Francesco Ruggiero
CERN, SL Division
CH-1211 Geneva 23
Switzerland
francesco.ruggiero@cern.ch

James Sebek
SSRL/SLAC
P.O. Box 4349, MS 99
Stanford, CA 94309
sebek@slac.stanford.edu

Elena Shaposhnikova
CERN, SL Division
CH-1211 Geneva 23
Switzerland
elena.chapochnikova@cern.ch

Gennady Stupakov
SLAC
P.O.Box 4349
Stanford, CA 94309
stupakov@slac.stanford.edu

Lee Teng
Argonne National Laboratory
Bldg. 401, 9700 S. Cass Ave.
Argonne, IL 60439
teng@aps.anl.gov

Takeshi Toyama
KEK, 1-1 Oho, Tsukuba
Ibaraki 305-0801
Japan
takeshi.toyama@kek.jp

Stephan Tzenov
SLAC
MS 26
Stanford, CA 94309
tzenov@SLAC.Stanford.edu

Ryuichi Ueno
KEK, 1-1 Oho
Tsukuba, Ibraaki 305-0801
Japan

Willem van Asselt
Brookhaven National Laboratory
Bldg. 911B
Upton, NY 11973-5000
vanasselt@bnl.gov

J.G. Wang
Oak Ridge National Laboratory/BNL
Bldg. 911B
Upton, NY 11973
jgwang@bnl.gov

T.F. Wang
LANL
LANSCE-1, MS H808
Los Alamos, NM 87545
twang@lanl.gov

Chris Warsop
Rutherford Appleton Laboraory
Chilton, Didcot, Oxfordshire
OX11 0QX, United Kingdom
c.m.warsop@rl.ac.uk

Jie Wei
Brookhaven National Laboratory
Bldg. 911B
Upton, NY 11973-5000
wei1@bnl.gov

Bill Weng
Brookhaven National Laboratory
Bldg. 911B
Upton, NY 11973-5000
weng@bnl.gov

Masahito Yoshi
KEK, 1-1 Oho
Tsukuba
Ibraaki 305-0801
Japan
masahito.yoshii@kek.jp

S.Y. Zhang
Brookhaven National Laboratory
Bldg. 911B
Upton, NY 11973-5000
syzhang@bnl.gov

Bruno Zotter
CERN, SL Division
CH-1211 Geneva 23
Switzerland
bruno.zotter@cern.ch

# INVITED TALKS

# Overview of beam instabilities

## A. HOFMANN

## 1  Introduction:

The motion of a single particle in a ring is determined by external guide fields (dipole and quadrupole magnets, RF-system) and initial conditions. The many particles in a high intensity beam represent themselves charges and currents which are sources of electromagnetic fields (self-fields). They are modified by the boundary conditions (impedance) of the beam surroundings (vacuum chambers, cavities, etc.) and act back on the beam. This can lead to a frequency shift (change of the betatron or synchrotron frequency), to an increase of a small perturbation of the beam, i.e. an instability, or to a change of the particle distribution, like bunch lengthening.

If the self-fields are small compared to the guide fields their effect is treated as a perturbation. In some cases, like bunch lengthening, a self-consistent distribution has to be found.

Multi-traversal effects require an impedance with memory, usually a narrow band cavity. Single traversal effects can be driven by a broad-band impedance. The two effects are usually treated separately, except for continuous (unbunched or coasting) beams. These instabilities can be longitudinal or transverse, involving synchrotron or betatron oscillations.

## 2  Longitudinal coupled bunch instability, Robinson instability

### 2.1  Longitudinal dynamics

The longitudinal dynamics is based on the relation between the deviations from nominal energy $E$ and revolution frequency $\omega_0$ expressed by the momentum compaction $\alpha_c$.

$$\frac{\Delta E}{E} = \beta^2 \frac{\Delta p}{p} = -\frac{\beta^2}{\eta_c} \frac{\Delta \omega_0}{\omega_0}, \text{ with } \eta_c = \alpha_c - \frac{1}{\gamma^2}$$

The RF-cavity oscillates with harmonic frequency $\omega_{RF} = h\omega_0$ giving a particle arriving at the synchronous time $t_s$ or phase $\phi_s = \omega_{RF} t_s$ an energy $U_s = e\hat{V} \sin\phi_s$

CP496, *Workshop on Instabilities of High Intensity Hadron Beams in Rings*, edited by T. Roser and S. Y. Zhang
© 1999 American Institute of Physics 1-56396-910-6/99/$15.00

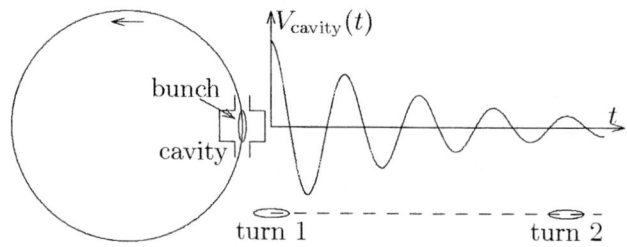

Figure 1: Induced field acting over one turn

per turn to compensate losses. Particles arriving before or after $t_s$ receive a different energy. This, together with the energy dependence of $\omega_s$, leads to longitudinal focusing. Particles execute synchrotron or energy oscillations with frequency $\omega_s$

$$V_{RF}(t) = \hat{V}\sin(h\omega_0 t)\,, \quad \omega_s^2 = -\omega_0^2 \frac{\eta_c he\hat{V}\cos\phi_s}{2\pi\beta^2 E} = Q_s^2 \omega_0^2.$$

Above transition energy we have $\eta_c > 0$ and $\cos\phi_s < 0$, below transition energy $\eta_c < 0$ and $\cos\phi_s > 0$.

## 2.2 Qualitative treatment

A bunch circulating in a storage ring induces a voltage in a cavity which can change its energy on the next turn, Fig. 1. This can increase the synchrotron oscillation amplitude and lead to an instability. We consider in the following a single, narrow band cavity resonance with shunt impedance $R_s$, resonant frequency $\omega_r \approx p\omega_0$ quality factor $Q$. An oscillating current $I(t) = \hat{I}\cos(\omega t)$ induces a voltage $V(t)$ in the cavity

$$V(t) = \hat{I}\left(Z_r \cos(p\omega_0 t) - Z_i \sin(p\omega_0 t)\right) = \hat{I}R_s \frac{\cos(\omega_p t) - Q\frac{\omega_r^2 - \omega_p^2)}{\omega_r \omega_p}\sin(\omega_p t)}{1 + Q^2\left(\frac{\omega_r^2 - \omega_p^2}{\omega_r \omega_p}\right)^2}$$

In Fig. 2 we consider now the circulating bunch with nominal revolution frequency $\omega_0$ which is executing a synchrotron oscillation $\Delta E = \widehat{\Delta E}\cos(\omega_s t)$, above transition, $\eta_c > 0$ and a cavity with its resonant frequency below the revolution harmonics $\omega_r < p\omega_0$, (A). At the moment when the bunch has an excess energy $\Delta E > 0$ its revolution frequency harmonic is below the nominal one and it sees a large cavity impedance giving a large energy loss. Half a synchrotron oscillation period later the energy is low $\Delta E < 0$ and the revolution frequency is above the nominal one. The bunch sees a low impedance and suffers a small energy loss. The energy loss

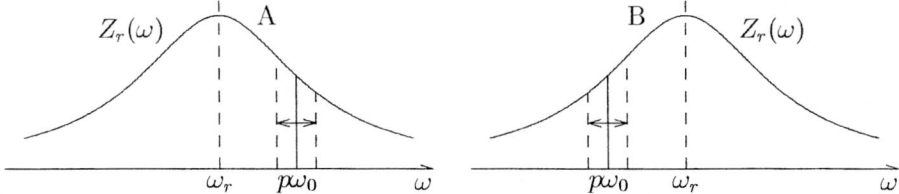

Figure 2: Qualitative understanding of the Robinson instability

of the bunch is large when its energy is high and the loss its small when the energy is low. This leads to damping of the energy oscillation.

We now tune the cavity to a resonant frequency being below the nominal revolution harmonics $\omega_r > p\omega_0$, (B). At the moment the energy of the bunch is high $\Delta E > 0$, its revolution frequency harmonic is below the nominal one $\Delta\omega_0 < 0$. Here, the impedance and the energy loss in the cavity are small. Half a synchrotron oscillation period later we have $\Delta E < 0$ and $\Delta\omega_0 > 0$. Here, impedance and energy loss in the cavity are large. The bunch looses more energy when it does not have enough and less when it has too much. This leads to an instability.

Below transition energy the situation is reversed.

## 2.3 Quantitative results

We consider now a bunch of average current $I_0$ executing as a whole a synchrotron oscillation shown on top of Fig 3. The arriving time at a certain location in the ring is modulated with the synchrotron oscillation frequency $\omega_s = \omega_0 Q_s$. Its deviation $\tau$ from the synchronous time in successive turns $k$ is

$$\tau(k) = \hat{\tau}\cos(2\pi Q_s k)$$

This phase modulation leads to a spectrum having sidebands at distance $\omega_s$ from the revolution harmonics $p\omega_0$. The current component at these sidebands, for a normalized oscillation amplitude, is given by the Fourier transform of the bunch current $I(t)$ at these frequencies.

$$\omega_{p\pm} = \omega_0(p \pm Q_s), \quad I_{p\pm} = \frac{\omega_0}{\sqrt{2\pi}}\tilde{I}(\omega_{p\pm}), \quad \tilde{I}(\omega) = \frac{1}{\sqrt{2\pi}}\int_{-\infty}^{\infty} I(t)e^{-\omega t}dt.$$

We assume a cavity with impedance having a resistive and reactive part expressed in complex notation $Z(\omega) = Z_r(\omega) + jZ_i(\omega)$ tuned close to a revolution harmonics $p\omega_0$ with a bandwidth sufficiently narrow that only one harmonic with sidebands is covered by the impedance. The sideband currents induce a voltage in the cavity impedance which has to be included in the longitudinal dynamics. This leads to a synchrotron oscillation with an exponential growth (or damping) rate and frequency shift

$$\Delta E = \widehat{\Delta E}e^{-\Delta\omega_{si}t}\cos\left((\omega_s + \Delta\omega_{sr})t\right)$$

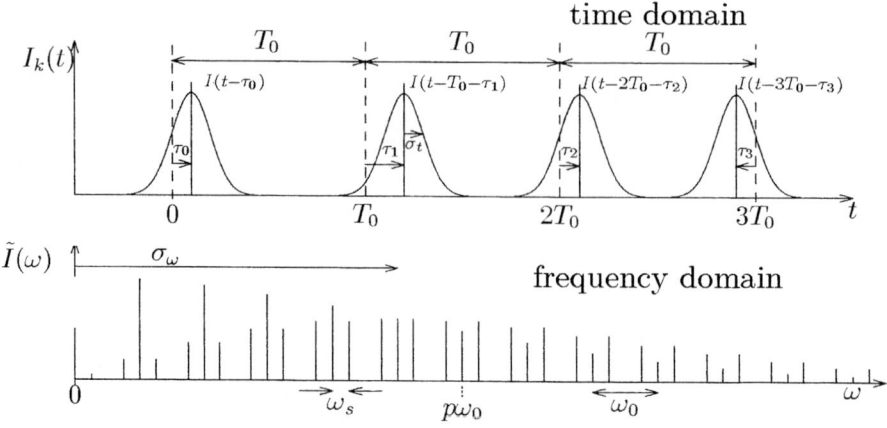

Figure 3: Oscillating bunch in time and frequency domain

$$\frac{1}{\tau_s} = \Delta\omega_{si} = \omega_s \frac{\left((p+Q_s)I_{p+}^2 Z_r(\omega_{p+}) - (p-Q_s)I_{p-}^2 Z_r(\omega_{p-})\right)}{8hI_0\hat{V}\cos\phi_s}$$

$$\Delta\omega_{sr} = \omega_s \frac{\left((p+Q_s)I_{p+}^2 Z_i(\omega_{p+}) + (p-Q_s)I_{p-}^2 Z_i(\omega_{p-})\right)}{8hI_0\hat{V}\cos\phi_s}$$

In agreement with the qualitative treatment we find stability above transition energy ($\cos\phi_s < 0$) if the impedance is lower at the higher frequency (upper sideband) and higher at the lower sideband.

The growth (or damping) rate depends on the difference in resistive impedance at the two sidebands as illustrated in Fig. 4. This can be understood qualitatively from Fig. 5 where we consider a synchrotron oscillation with the special tune $Q_s = 0.25$ shown in phase space with coordinates $\tau$ and $\epsilon = \Delta E/E$. This oscillation can be split into a stationary bunch plus a perturbation with no total charge but a dipole moment oscillating with $\omega_s$. The stationary part will induce a voltage at $p\omega_0$ which can lead to a fixed energy loss and, due to its slope, a change in synchrotron frequency of the particles inside the bunch. The oscillating dipole however induces fields at the two sidebands. We take a narrow band cavity and tune it first to the upper sideband and get the field shown in the upper curve of the figure. Tuning it to the lower sideband gives the induced field shown in the lower curve of the figure. We start with the case above transition energy. Here, the voltage induced by the oscillating bunch will be positive a turn later and enhance the energy deviation of the bunch. At the lower sideband the voltage after one turn is negative and reduces the energy deviation. The upper sideband has a destabilizing the lower on a stabilizing effect. Below transition energy the phase space motion goes in the other direction which reverses the situation. We have chosen here a special

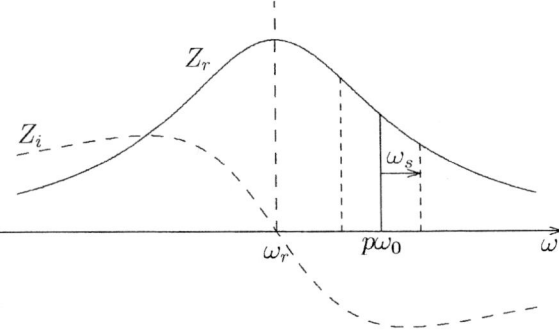

Figure 4: Growth rate given by the resistive impedance at the two sidebands

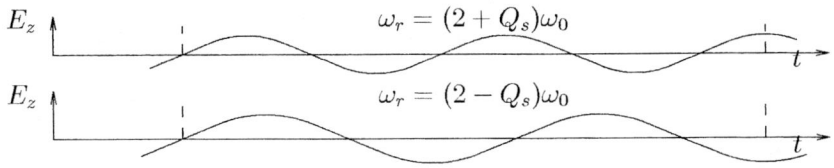

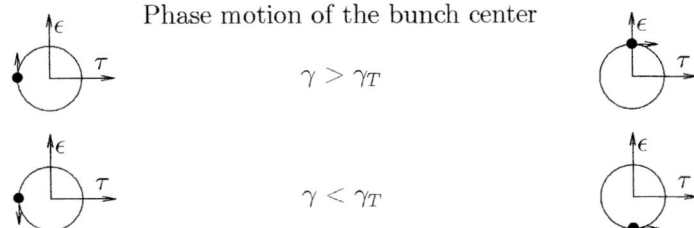

Figure 5: Effect of the induced voltage at the sidebands on stability

synchrotron frequency $\omega_s = 0.25\omega_0$. This makes the drawing and the explanation easier but it contains the physics of the general case.

## 2.4 Generalization

**General impedance:** We consider now a broader impedance which covers several revolution harmonics as shown in Fig. 6. Here, we have to take a sum over all impedance differences between the sidebands pairs.

$$\frac{1}{\tau_s} = \frac{\omega_s}{8hI_0\hat{V}\cos\phi_s} \sum_p \left((p+Q_s)I_{p+}^2 Z_r(\omega_{p+}) - (p-Q_s)I_{p-}^2 Z_r(\omega_{p-})\right) \quad (1)$$

and a corresponding expression for the frequency shift $\Delta\omega_{sr}$.

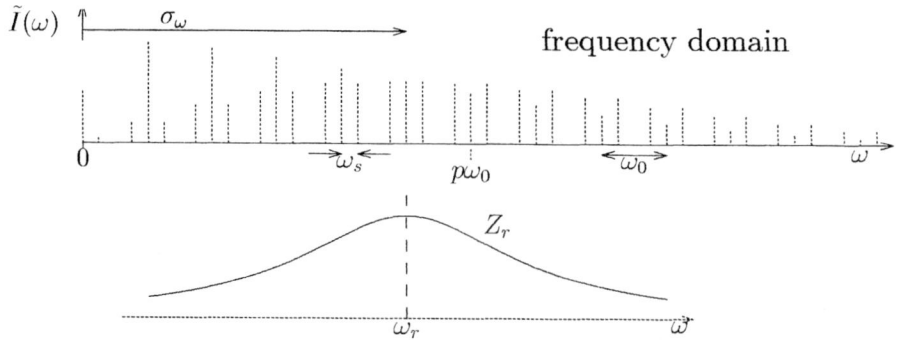

Figure 6: More general impedance covering many sidebands

These expressions become more compact using a complex notation for the impedance and frequency shift

$$Z(\omega) = Z_r(\omega) + jZ_i(\omega) \,, \quad \Delta\omega = \Delta\omega_{sr} + j\Delta\omega_{si} \,, \quad \text{with } e^{j\omega t}, \quad -\infty < \omega < \infty.$$

**Many bunches:** With $M$ equidistant bunches in the ring there are $M$ independent coupled bunch oscillations modes denoted by the mode number $0 \leq n \leq M-1$ related to the phase difference $\Delta\phi$ between the oscillations of adjacent bunches $n = \Delta\phi/(2\pi M)$. Each mode $n$ has one pair of sidebands in each frequency range of $M\omega_0$ at

$$\omega_{p\pm} = \omega_0(pM \pm (n + Q_s))$$

The growth rate is given by (1) using the above frequencies of each sideband pair.

**Bunch shape oscillations:** In addition to the rigid dipole modes ($m = 1$) there are bunch shape oscillations, quadrupole mode ($m = 2$), sextupole mode ($m = 3$), ... with the frequencies

$$\omega_{p\pm} = \omega_0(pM \pm (n + mQ_s)).$$

# 3 Transverse coupled bunch instabilities

## 3.1 Transverse impedance

Cavities can have transversely deflecting modes, Fig 7. They are excited by the beam through the longitudinal electric field which is converted into a deflecting magnetic field a quarter of an oscillation period $T_r/4$ later. The impedance is given by an integral around the ring over the Fourier component of the deflecting fields divided by the Fourier component of the dipole moment. The $'j'$ indicates that deflection and dipole moment are out of phase. The expression becomes real if we

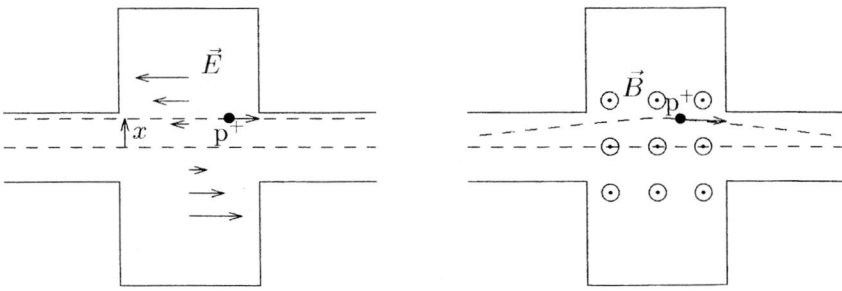

Figure 7: Transverse impedance

relate the deflection to the time derivative of the dipole moment

$$Z_T(\omega) = j\frac{\oint \left(\vec{E}(\omega) + [\vec{\beta}c \times \vec{B}(\omega)]\right)_T}{Iy(\omega)} = -\frac{\omega \oint \left(\vec{E}(\omega) + [\vec{\beta} \times \vec{B}(\omega)]\right)_T ds}{I\dot{y}(\omega)}$$

A single bunch executing a betatron oscillation will have a phase advance $2\pi Q_\beta$ each successive revolution where $Q_\beta$ is the betatron tune in the two planes. At a single location around the ring the integer part of the tune cannot be observed and we use $Q_\beta = \text{integer} + q$. Similar to the longitudinal case a single oscillating bunch excites a cavity at frequencies $\omega_0(p \pm q)$. For $M$ equidistant bunches there are $M$ coupled modes possible labeled by $0 \leq n \leq M-1$ with frequencies

$$\omega_{p\pm} = \omega_0 \left(pM \pm (n+q)\right)$$

A bunch passes with a displacement $x$ through the cavity and excites a field $\vec{E}$ which is converted after time intervals of $T_r/4$ into a magnetic field $-\vec{B}$, then an electric field $-\vec{E}$ and after again into a magnetic field $\vec{B}$ as shown in Fig. 8. The oscillating bunch has sidebands at $\omega_0(\text{integer} \pm q)$, (we take $q = 1/4$ in the figure). With the cavity tuned to the upper sideband the bunch will traverse it in the next turn at $t = T_r(k + 1/4)$ with a transverse velocity in the $-x$ direction and receive by the magnetic field a force in the opposite direction which damps the oscillation. With the cavity tuned to the lower sideband the bunch traverses at $t = T_r(k' + 3/4)$ with a negative velocity and receive a force in the same direction. Like in the longitudinal case the resistive impedances at the two sidebands have opposite effects leading to a similar expression for the growth or damping rate.

$$\frac{1}{\tau_s} \propto \sum_p \left(I_{p+}^2 (Z_T(\omega_{p+}) - I_{p-}^2 Z_T(\omega_{p-})\right) , \quad \omega_{p\pm} = \omega_0 \left(pM \pm (n+q)\right).$$

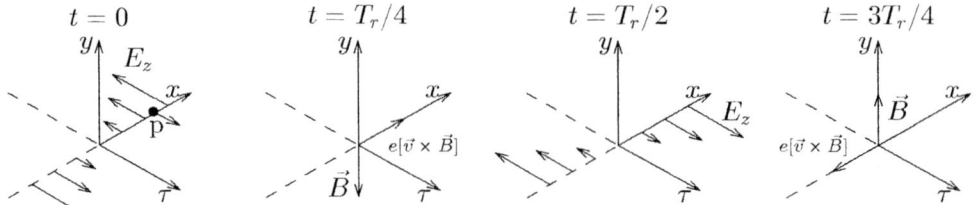

Figure 8: Development of the fields excited by a bunch as a function of time

## 4 Single traversal head-tail instability

The transverse head-tail instability involves a coherent transverse bunch motion combined with the incoherent longitudinal motion. The synchrotron motion of a particle represents an ellipse in the phase space of energy $\Delta E$ and time deviation $\tau$. A transverse oscillation is influenced through the chromaticity $Q' = dQ/(dp/p)$. A particle going from head to tail of the bunch due to a synchrotron oscillation has for ($\gamma_T > 0$) an excess energy and, for $Q' > 0$, a higher betatron frequency. This is reversed in the second half of the oscillation, going from tail to head. Consequently there is a betatron phase advance between head and tail for the coherent transverse oscillations. The detailed phase space motion is shown in Fig. 9 for a vertical oscillation of a bunch with eight particles having all the same synchrotron oscillation amplitude for a vanishing and for a finite chromaticity.

The chromatic phase along the bunch represents a frequency $\omega_\xi = \omega_0 Q'/\eta_c$ which shifts the envelope of the sidebands. This is illustrated in Fig. 10 first by using positive and negative frequencies with only upper sidebands and at the bottom with positive frequencies only and having two sidebands.

We consider a broad band transverse cavity mode which is excited by oscillating particles $A$ at the head of the bunch. Even if the mode is broad band, with a memory only a little longer than the bunch length, the particles B at the tail of the bunch can get now excited and oscillate with a certain phase $\Delta\phi$ compared to the head. After half a synchrotron oscillation period the particles $B$ are at the head and the particles $A$ at the tail oscillating with a phase $-\Delta\phi$ compared to $B$ assuming $Q' = 0$. The excitation by the head has now the wrong phase to excite the tail such that the oscillation keeps growing unless the chromaticity is finite and provides the correct phase shift between head and tail. In this case a head-tail instability can occur. It is usually calculated in frequency domain using the sideband spectrum shown in Fig. 10 with its envelope shifted by the frequency $\omega_\xi = Q'\omega_0/\eta_c$ giving the current components and the growth rate

$$I_{p\xi\pm} = \frac{\omega_0}{\sqrt{2\pi}} \tilde{I}(\omega_{p\pm} \pm \omega_\xi), \quad \omega_{p\pm} = \omega_0 (pM \pm (n+q))$$

$$\frac{1}{\tau_s} \propto \sum_p \left( I_{p\xi+}^2 (Z_T(\omega_{p+}) - I_{p\xi-}^2 Z_T(\omega_{p-}) \right).$$

Since the current components of adjacent sidebands can be very different even a broad band impedance can cause an instability. The memory necessary for an instability is now in the dynamic of the internal bunch motion and not in the impedance.

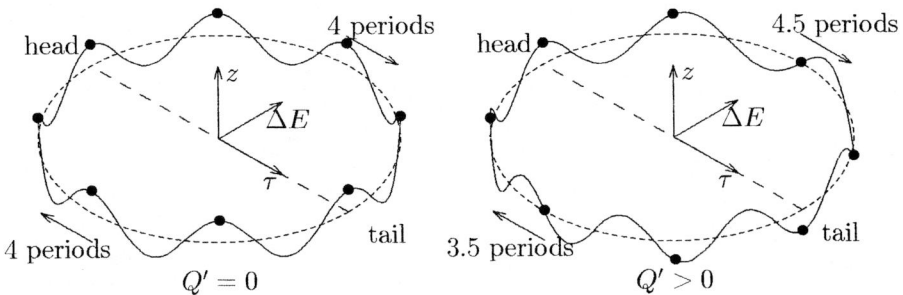

Figure 9: Combined transverse and longitudinal motion for two chromaticities

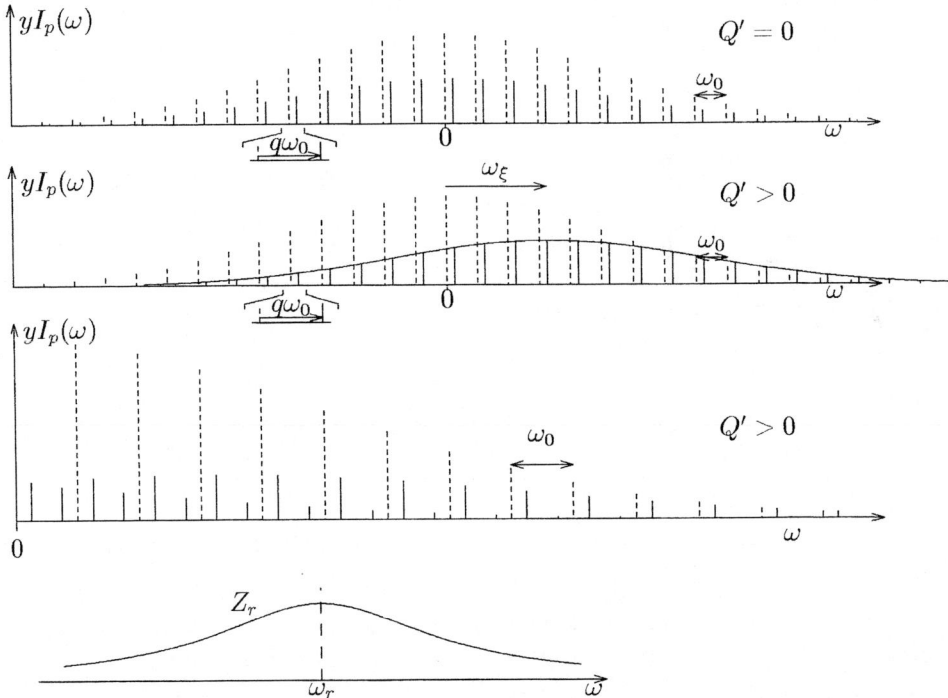

Figure 10: Spectrum of an oscillation bunch for vanishing and finite chromaticity

# 5 Coasting beam instabilities

## 5.1 Transverse coasting beam instabilities

The transverse betatron frequencies of a particle with nominal momentum are

$$\omega_{\beta f} = (n_f + Q)\omega_0 \ , \ \omega_{\beta s} = (n_s - Q)\omega_0.$$

Through

$$\frac{\Delta E}{E} = \beta^2 \frac{\Delta p}{p} = -\frac{\beta^2}{\eta_c} \frac{\Delta \omega_0}{\omega_0} \ , \text{ and } \Delta Q = Q'\frac{\Delta p}{p}.$$

they are affected by a momentum deviation

$$\Delta\omega_{\beta f} = (Q' - \eta_c(n_f + Q))\omega_0 \frac{\Delta p}{p} \ , \ \Delta\omega_{\beta s} = (Q' - \eta_c(n_s - Q))\omega_0 \frac{\Delta p}{p}.$$

Based on an energy distribution $f_0(\Delta E)$ of the particles in the beam, the above relation results in two frequency distributions $f(\omega_{\beta f})$, $f(\omega_{\beta s})$. We excite the beam by an acceleration $\hat{G}\exp(j\omega t)$ with frequency $\omega$ being close to $\omega_{\beta_f s}$ or $\omega_{\beta s}$ and get a velocity response of the center of mass of the distribution

$$\langle\hat{y}\rangle_f = -\frac{\hat{G}\omega}{2Q\omega_0} \int \frac{f(\omega_{\beta f})}{\omega_{\beta f} - \omega} d\omega_{\beta f} = -\frac{\hat{G}\omega}{2Q\omega_0}\left(\pi f(\omega) + jPV \int \frac{f(\omega_{\beta f})}{\omega_{\beta f} - \omega} d\omega_{\beta f}\right).$$

$$\langle\hat{y}\rangle_s = \frac{\hat{G}\omega}{2Q\omega_0} \int \frac{f(\omega_{\beta s})}{\omega_{\beta s} - \omega} d\omega_{\beta s} = \frac{\hat{G}\omega}{2Q\omega_0}\left(\pi f(\omega) + jPV \int \frac{f(\omega_{\beta s})}{\omega_{\beta s} - \omega} d\omega_{\beta s}\right),$$

where $PV\int$ is the "principle valued" integral.

The term $\pi f(\omega)$ is real, excitation and velocity response are in phase resulting in an absorbtions of energy and damping, called Landau damping. It is only present if the excitation frequency $\omega$ is within the frequency distribution of the individual particles. The second term is imaginary and gives the out-of-phase response being of less interest.

The oscillating beam can induce a voltage in a transverse impedance which in turn applies a self acceleration $G_s$ to the beam

$$Z_T(\omega) = -\frac{\omega}{I\hat{y}(\omega)} \oint \left(\vec{E}(\omega) + [\vec{\beta} \times \vec{B}(\omega)]\right)_T ds \ , \ \hat{G}_s = -\frac{eZ_T I\langle\hat{y}\rangle}{\gamma m_0 2\pi R\omega}.$$

If $\hat{G}_s = \hat{G}$ we can have a steady self sustained oscillation without external excitation, i.e. a threshold of an instability. Introducing this into the response we get for this threshold

$$1 = \frac{jecIZ_T(\omega)}{4\pi QE} \int \frac{f(\omega_{\beta s})}{\omega_{\beta f} - \omega} d\omega_{\beta f} = -\frac{ecIZ_T(\omega)}{4\pi QE}\left(\pi f(\omega) + jPV \int \frac{f(\omega_{\beta f})}{\omega_{\beta f} - \omega} d\omega_{\beta f}\right).$$

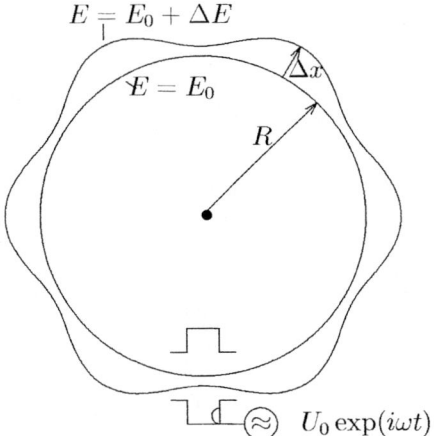

Figure 11: Excited longitudinal mode of a coasting beam

$$1 = \frac{-jecIZ_T(\omega)}{4\pi QE} \int \frac{f(\omega_{\beta s})}{\omega_{\beta s} - \omega} d\omega_{\beta s} = \frac{ecIZ_T(\omega)}{4\pi QE} \left( \pi f(\omega) + jPV \int \frac{f(\omega_{\beta s})}{\omega_{\beta s} - \omega} d\omega_{\beta s} \right).$$

These equations represent a relation between the complex impedance and the complex beam response to an excitation. It is represented by the so-called stability diagram shown on the left of Fig. 12 for a Gaussian distribution. If the impedance lies inside the central curve we have stability, outside an instability. The curve itself represents the threshold. Its shape is determined by the frequency distribution of the particles.

## 5.2 Longitudinal coasting beam instability

The longitudinal dynamics of a coasting or unbunched beam is governed by the relation between the deviations in momentum and revolution frequency

$$\frac{\Delta E}{E} = \beta^2 \frac{\Delta p}{p} = -\frac{\beta^2}{\eta_c} \frac{\Delta \omega_0}{\omega_0}, \text{ with } \eta_c = \alpha_c - \frac{1}{\gamma^2}.$$

The beam has an equilibrium energy distribution which translates into a distribution in revolution frequency

$$f_0(\Delta E) = \frac{1}{N} \frac{d^2 N}{d\theta dE} \rightarrow F_0(\Delta \omega_0) = \frac{1}{N} \frac{d^2 N}{d\theta d\omega_0}$$

A stable beam has a continuous current $I_0$, however, exciting it with $U_0 \exp(j\omega t)$ close to $n\omega_0$ gives a current perturbation

$$I_1(t) = \frac{-jNe^2\omega_0^3 U_0}{2\pi \beta^2 E} \int \frac{dF_0(\omega_0)/dt}{\omega - n\omega_0} d\omega_0 = \frac{Ne^2\omega_0^3 U_0}{2\pi \beta^2 E} \left( \pi \frac{dF_0}{d\omega_0}(\omega) - jPV \int \right).$$

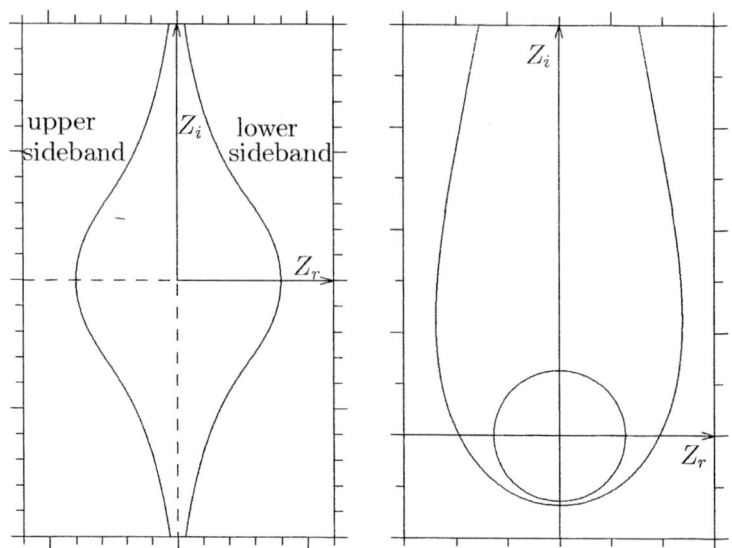

Figure 12: Transverse (left) and longitudinal (right) stability diagrams

This current $I_1$ can induce a voltage in an impedance $Z$. If it is as large or larger than $U_0$ it can replace the external excitation and keep the current modulation going or increase it. We get for this stability limit

$$1 = \frac{Ne^2\omega_0^3\eta Z(\omega)}{2\pi\beta^2 E}\left(\frac{\pi dF_0}{d\omega_0}(\omega) - jPV\int\frac{dF_0(\omega_0)/dt}{\omega - n\omega_0}d\omega_0\right).$$

This equation is a complex mapping which can be represented in form of a stability diagram, shown on the right of Fig. 12, which depends on the energy or revolution frequency distribution of the particles.

To separate the dependence on the form of the distribution from the one on physical parameters like $E$, $I_0$, $\Delta p/p$ and $\eta_c$ the stability diagram is normalized with the width the momentum spread. Taking many such diagrams and approximating them with a circle gives the (Keil-Schnell) stability criterion

$$\left|\frac{Z}{n}\right| \leq \frac{2\pi\beta^2 E\eta_c(\Delta p/p)^2}{eI_0}.$$

Important is the strong dependence on the momentum spread, or the connected frequency spread, which gives rise to Landau damping.

## 6 Bunch lengthening

The impedance of a ring consists often to a large part of many resonances with different frequencies $\omega_r$, shunt impedance $R_s$ and quality factors $Q$. At low frequencies

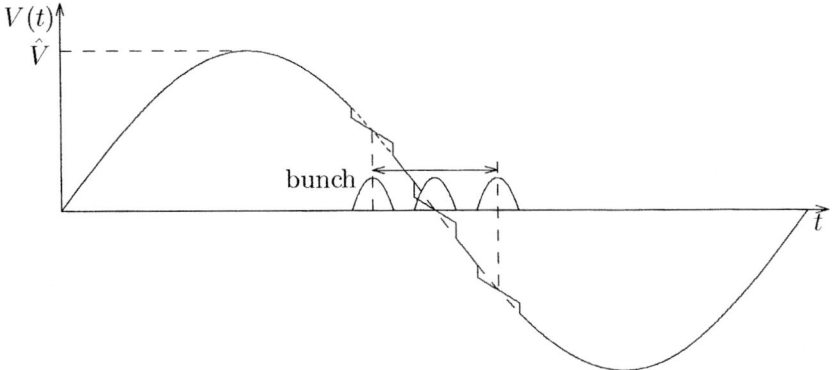

Figure 13: Potential distortion created by an oscillating bunch

$\omega \ll \omega_r$ their impedances are mainly inductive

$$Z(\omega) = R_s \frac{1 - jQ\frac{\omega^2-\omega_r^2}{\omega\omega_r}}{1 + \left(Q\frac{\omega^2-\omega_r^2}{\omega\omega_r}\right)^2} \approx j\frac{R_s\omega}{Q\omega_r} + \ldots$$

The sum of the low frequency impedance due to all these resonances divided by the mode number $n = \omega/\omega_0$ is called

$$\left|\frac{Z}{n}\right|_0 = \sum_k \frac{R_{sk}\omega_0}{Q_k\omega_{rk}} = L\omega_0.$$

with $L$ being the inductance. A bunch with current $I_b(t)$ induces a voltage $V_i = -L dI_b/dt$ which is added to the RF-voltage

$$V(t) = \hat{V}\sin(h\omega_0 t) - L\frac{dI_b}{dt}.$$

Developing around $t_s$, using $\tau = t - t_s$, $\phi_s = h\omega_0 t_s$ and using a single parabolic bunch with average current $I_0$

$$I_b(\tau) = \hat{I}\left(1 - \frac{\tau^2}{\tau_0^2}\right), \quad \frac{dI_b}{d\tau} = -\frac{3\pi I_0}{\omega_0 \tau_0^3}\tau$$

we get the voltage and synchrotron frequency shift of the particles in the bunch

$$V = \hat{V}\sin\phi_s + \hat{V}\cos\phi_s h\omega_0\tau\left(1 + \frac{3\pi|Z/n|_0 I_0}{h\hat{V}\cos\phi_s(\omega_0\tau_0)^3}\right), \quad \frac{\Delta\omega_s}{\omega_s} \approx \frac{3\pi|Z/n|_0 I_0}{2h\hat{V}\cos\phi_s(\omega_0\tau_0)^3}$$

Only the incoherent frequency of the individual particles in the bunch is changed (reduced for $\gamma > \gamma_T$, increased for $\gamma < \gamma_T$). The coherent dipole (rigid bunch) is not

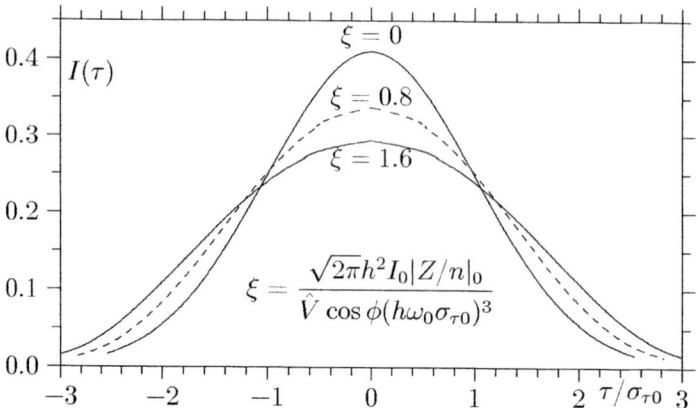

Figure 14: Potential well lengthening of a bunch with Gaussian energy distribution

affected by the inductive impedance since it carries the voltage distortion with it, Fig. 13. This can separate the coherent synchrotron frequency from the incoherent distribution and lead to a loss of Landau damping.

The reduction of the longitudinal focusing leads to an increase of the bunch length given by a 4th order equation for protons with constant phase space area

$$\left(\frac{\tau_0}{\tau_{00}}\right)^4 + \frac{3\pi |Z/n|_0 I_0}{h\hat{V}\cos\phi_s(\omega_0\tau_{00})^3}\left(\frac{\tau_0}{\tau_{00}}\right) - 1 = 0$$

The assumed parabolic bunch current is the projection of an elliptic phase space distribution. In this case the bunch form is not changed just its length increased. The situation is more complicated for other distribution like for the Gaussian shown in Fig. 14.

# Aspects of Beam Stability at ISIS

## G H Rees

*Rutherford Appleton Laboratory, CLRC, UK*

**Abstract.** Features of beam stability for the 50 Hz, 800 MeV, ISIS proton synchrotron are described. The issues discussed include the following: longitudinal coasting beam, longitudinal bunched beam, transverse coasting beam, transverse head-tail bunched beam, and the absence of the electron-proton instability.

## INTRODUCTION

The high intensity performance of the ISIS synchrotron is outlined in the reference [1]. The proton beam is more stable than initially predicted, and the operating parameters are set by the beam loss levels and not by the onset of instabilities. A maximum coasting beam of $4 \; 10^{13}$ protons has been obtained at the injection energy of 70 MeV with the limit set by the performance of the linac injector at the time of the experiment. The maximum beam accelerated to 800 MeV has been $2.7 \; 10^{13}$ protons per pulse, but a more typical operating level at 50 Hz is $2.5 \; 10^{13}$ per pulse. Stability issues to be discussed may be summarised as follows:

Longitudinal coasting beam resistive wall – not observed.

Longitudinal bunched beam – not observed in normal operation.

Transverse coasting beam resistive wall – cured by lowering $Q_v$.

Transverse bunched beam head-tail – cured by ramping $Q_v$.

Electron-proton – not observed throughout 70 to 800 MeV range.

These issues are now discussed in turn.

## LONGITUDINAL COASTING BEAM

The longitudinal space charge component of the beam coupling impedance is large, particularly at the 70 MeV injection energy, where the estimated value for Z/n is about 700 Ω. The value would have been over 1000 Ω if the vacuum chamber wall had not been profiled to give an approximately constant ratio of the beam-wall radii around the ring. A lower value of the space charge impedance eases the problem of the trapping of the unbunched circulating beam by the very low fields required by the radio frequency (rf) system.

Parameters during injection are such that operation is above the Keil-Schnell instability threshold for a coasting beam. For this situation, even though the ring operates below its transition energy, it had been predicted that the residue of the linac bunch structure (202.5 MHz and harmonics) would exhibit progressive self-bunching until saturation. An experimental check of this prediction has been made during the early stages of commissioning the ISIS ring.

A fast longitudinal monitor has been used to follow the coherent motion of the residual distribution of the linac bunch structure in the circulating beam during the injection interval and subsequently. The monitor signal has been processed in a spectrum analyser set at a harmonic of the linac bunch structure, or at the ring revolution frequency harmonic nearest to this frequency. The analogue output signal from the spectrum analyser has then been observed on an oscilloscope trace triggered just prior to injection. As each injected turn enters the ring, a small growth of the analogue signal has been observed but, immediately injection stops, rapid debunching has been seen to occur. Typical debunching times are less than 50 µs at maximum intensity, and up to 100 µs at lower intensities. The rapid damping observed, in place of anti-damping, appears to contradict conventional coasting beam longitudinal instability theory.

The observations are, however, interpreted as follows. Self-bunching, due to growth of quadrupole or higher order modes, causes bunch frequency modulation at the mode frequencies. Pairs of modulation sidebands of equal amplitude appear for symmetrical bunches. The longitudinal space charge fields are proportional to the derivatives of the sideband frequencies, so there are two potential damping/anti-damping terms for each sideband, one large and one small. The two large terms cancel when the sidebands are of equal amplitude, while the two small terms are additive. The additive signal corresponds to anti-damping above transition energy but to damping below, thus explaining the observations. This is in contrast to the analysis for conventional coasting beam theory, where the stability of only a single sideband is considered. The effect of a resistive beam coupling impedance is similarly influenced by the presence of two sideband frequencies. In this case, one sideband gives damping for the coherent motion and the second sideband, anti-damping, with no net effect on stability for equal amplitude sidebands, provided growth times are longer than oscillation periods.

## LONGITUDINAL BUNCHED BEAM

In its normal mode of operation, the ISIS synchrotron uses rf feed forward compensation to reduce the effect of the heavy beam loading. The rf field levels are set very low at the start of acceleration to obtain the correct conditions for trapping of an unbunched circulating beam. Pre-chopping at 50 Hz in the linac has not been included and would be advantageous only if larger H⁻ ion source currents were available. RF cavities are arranged in diametrically opposite pairs in the ring, with anti-phasing adopted to provide low net voltage per turn during injection. At the end of injection, pairs of cavities are swung into phase and the total voltage has to rise rapidly in the first millisecond of acceleration, typically from 2.5 to 60 kV per turn, reaching 140 kV per turn by mid-cycle. Application of feed forward control is necessary for all intensities above 10% of the normal operating level of $2.5 \; 10^{13}$ protons per pulse. Without feed forward, there is significant phase and amplitude modulation of the cavity fields, with resultant beam loss.

There is insufficient time for adiabatic trapping and there is strong filamentation during the early stage of acceleration. As a result, there is a pronounced non-uniformity in the longitudinal beam distribution, particularly at low intensity. Longitudinal space charge effects lead to smoother distributions at higher intensity, but the beam does not reach an equilibrium state by the end of the 10 ms period of acceleration. There is no evidence of any instabilities due to the lumpiness of the distributions. Controls include beam radial and phase control loops, cavity field amplitude control and a cavity frequency tuning loop, in addition to the feed forward of the fundamental component of beam current, after a one turn delay. Digital control techniques are used to optimise the performance of the tuning loop, which adjusts the biasing current to set the permeability of the cavity ferrite. Initially, in the first 2 ms, the cavity is held on tune at its actual resonant frequency, but there is then a smooth transition to the correct detuning condition for reactive beam loading compensation. The initial stage is necessary to provide adequate response in the control loops while the cavity fields are low. Feed-forward control is removed gradually after 2 ms.

Quadrupole mode instabilities may develop towards the end of the acceleration interval if the tuning loop has not been optimised. This has been the only instability observed and may be cured by re-setting the tuning loop, or by using quadrupole mode feedback, or both. Typical acceleration efficiencies approach 90%, with the loss due to the nature of the non-adiabatic trapping, and with high efficiency loss collection obtained at a momentum collimator.

## TRANSVERSE COASTING BEAM

A maximum coasting beam of $4 \; 10^{13}$ protons has been obtained in a storage ring mode of ISIS at the injection energy of 70 MeV, with the limit set by the performance of the linac injector at the time of the experiment. To achieve the maximum intensity, the betatron tunes had to be adjusted, with the vertical tune decreased from its design value of 3.8 to 3.7. A vertical resistive wall instability develops at the higher tune value, with a growth time of a few milliseconds. The unstable mode observed has been at a frequency of approximately 130 kHz, corresponding to the lowest resistive wall mode, $4-Q_v$.

Confirmation of a resistive wall mode was made by observation of the growth rate as a function of the vertical tune. The growth rate decreased as the distance of the tune from the integer value 4 was increased, with the beam becoming stable for tunes less than or equal to 3.7. No accompanying electron-proton type instability has been observed. Results have been obtained with the ring operating at the natural value of its chromaticity and with an injected beam fractional momentum spread of about $\pm 2\ 10^{-3}$. The corresponding chromatic tune spreads are $\pm 0.01$, small compared to the transverse space charge tune shifts and spreads of 0.2.

The chromatic tune spreads are less than the values predicted as necessary to stabilize the instability. There has been no evidence of any similar coherent betatron growth during the normal operation of the ring. The rf fields are applied at low level shortly after injection commences so that the beam begins to bunch and hence to disturb the development of any coasting beam mode.

## TRANSVERSE BUNCHED BEAM HEAD-TAIL MOTION

The two proton bunches in the ISIS synchrotron exhibit some coherent vertical growth during the 2 to 4 ms interval of the 0 to 10 ms acceleration period. The instability develops at beam intensities above $2\ 10^{12}$ and may lead to complete loss of beam, but the growth rate tends to decrease at higher intensities when the space charge tune depressions are larger. Coherent growth is suppressed almost entirely when operating at the natural negative chromaticity by ramping the vertical tune over the 2 to 4 ms interval, away from the nearby integer value. The 2 ms ramping is achieved by programming of separate trim quadrupoles, adjacent to each main quadrupole. There is no observation of instability in the 0 to 2 ms interval, nor in the 4 to 10 ms interval, suggesting the bunches reach an optimal length during the intervening interval for the development of the instability.

The pattern of beam behaviour is highly repetitive, with a single vertical displacement node for the coherent motion at each bunch centre. This is the signature for the m = 1, head-tail, coherent mode; it had not been expected at ISIS because the head-tail chromatic phase shifts favour the development of modes m >1. For the 2 to 4 ms interval, the expected head-tail mode is that with m = 2, not m = 1, but the m = 2 mode pattern has never been observed. The value for the head-tail phase shift is approximately 2.8 $\pi$ and, for such a value, it is difficult to envisage how the traditional perturbation distribution assumed for head-tail motion may explain the mechanism of anti-damping for m = 1 motion. A modified distribution [2] has therefore been proposed in an attempt to explain the observations at ISIS.

In [2], the beam oscillation pattern is analysed to find the sideband amplitudes in the spectra of the differential wall currents and the associated resistive wall forces. It is found that the largest resisitive wall forces are not at the lowest (n - $Q_v$) sideband frequency but at some of the (n + $Q_v$) sidebands of low n value, where n defines the revolution frequency harmonic involved. The perturbed distribution of the bunches must allow anti-damping for (n + $Q_v$) frequencies over the full bunch extents.

The standing wave pattern for the m = 1 mode has a single central displacement node which is traditionally assumed to arise due to the cancellation between the coherent betatron motions of the positive and negative off-momentum particles in the bunches. The alternative assumption, made for ISIS, is that the central node is caused by the cancellation between the coherent betatron motions of small and large synchrotron amplitude particles. The protons of small synchrotron amplitude then experience continuous anti-damping coherent betatron forces, while those of large amplitude experience damping when they are near the bunch centre and anti-damping when they are not, being anti-damped on average. Experimentally it is confirmed that for the natural values of the chromaticity, the dominant sidebands are those at $(n + Q_v)$. More details are given in [2].

## NON-APPEARANCE OF ELECTRON-PROTON INSTABILITY

The electron-proton instabilty is not observed at ISIS for either a coasting or a bunched beam, in contrast to the situation at the PSR at LANL [3]. Beam densities are lower in the case of ISIS, but not significantly so for the case of a coasting beam, where $4 \ 10^{13}$ protons have been stored at 70 MeV, at transverse rms emittances of approximately 50 $\pi$ μrm. No high frequency, coherent, transverse motion is observed in either transverse plane, even though transverse coasting beam stability criteria are exceeded.

It is of interest to speculate why no coasting beam electron-proton instability is observed at ISIS while the instability limits the coasting beam threshold at the PSR to $8 \ 10^{12}$ protons per pulse. The 1-D and 2-D beam densities are comparable, though the beam emittances and acceptances are larger at ISIS. Since the instability involves the trapping of electrons in the potential well of the proton beam, and the electrons drift towards the regions of deepest potential well, the influence of the vacuum chamber walls and of the beam distributions may be significant. ISIS has a smoothly contoured vacuum chamber wall, with shielding of the bellows sections. The PSR does not, so the associated electric fields are larger, particularly near the discontinuities of the bellows sections, resulting in a larger potential well for trapped electrons.

Of interest is the design of next generation spallation neutron sources and, in particular, how to enhance growth times for possible electron-proton and fast transverse instabilities. It appears prudent to shield all bellows sections and to consider reducing the resistivity of the vacuum chamber wall. Extraction kicker modules should also be designed for low transverse coupling impedances.

## REFERENCES

[1] G H Rees, High Intensity Performance of the ISIS Synchrotron, Proceedings of the Workshop on Beam Instabilities in Storage Rings, Hefei, China, July 1994, pp 198 – 205.
[2] G H Rees, Interpretation of the Higher Mode, Head-Tail Motion Observed on ISIS, Particle Accelerators, Vol 39, 1992, pp 159 – 168.
[3] R J Macek, PSR Experience with Beam Losses, Instabilities and Space Charge Effects, Shelter Island Workshop, 1998, AIP Proceedings 448, pp 116 - 127.

# High Intensity Beam Operation of the Brookhaven AGS*

Thomas Roser

*AGS Department, Brookhaven National Laboratory
Upton, NY 11973-5000*

**Abstract.** For the last few years the Brookhaven AGS has operated at record proton intensities. This high beam intensity allowed for the simultaneous operation of several high precision rare kaon decay experiments. The record beam intensities were achieved after the AGS Booster was commissioned and a transition jump system, a powerful transverse damper, and an rf upgrade in the AGS were completed. The intensity is presently limited by space charge effects at both Booster and AGS injection and transverse instabilities in the AGS.

## THE AGS ACCELERATOR COMPLEX

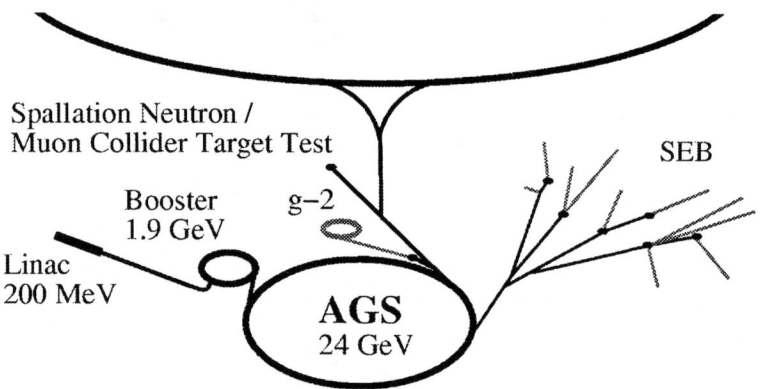

**FIGURE 1.** The AGS-RHIC accelerator complex.

Fig. 1 shows the present layout of the AGS-RHIC accelerator complex. The high intensity proton beam of the AGS is used both for the slow-extracted-beam (SEB) area with many target station to produce secondary beams and the fast-extracted-beam (FEB) line used for the production of muons for the g-2 experiment and for

---

*) Work performed under the auspices of the U.S. Department of Energy

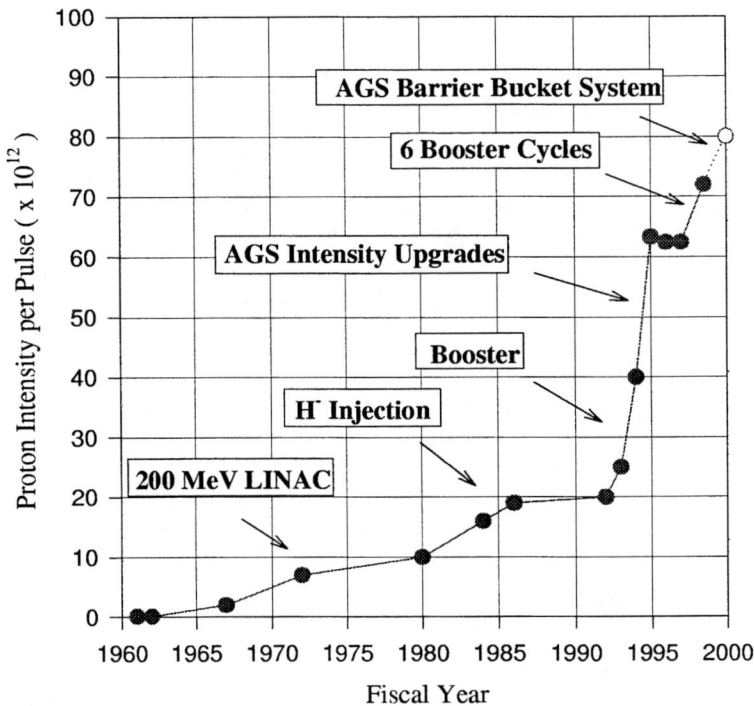

**FIGURE 2.** The history of the evolution of the proton beam intensity in the Brookhaven AGS.

high intensity target testing for the spallation neutron sources and muon production targets for the muon collider. The same FEB line will also be used for the transfer of beam to RHIC.

The proton beam intensity in the AGS has increased steadily over the 40 year existence of the AGS, but the most dramatic increase occurred over the last few years with the addition of the new AGS Booster. In Fig. 2 the history of the AGS intensity improvements is shown and the major upgrades are indicated. The AGS Booster has one quarter the circumference of the AGS and therefore allows four Booster beam pulses to be stacked in the AGS at an injection energy of 1.5 − 1.9 $GeV$. At this increased energy, space charge forces are much reduced and this in turn allows for the dramatic increase in the AGS beam intensity.

# HIGH INTENSITY OPERATION OF THE LINAC AND BOOSTER

The 200 MeV LINAC is being used both for the injection into the Booster as well as an isotope production facility. A recent upgrade of the LINAC rf system made it possible to operated at an average H$^-$ current of 150 $\mu A$ and a maximum of $12 \times 10^{13}$ H$^-$ per 500 $\mu s$ LINAC pulse for the isotope production target. Typical beam currents during the 500 $\mu s$ pulse are about 80 $mA$ at the source, 60 $mA$ after the 750 $keV$ RFQ, 38 $mA$ after the first LINAC tank (10 $MeV$), and 37 $mA$ at end of the LINAC at 200 $MeV$. The normalized beam emittance is about 2 $\pi$ $mm$ $mrad$ for 95 % of the beam and the beam energy spread is about $\pm 1.2$ $MeV$. A magnetic fast chopper installed at 750 $keV$ allows the shaping of the beam injected into the Booster to avoid excessive beam loss.

The achieved beam intensity in the Booster surpassed the design goal of $1.5 \times 10^{13}$ protons per pulse and reached a peak value of $2.3 \times 10^{13}$ protons per pulse. This was achieved by very carefully correcting all the important nonlinear orbit resonances especially at the injection energy of 200 $MeV$ and by using the extra set of rf cavities that were installed for heavy ion operation as a second harmonic rf system. The second harmonic rf system allows for the creation of a flattened rf bucket which gives longer bunches with lower space charge forces. The fundamental rf system operated with 90 $kV$ and the secondary harmonic with 30 $kV$. The typical bunch area was about 1.5 $eVs$. Even with the second harmonic rf system the incoherent space charge tune shift can reach one unit right at injection ($3 \times 10^{13}$ protons, norm. 95 % emittance: 50 $\pi$ $mm$ $mrad$, bunching factor: 0.5). Of course, such a large tune shift is not sustainable, but the beam emittance growth and beam loss can be minimized by accelerating rapidly during and after injection. Best conditions are achieved by ramping the main field during injection with 3 T/s increasing to 9 T/s after about 10 ms. The quite large non-linear fields from eddy currents in the Iconel vacuum chamber of the Booster are passively corrected using correction windings on the vacuum chamber that are driven by backleg windings [1].

Recently the Booster rf system was modified to allow for the acceleration of a single bunch. At this lower frequency 45 kV was available for the first harmonic and 22 kV for the second harmonic. This results in the same available total bucket area. With only one bunch in the Booster a peak intensity of about $1.8 \times 10^{13}$ was reached. Single bunch operation in the Booster allowed for the transfer of six Booster loads into the AGS reducing the need for very high intensity in the Booster. Nevertheless, the lower peak intensity was a surprise which was eventually traced to the factor of two lower synchrotron frequency. During the approximately $250\mu s$ long H$^-$ injection period the synchrotron motion is now not providing enough dilution and high peak line densities are developing. More elaborate longitudinal painting schemes were studied which may eventually result in higher intensity even with single bunch operation.

# AGS HIGH INTENSITY OPERATION

The AGS itself also had to be upgraded to be able to cope with the higher beam intensity. During beam injection from the Booster, which cycles with a repetition rate of 7.5 $Hz$, the AGS needs to store the already transferred beam bunches for about 0.4 seconds. During this time the beam is exposed to the strong image forces from the vacuum chamber which causes beam loss from resistive wall coupled bunch beam instabilities within as short a time as a few hundred revolutions. An example of such an instability for relatively low beam intensity is shown in Fig. 3. Even though the eight bunches individually show a rather complicated growth pattern of the vertical displacement a Fourier analysis shows that the coupled bunch modes develop as expected for a resistive wall instability with the lowest frequency mode growing the fastest.

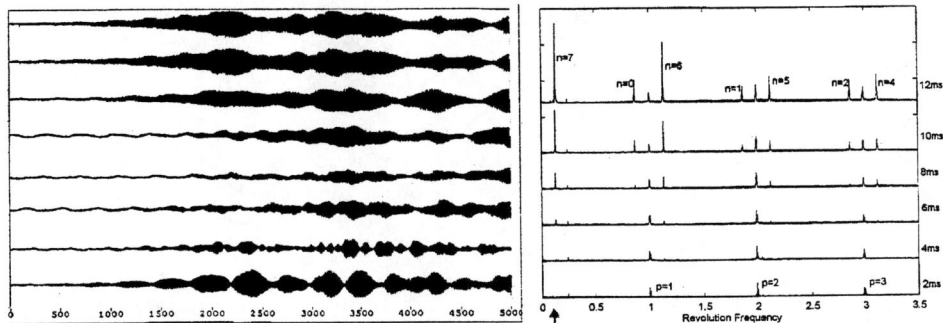

**FIGURE 3.** The left figure shows the growth of a vertical coupled bunch instability at AGS injection energy of 1.5 GeV for the eight bunches. The horizontal scale is in turns. The right graph shows the evolution of the frequency components during the growth of the instability.

A very powerful feedback system was installed that senses any transverse movement of the beam and compensates with a correcting kick. This transverse damper can deliver $\pm 160\ V$ to a pair of 50 $\Omega$, one-meter-long strip-lines. A recursive digital notch filter is used in the feed-back circuit to allow for accurate determination of the average beam position and increased sensitivity to the unstable coherent beam motion. This filter design is particular important for the betatron tune setting of about 8.9 which is required to avoid non-linear octupole stopband resonance at 8.75. With an incoherent tune shift at the AGS injection energy of 0.1 to 0.2 it is still necessary, however, to correct the octupole stopband resonances to avoid excessive beam loss.

To reduce the space charge forces further the beam bunches in the AGS are lengthened by purposely mismatching the bunch-to-bucket transfer from the Booster and then smooth the bunch distribution using a high frequency 100 $MHz$ dilution cavity. The resulting reduction of the peak current helps both with coupled bunch instabilities and stopband beam losses. Fig. 4 shows the evolution of a mismatched bunch being diluted.

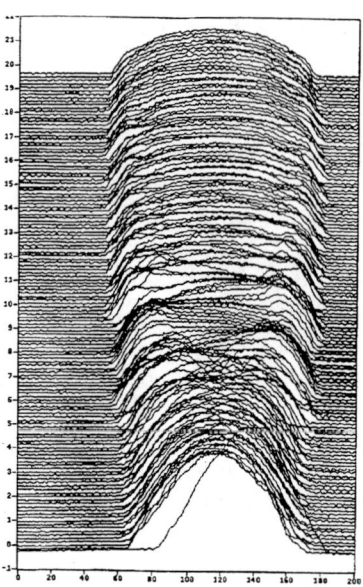

**FIGURE 4.** Evolution of a mismatched bunch at AGS injection under the influence of the 100 MHz dilution cavity. The traces show the intensity of the same bunch every revolution. The horizontal scale is in nanoseconds

A large part of the injection losses at the AGS are due to a relative slow loss during the first millisecond the transferred bunches circulate in the AGS. No direct cause for this loss could be identified but it is correlated with a sustained transverse coherent beam oscillation shown in Fig. 5. The coherent oscillations result from miss-matched beam injection to blow-up the transverse emittance and therefore reduce the space charge tune shift. Although the coherence persists over a whole millisecond the middle part of the beam bunch has a coherent space charge tune shift of about 0.1 and therefore very high frequency vertical modulations appear. The bunch intensity is about $1.3 \times 10^{13}$ protons.

At bunch intensities above $1.3 \times 10^{13}$ protons a vertical head-tail instability develops with a single bunch in the AGS. Fig. 6 shows both the bunch shape as well as the vertical modulation. The modulation develops towards the tail of the bunch and is very asymmetric. The growth rate of about 50 ms is quite slow consistent with a weak head-tail instability. However, the observed asymmetry suggests a very short wake or possibly an electron-proton instability. The instability can be cured by changing the vertical tune.

The large space charge forces during AGS injection can be mitigated by maintaining a practically debunched beam during the AGS injection process. To accomplish this, cavities that produce isolated rf buckets can be used to maintain an empty gap for injection of additional Booster beam pulses. Isolated bucket cavities, also called Barrier Bucket cavities, have been used elsewhere [3]. However, for this stacking

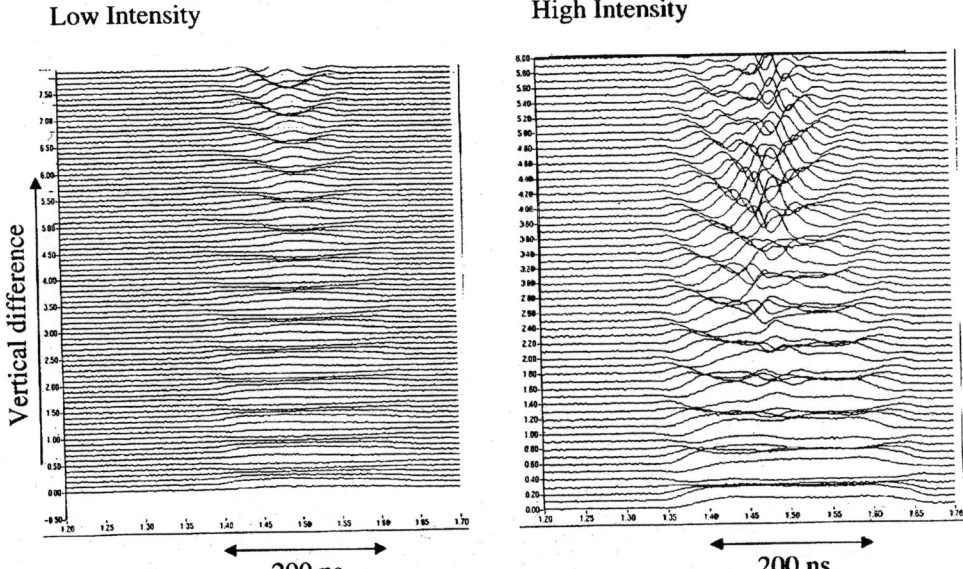

**FIGURE 5.** Evolution of the vertical miss-match at AGS injection. The traces show the vertical displacement of the same bunch every revolution. The horizontal scale is in microseconds

scheme, a high rf voltage will be needed to contain the large bunch area of the high intensity beam. A successful test of this scheme has recently been completed using two 40 kV Barrier cavities [4].

During acceleration the AGS beam has to pass through the transition energy after which the revolution time of higher energy protons becomes longer than for the lower energy protons. This potentially unstable point during the acceleration cycle was crossed very quickly with a new powerful transition energy jump system with only minimal losses even at the highest intensities [2]. The large lattice distortions introduced by the jump system prior to the transition crossing severely limits the available aperture of the AGS in particular for momentum spread. Efforts to correct the distortions using sextupoles have been partially successful. After the transition energy, a very rapid, high frequency instability developed which could only be avoided by purposely further increasing the bunch length using again the high frequency dilution cavity.

The peak beam intensity reached at the AGS extraction energy of $24\,GeV$ was $7.2 \times 10^{13}$ protons per pulse also exceeding the design goal for this latest round of intensity upgrades. It also represents a world record beam intensity for a proton synchrotron. Individual bunches with an intensity of $1.2 \times 10^{13}$ had a bunch area of about 3 eVs at AGS injection but were diluted to about 10 eVs during acceleration.

At maximum beam intensity about 30 percent of the beam is lost at Booster injection ($200\,MeV$), 20 percent at injection into the AGS ($1.9\,GeV$), which includes

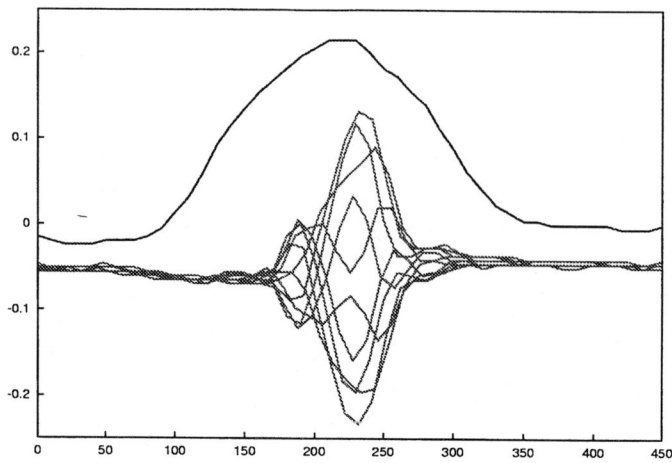

**FIGURE 6.** Bunch shape (top) and vertical modulation of a high intensity bunch at AGS injection energy of 1.9 GeV during a single bunch head-tail instability. The vertical scale is in arbitrary units and the horizontal scale is in ns.

losses during the 0.4 second storage time in the AGS, and about 3 percent is lost at transition (8 $GeV$). Although activation levels are quite high all machines can still be manually maintained and repaired in a safe manner.

# HIGH INTENSITY OPERATION FOR FAST-EXTRACTED BEAM

During fast-extracted beam operation single bunches are extracted multiple times from the AGS at 24 GeV. The bunch length at extraction has to be as short as possible. This requirement conflicts with the effort to dilute the beam to stabilize it at high intensity. In particular, without additional bunch dilution after transition crossing, fast transverse single bunch instabilities develop leading to beam loss or beam blow-up. A technique to shorten the bunches before extraction was developed using adiabatic quadrupole pumping. With this technique a coherent quadrupole excitation as shown on the left side of Fig. 7 could be maintained for the multiple extractions [5]. To further shorten the bunches and also to limit the amount of beam per single extraction to less than the target station limit of $7 \times 10^{12}$ protons the six bunches were adiabatically split into twelve bunches. This is shown on the right side in Fig.7.

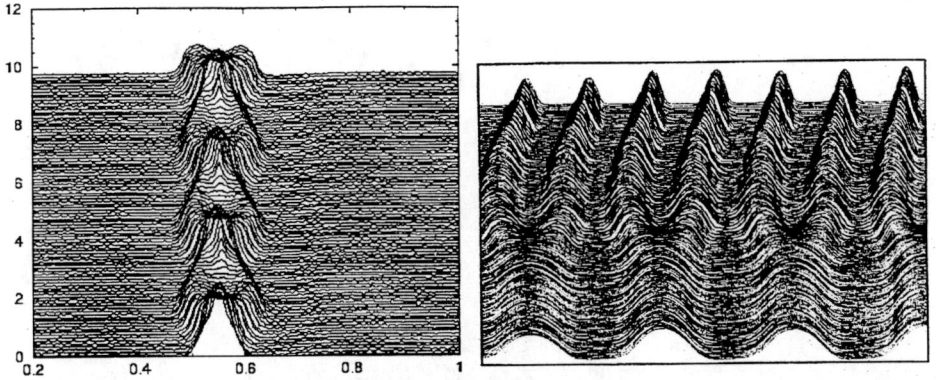

**FIGURE 7.** Evolution of the shape of a high intensity proton bunch during adiabatic quadrupole excitation (left) and adiabatic bunch splitting (right)

# REFERENCES

1. G.T. Danby and J.W. Jackson, Part. Accel. **27** (1990) 33
2. W.K. van Asselt et al., The Transition Jump System for the AGS, 1995 Particle Accel. Conf., Dallas, Texas, May 1995, p. 3022., L.A. Ahrens et al., Performance of the AGS transition jump system, 1999 Particle Accel. Conf., New York, to be published.
3. J.E. Griffin et al., IEEE Trans. on Nucl. Sc. Vol. NS-30, No. 4, (1983) 3502
4. M. Blaskiewicz et al., Barrier cavities at the Brookhaven AGS, 1999 Particle Accel. Conf., New York, to be published; M. Fujieda et al., Magnetic alloy loaded rf cavity for barrier bucket experiment at the AGS, 1999 Particle Accel. Conf., New York, to be published.
5. M. Bai et al., Adiabatic excitation of longitudinal bunch shape oscillations, 1999 Particle Accel. Conf., New York, to be published.

# Proton Driver Study at Fermilab

Weiren Chou

*Fermi National Accelerator Laboratory*[1]
*P.O. Box 500, Batavia, IL 60510*

**Abstract.** Fermilab has started the design work of a high intensity proton source called the proton driver. It would provide a 4 MW proton beam to the target for muon production. This paper discusses the basic features of this machine and the associated accelerator physics and design issues.

## INTRODUCTION

Since about 1996, a Muon Collider Collaboration has been formed within the high energy physics (HEP) community to study the feasibility of a future collider using muon beams. Recently, this collaboration has turned its attention to a relatively cheaper and easier muon storage ring called the neutrino factory. Either a collider or a storage ring, it requires muon beams whose intensity is several orders of magnitude higher than that in any existing muon source. In order to produce such intense muon beams, a high intensity proton source, called the proton driver, is needed.

The proton driver is a high intensity rapid cycling proton synchrotron. Its primary function is to deliver intense short proton bunches to the target for muon production. These muons will be captured, phase rotated, cooled, accelerated and finally, injected into either a storage ring for neutrino experiment (a $\nu$-Factory) or a collider for $\mu\mu$ collision. In this sense, the proton driver is *the front end* of a muon facility.

There are two primary requirements of the proton driver:

1. High beam power: $P_{\text{beam}} = 4$ MW.
   This requirement is similar to other high intensity proton machines that are presently under design, *e.g.*, the SNS, ESS and JHF.

2. Short bunch length at exit: $\sigma_b = 1\text{-}2$ ns.
   This requirement is *unique* for the proton driver. It brings up a number of interesting and challenging beam physics issues that we will discuss in this paper.

---

[1] Fermilab is operated by Universities Research Association, Inc. under contract with the US Department of Energy.

The bunch length is related to the longitudinal emittance $\epsilon_L$ and momentum spread $\Delta p$ by:

$$\sigma_b \propto \frac{\epsilon_L}{\Delta p}$$

In order to get short bunch length, it is essential to have:

- longitudinal emittance preservation (no intentional blow-up);
- large momentum acceptance (in both rf and lattice);
- bunch rotation at the end of the cycle.

It is interesting to compare the proton driver with the former SSC. The SSC required proton beams very bright in the transverse plane. In order to reach the design luminosity, the design value of the transverse emittance $\epsilon_T$ was 1 $\mu$m, which was several times smaller than that in any existing high energy proton colliders. In the longitudinal plane, however, $\epsilon_L$ would be blown up by two orders of magnitude in the injector chain in order to avoid instability and intrabeam scattering problem. The proton driver, on the contrary, requires high brightness in the longitudinal plane because of short bunch length, whereas $\epsilon_T$ would be diluted by painting during the injection from the linac to the ring in order to reduce the space charge effect.

## CHOICE OF THE PRIMARY PARAMETERS

The beam energy is the product of three parameters: proton energy $E_p$, number of protons per cycle $N_p$ and the repetition rate (rep rate) $f_{\text{rep}}$:

$$P_{\text{beam}} = f_{\text{rep}} \times E_p \times N_p$$

The rep rate is chosen to be 15 Hz. There are two reasons: (1) Muons decay quickly. So the collider needs to get re-fill quickly. The life time of a 2 TeV muon is about 40 ms. The rep rate should be comparable to the muon decay rate. (2) Fermilab is experienced in operating 15 Hz linac and booster.

Given the beam power and rep rate, the product $E_p \times N_p$ is determined. At this time, we choose 16 GeV and $1 \times 10^{14}$ protons per pulse (ppp). This is based on the following trade-off considerations: (1) Higher $E_p$ would give lower longitudinal phase space density $N_b/\epsilon_L$, lower space charge tune shift $\Delta Q$ at top energy, and would give smaller momentum spread $\frac{\Delta p}{p}$. (2) However, higher $E_p$ would also mean higher cost (e.g., $V_{rf} \propto E_p^2$) and higher radiation power to the environment. The choice of 16 GeV is a reasonable compromise. There are two other important issues related to the choice of $E_p$:

- The muon yield per proton $N_\mu/N_p$ at the beginning of the decay channel is believed to be proportional to $E_p$. However, the effective muon yield (i.e., $N_\mu/N_p$ at the exit of the decay channel) as a function of $E_p$ is yet to be studied.

- For given $P_{\text{beam}}$, the total energy deposition on the target is a constant (about 10%) in the range of $E_p$ from 8 GeV to 30 GeV. However, the deposited energy is not distributed uniformly. A crucial parameter in the target design is the maximum density of energy deposition on the target. It needs to be simulated as a function of $E_p$.

Table 1 is a comparison of these parameters in high beam power proton machines.

**TABLE 1.** High Beam Power Proton Machines

| Machine | Protons per Cycle | Repetition Rate (Hz) | Protons per Second | Beam Energy (GeV) | Beam Power (MW) |
|---|---|---|---|---|---|
| *Existing:* | | | | | |
| BNL AGS | $8 \times 10^{13}$ | 0.5 | $4 \times 10^{13}$ | 30 | 0.2 |
| FNAL Booster | $5 \times 10^{12}$ | 15 | $7.5 \times 10^{13}$ | 8 | 0.1 |
| RAL ISIS | $2.5 \times 10^{13}$ | 50 | $1.25 \times 10^{15}$ | 0.8 | 0.16 |
| LANL PSR | $2.5 \times 10^{13}$ | 20 | $5 \times 10^{14}$ | 0.8 | 0.064 |
| *Planned:* | | | | | |
| Proton Driver | $1 \times 10^{14}$ | 15 | $1.5 \times 10^{15}$ | 16 | 4 |
| Japan JHF | $2 \times 10^{14}$ | 0.3 | $0.6 \times 10^{14}$ | 50 | 0.5 |
| ORNL SNS | $2 \times 10^{14}$ | 60 | $1.2 \times 10^{16}$ | 1 | 2 |
| Europe ESS | $2.34 \times 10^{14}$ | 50 | $1.2 \times 10^{16}$ | 1.334 | 2.5 |

The layout of the proton driver was described in Ref. [1]. It consists of three accelerators: a 1 GeV linac, a 3 GeV pre-booster and a 16 GeV booster. At present, Fermilab has a 400 MeV linac and a 8 GeV booster. The proton driver would increase the beam intensity by a factor of 20 and beam power a factor of 40.

The proton driver would be built in two phases. In Phase I, a 16 GeV new booster will be built in a new tunnel, using the present 400 MeV linac as its injector. The beam intensity will be increased by a factor of 5. In Phase II, a 1 GeV linac and a 3 GeV pre-booster will be added, bringing up the beam intensity by another factor of 4. The parameters in these stages are listed in Table 2.

# BEAM PHYSICS

## Longitudinal beam dynamics

1. High longitudinal phase space density – Keep $\epsilon_L$ small:
   Table 3 is a comparison of the longitudinal brightness $N_b/\epsilon_L$ in existing as well as planned proton machines. The proton driver requires $12.5 \times 10^{12}$ particles per eV-s, which is almost an order of magnitude (or more) higher than most

**TABLE 2.** Parameters of Present, Phase I and Phase II

|  | Present | Phase I | Phase II |
|---|---|---|---|
| **Linac** (operating at 15 Hz) | | | |
| Kinetic energy (MeV) | 400 | 400 | 1000 |
| Peak current (mA) | 40 | 45 | 80 |
| Pulse length ($\mu$s) | 25 | 90 | 200 |
| $H^-$ per pulse | $6.3 \times 10^{12}$ | $2.5 \times 10^{13}$ | $1 \times 10^{14}$ |
| **Pre-booster** (operating at 15 Hz) | | | |
| Extraction kinetic energy (GeV) | | | 3 |
| Protons per bunch | | | $2.5 \times 10^{13}$ |
| Number of bunches | | | 4 |
| Total number of protons | | | $1 \times 10^{14}$ |
| Norm. transverse emittance (mm-mrad) | | | $200\pi$ |
| Longitudinal emittance (eV-s) | | | 2 |
| RF frequency (MHz) | | | 7.5 |
| **Booster** (operating at 15 Hz) | | | |
| Extraction kinetic energy (GeV) | 8 | 16 | 16 |
| Protons per bunch | $6 \times 10^{10}$ | $3 \times 10^{11}$ | $2.5 \times 10^{13}$ |
| Number of bunches | 84 | 84 | 4 |
| Total number of protons | $5 \times 10^{12}$ | $2.5 \times 10^{13}$ | $1 \times 10^{14}$ |
| Norm. transverse emittance (mm-mrad) | $15\pi$ | $50\pi$ | $200\pi$ |
| Longitudinal emittance (eV-s) | 0.1 | 0.1 | 2 |
| RF frequency (MHz) | 53 | 53 | 7.5 |
| Extracted bunch length $\sigma_t$ (ns) | 0.2 | 0.2 | 1 |
| Target beam power (MW) | 0.1 | 1 | 4 |

of the existing machines except the PSR and ISIS, which are low energy (800 MeV) machines.

In order to achieve such a high longitudinal brightness, one has to preserve $\epsilon_L$, which is in contrast to the "conventional wisdom" of blowing up $\epsilon_L$ to keep beam stable. The following measures are taken for $\epsilon_L$ preservation:

- Avoid transition crossing in the lattice design. This would eliminate a major source of emittance dilution.

- Avoid longitudinal microwave instability by keeping the beam always below transition and keeping the resistive impedance small in the machine design. Experience shows that, below transition, the microwave instability is much less likely to occur when the large capacitive space charge impedance is dominant.

- Avoid coupled bunch instability by using low Q rf cavity.

- Apply inductive insert to compensate the potential well distortion due to the space charge.

**TABLE 3.** Longitudinal Brightness of Proton Machines

| Machine | $E_{max}$ (GeV) | $N_{tot}$ ($10^{12}$) | $N_b$ ($10^{12}$) | $\epsilon_L$ (eV-s) | $N_b/\epsilon_L$ ($10^{12}$/eV-s) |
|---|---|---|---|---|---|
| *Existing:* | | | | | |
| CERN SPS | 450 | 46 | 0.012 | 0.5 | 0.024 |
| FNAL MR | 150 | 20 | 0.03 | 0.2 | 0.15 |
| KEK PS | 12 | 3.6 | 0.4 | 2 | 0.2 |
| FNAL Booster | 8 | 4 | 0.05 | 0.1 | 0.5 |
| PETRA II | 40 | 5 | 0.08 | 0.12 | 0.7 |
| DESY III | 7.5 | 1.2 | 0.11 | 0.09 | 1.2 |
| FNAL Main Inj | 150 | 60 | 0.12 | 0.1 | 1.2 |
| CERN PS | 14 | 25 | 1.25 | 0.7 | 1.8 |
| BNL AGS | 24 | 63 | 8 | 4 | 2 |
| LANL PSR | 0.797 | 23 | 23 | 1.25 | 18 |
| RAL ISIS | 0.8 | 25 | 12.5 | 0.6 | 21 |
| *Planned:* | | | | | |
| Proton Driver | 16 | 100 | 25 | 2 | 12.5 |
| Japan JHF | 50 | 200 | 12.5 | 5 | 2.5 |
| AGS for RHIC | 25 | 0.4 | 0.4 | 0.3 | 1.3 |
| PS for LHC | 26 | 14 | 0.9 | 1.0 | 0.9 |
| SPS for LHC | 450 | 24 | 0.1 | 0.5 | 0.2 |

2. Bunch rotation:
A bunch rotation is needed at the end of the cycle in order to shorten the bunch to 1-2 ns. There are three possible ways to do this gymnastics:

- RF amplitude jump
- RF phase jump
- $\gamma_t$ manipulation

The first two methods have been in use for many years. The third one is a new idea first suggested by J. Norem and has been partially demonstrated at an experiment at the AGS. In either of these methods, the bunch length compression ratio is about 3-4.

Our simulation study is focused on the rf amplitude jump. Although Fermilab has years of experience with this operation, the high bunch intensity poses new problems:

(a) During the debunching process, how low can the rf voltage be?
Lower $V_{rf}$ means lower $\frac{\Delta p}{p}$, which in turn gives larger compression ratio. At a high intensity bunch rotation experiment at the AGS, it was found that the minimum $V_{rf}$ is limited by the beamloading effect rather than beam instabilities.

(b) What is the effect of the higher order terms of the momentum compaction factor $\alpha_1$ and $\alpha_2$ in bunch rotation?

In a regular bunch rotation simulation, the momentum compaction is assumed to be a constant $\alpha_0$. However, the proton driver lattice is nearly isochronous ($\alpha_0 \approx 0$). The higher order terms become important. In other words, particles with different $\frac{\Delta p}{p}$ would have different path length $\Delta L$. We are still in the process to understand this effect. Preliminary simulations show that, if $\alpha_1$ is not properly chosen, the bunch rotation could fail.

(c) What is the effect of the transverse tune shift in bunch rotation?

This is an effect similar to the above but from the transverse plane. Due to short bunch length, the tune shift $\Delta Q$ from direct space charge and image charge remains large even at 16 GeV. This $\Delta Q$ also gives different path length $\Delta L$, which would affect bunch rotation. In other words, the path length of each particle depends not only on its longitudinal position but also on its transverse amplitude. One difficulty in studying this problem is how to mimic this space charge effect. Is it conceptually correct simply changing the lattice quad strength to estimate $\Delta L$? Is a 6D simulation necessary to study this complicated transverse-longitudinal coupling problem? It will take a while for us to fully understand this problem.

At this workshop, it was proposed to have a 5-lab "contest." Namely, the machines in the five labs – BNL-AGS, FNAL-MI, KEK-PS, CERN-PS and SPS, and Indiana University-IUCF – would carry out two experiments: (1) bunch rotation, and (2) longitudinal microwave instability below transition. The competing items are: i. maximum peak current $I_{\text{peak}}$; ii. maximum longitudinal phase space density $N_b/\epsilon_L$; and iii. maximum compression ratio.

## Space charge and instabilities

As in all other high intensity machines, the following measures are taken to reduce the Laslett tune shift $\Delta Q$ at injection: high injection energy (3 GeV), large transverse emittance ($200\pi$ mm-mrad, normalized), painting and a 2nd harmonic rf.

In addition to these, it is also planned to use inductive inserts to reduce the potential well distortion from the space charge. Although this idea was proposed many years ago, nobody ever tried it on a real machine until recently. There are two on-going experiments: one at the LANL-PSR using ferrite inserts, another at the KEK-PS using Finemet inserts. The data are being analyzed. A third experiment at the ANL-IPNS is also being discussed. Simulations show that inductive compensation helps during injection, acceleration and bunch rotation. However, because these inserts also introduce additional resistive impedance, one needs to be careful so that it would not cause any instability problem.

The "conventional" type of instabilities, namely, those we have studied for decades, include the impedance budget, resistive wall, slow head-tail, Robinson, coupled bunch, *etc*. These are by no means trivial. However, it is believed that one knows how to deal with them.

More difficult is another type of problems, the "non-conventional" ones, which become important because of the special requirements of the proton driver. They are yet to be understood.

- Longitudinal microwave instability below transition:
  Below transition, the validity of the Keil-Schnell criterion is questionable. There are a number of cases in which this criterion is violated. For example, the beam intensity in the RAL-ISIS is 10 times higher than the calculated threshold.
  
  In fact, there is no report on observation of microwave instability in any existing proton machines when they operate below transition. Only in some special machine experiments, which introduce large resistive impedance on purpose, self-bunching and perhaps also instability were seen in a coasting beam.
  
  More theoretical, simulation and experimental studies are needed on this subject.

- Fast head-tail (transverse mode-coupling) in the presence of strong space charge:
  This type of instability is clearly observed in electron machines. Moreover, the calculated threshold and growth rate agree well with the measurements. However, it has never been observed in any proton machine. There are two possible explanations:
  
  1. If the betatron tune spread $\Delta Q_\beta$ in a proton machine is many times larger than the synchrotron tune $Q_s$, then the mode lines ($m = 0, \pm 1, ...$) would get smeared and there won't be any coupling.
  2. In low- and medium-energy proton machines, the space charge force is significant. It would shift $m = -1$ mode downward as the beam intensity increases. Meanwhile, the inductive broadband wall impedance would shift this mode upward. Thus, they intend to cancel each other. This makes the coupling between the mode $m = 0$ and $m = -1$ more difficult.
  
  These claims need support from more careful analytical and numerical study.

- Synchro-betatron resonance due to dispersion in rf section:
  Due to the compact size of the proton driver, some rf cavities may have to be installed in the dispersion region. Would this be a problem? The concern is about the synchro-betatron resonance $kQ_\beta \pm mQ_s = n$. In previous studies, the case $k = 1$ has been fully analyzed. [2] However, the cases of $k = 2, 3,$ ... remain open. It is not clear at this moment if this would be a problem, although experiences tell us that betatron resonances up to the 3rd order could still be important even in a rapid cycling synchrotron.

## Particle loss, collimation and shielding

The tolerable particle loss is an important issue in high intensity proton machine design. One main concern is the hands-on maintenance, which requires the residual dose below certain level before one may proceed to do any repair work. Using the preliminary lattice and magnet design of the proton driver, Monte Carlo simulations using the code MARS show that, at an average particle loss rate of 1 W/m, the residual dose after 30 days irradiation and 4 hours cool down would be below 100 mrem/hr. This result agrees with that obtained at LANL and ORNL.

In the meantime, a collimation system has been incorporated into the lattice. Even with an assumed 10% loss at 3 GeV (which gives 72 kW), simulation shows that this system would confine more than 99% of the losses in a local section, leaving most of the ring (the so-called "quiet" area) below 5 W/m.

The MARS code was also used for radiation shielding calculation. The needed dirt thickness for shielding 1-hour accidental full beam loss is 29 feet. It is close to the result obtained from the simplified scaling formula (the Dugan criterion), which gives 32 feet.

## Transient beamloading

This problem is crucial to the intense short bunch operation. The single bunch intensity ($2.5 \times 10^{13}$) gives a charge $q = 4$ $\mu$C. For a 20 kV cavity and a gap capacitance $C = 400$ pF, the single pass beamloading voltage $q/C$ would reach 10 kV, which has to be compensated. However, because the bunch is very short ($\sigma_b = $ 1-2 ns), how to inject a short current pulse to do the compensation is challenging. This is a high priority item in the proton driver study. At this moment, the plan is to use an rf feedforward system for global compensation and an rf feedback system for reducing bunch-to-bunch and turn-by-turn variations.

The multi-pass beamloading voltage has a rich Fourier spectrum when a low Q cavity is used. An rf feedback system would have to provide several harmonics for sufficient compensation.

## Lattice

The constraints and requirements of the lattice design are:

- $B_{max} \leq 1.5$ Tesla

- No transition crossing (which excludes the FODO lattice)

- Large dynamic aperture ($\epsilon_N = 200\pi$ mm-mrad)

- Large momentum acceptance: $\frac{\Delta p}{p} = \pm 2.5\%$

- Dispersion free straight sections for rf, injection and extraction

- Suitable locations for a collimation system

There are two FMC (flexible momentum compaction) lattice candidates, one is triangular, another racetrack. Both give large or imaginary $\gamma_t$ and use sextupoles to increase the momentum acceptance. The same sextupoles can also be used to control the slope ($\alpha_1$) of the momentum compaction factor if a compromise in chromaticity control is acceptable.

It turns out to be impractical to design a 16 GeV lattice that meets all the above requirements while keeps the same size (474 m) of the present booster. The new booster would be larger.

## TECHNICAL SYSTEMS

### RF

The required rf voltage is about 1.2 MV. Due to small size of the machine, the cavity needs to have high gradient ($> 20$ kV/m). A new type of magnetic alloy called the Finemet will be used. Thanks to the US-Japan collaboration, a 7.5 MHz, 20 kV prototype cavity is under construction. After high power bench test, it will be installed in the Main Injector for beam test.

### Magnets

These are big magnets. The dipole has an aperture of 5" × 13" and weighs about 10 tons per meter. The peak field is 1.5 Tesla. Simulation shows the field quality is good: $\frac{\Delta B}{B} < 10^{-3}$ within ±4". One important parameter is the ac loss of these magnets. The data from vendors' catalogs are not applicable, because they are measured at 60 Hz and without dc bias. Our measurement shows that, at 15 Hz and with dc bias current, the as loss is about 1/15 of that in the catalog.

### Power supply

There are two proposals. One is a programmable power supply using fast switching IGBT (about 7 kHz). The reactive power will be stored in a capacitor bank. Another is a resonant power supply. The latter has three variations: (1) single 15 Hz resonance circuit (as in the present booster), (2) dual-resonance circuit (15 Hz plus 12.5% 30 Hz component), and (3) dual-frequency circuit (up-ramp 10 Hz, down-ramp 30 Hz, using IGBT to switch the frequency). Both (2) and (3) can save significant rf power and rf voltage. One concern about (3) is the ripple effect during injection.

An attractive feature of the programmable power supply is its feasibility, namely, to allow different ramp rate, a flat top and a flat bottom. But it is more expensive.

## Vacuum pipe

The present design is a thin Inconel pipe (5" × 9" × 50 mils) with water cooling. Compared with stainless steel, Inconel has high strength and high electric resistivity. Its eddy current is 4 times smaller than that in stainless steel. Compared with the ceramic pipe (as used in the ISIS), Inconel reduces the vertical magnet aperture by 1.5-2 inches. The main concerns about an Inconel pipe are:

- Large deflection under vacuum: $\Delta y = -1"$, $\Delta x = 0.7"$
- Eddy current heating: $\sim 3$ kW/m
- Eddy current induced error field: At $\dot{B}_{max}$, 1" reference point, the harmonic components are $b_1 = 93$ unit, $b_3 = 2.6$ unit.

Several pieces of prototype pipes are being fabricated. The water cooling tubes are glued to the beam pipe by some special epoxy, which is electrically an insulator but has good thermal conductivity. The pipes will undergo vacuum and heating tests.

Another design using ultra thin (5 mils) Inconel or Ti-Al alloy is under investigation. It uses ribs for mechanical stability. Because the heat load would be reduced by a factor of 10, the cooling system could be eliminated.

## RF chopper

A new type of chopper, which is a pulsed beam transformer using Finemet and can be placed in front of the RFQ in the linac, has been designed and built in collaboration with the KEK. The beam test will be performed at the HIMAC in Japan.

## ACKNOWLEDGEMENT

This paper is a report of the work carried out by the Fermilab Proton Driver Design Team.

## REFERENCES

1. S. Holmes, editor, "A Development Plan for the Fermilab Proton Source," FERMILAB-TM-2021 (1997).
2. T. Suzuki, Particle Accelerators, Vol. 18, pp. 115-128 (1985).

# Collective Instabilities in the LHC: Electron Cloud and Satellite Bunches

Francesco Ruggiero and Xiaolong Zhang

*CERN, Geneva, Switzerland*

**Abstract.** We shortly review electron cloud effects in the CERN Large Hadron Collider. In particular, we discuss recent simulations showing a significant reduction of the beam induced heat load on the cold beam screen with weak satellite bunches at 5 ns from the main proton bunches. This can be an effective solution to shorten the conditioning period required to lower the secondary electron yield of the screen surface.

## INTRODUCTION

The CERN Large Hadron Collider (LHC) has a design energy of 7 TeV per beam and an unprecedented design luminosity of $10^{34}$ cm$^{-2}$ s$^{-1}$. For cost optimisation it makes use of the existing injector chain and has to be accommodated in the LEP tunnel. Such design objectives are accompanied by several technological and beam dynamics challenges [1], ranging from the 8.3 T superconducting dipole magnets operating in superfluid helium at 1.9 K to the large number of low-emittance, intense proton bunches to be injected in the two magnetic channels at 450 GeV, safely accelerated and collided at top energy. A list of LHC parameters is shown in Table 1.

The heat load of about 1 W/m, mainly due to synchrotron radiation and to beam image currents [2], has to be taken by a cold dipole beam screen, with a 50 $\mu$m copper coating, operated between 5 and 20 K. At this temperature the cryopumping capacity is strongly reduced and about 4% of the screen surface will be covered by pumping slots, so that the cold bore at 1.9 K can pump away the gas, while being protected from synchrotron radiation. A summary of conventional collective effects in the LHC, including parasitic losses and coherent tune shifts, mode coupling and resistive wall instability, can be found in Refs. [3,4], while more recent results on head-tail instability, Landau damping and octupole correctors have been presented in [5]. Coaxial wire measurements, simulations and analytic estimates of RF-penetration through thin metallic layers, such as metallized ceramic chambers for kicker magnets, are discussed in [6] and RF-screening will be the subject of a forthcoming experiment in a CERN EPA beam line. Direct space charge effects have been recently studied in [7] and found to have modest consequences on the

**TABLE 1.** List of LHC parameters.

| | | |
|---|---|---|
| Energy | (TeV) | 7.0 |
| Dipole field | (T) | 8.3 |
| Coil aperture | (mm) | 56 |
| Distance between apertures | (mm) | 194 |
| Luminosity | (cm$^{-2}$ s$^{-1}$) | $10^{34}$ |
| Beam-beam parameter | | 0.0032 |
| Injection energy | (GeV) | 450 |
| Circulating current/beam | (A) | 0.530 |
| Bunch spacing | (ns) | 25 |
| Protons per bunch | | $1.05 \times 10^{11}$ |
| Stored beam energy | (MJ) | 332 |
| Normalised transverse emittance | ($\mu$m) | 3.75 |
| r.m.s. bunch length | (m) | 0.075 |
| r.m.s. bunch radius (arcs) | (mm) | 0.303 |
| Vertical aperture of dipole beam screen | (mm) | 36 |
| Beta values at I.P. | (m) | 0.5 |
| Full crossing angle | ($\mu$rad) | 300 |
| Beam lifetime | (h) | 22 |
| Luminosity lifetime | (h) | 10 |
| Energy loss per turn | (keV) | 6.9 |
| Critical photon energy | (eV) | 44.1 |
| Total radiated power per beam | (kW) | 3.7 |

LHC dynamic aperture and on the beam emittance (through coherent quadrupole oscillations). Finally, a review of incoherent and coherent beam–beam effects in the LHC can be found in Ref. [8].

In the following we shortly review electron cloud effects in the LHC arcs: a fairly complete list of publications is available at the web address wwwslap.cern.ch/collective/electron-cloud/. In particular, the reduction of beam induced heat load by a dipole beam screen with ribbed surface and low reflectivity was first discussed in [9], while beam scrubbing scenarios and the role of satellite bunches have been recently discussed in [10]. We summarise the physical mechanism on which the clearing effect of satellite bunches is based and present some recent simulation results. Satellite bunches may turn out to be an effective cure also for electron cloud effects recently observed in the CERN SPS with LHC-type of beam [12].

## MECHANISM OF THE ELECTRON CLOUD BUILD-UP

A sketch of the electron cloud build-up in the LHC is presented in Fig. 1. The physical mechanism is the following:

- Photoelectrons created at the pipe wall are accelerated by proton bunches up to 200 eV and cross the pipe in about 5 ns, i.e. significantly less than the 25 ns bunch spacing. In the strong dipole field, only the vertical component of the beam force is effective in accelerating the electrons. Indeed they spiral along the vertical magnetic field lines with typical Larmor radii of a few $\mu$m and perform about a hundred cyclotron rotations during a bunch passage.

- Slow secondary electrons with energies below 10 eV have a time-of-flight longer than 20 ns and survive until the next bunch. When the next bunch arrives, there is a relatively uniform distribution of photoelectrons plus secondary electrons in the screen cross section: the energy gain can reach a few keV and these fast particles hit very quickly the screen walls, producing again low-energy secondary electrons.

- If the maximum secondary electron yield $\delta_{\max}$ is larger than a critical value, weakly dependent on the horizontal position along the pipe cross section, this process may lead to an electron cloud build-up (ultimately limited by space charge effects) with potential implications for beam stability and heat load on the LHC beam screen. The critical yield is around 1.3 for nominal LHC parameters.

Secondary emission can be reduced by special coatings or by an adequate electron dose. An electron dose of 1 mC/mm$^2$ is sufficient to lower the maximum secondary yield below the critical value of 1.3. To condition the screen surface in the shortest possible time, while keeping the heat deposition within acceptable bounds, it is possible to use electrons accelerated by a special proton beam, either with increased bunch spacing or with weak satellite bunches at 5 ns from the main bunches. These behave as 'clearing bunches' for slow secondary electrons.

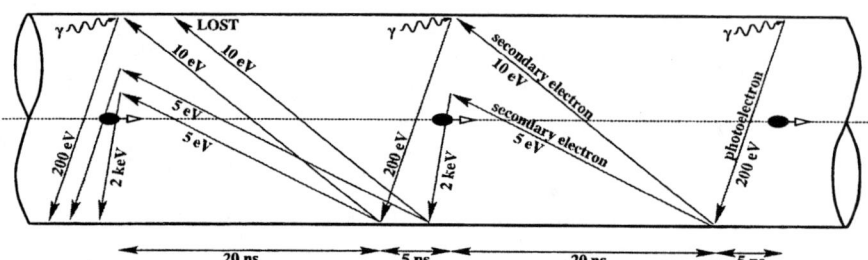

**FIGURE 1.** Sketch of the electron cloud build-up in the LHC.

## Electron Cloud Instability

Simulations performed in 1997 [13] assuming a maximum secondary electron yield $\delta_{\max} = 1.5$, high reflectivity $R = 1$ and large photoelectron yield $\delta_{\gamma e} = 1$ gave an instability rise time of 50 msec horizontally and 130 msec vertically. More recent simulations by X. Zhang with $R = 10\%$ and $\delta_{\gamma e} = 0.2$ gave a slower instability with 340 msec horizontal and 170 msec vertical rise times.

According to the 1997 simulations, lowering $\delta_{max}$ from 1.5 to 1.1 reduces the horizontal wake by less than a factor of 2. If the beam current is a factor 2 smaller than the design value, the effective horizontal wakefield increases by about 25%; the wakefield does not decrease in proportion to the beam current, because, for lower current, the high-density region of the electron cloud is closer to the beam pipe centre.

The vertical wake is instead very sensitive to the initial secondary-electron energy and may even change sign for maximum initial energies above 25 eV, when more than half of the secondaries cross the centre of the beam pipe before the arrival of the next bunch.

## SIMULATION RESULTS WITH SATELLITE BUNCHES

Satellite bunches may increase the critical secondary electron yield and at the same time modify the energy distribution of the electrons hitting the pipe wall. A proper choice of satellite intensity and spacing from the forerunner nominal bunches can significantly reduce the head load on the beam screen.

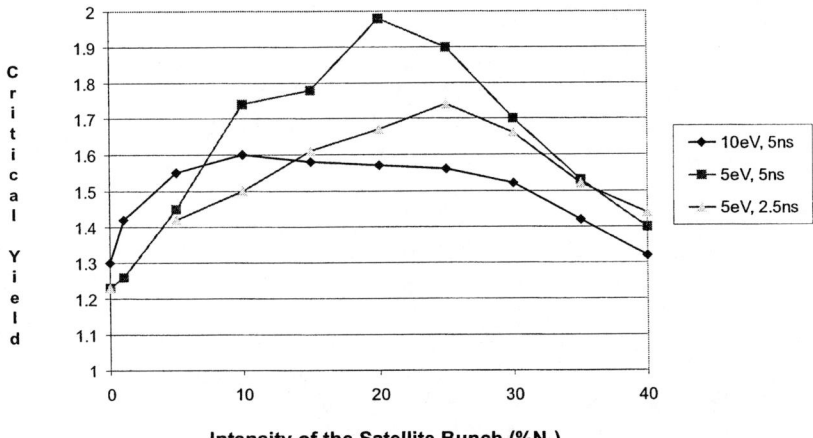

**FIGURE 2.** Critical value of the maximum secondary electron yield $\delta_{max}$ versus the relative intensity of satellite bunches following the nominal bunches at a spacing of 2.5 ns (one LHC RF bucket) or 5 ns (two LHC RF buckets). We assume a highly reflective beam screen surface, with $R \simeq 1$, and a half-gaussian secondary electron energy distribution with 5 eV or 10 eV r.m.s. width.

# Critical secondary electron yield

To find the critical secondary electron yield with different satellite bunch patterns, we neglect space charge effects and generate photoelectrons only for the first bunch. We observe the evolution of the number of electrons per unit length over 60 bunch passages for different values of $\delta_{max}$ (corresponding to a primary electron energy of 300 eV). We look for the critical value below which the electron density will decrease and above which the electron density will grow exponentially.

In Fig. 2 we plot the critical value of $\delta_{max}$ for different intensities of the satellite bunches. The legend for each curve indicates the initial energy distribution $\sigma_{se}$ of the secondary electrons and the spacing between each satellite bunch and its forerunner nominal bunch. We find that the highest critical value of $\delta_{max}$ is about 2, for $\sigma_{se} = 5$ eV, a spacing of 5 ns and a satellite bunch intensity of 20% of the nominal intensity. Therefore, satellite bunches with such intensity have the best clearing effect for secondary electrons. In the case of 2.5 ns spacing (one LHC RF bucket), the effect of the satellite bunch is mainly to decelerate the photoelectrons rather than clearing the secondary electrons, since most photoelectrons just pass the centre of the vacuum chamber around this time. In this case the satellite bunch reduces the energy of the hitting electrons and in turn can again increase the critical value of $\delta_{max}$.

Fig. 3 shows the critical secondary yield for different spacings between satellite and nominal bunches. The intensity of the satellite bunch is 20% of the nominal bunch, with $\sigma_{se} = 5$ eV. We see that the filling pattern of 5 ns bunch spacing represents the most favourable case.

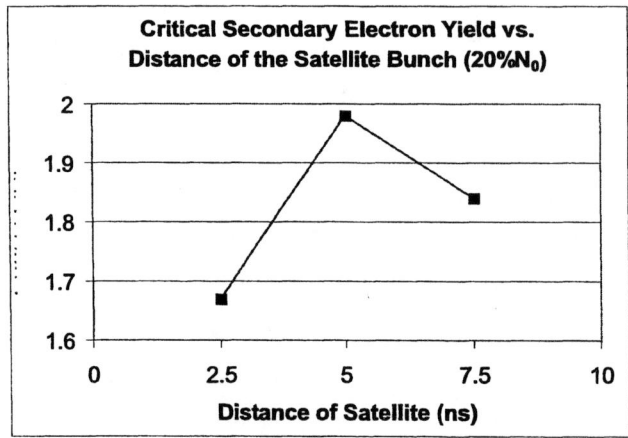

**FIGURE 3.** Critical value of the maximum secondary electron yield $\delta_{max}$ for satellite bunches having 20% of the nominal bunch intensity and following nominal LHC bunches at various distances. We assume a highly reflective beam screen surface, with $R \simeq 1$, and half-gaussian secondary electron energy distribution with 5 eV r.m.s. width.

# Heat load and scrubbing time

The simulation results shown in Fig. 4 refer to the LHC dipole beam screen and have been performed assuming a photoelectron yield $\delta_{\gamma e} \simeq 0.2$ and a surface reflectivity of 10%. This means that 10% of the photoelectrons are uniformly distributed around the beam screen, while the remaining 90% have a Gaussian angular distribution with an r.m.s. angle of 22.5° from the horizontal plane. The maximum secondary electron yield corresponds to a primary electron energy of 300 eV and secondary electrons have a Gaussian energy distribution with 5 eV r.m.s. value and cut-off at 5 sigma. For an accurate modelling of the electron dynamics, there are 50 slices per bunch (both for nominal and satellite bunches, having the same r.m.s. length of 7.5 cm) and again 50 slices for each inter-bunch gap. Space charge effects are included in the simulation.

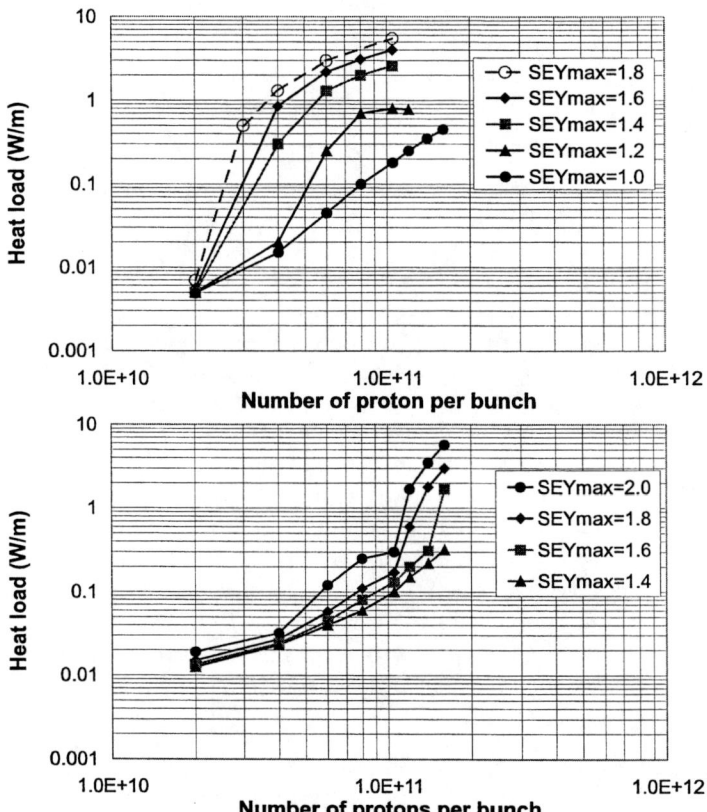

**FIGURE 4.** Heat load versus bunch population for different values of $\delta_{\max}$ and 10% reflectivity: without (top) and with (bottom) satellite bunches having 20% of the nominal bunch intensity and a spacing of 5 ns.

With nominal LHC bunch intensity and spacing, but with satellite bunches at a distance of 2 RF buckets, the heat load for $\delta_{max} = 1.8$ is 180 mW/m and the resulting scrubbing time is 36 hours, assuming that only electrons hitting the screen wall with more than 100 eV be effective for surface conditioning. As shown in Fig. 5 there is a window around 15-20% for the relative intensity of satellite bunches, where the heat load is significantly reduced; the corresponding critical value of $\delta_{max}$ is large (above 1.8). For lower intensities of the satellite bunches, the effect of space charge repulsion is reduced and the heat load increases. For a reduced reflectivity of 2% and a photoelectron yield of 0.1, the heat load becomes only 18 mW/m and the corresponding scrubbing time increases to about 310 hours. This is only slightly shorter than the scrubbing time of 360 hours estimated by taking into account only photoelectrons.

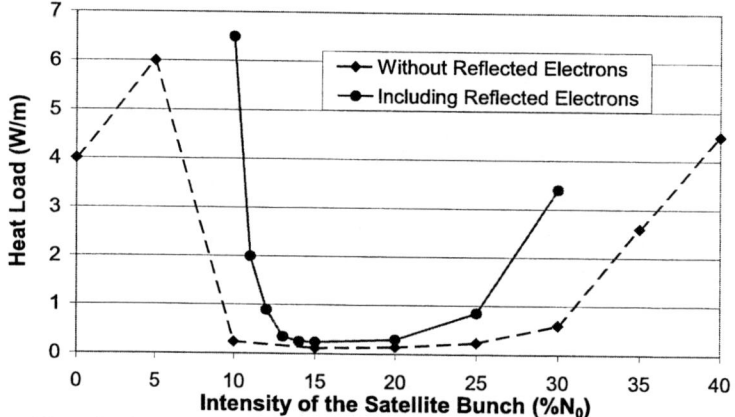

**FIGURE 5.** Heat load versus relative intensity of satellite bunches, following nominal LHC bunches at 2 RF wavelengths, with (solid line) or without (dashed line) elastic electron reflection as described in Ref. [11], with $\delta_{max} = 1.6$ and 10% reflectivity.

It is possible to get an independent estimate of the minimum time required for surface conditioning, without any assumption on reflectivity and photoelectron yield[1]. Let us assume a maximum heat load of 200 mW/m, compatible with cooling, and an average electron energy around 200 eV. As shown in Fig. 6, this is consistent with simulation results for a nominal LHC proton beam with satellite bunches. The corresponding linear flux of electrons bombarding the screen surface is $6 \times 10^{15}$ s$^{-1}$m$^{-1}$. Since a meter of LHC beam screen has a surface of $1.25 \times 10^5$ mm$^2$, the electron dose accumulated per hour is $\frac{200\,\text{mW/m}}{200\,\text{eV}} \frac{\text{m}}{1.25 \times 10^5 \,\text{mm}^2} 1.6 \times 10^{-19} C \simeq 8 \times 10^{-9} \frac{C}{\text{mm}^2 \text{s}}$ and the beam scrubbing time necessary to accumulate the required electron dose of 1 mC/mm$^2$ is about 35 hours. Some uncertainty remains in connection with the scrubbing efficiency of electrons bombarding the screen surface with energies lower than 100 eV, although experimental data indicate that the required

---

[1] This estimate has been suggested by C. Benvenuti, P. Chiggiato and V. Rouzinov.

dose is about the same for primary electron energies ranging from 500 eV down to 100 eV [10].

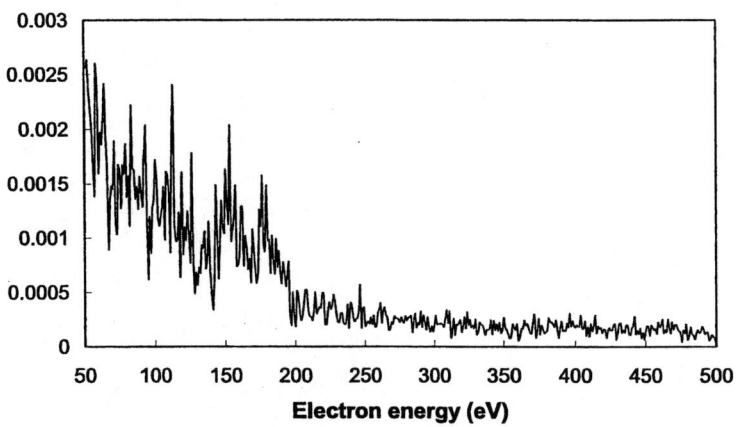

**FIGURE 6.** Integrated energy distribution of electrons hitting the dipole screen surface with satellite bunches having 15% of the nominal bunch intensity and following nominal LHC bunches at 2 RF wavelengths. We assume reflectivity $R = 0.1$, photoelectron yield $\delta_{\gamma e} \simeq 0.2$ and half-gaussian secondary electron energy distribution with 5 eV r.m.s. width.

# REFERENCES

1. Evans, L.R., LHC Project Report 303, CERN, Geneva, Switzerland (April 1999), in *Proc. Particle Accelerator Conference, New York, 1999*, edited by Luccio, A. and MacKay, W. (IEEE, New York, 1999), pp. 21-25.
2. Caspers, F., Morvillo, M., Ruggiero, F., and Tan, J., LHC Project Report 307, CERN, Geneva, Switzerland (August 1999).
3. Ruggiero, F., Tech. Rep. SL-95-09-AP, CERN, Geneva, Switzerland (February 1995), in *Proc. Workshop on Collective Effects in Large Hadron Colliders, Montreux, 1994*, edited by Keil, E. and Ruggiero, F. (Gordon and Breach, New York, 1995), *Part. Accel.* **50**, pp. 83-104.
4. Ruggiero, F., *et al.*, LHC Project Report 120, CERN, Geneva, Switzerland (June 1997), in *Proc. Particle Accelerator Conference, Vancouver, 1997*, edited by Comyn, M., Craddock, M. K., Reiser, M., and Thomson, J. (IEEE, New York, 1998), pp. 107-109. See also Berg, J.S., LHC Project Report 16, CERN, Geneva, Switzerland (July 1996).
5. Gareyte, J., Koutchouk, J.-P., and Ruggiero, F., LHC Project Report 91, CERN, Geneva, Switzerland (revised version, April 1997). See also Berg, J.S. and Ruggiero, F., Tech. Rep. SL-96-71-AP, CERN, Geneva, Switzerland (December 1996), Berg, J.S., LHC Project Report 100, CERN, Geneva, Switzerland (April 1997) and Koutchouk, J.-P. and Ruggiero, F., LHC Project Note 163, CERN, Geneva, Switzerland (October 1998).

6. Caspers, F., *et al.*, LHC Project Report 300, CERN, Geneva, Switzerland (April 1999), in *Proc. Particle Accelerator Conference, New York, 1999*, edited by Luccio, A. and MacKay, W. (IEEE, New York, 1999), pp. 1408-1410.
7. Ruggiero, F. and Zimmermann, F., LHC Project Report 295, CERN, Geneva, Switzerland (April 1999), in *Proc. Particle Accelerator Conference, New York, 1999*, edited by Luccio, A. and MacKay, W. (IEEE, New York, 1999), pp. 2626-2628.
8. Poole, J. and Zimmermann, F. (editors), *Proc. Workshop on Beam-Beam Effects in Large Hadron Colliders, CERN, April 1999*, Tech. Rep. SL-99-039-AP, CERN, Geneva, Switzerland (June 1999).
9. Baglin, V., *et al.*, LHC Project Report 188, CERN, Geneva, Switzerland (June 1998), in *Proc. 6th European Particle Accelerator Conference, Stockholm, 1998*, edited by Myers, S., Liljeby, L., Petit-Jean-Genaz, Ch. *et al.* (IOP, Bristol, 1999), pp. 359-361.
10. Brüning, O., *et al.*, LHC Project Report 290, CERN, Geneva, Switzerland (April 1999), in *Proc. Particle Accelerator Conference, New York, 1999*, edited by Luccio, A. and MacKay, W. (IEEE, New York, 1999), pp. 2629-2631.
11. Furman, M. and Lambertson, G.R., Tech. Rep. LBNL-41123/CBP Note-246/PEP-II AP Note AP 97.27 (January 1998), in *Proc. Int. Workshop on Multibunch Instabilities in Future Electron and Positron Accelerators (MBI97), Tsukuba, 1997*, edited by Chin, Y.H., KEK Proceedings 97-17, KEK, Tsukuba, Japan (December 1997), pp. 170-199. See in particular Eqs. (4.10) and (4.11).
12. Höfle, W. and Jimenez, J., private communication (CERN, August 1999).
13. Zimmermann, F., LHC Project Report 95, CERN, Geneva, Switzerland (February 1997).

# Stability Issues of Low-Energy Intense Beams

K.Y. Ng and A.V. Burov

*Fermi National Accelerator Laboratory,[1] P.O. Box 500, Batavia, IL 60510*

**Abstract.** Some stability issues of low-energy intense beams are discussed. These include inductor tuners for the cancellation of the longitudinal space-charge induced potential-well distortion and their consequences, transient beamloading and possible feed-forward alleviation, coherent and incoherent transverse tune shifts, as well as the impact of transverse space charge on transverse mode-coupling instability.

## I INTRODUCTION

Several relatively low-energy and very high-intensity proton rings, such as the storage rings of the U.S. and European neutron spallation sources, the booster of the Japan Hadron Project, and the low-energy ring of the Fermilab future booster, are under design [1–4]. These rings have circumferences around 150 to 200 m, containing $\sim 1 \times 10^{14}$ protons per cycle. High intensity and low energy imply large space-charge forces in the longitudinal and transverse directions. The longitudinal space-charge force will counteract significantly the rf focusing force giving rise to a large potential-well distortion. To cope with this distortion, one method is to insert inductor tuners to cancel the longitudinal space-charge force. This insertion together with its consequences will be discussed. The intense charge density will also impact large transient beamloading onto the rf cavities. A feed-forward scheme to alleviate the beamloading voltage is addressed. Transversely, the importance of the coherent and incoherent space-charge tune shifts for an intense low-energy beam is reviewed. Finally, the effect of space charge on transverse mode-coupling instability is investigated.

## II LONGITUDINAL SPACE CHARGE

Let us take, for example, an older design of the low-energy ring in the Fermilab future booster, which accelerates 2 bunches each containing $N_b = 5.0 \times 10^{13}$ protons from kinetic energy 1.0 GeV to 4.5 GeV at 15 Hz. The ring has a circumference of 180.649 m, rf harmonic $h = 2$, and transition gamma $\gamma_t = 7$. The 95% bunch area is $A = 1.0$ eV-s and 95% normalized emittance $\epsilon_{95} = 200 \times 10^{-6}$ $\pi$m. The average current in the ring is $I_{\text{av}} = 23.27$ A and the peak current is $I_{\text{pk}} = I_{\text{av}}/B = 93.06$ A, where $B = 0.25$ is the *bucket bunching factor*. Assuming parabolic distribution, the half bunch length is $\hat{\tau} = 3eN_b/(4I_{\text{pk}}) = 64.56$ ns. The half momentum spread is therefore $\hat{\delta} = A/(\pi\hat{\tau}) = 3.322 \times 10^{-3}$. The average betatron function and average dispersion of the ring are, respectively, $\langle\beta\rangle = 25$ m and $\langle D\rangle = 1.8$ m. Thus, the average beam radius is about $a = [\epsilon_{95}\langle\beta\rangle/(\gamma\beta) + (\langle D\rangle\hat{\delta})^2]^{1/2} = 5.29$ cm. A beam

---

[1] Operated by the Universities Research Association, Inc., under contract with the U.S. Department of Energy.

CP496, *Workshop on Instabilities of High Intensity Hadron Beams in Rings*,
edited by T. Roser and S. Y. Zhang
© 1999 American Institute of Physics 1-56396-910-6/99/$15.00

pipe of radius $b \sim 8$ cm will be recommended. We can therefore estimate the longitudinal space-charge impedance of the ring [5]:

$$\left.\frac{Z_0^\parallel}{n}\right|_{\text{spch}} = -j\frac{Z_0}{2\gamma^2\beta}\left(1 + 2\ln\frac{b}{a}\right) = 92.11\ \Omega\ ,  \qquad (2.1)$$

where $\gamma = 2.0658$ and $\beta = 0.8750$ are the Lorentz factors at injection while $Z_0 \approx 377\ \Omega$ is the free-space impedance. This is not bad for microwave instability, because the operation is below transition and the Keil-Schnell stability limit is

$$\left|\frac{Z_0^\parallel}{n}\right| \leq F_\parallel \frac{E_0|\eta|}{e\beta^2 I_{\text{pk}}}\left[\frac{\Delta E}{E_0}\right]^2_{\text{FWHM}} = 78.86\ \Omega\ ,  \qquad (2.2)$$

where $E_0$ is the total energy of the synchronous particle and $F_\parallel = 1.047$ is the form factor for a bunch with parabolic momentum distribution.

## A  Potential-Well Distortion and Inductor Insertion

Ignoring coupling impedances, for a bunch with the half length and half momentum spread specified above, the rf bucket holding the bunch must have a synchrotron tune $\nu_s$ and a rf voltage $V_{\text{rf}}$,

$$\nu_s = \frac{|\eta|\hat\delta}{\omega_0\hat\tau} = 1.207\times 10^{-3}\ ,\qquad V_{\text{rf}}\cos\phi_s = \frac{2\pi\beta^2 E_0\nu_s^2}{|\eta|h} = 31.73\ \text{kV}\ , \qquad (2.3)$$

where $\omega_0/(2\pi) = 1.452$ MHz is the revolution frequency and $\phi_s$ is the synchronous angle. Therefore, a particle with time advance $\tau$ ahead the synchronous particle will see the relative rf voltage

$$V_{\text{rf}}[\sin(\phi_s - h\omega_0\tau) - \sin\phi_s] \approx -V_{\text{rf}}\cos\phi_s\left[\frac{3\pi B}{2}\right]\frac{\tau}{\hat\tau} = -37.38\frac{\tau}{\hat\tau}\ \text{kV}\ , \qquad (2.4)$$

where the rf sine wave has been linearized. However, the intense beam creates on the particle a strong repulsive longitudinal electric field

$$E_z^{\text{spch}} = -\frac{eZ_0}{4\pi\gamma^2\beta^2 c}\left[1 + 2\ln\frac{b}{a}\right]\frac{d\lambda}{d\tau}\ , \qquad (2.5)$$

where $c$ is the velocity of light. Assuming the linear parabolic bunch distribution $\lambda(\tau) = 3eN_b/(4\hat\tau)(1 - \tau^2/\hat\tau^2)$, the space-charge voltage seen per turn is

$$V_{\text{spch}} = E_z^{\text{spch}} C = \frac{3\pi I_b}{(\omega_0\hat\tau)^2}\left.\frac{Z_0^\parallel}{n}\right|_{\text{spch}}\frac{\tau}{\hat\tau} = +29.11\frac{\tau}{\hat\tau}\ \text{kV}\ , \qquad (2.6)$$

where $C$ is the circumference of the accelerator ring. As a result, to maintain the bunch shape, such as the length and momentum spread, the rf voltage required must be increased to $V_{\text{rf}} = (37.38 + 29.11)31.73/37.38 = 56.43$ kV.

If we do not want such a large rf voltage brought about by the space-charge force, we need to cancel the space charge by, for example, inductor insertion. Two such experiments have been performed lately.

## 1. Fermilab-Los Alamos Collaboration

In 1997, the Los Alamos PSR was running at 797 MeV with an intensity of $3 \times 10^{13}$ protons in the beam. The space-charge force was intense and an rf voltage of 10 kV was required to bunch the beam so that the injection-extraction gap could be kept clean. Two ferrite tuners designed to cancel $\frac{2}{3}$ of the space charge were built, each $\sim 76$ cm long consisting of 30 Toshiba $M_4C_{21A}$ cores (12.7 cm I.D., 20.3 cm O.D., and 2.54 cm thick). The relative magnetic permeability is $\mu = 50$ to 70 over a modest temperature range. These properties remain approximately constant up to 30 MHz, after which $\mu$ rolls off. A solenoid was wound outside so that relative magnetic permeability could be decreased through perpendicular biasing.

The experiment was performed in August of 1997 [6]. First, bunch lengthening was observed when the ferrite was biased (Fig. 1 left) as was expected with the decrease of the inductance. Second, the rf voltage required for bunching was reduced by about $\frac{1}{3}$, indicating that the space-charge force had been cancelled partially by the inductance of the ferrite (Fig. 1 right). Third, the injection-extraction gap during the experiment was probably the cleanest ever observed.

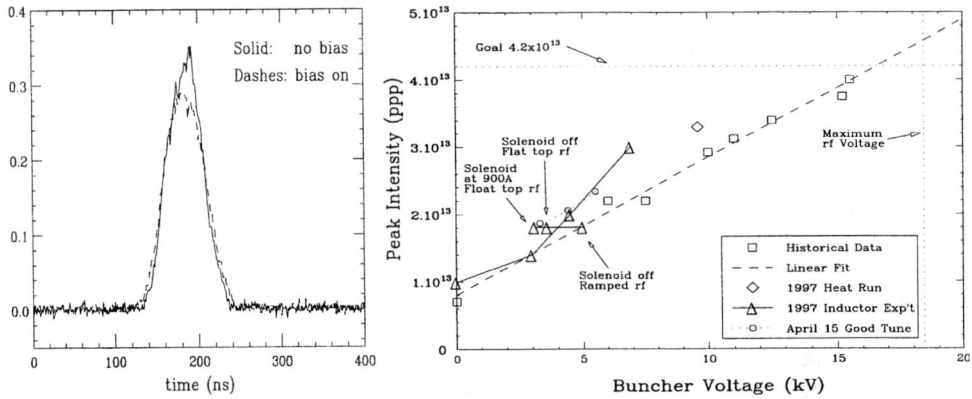

**FIGURE 1.** Left: PSR beam bunch shapes with unbiased (solid) and biased (dashes) ferrite compensation. Right: Stability threshold versus rf bunching voltage. Results of this experiment are depicted by triangles. (Reproduced from Ref. 6).

## 2. KEK experiment

A similar experiment started at the KEK PS Main Ring in 1997, but with a much lower intensity of 2 to $9 \times 10^{11}$ protons per bunch [7]. The beam kinetic energy was 500 MeV with a space-charge impedance $Z_0^{\parallel}/n = -j310\ \Omega$. Instead of ferrite, a Met-Glass-like material called Finemet was used. The inductor tuner consisted of 12 Finemet cores (14.0 cm I.D., 34.0 cm O.D., and 2.54 cm thick), without current biasing. The control of the relative permeability was achieved by installing copper short bars across the Finemet cavities. The coherent frequency of the quadrupole synchrotron oscillation was measured as a function of bunch intensity. As shown in Fig. 2, with the inductor tuner on, the coherent frequency was less dependent on intensity, indicating that the space-charge force had been partially cancelled.

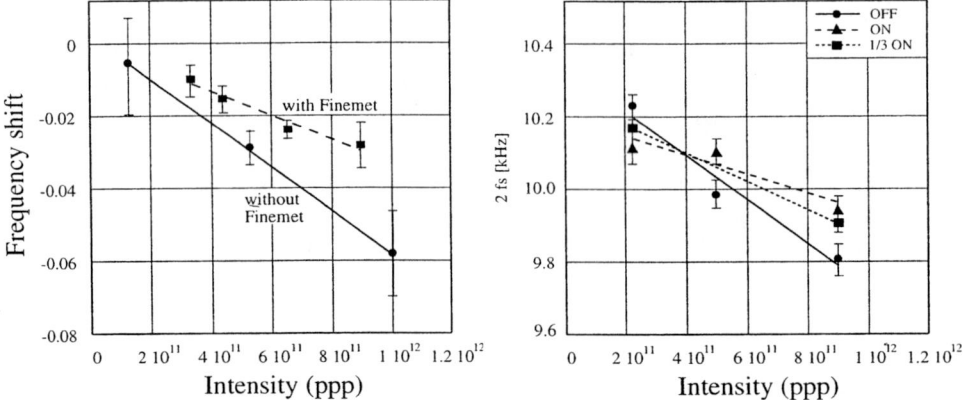

**FIGURE 2.** Left: Measured frequency shifts of the quadrupole oscillations versus beam intensity at KEK with and without Finemet. Right: New KEK results of quadrupole oscillation frequency versus beam intensity with Finemet tuners on, $\frac{1}{3}$ on, and off. (Reproduced from Ref. 7).

## B  Power Loss to Ferrite or Finemet

To incorporate loss, the relative permeability can be made complex: $\mu \to \mu' - j\mu''$. The impedance of the ferrite is therefore

$$\frac{Z_0^\parallel}{n} = j(\mu' - j\mu'')\omega_0 L ,  \qquad (2.7)$$

where $L$ denotes the inductance of the ferrite or Finemet required to compensate for the space charge of the bunch. It is clear that $\mu'$ and $\mu''$ must be frequency-dependent. Their general behaviors are shown in the left plot of Fig. 3. For the Toshiba $M_4C_{21A}$ ferrite, $\mu'$ is roughly constant at $\mu'_L \sim 50$ at low frequencies and starts to roll off around $\omega_r/(2\pi) \sim 30$ MHz, while $\mu''$, being nearly zero at low frequencies, reaches a maximum $\mu''_R$ near $\omega_r/(2\pi)$. For the Finemet this roll-off frequency can be at very much lower frequency, around 1 to 10 MHz. Thus the power loss to the inductor tuners from the beam may become very large and may not be ignored. First, the energy lost by the beam has to be compensated by the rf system. Second, the ferrite or Finemet can become too warm. Third, a large $\mathcal{R}e\, Z_0^\parallel$ of the inductor can lead to microwave instability. In fact, such an instability had been already observed in the 1997 Los Alamos experiment, and this instability has been much more serious at the present moment when the PSR beam intensity has been upgraded to $\sim 5.0 \times 10^{13}$ protons as indicated in right plot of Fig. 3.

Essentially, the loss is the overlap integral of the bunch spectrum and $\mathcal{R}e\, Z_0^\parallel$ of the inductor tuners. To compute this, an impedance model for the inductor tuners is necessary. The simplest 2-parameter model consists of an *ideal* inductance $L$ and an *ideal* resistor $R$ in parallel, which gives

$$Z_0^\parallel(\omega) = j\omega L \frac{1 - j\omega/\omega_r}{1 + \omega^2/\omega_r^2} \propto j\omega(\mu' - j\mu'') , \qquad \omega_r = \frac{R}{L} . \qquad (2.8)$$

The corresponding longitudinal wake potential is $W(t) = R[\delta(t) - \omega_r e^{-\omega_r t}]$. A 3-parameter model is the broadband parallel-$RLC$ resonance:

$$Z_0^\|(\omega) = \frac{R}{1 + jQ\left(\dfrac{\omega}{\omega_r} - \dfrac{\omega_r}{\omega}\right)}, \tag{2.9}$$

where $\omega_r$ is roughly where $\mu''$ peaks. The other two parameters $R$ and $Q$ can be obtained in terms of $\mu'_L$, the value of $\mu'$ at low frequencies, and $\mu''_R$, the value of $\mu''$ at resonant frequency $\omega_r/(2\pi)$. From Eq. (2.7), we obtain

$$\left|\frac{Z_0^\|}{n}\right|_{\text{ind}} = \mu'_L \omega_0 L \quad \text{and} \quad \mathcal{R}e\, Z_0^\|(\omega_r) = \mu''_R \omega_r L, \tag{2.10}$$

where $|Z_0^\|/n|_{\text{ind}}$ is the inductor impedance per harmonic. From Eq (2.9), we obtain

$$\left|\frac{Z_0^\|}{n}\right|_{\text{ind}} = \frac{\omega_0 R}{Q\omega_r} \quad \text{and} \quad \mathcal{R}e\, Z_0^\|(\omega_r) = R. \tag{2.11}$$

Thus, we can solve for $\mu''_R = Q\mu'_L$. Note that $Q$ here is the quality factor describing the $\mu''$ peak. It relates the values of $\mu'$ and $\mu''$ at *different* frequencies, and is not the usual industrial-quoted $Q$ which relates them at the *same* frequency.

The energy the particle lost to the inductor in one passage can now be computed:

$$\mathcal{E} = \frac{3e^2 N_b \tau}{2\omega_0 \hat{\tau}^3} \left|\frac{Z_\|}{n}\right|_{\text{ind}} + \frac{3e^2 N_b}{2Q\omega_0 \omega_r \hat{\tau}^3} \left|\frac{Z_\|}{n}\right|_{\text{ind}}, \tag{2.12}$$

where a parabolic bunch distribution has been used and $Q\omega_r\hat{\tau} \gg 1$ has been assumed. The first term is the linear force from the inductive impedance $Z_\|/n|_{\text{ind}} = j\omega_0 L$, which is supposed to cancel the space-charge force, leaving behind the second

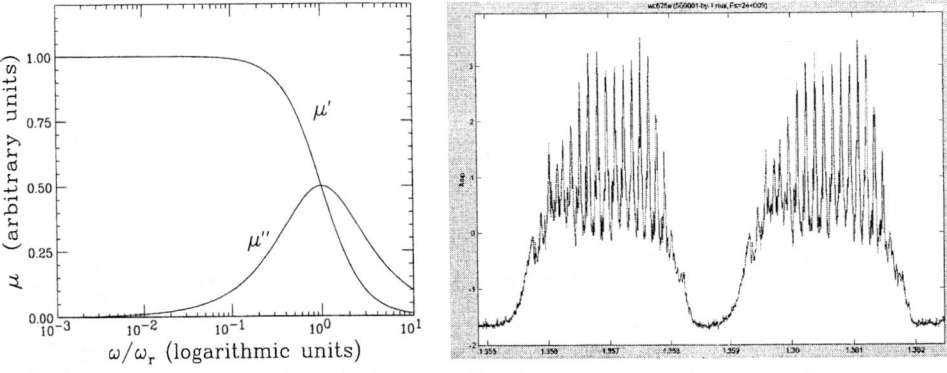

**FIGURE 3.** Left: A typical plot of $\mu'$ and $\mu''$ as functions of frequency. Right: Longitudinal microwave instability observed in a chopped coasting beam (for 2 turns) at the Los Alamos PSR. The collective frequency of the instability is around 75 MHz. Because of the increase in momentum spread, some protons will be rotated into the injection-extraction gap through synchrotron oscillations. They trap electrons leading to transverse e-p instability.

term, which is the actual energy lost to the insertion. For the RL model under the same assumption, exactly the same result is obtained if we make the substitution $Q = 1$. Thus for $n_b$ bunches the total power lost to the insertion becomes

$$P = \frac{3e^2 n_b N_b^2}{4\pi Q \omega_r \hat{\tau}^3} \left| \frac{Z_\parallel}{n} \right|_{\text{ind}}. \tag{2.13}$$

Numerical evaluations in various rings are listed in Table 1, where a full compensation by ferrite or Finemet and $Q = 1$ in the resonance model of Eq. (2.9) are assumed. Although the energy lost by a particle per turn is in general small, the total power lost to the ferrite can become big when the bunch is short and intense. For example, in the new design of the Fermilab future booster, the loss of 361 kW to the ferrite is large, although it is tolerable with water cooling. However, for some future machines with, for example, bunch length shortened by another factor of 10, this power loss will increase by a factor of 1000, and will certainly become intolerable.

**TABLE 1.** Energy loss per turn and total power loss to ferrite or Finemet at various rings.

|  | KEK | PSR | Fermilab Future Booster | |
|---|---|---|---|---|
|  |  |  | Old design | New design |
| $N_b$ | $2 - 9 \times 10^{11}$ | $3.0 - 5.0 \times 10^{13}$ | $5.0 \times 10^{13}$ | $2.5 \times 10^{13}$ |
| $n_b$ | 1 | 1 | 2 | 4 |
| $\hat{\tau}$ (ns) | 44 | 100 | 64.56 | 28.25 |
| $Z_0^\parallel/n|_{\text{spch}}$ ($\Omega$) | 310 | 200 | 100 | 100 |
| Ferrite: $f_r = 30$ MHz |  |  |  |  |
| $\mathcal{E}$ (keV) | 0.221–0.992 | 0.183–0.344 | 2.60 | 13.6 |
| $P$ (kW) | 0.0047–0.096 | 2.45–6.82 | 60.4 | 361 |
| Finemet: $f_r = 5$ MHz |  |  |  |  |
| $\mathcal{E}$ (keV) | 1.32–5.95 | 1.10–1.83 | 15.6 | 81.4 |
| $P$ (kW) | 0.0028–0.058 | 14.7–40.9 | 362 | 2160 |

## C  Perpendicular Bias to Saturation

The loss to the ferrite is mainly hysteresis effect. In the hysteresis $B$-$H$ plot on the left side of Fig. 4, the loss due to one complete oscillation of the ac magnetic field $H_1$ produced by the beam is proportional to the enclosed area marked 1. If the ferrite is biased with a dc magnetic field $H$ at Point 2, the hysteresis area will be smaller and so is the loss. If we bias at Point 3 at saturation, there will not be any hysteresis loop due to the ac magnetic field $H_1$ and therefore all hysteresis loss will be eliminated. This strong dc bias field $H_c$ will leave the magnetization $\vec{M}$ precessing around it, and the only loss will be due to the spin wave inside the ferrite, which is small. The price we need to pay here is a much lower relative permeability, which is equal to the slope the line joining the origin to Point 3. Thus more ferrite cores will be needed to compensate for the same amount of space charge. Such a bias scheme can be achieved by encircling the ferrite rings with a solenoid, where the dc biasing field $H_c$ is along the beam direction and is perpendicular to the ac field $H_1$ carried by the beam. In fact, such a solenoid is always needed so that the relative permeability can be suppressed as the space charge decreases while the beam particles are ramped. A schematic picture of the precessing magnetization is shown in the right plot of Fig. 4.

Without the hysteresis loss, there will not be any broadband loss peak. There is still a resonance at the much higher gyromagnetic circular frequency of $\omega_c = \gamma_g H_c$, where $\gamma_g = 2\pi \times 2.80$ MHz/Oersted is the gyromagnetic ratio. By choosing the suitable biasing field $H_c$, this resonant frequency can be made at least 10 times higher than $\omega_r$ where the broadband loss peaks but is absent now in the presence of the saturated biasing. This resonance peak is very much narrower and the quality factor is $Q \lesssim 10$. Thus, we can see from Eq. (2.13) that the power loss can be reduced by at least a factor of $\sim 100$.

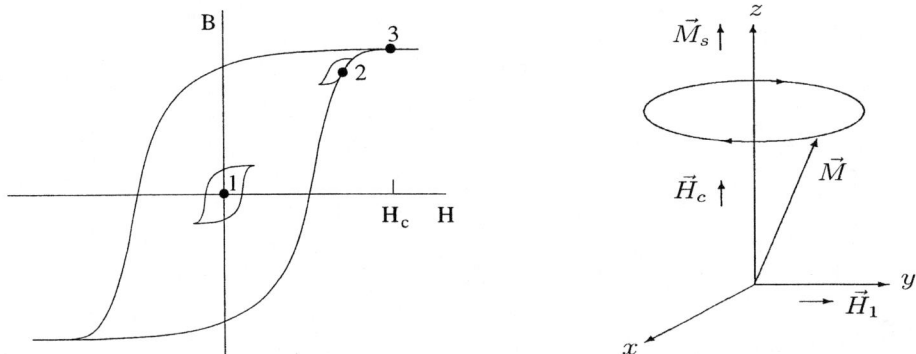

**FIGURE 4.** Left: Hysteresis $B$-$H$ plot. At zero bias, the loss is proportional to the enclosed area marked 1. With dc bias field at Point 2, the hysteresis loss will be smaller. At saturated bias at Point 3, all hysteresis loss will be eliminated. Right: System with saturated perpendicular bias $H_c$ in the z-direction. With the application of the ac field $\vec{H}_1$ in the y-direction, the magnetization $\vec{M}$ acquires an ac component in the x-y plane precessing about the z-axis.

## D  Microwave Instability

Actual area of beam stability in the complex $Z_0^\parallel$-plane (or the traditional $U'$-$V'$ plane) is somewhat different from the commonly quoted Keil-Schnell estimation. In Fig. 5, the heart-shape solid curve, denoted by 1, is the threshold curve for parabolic distribution in momentum spread, where the momentum gradient is discontinuous at the ends of the spread. Instability develops and a smooth momentum gradient will result at the ends of the spread, changing the threshold curve to that of a distribution represented by 2, for example, $\frac{15}{16}(1-\delta^2/\hat{\delta}^2)$. Further smoothing of the momentum gradient at the ends of the spread to a Gaussian distribution will change the threshold curve to 3. On the other hand, the commonly known Keil-Schnell threshold is denoted by the circle of unit radius in dots. This is the reason why in many low-energy machines the Keil-Schnell limit has been significantly overcome by a factor of about 5 to 10. In this case, the space charge is almost the only source of the impedance, the real part of the impedance can be typically orders

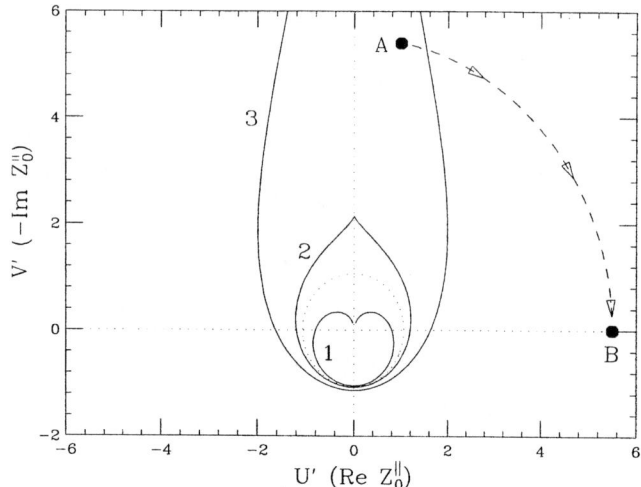

**FIGURE 5.** Microwave instability threshold curves in the complex $Z_0^\|$ (or $U'$-$V'$) plane, for (1) parabolic momentum distribution, (2) distribution with a continuous momentum gradient, and (3) Gaussian momentum distribution. The commonly quoted Keil-Schnell threshold criterion is denoted by the circle in dots. An intense space-charge beam has impedance denoted by Point A outside the Keil-Schnell circle. A ferrite tuner compensating the space charge completely will have a resistive impedance roughly at Point B and is therefore unstable.

of magnitude smaller. As an example, if the impedance of the Los Alamos PSR is at Point A, the beam is within the microwave stable region if the momentum spread is Gaussian like, although it exceeds the Keil-Schnell limit. Now, if we compensate the space-charge potential-well distortion by the ferrite inductance, the ferrite required will have an inductive impedance at low frequency equal to the negative value of the space charge impedance at A, for example, about $-5.5$ units according to Fig. 5. However, the ferrite also has a resistive impedance or $\mathcal{R}e\, Z_0^\|$. Although $\mathcal{R}e\, Z_0^\|/n$ is negligible at low frequencies, it reaches a peak value near $\omega_r/(2\pi)$ (about 50 to 80 MHz for the Toshiba $M_4C_{21A}$) with the peak value the same order of magnitude as the low-frequency $\mathcal{I}m\, Z_0^\|$. Actually, according to the $RLC$ model discussed above, we get

$$\frac{\mathcal{R}e\, Z_0^\|/n|_{\mathrm{pk}}}{\mathcal{I}m\, Z_0^\|/n|_{\omega\to 0}} \approx \frac{Q^2+Q+1}{Q+2} = \begin{cases} Q & \text{if } Q \gg 1 \\ 1 & \text{if } Q \sim 1 \\ \frac{1}{2} & \text{if } Q \ll 1 \end{cases} \geq \frac{1}{2}. \qquad (2.14)$$

The $RL$ model gives the same impedance ratio of $\frac{1}{2}$ as the low-$Q$ case of Eq.(2.14). Thus the ferrite will contribute a resistive impedance denoted roughly by Point B ($\sim 5.5$ units) when $Q \sim 1$ or at least one half of it when $Q \ll 1$. This resistive impedance of the ferrite will certainly exceed the threshold curve and we believe that the longitudinal instability observed at the Los Alamos PSR is a result of this consideration. It follows from here that such low-frequency compensation of an intense space-charge induced potential-well distortion will definitely result in the microwave instability at high frequencies, $\omega \simeq \omega_r$. In other words, the strong space-charge potential-well distortion can only be compensated by the ferrite or Finemet inductance to a small extent to ensure that the resistive part of the ferrite or Finemet is kept below the microwave instability threshold.

For the transverse bias of the ferrite to saturation discussed in the previous section, although the power dissipation in the ferrite can be reduced to a large extent, the sharp gyromagnetic resonance resulting at higher frequency can become more susceptible to microwave instability, unless the gyromagnetic resonance is very much narrower the width of the bunch spectrum.

## E  Transient Beamloading

The intense proton beam will excite modes of oscillation in the rf cavities when passing through them. This is called transient beamloading. For the new design of the Fermilab low-energy booster ring, a total rf voltage of $\sim 185$ kV is required. Since the rf frequency is low, about 7.5 MHz, one needs to split the rf system into 10 cavities, each with $V_{\rm rf} = 18.5$ kV. Each cavity is loaded with 30 cm of ferrite cores ($\mu' = 21$) with inner/outer radii 20/35 cm. This gives an inductance of $L \sim 0.61$ $\mu$H and capacitance $C \sim 820$ $\mu$F. Then for a point bunch containing $N_b = 2.5 \times 10^{13}$ protons, the transient beam loading voltage is

$$V_{t0} = \frac{eN_b}{C} = 5.4 \text{ kV}, \tag{2.15}$$

which is an appreciable fraction of the $V_{\rm rf} = 18.5$ kV supplied by the klystron. For a longer bunch, the transient beamloading will be less. In fact, this is just the wake potential seen by a particle at time $\tau$ ahead the bunch center due to the wake of a cavity gap, or for a Gaussian bunch,

$$V_t(\tau) = e\int_\tau^\infty d\tau' \lambda(\tau') W(\tau' - \tau) = -\frac{eN_b \omega_r R_\|}{2Q\cos\phi_0} \mathcal{R}e\, e^{j\phi_0 - \tau^2/(2\sigma_\tau^2)} w\left[\frac{\sigma\tau\omega_r e^{j\phi_0}}{\sqrt{2}} - \frac{j\tau}{\sqrt{2}\sigma_\tau}\right], \tag{2.16}$$

where $\omega_r/(2\pi)$ is the resonant frequency, $R_\|$ the shunt impedance, and $Q$ the quality factor of the cavity mode excited, $\phi_0 = \sin^{-1}\frac{1}{2Q}$, and $w$ is the complex error function. It is easy to show that as the bunch length $\sigma_\tau \to 0$, $V_t(\tau)$ approaches the point-bunch limit $V_{t0}$ in Eq. (2.15).

Let us understand how the transient beamloading originates. As a bunch of protons passes through the cavity gap, a negative charge equal to that carried by the bunch will be left by the image current at the upstream end of the cavity gap. Since the negative image current will resume from the downstream end of the cavity gap following the bunch, an equal amount of positive charge will accumulate there. Thus, a voltage will be created at the gap opposing the beam current and this is the transient beamloading voltage as illustrated in Fig. 6 left. Griffin [8] suggested to use a feed-forward system, which will monitor the linear charge distribution of the bunch and deliver via a tetrode the same amount of negative charge density to the downstream end of the cavity gap so as to cancel the positive charge there and thus alleviating the transient beamloading, as illustrated in Fig. 6 right.

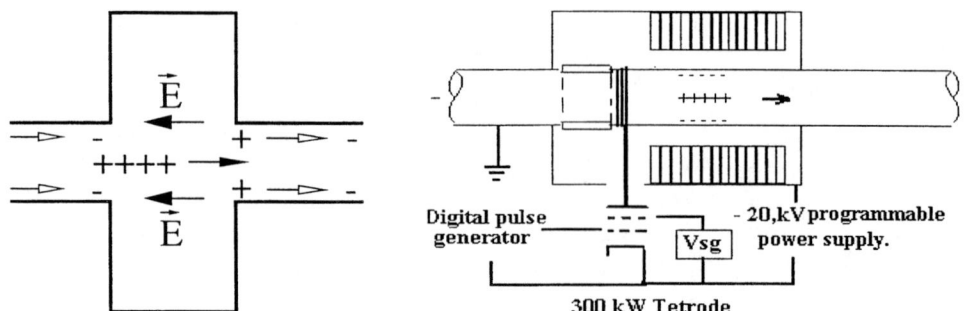

**FIGURE 6.** Left: As a positively charged bunch passes through a cavity, the image current leaves a negative charge at the upstream end of the cavity gap. As the image current resumes at the downstream side of the cavity, a positive charge is created at the downstream end of the gap because of charge conservation, thus setting up an electric field $\vec{E}$ and therefore the transient beamloading voltage. Right: The bunch density is monitored and a negative charge density is fed-forward through a tetrode to the downstream side of the cavity gap to cancel the positive charge left there, thus eliminating the transient beamloading.

## III TRANSVERSE SPACE-CHARGE EFFECTS

### A Coherent and Incoherent Tune Shifts

Usually, people say that a large incoherent space-charge tune spread will encompass a lot of parametric resonances and lead to instability. The common rule of thumb is that incoherent self-field tune spread should not exceed $\sim 0.40$. Both rings of the future Fermilab booster are designed to have normalized 95% emittances equal to $2.0 \times 10^{-4}$ $\pi$m, so that such tune spreads can be below 0.40. However, this self-field tune spread at injection has never been a well-measured beam parameter. It is difficult to measure because low-energy rings are usually ramped very rapidly.

Machida and Ikegami [9] pointed out at the space-charge workshop at Shelter Island that it is the *coherent* rather than the *incoherent* tune shifts that determine the instability of a beam. In fact, this is quite reasonable. When the bunch is oscillating at an integer coherent tune, we have the usual integer resonance. This leads to an instability because all particles are performing betatron oscillations with a tune component that is at an integer. The whole beam will become unstable. On the other hand, if the incoherent tune spread covers an integer resonance, only *a small amount* of particles are hitting the integer resonance; thus the whole beam may not be unstable. The coherent betatron tune is not affected by space charge when the image forces are small. This is because the centroid of the bunch does not see any space-charge force. On the other hand, the coherent quadrupole betatron tune and coherent sextupole betatron tune will be affected by space charge. Therefore, when they hit a resonance, there will be instability. This is demonstrated by the simulation of Machida and Ikegami in Fig. 7. In the simulation, the horizontal coherent quadrupole tune hits the integer of 13 when the beam intensity reaches $\sim 15$ A. We do see that the horizontal emittance increases rapidly around the beam intensity of 15 A. The vertical coherent quadrupole tune hits the integer 11 when the beam intensity is raised to around 13 to 15 A. The vertical emittance increases also around those intensities. However, we do not see any growth of emittance when the coherent quadrupole tunes cross half integers.

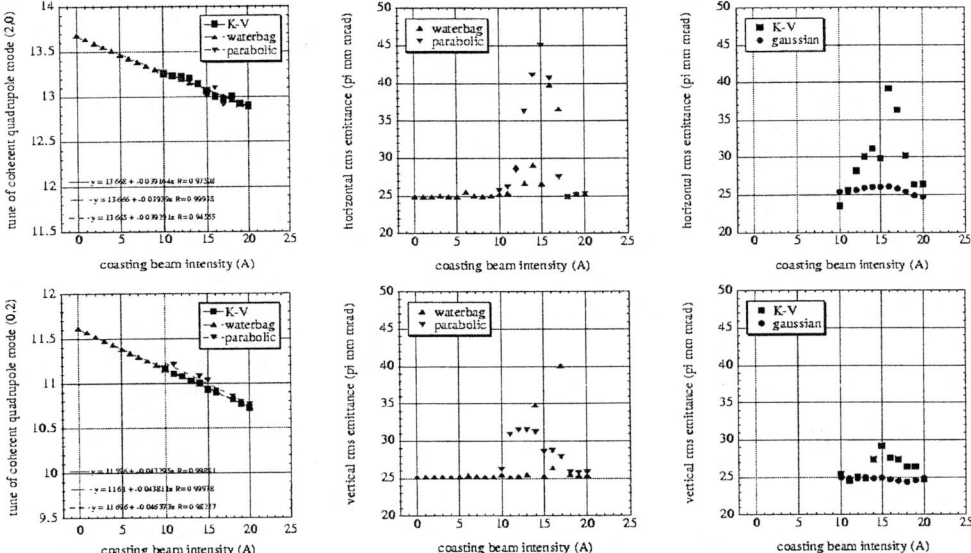

**FIGURE 7.** Tune of coherent quadrupole mode (left) and rms emittance at 512 turns after injection (center and right) versus beam intensity. Upper figures show horizontal results and lower ones vertical. Rms emittance growth is observed when either the horizontal or vertical coherent quadrupole tune becomes integer. (Reproduced from Ref. 9).

## B  Space Charge and TMCI

It was reported in a recent paper of Blaskiewicz [10] that the space-charge tune shift can strongly damp the transverse mode coupling instability (TMCI), which is also known as strong head-tail instability. The investigation was made on the basis of particle tracking and the analytically solvable *square-well air-bag model* [11], with the bunch distribution in the longitudinal phase space,

$$\Psi(\phi, \Delta E) = \tfrac{1}{2}\rho(\phi)[\delta(\Delta E - \widehat{\Delta E}) + \delta(\Delta E + \widehat{\Delta E})] \,, \qquad (3.1)$$

where $\rho(\phi) = 1/(2\pi)$ is the linear distribution or the projection onto the longitudinal axis. In this model the synchrotron phase $\phi$ ranges from $-\pi$ to 0 at $\Delta E = -\widehat{\Delta E}$ and from $\pi$ to 0 at $\Delta E = \widehat{\Delta E}$, with $\phi = 0$ representing the head of the bunch.

What is going to be presented here is a qualitative explanation why the space charge helps TMCI. Without space charge, the bunch starts to be unstable when two neighboring synchro-betatron modes merge under the influence of the wake forces. Typically, the pure betatron mode (the azimuthal or synchrotron harmonic 0 mode, also known as the rigid-bunch mode) is affected by the wake force and shifts downward, while the other azimuthal modes are not much affected, at least at low intensity. The transverse wake force produced by an off-axis beam has the polarity that deflects the beam further away from the pipe axis. This force acts as a defocusing force for the rigid beam mode, and therefore the frequency shifts downward. Such a down-shift of the betatron frequency is routinely observed in electron rings and serves as an important tool of probing the impedance. As a result, the instability threshold is determined by the coupling of the 0 and −1 modes, as illustrated in the left plot of Fig. 8, (see below for definitions of parameters).

The space charge by itself also shifts all the frequencies downward, as illustrated in the right plot of Fig. 8. The only exception is the azimuthal 0 mode, which describes the motion of the bunch as a whole, and, therefore, is not influenced by the space charge at all. Thus, in the presence of space charge, the 0 mode will couple with the −1 mode at a higher current intensity and therefore the threshold is raised in the presence of space charge. This is illustrated in the left plot of Fig. 9.

Let us go in more details with mathematics. The transverse displacement $x(\phi)$ of a particle at the synchrotron phase $\phi$ satisfies the equation of motion:

$$\frac{d^2 x(\phi)}{dt^2} + \omega_\beta^2 x(\phi) = F(\phi) + S\rho(\phi)[x(\phi) - \bar{x}(\phi)] , \qquad (3.2)$$

where $\omega_\beta/(2\pi)$ is the unperturbed betatron frequency and the smooth approximation for the betatron oscillations has been applied. To incorporate synchrotron oscillation, the full time derivative takes the form

$$\frac{d}{dt} = \frac{\partial}{\partial t} + \omega_s \frac{\partial}{\partial \phi} , \qquad (3.3)$$

with $\omega_s/(2\pi)$ being the synchrotron frequency. The right-hand side of Eq. (3.2) contains the transverse driving forces. The first term is the transverse wake force

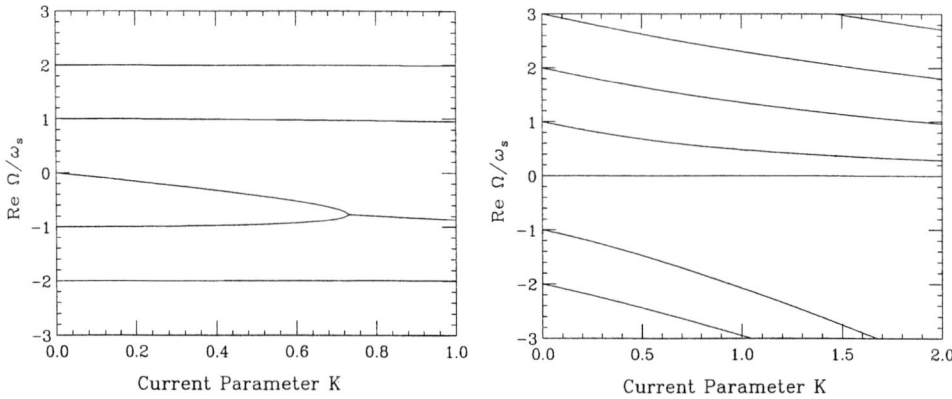

**FIGURE 8.** Left: The transverse wake force shifts mostly the azimuthal 0 mode downward but not the other modes. Instability occurs when the 0 and −1 modes meet with each other. Right: The space-charge force in the absence of the wake forces shifts all modes downward with the exception of the 0 mode.

$$F(\phi) = \frac{N_b e^2 c^2}{E_0 C} \int_0^{|\phi|} W_1[z(\phi') - z(\phi)]\rho(\phi')\bar{x}(\phi')d\phi' , \qquad (3.4)$$

where $N_b$ is the number of particles in the bunch, $W_1$ the transverse wake function, $z(\phi)$ the longitudinal position of the beam particle. The second term is the space-charge contribution. It is proportional to the linear density $\rho(\phi)$ and the displacement relatively the local beam center $x(\phi) - \bar{x}(\phi)$, with the constant $S$ representing the space-charge strength.

To solve the problem quantitatively, we expand the offset into the synchrotron harmonics (or azimuthals):

$$x(\phi, t) = e^{-i\omega_\beta t - i\Omega t} \sum_{n=-\infty}^{\infty} x_n e^{in\phi} , \qquad (3.5)$$

where $\Omega/(2\pi)$ is the collective frequency shift. In this air-bag model, all particles reside at the edge of the bunch distribution in the longitudinal phase space. Note that because of the square-well air-bag model, these synchrotron azimuthals are slightly different from the conventional ones. The average offset at the synchrotron phase $\phi$ is therefore given by

$$\bar{x}(\phi, t) = \tfrac{1}{2}[x(\phi, t) + x(-\phi, t)] = e^{-i\omega_\beta t - i\Omega t} \sum_{n=-\infty}^{\infty} x_n \cos n\phi . \qquad (3.6)$$

Following basically Ref. [12], Eq. (3.2) transforms into an eigenvalue equation,

$$\left(\frac{\Omega}{\omega_s} - n\right) x_n = -K \sum_{m=-\infty}^{\infty} x_m (\mathcal{W}_{nm} + \xi \mathcal{Q}_{nm}) . \qquad (3.7)$$

Here, the current parameter is written as

$$K = \frac{N_b e^2 c^2 W_0}{2\pi^2 \omega_\beta \omega_s C E_0} . \qquad (3.8)$$

The wake matrix elements are then given by

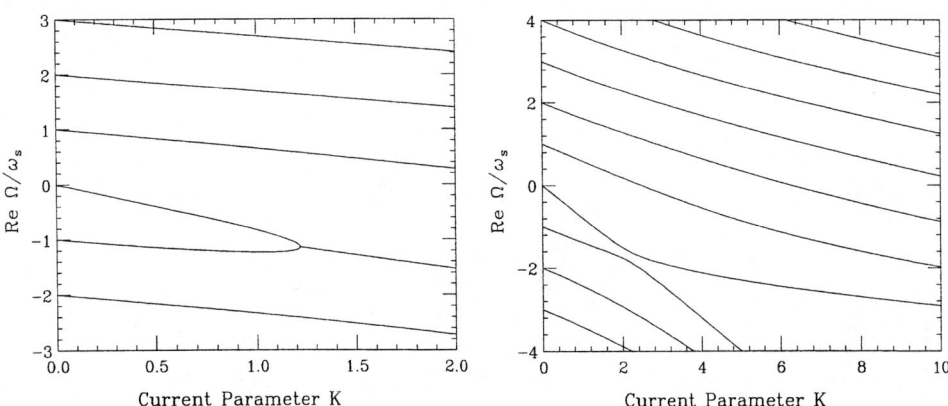

**FIGURE 9.** Left: With the transverse space-charge force added to the wake forces, all modes except the 0 mode are shifted downward, thus requiring the 0 and $-1$ modes to couple at a much higher current threshold. Right: When space charge reaches the critical value of $\xi = 5$, the $-1$ mode is shifted away from the 0 mode by so much that they do not couple anymore.

$$\mathcal{W}_{nm} = \int_0^\pi d\phi \int_0^\phi d\phi' w[z(\phi') - z(\phi)] \cos(n\phi) \cos(m\phi') , \qquad (3.9)$$

where the wake function is presented as $W(z) = W_0 w(z)$ with $W_0$ serving as a normalizing constant. The space-charge parameter

$$\xi = \frac{\Delta\omega_\beta}{2K\omega_s} \qquad (3.10)$$

is a current-dependent ratio of the incoherent tune shift

$$\Delta\omega_\beta = \frac{S\rho}{2\omega_\beta} \qquad (3.11)$$

to the current parameter $K$. The space-charge matrix elements are

$$\mathcal{Q}_{nm} = \delta_{nm} - \delta_{n,-m} \qquad (3.12)$$

in the assumed air-bag distribution.

Without wake forces, the eigenvalue equation leads to the mode behavior presented in the right plot of Fig. 8. For the simplest step-like wake function $w(z) = \theta(z)$ and without space charge ($\xi = 0$), the mode coupling is shown in the left plot of Fig. 8, where the threshold is $K = 0.73$. Now space charge is introduced with the space-charge parameter $\xi = 4$. We do see in the left plot of Fig. 9 that, because the $-1$ mode is shifted downward by the space charge, the instability threshold has been pushed up to $K = 1.25$ as compared with the left plot of Fig. 8.

Further increasing the space-charge parameter to $\xi = 5$, we see in the right plot of Fig. 9 that modes 0 and $-1$ do not merge any more. What is not shown in the plot is a much higher new threshold where the 0 mode couples with the 1 mode instead. This new threshold is very much model dependent. In the present model, it depends strongly on the number of modes included in the truncated matrix. For truncation at modes $|n| = 32$, this new threshold is at least a factor of 30 higher than when space charge is absent. A dependence of the calculated threshold $K_{th}$ on the mode truncation number $|n|$ was found as $K_{th} \propto |n|^{1/2}$ for $|n| \leq 10$ and even weaker,

$$K_{th} \propto |n|^{1/3} , \qquad (3.13)$$

for $10 \leq |n| \leq 32$. The divergence is caused by the fact that the Fourier components of the space charge in Eq. (3.12) do not roll off at high frequencies. Taking into account the finite value of the ratio of transverse bunch size $\sigma_\perp$ to longitudinal bunch size $\sigma_\parallel$, we estimate this roll off limit as $|n| \simeq \sigma_\perp/\sigma_\parallel \simeq 200$ to 1000 for typical hadron bunches. Extrapolation of the dependence Eq. (3.13) into this area brings to a conclusion that the actual threshold can be 2 to 3 times higher than the result reported for $|n| = 32$. So for this simplified wake-beam model, the space charge is found to be able to increase the TMCI threshold by a factor of 50 to 100.

Unlike the longitudinal mode-coupling instability where the bunch may just lengthen as the beam becomes unstable essentially without losing beam particles, this transverse instability is devastating; as soon as the threshold is reached, the bunch disappears. TMCI in electron machines are usually damped with a *reactive* feedback system; i.e., the kicker is located at an even multiple of $90°$ from the pickup [13]. This implies the addition of a term $G\bar{x}(\phi)$ to the right-hand side of Eq. (3.2), where $G$ is the gain of the feedback system. Notice that the reactive feedback acts on the center of the bunch and is *in phase* with the particle displacements; hence the term reactive. It therefore modifies the betatron tune by introducing a tune shift. Thus, only the 0 mode is affected but not the other modes. The instability threshold can then be raised by properly choosing the strength and sign of the feedback gain $G$ so that the 0 mode has a *positive* shift. The space-charge tune shift in a proton machine, as discussed above, constitutes a natural inverse reactive feedback.

One of the authors (A. Burov) expresses his gratitude to Slava Danilov and Mike Blaskiewicz for fruitful discussions.

# REFERENCES

1. J.R. Alonso, J.R., *Proceedings of the Sixth European Particle Accelerator Conference*, p.493 (Stockholm, Sweden, June 22-26, 1998).
2. *The JHF Accelerator Design Study Report*, Sept. 1997, KEK.
3. *ESS, A Next Generation Neutron Source for Europe*, Vol. 3, The ESS Technical Study, ISBN 090 237 6 500 090 237 6 659.
4. *A Development Plan for the Fermilab Proton Source*, Ed. S. Holmes, Fermilab Report TM-2021.
5. Keil, E., and Schnell, W., CERN Report TH-RF/69-48, 1969; Neil, V.K., and Sessler, A.M., *Rev. Sci. Instr.*, **36**, 429 (1965); Boussard, D., CERN Report Lab II/RF/Int./75-2, 1975.
6. Plum, M.A., Fitzgerald, D.H., Langenbrunner, J., Macek, R.J., Merrill, F.E., Neri, F, Thiessen, H.A., Walstrom, P.L., Griffin, J.E., Ng, K.Y., Qian, Z.B., Wildman, D., and Prichard, B.A. Jr., *Phys. Rev. ST Accel. Beams*, **2**, 064201 (1999).
7. Koba, K., Machida, S., and Mori, Y., KEK Note, 1997 (unpublished); Koba, K., these proceedings; Koba, K., *et al, Phys. Sci. Instr.*, **70**, 2988 (1999).
8. Griffin, J.E., *RF System Considerations for a Muon Collider Proton Driver Synchrotrons*, Fermilab report FN−669, 1998.
9. Machida, S., and Ikegami, M,, *Proceedings of Workshop on Space Charge Physics in High Intensity Hadron Rings*, p.73, Ed. Luccio, A.U., and Weng, W.T., (Shelter Island, New York, May 4-7, 1998).
10. Blaskiewicz, M., *Fast Head-tail Instability with Space Charge*, Phys. Rev. ST Accel. Beams, **1**, 044201 (1998).
11. Danilov V., and Perevedentsev, E., *Strong Head-Tail Effect and Decoupled Modes in the Space-Time Domain*, Proceedings of XVth International Conference on High Energy Accelerators, p.1163 (Hamburg, 1992).
12. Danilov, V., and Perevedentsev, E., *Feedback system for elimination of the TMCI*, Nucl. Instr. and Methods, **A391**, 77 (1997).
13. Ruth, R., CERN Report LEP-TH/83-22, 1983; Myers, S., *Proceedings of IEEE Particle Accelerator Conference*, p.503 (Washington, 1987); Zotter, B., *IEEE Trans. Nucl. Sci.*, **NS-32**, 2191 (1985); Myers, S., CERN Report LEP-523 1984.

# Impedance Issues in the CERN SPS

T. Linnecar

*CERN, Geneva, Switzerland*

**Abstract**

The future use of the CERN SPS accelerator as injector for the Large Hadron Collider, LHC, and the possible use of the SPS as a neutrino source for the Gran Sasso experiment are pushing the maximum intensity requirements of the accelerator much higher than achieved up to now. At the same time the requirements on beam quality are becoming far more stringent. The SPS machine, built in the 70's, is not a "smooth" machine. It contains many discontinuities in vacuum chamber cross-section and many cavity-like objects, as well as the 5 separate RF systems at present installed. All these lead to a high impedance, seen by the beam, spread over a wide frequency range. As a result there is a constant fight against instabilities, both single and multi bunch, as the intensity increases. A programme of studies is under way in the SPS to identify, reduce, and remove where possible the sources of these impedances.

# 1 Introduction

To fulfil the different fixed target physics requirements the total intensity of the CERN SPS accelerator has increased from its design value of $10^{13}$ protons/pulse, achieved soon after commisioning in 1976, to its present peak value of $4.84 \times 10^{13}$ attained in 1997. In the early 80's, the SPS became the Sp$\bar{\text{p}}$S collider where 6 proton bunches were accelerated and then collided with 6 antiproton bunches. The single bunch intensity was increased during this time from the $10^{10}$ in fixed target operation to $1.6 \times 10^{11}$. To obtain these total and single bunch intensity increases a continuing battle has had to be waged against instabilities, leading to many modifications and upgrades to the machine hardware. Here are a few examples. Already in the first two years of operation the impedance of two higher order modes, one transverse at 460 MHz and the other longitudinal at 628 MHz [1] in the main RF cavities, were identified as causing instability and intensity limitation and had to be passively damped. When preparing for Sp$\bar{\text{p}}$S, bunch intensities were limited when crossing transition by both a longitudinal negative mass instability and a transverse head-tail instability to $2 \times 10^{10}$ protons [2]. This limitation was removed by raising the injection energy to E = 26 GeV, above transition, $\gamma_{tr} = 23.4$. Single turn RF feedback has been required for the main 200 MHz travelling wave RF system to raise intensities above $3 \times 10^{13}$. This type of feedback has also had to be added in addition to the strong short delay RF feedback around the 352 MHz superconducting cavities used for lepton acceleration, but also seen by the intense proton beam. Together they reduce the impedance at the synchrotron satellites around each revolution frequency line from many G $\Omega$ to $\sim$ 100k$\Omega$ per cavity.

Future uses of the SPS require much higher intensities. At the same time much improved beam quality is demanded. The presence or not of instabilities will clearly be a dominant factor in the success of our programme. Identification and control of the different impedance sources in the machine is and will remain a very hot issue.

## 2 Future beam requirements for the SPS

The SPS machine is now being prepared for use as injector for the Large Hadron Collider (LHC) [3], at present under construction at CERN and which will come into operation in 2005. To obtain the highest luminosities in the LHC, the SPS must accelerate beams with single bunch intensities similar to that in Sp$\bar{\text{p}}$S operation, $N = 1.1 \times 10^{11}$, and batch intensities much higher than seen before in the SPS, even though the total intensity will be lower than the record value due to the incomplete filling of the ring, $2.4 \times 10^{13}$ protons in 3/11 of the ring. These performances must be achieved with small longitudinal and transverse emittances, both for luminosity reasons, but also in the longitudinal case to ease transfer to the LHC where beam losses at injection must be kept very small [4].

| Operating Mode | Energy inj., top (GeV) | Total intensity ($\times 10^{12}$) | Bunch intensity ($\times 10^{10}$) | Batch current (A) | $\varepsilon_l$ inj., top (eVs) | $\varepsilon_{H,V}$ top ($\mu m$) |
|---|---|---|---|---|---|---|
| Sp$\bar{\text{p}}$S collider (past) | 26, 315 | 0.2 | 20 | | 0.6, 0.8 | 2.75, 2.75 |
| Fixed target (present) | 14, 450 | 48.4 | 1 | 0.33 | 0.2, 2 | 10, 7 |
| SPS for LHC (project) | 26, 450 | 24 | 11-17 | 0.7-1.1 | 0.35, 0.6-1 | 3.5, 3.5 (3.0 inj.) |
| Gran Sasso (possible) | 14, 400 | $\geq 70$ | 2 | 0.5 | 0.2, $\leq 2$ | 10, 10 |

Table 1: Parameters for present and future operation of the SPS

Table 1 gives various parameters for this mode of operation, and also for the present fixed target physics operation and the past Sp$\bar{\text{p}}$S operation. The tight emittance budget for LHC beams in the SPS is evident. To allow going to the higher value of longitudinal emittance at extraction, $\varepsilon_l = 1$ eVs, it will be necessary to install a separate RF system at 200 MHz in the LHC for capture [5]. Note also that present fixed target operation, with its lower batch current, has a far higher emittance at extraction than can be tolerated for LHC injection. This tight specification for the beam is supplemented by the further demand from the project to send a neutrino beam to Gran Sasso, Italy. This project requires maximising the total number of protons that can be produced by the SPS in a given operation period, say 200 days [6]. The optimisation of the accelerator chain is under study, but one parameter is clearly the maximum intensity per cycle that can be accelerated in the SPS. A desirable increase to $\geq 7 \times 10^{13}$ from the present maximum of $4.84 \times 10^{13}$ is hoped for. Although the emittances are less of a problem here, considerable care must be taken to prevent micro-losses during the acceleration cycle and in the extraction channel.

# 3 What type of problems do we expect?

The microwave instability is the single bunch instability of concern in the longitudinal plane. It was conjectured as early as 1976 [1] that the large number of discontinuities in the machine vacuum chamber could be the cause of the microwave instability responsible for blow-up during the debunching and recapture process at 10 GeV in the SPS. The problem was solved at that time by installing a 200 MHz RF bunching system in the PS (the SPS injector) and injecting bunch into bucket.

In $Sp\bar{p}S$, the microwave instability limited the bunch intensity at injection at 26 GeV, the limit being raised by installing a 100 MHz RF capture system in the SPS to allow transfer of longer bunches (5.0ns). It is believed from our present knowledge of the impedances responsible for this instability (see below) that the nominal LHC bunch of 4 ns, 0.35 eVs, and $1.1 \times 10^{11}$ protons, may be unstable. Rather than using a new RF system this time to reduce the danger, more radically the problem will be attacked at the source.

Multibunch instabilities in the longitudinal plane have been observed and studied on the operational fixed target beams and on batch type beams in machine study periods. Some conclusions from these studies can be drawn:

• The beam is already unstable at $4 \times 10^{12}$, becoming more and more unstable towards higher energies.

• The emittance at 445 GeV for $4 \times 10^{13}$ is ~2 eVs for the transfer voltage needed for LHC injection, but can be lower with voltage programming.

• Bunch intensity seems to be the dominant parameter for emittance blow-up. However some influence from the large gap between the batches suggests a wakefield covering several bunches.

The first point can be explained by an analysis of the threshold intensity through the cycle. The threshold curves for the normal operation cycle and various resonator frequencies are shown in Fig.1 taken from Ref.[7]. Whatever the source frequency, the threshold decreases significantly as energy increases. The spectrum sometimes shows specific lines growing but in general a broad spectrum filling the space between the bunch spacing frequency lines is seen.

In the transverse plane the head-tail instability, which is of greatest concern for single bunches, can be adequately controlled by careful chromaticity adjustment for the lowest modes. For the higher modes the frequency spread in the beam should be sufficiently large to give adequate Landau damping. The transverse mode coupling instability has

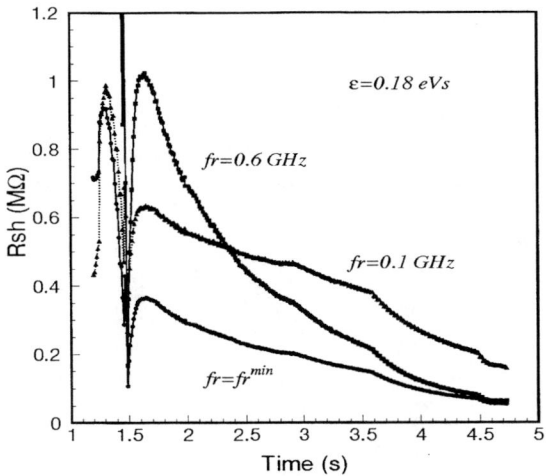

Figure 1: Coupled bunch instability threshold during normal operation in fixed target cycle for a beam intensity $4.2 \times 10^{13}$ and different resonant frequencies $f_r$.

not been observed with protons. Multi-bunch modes at high mode numbers are not at present observed, several transverse HOMs in the different RF cavities having been already damped. With increasing intensity this may no longer be the case. The resistive wall instability is of concern. For a given mode the risetime of this instability is modified by the batch structure of the beam, as is the mode number having the highest growth rate [8].

## 4 Identifying the sources

### 4.1 Impedances leading to a single bunch instability

The microwave instability threshold has been determined in the past and more recently [9] by measuring the variation in captured bunch length as a function of injected intensity. The threshold can also be found by observing the high frequency signals generated. From the threshold intensity the parameters for a broad-band impedance model can be defined and this is often used as a machine parameter for scaling purposes. Estimations of this impedance depend on the spectral bandwidth of the particular bunch available for the measurement. The low-frequency inductive part can also allow this impedance to be measured and this can be found from bunch lengthening or by observation of the change in debunching time as a function of intensity [10].

However the the broad-band model is far from reality, making prediction for future operation modes from scaling very difficult. As is now clear, a more realistic model consists of a number of resonant peaks. Attempts to determine a model by making an inventory of the different elements in the machine and computing the longitudinal impedance have been made [11] as well. We have developed a technique based on the measurement of the spectrum of unstable bunch modes to measure the fine structure of the machine impedance [12] and determine the dominant impedances for the model. Here we outline the procedure used and give the results.

Single, high-intensity proton bunches are injected into the machine with RF off and the spectrum is observed during very slow debunching. If the bunch intensity is above threshold then the different resonant impedances in the machine lead to a modulation of the line density at the resonant frequencies. The debunching time, $t_{deb}$, must be slow enough that the modulation grows and saturates before the bunch length has significantly increased. This is a function of the machine parameters, slip factor $\eta$, and the maximum momentum spread $\Delta p/p$ in the bunch. The instability reaches some maximum modulation amplitude which is recorded as a function of frequency. At each frequency point, the amplitude from many bunches is measured and averaged. The spectrum of an unstable bunch mode associated with a given narrow band impedance is centred at the resonant frequency and has a spectrum width given by the bunch distribution, proportional to $1/\tau_b$, $\tau_b$ being the bunch length. Hence it is interesting to use very long bunches which then allow the fine structure of the machine impedance to be resolved. While exact analytical formulae for saturation are unknown, numerical simulation [9] shows that for a narrow band impedance source with bandwidth $\Delta f_r < 1/\tau_b$, the maximum amplitude is a function of the $R_{sh}/Q$ of the cavity and N.

In our experiment the machine and bunch parameter were as in Table 2:

| E | $\eta$ | $\tau_b$ | $\varepsilon_l$ | $t_{deb}$ |
|---|---|---|---|---|
| 26 GeV | $5.257 \times 10^{-4}$ | 25 ns | 0.25 eVs | 80 ms |

Table 2: Experimental parameters for measuring the spectrum of unstable bunches.

We used two techniques to analyse the data, which was obtained from a wide-band wall-current monitor. In the first one we used a spectrum analyser with a bandwidth of 3MHz as a receiver and scanned the range from 100 MHz to 4 GHz. The second technique uses a digital oscilloscope to acquire the data in time domain, a Fourier transform giving the spectrum. This is much faster but is limited in any case to 2 GHz by the maximum sampling frequency. The results obtained by the first technique are shown in Fig.2, where numerous peaks corresponding to the different impedance in the machine can be seen.

We have succeeded in identifying most of the peaks in the figure. The peaks at 200 MHz and 800 MHz are the travelling wave RF fundamental modes and the peak at 400 MHz is due to the cavities formed by the vacuum chambers containing the extraction septa. All the peaks above 1.4 GHz can be explained by the resonances of the 800 or so cavity-like vacuum ports. All these latter accidental cavities contain damping resistors,

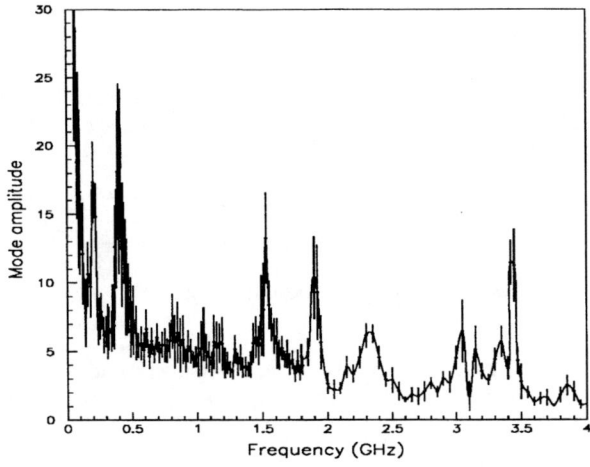

Figure 2: Measured spectrum of unstable bunches

which is important for coupled bunch instabilities but which does not change the $R_{sh}/Q$.

We have used bunch lengthening measurements to check our new impedance model which contains the four sets of elements described above with $R_{sh}/Q$ and Q values found from calculation or laboratory experiment. Results of simulations of bunch lengthening using this model can be compared with experimental results and show very close agreement. It seems that we may still be missing approximately 10% of the impedance in our model.

## 4.2 Impedances leading to multi-bunch instabilities

Our attempts to identify the sources of coupled bunch instabilities were less successful, so far, than for the sources of single bunch instabilities. There is evidence from instability spectrum measurements at high energies of a source in the frequency range 900 MHz to 1200 MHz which could come from the next passband of the main 200 MHz RF system. There is also some indication for the low Q, high frequency resonances in the pumping ports as being a possible source of multi-bunch instabilities at high energies.

The spectrum of the growing instability is in general broad band as mentioned before, and does not lend itself to easy interpretation. Methods to analyse these spectra are being investigated at the present time [13].

We have compiled a list of the known HOM's in the various RF cavities in the ring and have searched for evidence of excitation by direct observation in the cavities. At the moment we have 5 different types of cavities in the SPS. A résumé of the situation as known is given in Table 3.

| Source | Freq. range (GHz). | Comment |
|---|---|---|
| HOM in RF cavities | 0.1 - 1.109 | 25 long. modes damped |
| | 0.286 - 1.14 | 25 trans. modes damped |
| | 0.912 - 2.78 | 13 modes seen, undamped |
| Pumping ports | 1.39 - 4.23 | Max. long. $R_{sh}/Q$ at 1.55, 1.91, 2.14, 2.96 GHz |
| 400 MHz band, (septa) | 0.343 - 0.49 | 16 septa installed now, 8 later |

Table 3: Known impedance sources

For the transverse plane the resistive wall impedance and its sources are well known. High frequency transverse modes in the cavities have been measured, see Table 3. The pumping ports also have transverse modes which were calculated but for which we do not at the moment expect problems.

## 5 Reducing the machine impedance

### 5.1 Elements with high $R_{sh}/Q$

#### 5.1.1 Pumping ports

The SPS machine designed in the early seventies has a FODO type lattice, the bending and focussing functions being separated in different magnets. There are six long straight sections each assigned to a different function, injection, extraction, RF, etc...; and 216 short straight sections, adjacent to quadrupoles, containing corrector magnets position pick ups, and other instrumentation [14]. There are five main vacuum chamber types, but different cross-sections are found in the instrumentation, etc. The result is an extremely large number of abrupt changes in vacuum pipe dimension. In a significant number of cases, the different elements are connected by the insertion of a pumping port. This is in two halves, connected by a vacuum joint, one half being a solid cylinder, the other a bellows. Unfortunately this item has a larger diameter than any vacuum chamber and creates an RF cavity with, as we have seen, significant $R_{sh}/Q$. RF screens are being designed to lower the impedance of these ~1000 items [15].

The design of this screen must fulfil many difficult requirements:

• It must be easy to instal and align. It will be necessary to remove one out of two magnets, ~300, to obtain access.

• A mechanism must be provided so that the RF contacts in the centre can be retracted and the shield divided in two, when and if a magnet must be removed or replaced.

• The pumping-down time for the machine and the residual pressure should not change significantly.

• The shielding at GHz frequencies must be good.

Fig. 3 gives a 3-d cut-away view of the present design.

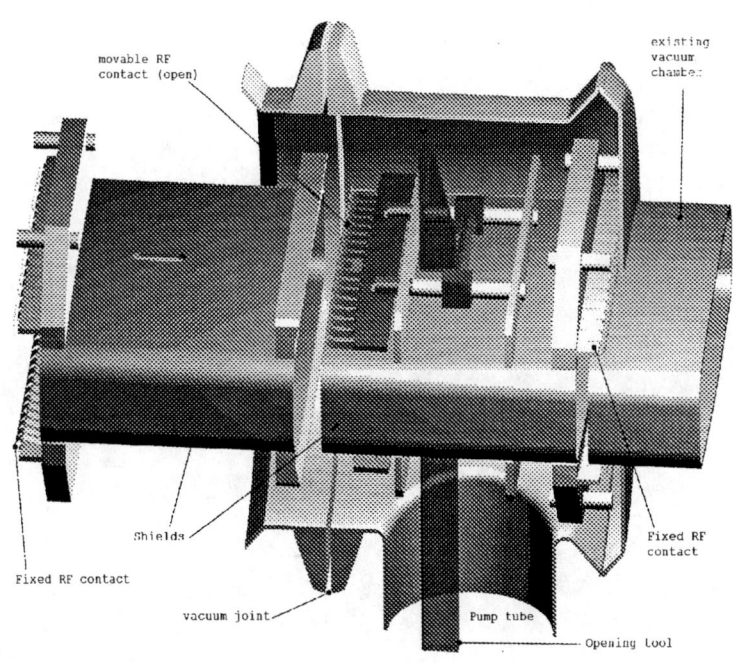

Figure 3: Pumping port screening, cut-away view.

Critical points are:
- The RF contacts at each end, which are near the uneven weld between the vacuum chamber and the end flange.
- The pumping holes which must be large enough for pumping while not affecting screening. Keeping the holes towards the side is optimum for the longitudinal modes. Care must be taken with their position so as not to aggravate any the transverse modes.
- The RF contact in the centre must be retractable. The mechanism to do this is by inserting a tool from beneath through slots in the vacuum chamber, pushing the contacts back. This contact must also be sufficiently flexible to take up significant movement, vertical and horizontal, $\sim \pm 2$ mm, during alignment and also position and angle changes when installed at the ends of bending magnets.
- Note also the two resistors, aluminium oxide on ceramic, which are inserted to lower the Q of any remaining resonances in the outer chamber.

Prototypes are being manufactured and will be installed in the machine in late summer 1999 for tests. The final installation of these screens will be spread over two long shutdowns, 1999/2000 and 2000/2001, the latter being associated with the closure of LEP.

### 5.1.2 Septum screens

The single bunch impedance measurement also identified the magnetic septa vacuum chambers as a significant source of impedance. Here, complete screening of this large vacuum tank is not possible due to the presence of the septum magnet and we have opted for a screen of diameter 28 cm, (the diameter of the adjacent vacuum chambers), covering half the circumference. A photo is shown in Fig.4.

Figure 4: RF shield in the magnetic septum vacuum tank.

This is fitted along the length of the chamber, 2.67 m with RF contacts at each end. A steel sheet perforated with holes 6 mm in diameter is used. At each end two damping resistors are inserted, capacitively connected to the floating kicker element, to damp out resonances. Measurements in the laboratory have shown a reduction of $\geq 10$ dB in impedance. A prototype was installed in the ring last year. Temperature measurements showed that heating due to beam image currents was negligible. This last shutdown half of the existing elements were shielded, the remainder will be shielded in the next.

## 5.2 Elements with high $R_{sh}$

For coupled bunch instabilities clearly the RF cavities are primary suspects in the search for devices with high $R_{sh}$. However, the known HOMs up to 900 MHz are already passively damped. If it is confirmed that the the next higher passband in the travelling wave cavities is a problem then we will examine ways to damp this. As mentioned before, there is also some evidence for the low Q, high frequency resonances in the pumping ports as being a possible source at high energies. Obviously screening will help here. When the LEP machine stops, the lepton acceleration systems in the SPS will be removed. This means the 100 MHz and 200 MHz standing wave cavity systems and the

352 MHz superconducting cavity system - a total of 28 cavities in all. In addition we will remove all other elements like the wiggler, some instrumentation etc., dedicated to lepton acceleration, to clean the machine as much as possible. Nonetheless the possibility remains that there are other elements in the machine that may have significant impedance. Our search for sources will continue.

All these changes are aimed primarily at the longitudinal impedances, though the screening of objects and the removal of unnecessary equipment will help in the transverse plane as well.

# 6 Other means of controlling the instabilities

While trying as far as possible to reduce the impedance sources responsible for the different instabilities, we are also pursuing other methods to control the problem. For the single bunch instability one possibility is to raise the threshold by increasing the slip-factor $\eta$. This can be done by lowering the value of the transition energy. In the SPS with injection at 26 GeV above and reasonably close to transition energy an improvement in threshold intensity by a factor 2.6 can be obtained by changing from $\gamma_{tr} = 23.4$ (normal operation) to $\gamma_{tr} = 19.5$. In principle either a dedicated set of quadrupoles would be installed to do this or the machine tunes could be significantly lowered. For tests [16] we have used the resonant behaviour of the dispersion function when operated at tune values close to a multiple of the machine superperiodicity of six. The result on the beam is measured by the unstable single bunch procedure outlined above. The results are given in Fig.5.

This shows that the effect of the machine impedance on the beam is reduced by approximately a factor two for all frequencies except at 200 MHz, the fundamental RF frequency, and 1.3 GHz. Note that lowering the injection energy requires a larger matching voltage at injection which is beneficial for beam loading problems.

For multi-bunch problems, we are attacking on various fronts:
- Improved RF feedback around the main RF cavities. This is under construction.
- Feedforward on the main RF cavities. This is under construction.
- Landau damping using the existing 800 MHz cavities. Analysis [17] has shown that the bunch shortening mode of operation (BS) is preferable to the bunch lengthening mode (BL) in practical applications. This is primarily due to the significantly reduced requirement on phase accuracy between the two RF systems in BS mode. This analysis confirms the operational experience of both HERA and the SPS where the BS mode is the only useful mode for Landau damping.
- Longitudinal feedback using recuperated lepton cavities. Preliminary designs are under way. Here the delicate problem is the coupling between the amplifier and cavity when bunch-by-bunch feedback (25 ns bunch spacing) is necessary.

Multi-bunch instabilities in the transverse plane are dominated by the resistive wall instability due to the vacuum chamber impedance, and the HOM modes in the RF cavities. The latter are already damped where known. The new transverse damper design

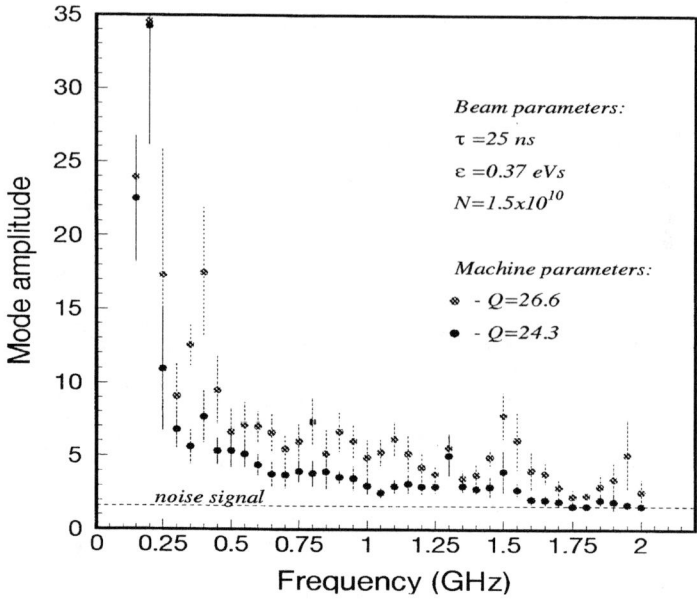

Figure 5: Spectrum of unstable modes for two values of $\gamma_{tr}$

will have a bandwidth extending to beyond 20 MHz, which will take care hopefully of all instabilities for the LHC type beams having a bunch spacing corresponding to 40 MHz [8]. However it will not cover all modes possible with the Gran Sasso high intensity beam, where the bunch spacing corresponds to 200 MHz. Here we will have to rely on Landau damping from octupoles. The transverse emittance for this beam is however less critical. The main single bunch instability is the lowest mode head-tail and we hope to control this via the chromaticity.

# 7 Conclusion

The future uses of the SPS accelerator at high intensities have created the need for an exciting programme of machine studies and hardware development projects. While standard techniques such as Landau damping and feedback are being explored to help control the beams, a vigorous programme to reduce the machine impedance, the source of the instabilities, is underway.

# Acknowledgements

This report summarises the work of a team of people in the SL division. A large number is contributing to the design of the RF shields, in particular, F. Caspers, C. Dalmas, A. Rizzo, J. Ramillon and A. Spinks. I would like to thank E. Shaposhnikova for reading the paper and helpful comments.

# References

[1] D. Boussard, G. Dome, T. Linnecar, A. Millich, Longitudinal Phenomena in the CERN SPS, IEEE Trans. on Nuclear Science, Vol. NS-24, No. 3, 1977

[2] D. Boussard, L. Evans, J. Gareyte, T. Linnecar, W. Mills, E.J.N. Wilson, Acceleration and Storage of a Dense Single Bunch in the CERN SPS, IEEE Trans. on Nuclear Science, Vol. NS-26, No. 3, 1979

[3] The SPS as Injector for LHC - Conceptual design, ed. P. Collier, CERN-SL-97-07 DI.

[4] J. Gareyte, Requirements of the LHC on its Injectors. Proc. workshop on LEP-SPS performance, Chamonix IX, Jan. 1999. CERN-SL-99-007 DI

[5] J. Tuckmantel, The SPS/LHC Longitudinal Interface. Proc. workshop on LEP-SPS performance, Chamonix IX, Jan. 1999. CERN-SL-99-007 DI

[6] E. Weisse, The CERN Neutrino Beam to Gran Sasso. Proc. workshop on LEP-SPS performance, Chamonix IX, Jan. 1999. CERN-SL-99-007 DI

[7] E. Shaposhnikova, Longitudinal Instabilities in the SPS. Proc. workshop on LEP-SPS performance, Chamonix IX, Jan. 1999. CERN-SL-99-007 DI

[8] W. Hofle, Towards a Transverse Feedback System and Damper for the SPS in the LHC Era. Particle Accelerators, 1997, Vol. 58, pp 269-279.

[9] T. Linnecar, E. Shaposhnikova, Microwave instability and Impedance Measurements in the CERN SPS. Particle Accelerators, 1997, Vol. 58, pp 241-255.

[10] T. Linnecar, E. Shaposhnikova, Another Method to Measure the Low-Frequency Machine Impedance. EPAC96, June 1996, Sitges.

[11] L. Vos, Computer Calculation of the Longitudinal Impedance of Cylindrically Symmetric Structures and its Application to the SPS. CERN SPS/86-21(MS).

[12] T. Bohl, T.P.R. Linnecar, E. Shaposhnikova, Measuring the Resonance Structure of Accelerator Impedance with Single Bunches. Phys. Rev. Letts. 1997, Vol. 78, No. 16.

[13] E. Shaposhnikova, Analysis of coupled bunch instability spectra. These proceedings.

[14] P. Collier, A.Spinks, Survey of the Short Straight Sections in the SPS for the Impedance Reduction Programme. CERN SL-Note-99-025 SLI.

[15] A. Spinks. Private communication.

[16] T. Bohl, M. Lamont, T. Linnecar, W. Scandale, E. Shaposhnikova, Measurement of the Effect in Single Bunch Stability of Changing Transition Energy in the CERN SPS. EPAC98, June 1998, Stockholm, and CERN SL-98-024 RF.

[17] T. Bohl, T. Linnecar, E. Shaposhnikova, J. Tuckmantel, Study of Different Operating Modes of the 4th RF Harmonic Landau Damping System in the CERN SPS. EPAC98, June 1998, Stockholm, and CERN SL-98-026 RF.

# Analytic Methods for Impedance Calculations

R.L. Gluckstern and A.V. Fedotov

*Physics Department, University of Maryland, College Park, Maryland 20742*

## I  INTRODUCTION

A beam bunch traveling along the axis of beam pipe produces a wakefield which can cause disruption and/or instability within the bunch or in the following bunches. Important measures of this disruption are the longitudinal and transverse coupling impedances produced by obstacles along the beam pipe. These impedances are the Fourier transforms of the wake functions.

In this paper we will review the analytic calculation of the longitudinal impedance of a variety of obstacles/obstructions in a typical beam pipe. Although most of these analyses have been for ultrarelativistic point charges, the focus of this workshop is on impedance issues relevant to the SNS storage ring. For this reason, we will indicate the changes to be expected when $v \neq c$ wherever they have been obtained. The reader should be aware of two excellent books on the subject [1,2] which provide a great deal of analytic and numerical information on wakefields and impedances in a variety of environments. An overview of impedance calculations can be found in [3]. Also some studies for a beam velocity $v < c$ were recently presented [4,5].

## II  IMPEDANCE OF A PERFECTLY CONDUCTING BEAM PIPE

We consider a charge $Q$ in the form of a uniformly charged thin disk of radius $a$, traveling with velocity $v$ along the $z$-axis of a perfectly conducting beam pipe of radius $b$. The frequency domain charge and current densities can then be written as

$$\rho(r, z; \omega) = \frac{Q}{\pi a^2 v} e^{-j\omega z/v}, \qquad (1)$$

$$J_z(r, z; \omega) = \frac{Q}{\pi a^2} e^{-j\omega z/v}, \qquad (2)$$

where we need to multiply by

---

CP496, *Workshop on Instabilities of High Intensity Hadron Beams in Rings,*
edited by T. Roser and S. Y. Zhang
© 1999 American Institute of Physics 1-56396-910-6/99/$15.00

$$\frac{1}{2\pi}\int_{-\infty}^{\infty}d\omega e^{j\omega t} \tag{3}$$

to obtain all quantities in the time domain. We now write the longitudinal impedance as a volume integral (over the transverse distribution of the beam) as [6]

$$Z_{\parallel}(\omega) = -\frac{1}{Q^2}\int d^3v \boldsymbol{E} \cdot \boldsymbol{J}^*. \tag{4}$$

The longitudinal impedance corresponding to the source charge and current for a length $\mathcal{L}$ of perfectly conducting beam pipe can then be written in terms of Bessel functions of imaginary arguments as

$$\frac{Z_{\parallel}(\omega)}{Z_0} = -\frac{j\mathcal{L}}{\pi a^2 \omega}[1 - 2F_1(\sigma a)I_1(\sigma a)], \tag{5}$$

where $Z_0 = (\mu/\epsilon)^{1/2}$, $\sigma = \omega/v\gamma$, and where

$$F_1(x) = K_1(x) - \frac{K_0(\sigma b)}{I_0(\sigma b)}I_1(x). \tag{6}$$

For $\sigma a \ll 1$, Eq. (5) can be written as

$$\frac{Z_{\parallel}(\omega)}{nZ_0} \simeq -\frac{j}{\beta\gamma^2}[\ln\frac{b}{a} + \frac{1}{4} + p(\sigma b)]. \tag{7}$$

For a ring of circumference $\mathcal{L}$, $n = \omega/\omega_0 = \omega\mathcal{L}/2\pi v$ is the harmonic number and

$$p(y) = \ln\left(\frac{2}{y}\right) - C - \frac{K_0(y)}{I_0(y)}, \tag{8}$$

where $C = 0.5772$ is Euler's constant. The 1/4 in Eq. (7) reflects the average over the uniform transverse charge density of the beam. For a charge density proportional to $r^n$, this constant is $(4+n)^{-1}$. Also, if the beam radius is $a$ but the integral over $E_z$ is taken only on the $z$-axis one obtains 1/2 instead of 1/4.

If we also have $\sigma b \ll 1$, which would be correct either for an ultrarelativistic beam ($\gamma \to \infty$, $\beta \to 1$) or in the low frequency limit, $p(\sigma b) \to 0$, Eq. (7) reduces to the familiar result [1,2,7].

## III IMPEDANCE OF A SMALL HOLE IN THE BEAM PIPE

An obstacle, such as a hole in the beam pipe wall, interrupts the image current flow in the wall causing the propagation of wave guide modes away from the obstacle. Using Maxwell's equations, the additional contribution to the impedance due to the hole can be written as the surface integral over the hole area at the pipe radius $r = b$:

$$Q^2 Z_\|(\omega) = \int_{\text{hole}} dS(E_z \times H_{1\theta}^*), \qquad (9)$$

where $H_{1\theta}(b, \theta; \omega)$ is the magnetic field in the absence of the hole and $E_z(b, \theta; \omega)$ is the axial electric field in the presence of the hole. For a small hole, Eq. (9) can be written in terms of the magnetic susceptibility $\psi$ and electric polarizability $\chi$ of the hole as

$$Q^2 Z_\|(\omega) = \frac{j\omega\mu\psi}{2}|H_\theta|^2 - \frac{j\omega\epsilon\chi}{2}|E_r|^2, \qquad (10)$$

valid for any value of $\beta$ [8]. The dependence on velocity comes from $E_r$ and $H_\theta$ evaluated at $r = b$. These can be easily obtained for the perfectly conducting beam pipe by a Lorentz transformation from the frame moving with the charge to the laboratory frame, and are given, in the frequency domain, by

$$Z_0 H_\theta = \beta E_r = \frac{QZ_0}{2\pi b}\frac{e^{-j\omega z/v}}{I_0(\sigma b)} \qquad (11)$$

leading to

$$\frac{Z_\|(\omega)}{Z_0} = \frac{j\omega}{8\pi^2 b^2 v}\frac{(\beta\psi - \chi/\beta)}{[I_0(\sigma b)]^2}. \qquad (12)$$

The result in Eq. (12) agrees with earlier results for $\beta = 1$. But for $\beta < 1$, it indicates that the sign of the imaginary impedance will change when $\beta^2 = \chi/\psi$, (which is always less than 1). Expressions for $\psi$ and $\chi$ for holes of different geometry can be found in [3]. In fact, the usual expectation that $Z_\|$ vanishes for a long thin slot is no longer correct for $\beta < 1$ where the induced electric dipole covers an extended axial region.

Equation (12) was derived for a hole in a thin wall. The same expression is valid for a hole whose thickness is comparable to its width and length. Specifically, $\psi$ and $\chi$ need to be replaced by $\psi_{\text{in}}$ and $\chi_{\text{in}}$, the inside susceptibility and polarizability defined and evaluated by Gluckstern and Diamond [9]. Also, the result in Eq. (12) can be generalized to be valid for a small obstacle which protrudes into the beam pipe [10].

For the transverse coupling impedance it is useful to write the impedance in terms of a surface integral. In the limit of a small beam displacement $\Delta_x$ we have

$$kQ^2\Delta_x^2 Z_x(\omega) = \int_{\text{hole}} dS(E_z \times H_{1\theta}^*). \qquad (13)$$

The form of the coupling integral in Eq. (13) is identical to Eq. (9). We then immediately obtain the transverse counterpart to Eq. (10):

$$kQ^2\Delta_x^2 Z_x(\omega) = \frac{j\omega\mu\psi}{2}|H_\theta|^2 - \frac{j\omega\epsilon\chi}{2}|E_r|^2. \tag{14}$$

The only remaining step is to find the new components of $E_r$, $H_\theta$ produced by the dipole source. With

$$Z_0 H_\theta = \beta E_r = \frac{QZ_0\Delta_x \cos\theta}{2\pi} \frac{\sigma/b}{I_1(\sigma b)} e^{-j\omega z/v} \tag{15}$$

we finally obtain

$$\frac{Z_x(\omega)}{Z_0} = \frac{j\cos^2\theta}{2\pi^2 b^4}\left[\frac{\sigma b}{2I_1(\sigma b)}\right]^2 \frac{(\beta^2\psi - \chi)}{\beta^2}. \tag{16}$$

## IV IMPEDANCE NEAR THE CUTOFF OF THE BEAM PIPE

In an earlier calculation for a small azimuthally symmetric pillbox cavity of height $\Delta b$ and width $g$ [11] we obtained the following expression for the admittance of the cavity for $\beta = 1$:

$$Z_0 Y_\parallel(k) = 2\pi kb\left[-\frac{j}{k^2 g\Delta b} + \frac{j 2\ln 2}{\pi} + \sum_{s=1}^{\infty}\frac{e^{-jb_s g/b}}{b_s}\right], \tag{17}$$

where $k = \omega/c$ and $b_s^2 = k^2 a^2 - p_s^2$, with $p_s$ being the roots of $J_0(p_s) = 0$. One sees in Eq. (17) that the real part of the admittance is independent of all features of the pillbox for $g \ll b$. It can be shown that this term corresponds to the energy which is lost as the pillbox generates outgoing propagating modes in the pipe. Apparently, the reactive part arises from the evanescent pipe modes generated by the pillbox. Equation (17) is valid for $kb \gtrsim 1$ as long as $k\Delta b \ll 1$, $kg \ll 1$, and accurately describes the broad resonance character of the impedance in this frequency range. For a small obstacle, the first term in the bracket [ ] is dominant and corresponds to the magnetic susceptibility found in Eq. (12). Thus, for low frequencies, we can rewrite the impedance of a small pillbox as

$$Z_\parallel = j\omega L, \tag{18}$$

where the inductance $L$ corresponds only to the magnetic portion of the problem. The immediate implication of our result given by Eq. (12) is that $Z_\parallel$ for a pillbox is unchanged for $\beta < 1$, except for the factor $1/(I_0^2(\sigma b))$. Of course, if one rewrites it in the form $Z_\parallel/n$ with $n = \omega/\omega_0 = \omega R/v$, then Eq. (18) becomes:

$$\frac{Z_\parallel}{n} = \frac{j\beta cL}{R[I_0(\sigma b)]^2}. \tag{19}$$

where $R = \mathcal{L}/2\pi$ is the radius of the accelerator ring.

We have recently repeated this analysis for a hole in a thin liner wall [12] with the result that the impedance separates into an electric term whose admittance is

$$Z_0 Y_\parallel^{(\text{el})}(k) = \frac{j8\pi^2 b^2}{k}\left[\frac{1}{\chi} - \tilde{W}\right] \tag{20}$$

and a magnetic term whose admittance is

$$Z_0 Y_\parallel^{(\text{mag})}(k) = -\frac{j8\pi^2 b^2}{k}\left[\frac{1}{\psi} + \tilde{V}\right], \tag{21}$$

where $\tilde{W}$, $\tilde{V}$ are complicated complex functions of the pipe geometry. We have thus far been unable to obtain explicit expressions for $\tilde{W}$ and $\tilde{V}$ but have confirmed Eqs. (20) and (21) numerically, obtaining approximate results for $\tilde{W}$ and $\tilde{V}$ for $kb \lesssim 1$. Once again the additional terms involving $\tilde{W}$ and $\tilde{V}$ are related to the outgoing modes in the pipe (and in [12] also the coax between the pipe and outer wall). Although these expressions were derived for $\beta = 1$, we believe that Eq. (20) should be multiplied by $\beta^2 I_0^2(\omega b/v\gamma)$ and that Eq. (21) should be multiplied by $I_0^2(\omega b/v\gamma)$, as suggested by Eq. (12). For convenience, we also give the result for the impedance, when $\chi\tilde{W} \ll 1$, $\psi\tilde{V} \ll 1$:

$$\frac{Z_\parallel(\omega)}{Z_0} = \frac{j\omega}{8\pi^2 b^2 v[I_0(\sigma b)]^2}\left[\beta\psi - \frac{\chi}{\beta} - \left(\beta\psi^2\tilde{V} + \frac{\chi^2\tilde{W}}{\beta}\right)\right]. \tag{22}$$

The real part of the impedance is contained in the terms proportional to $\text{Im}\tilde{V}$, $\text{Im}\tilde{W}$.

The simple form for the admittance in Eqs. (20), (21) is encountered frequently in several analyses for the impedance in different situations (small obstacles, periodic obstacles, etc.). It most likely originates from the most logical construct of an equivalent circuit. But in the present case, the separation into an electric and a magnetic term is broken when the thickness of the beam pipe wall is finite [12].

## V INTEGRAL EQUATION FOR THE LONGITUDINAL IMPEDANCE

An alternate approach for the calculation of the longitudinal impedance is by solving the integral equation for the longitudinal electric field at the border between the beam pipe and the obstacle [13]. To illustrate this for an azimuthally symmetric geometry, we assume that the obstacle is a pillbox cavity and write for the fields within the pipe ($r \leq b$):

$$E_z(r, z; \omega) = \int_{-\infty}^{\infty} dq\, e^{-jqz} A(q)\frac{J_0(\kappa r)}{J_0(\kappa b)} + E_z^{(s)}(r, z; \omega), \tag{23}$$

$$E_r(r, z; \omega) = j \int_{-\infty}^{\infty} q dq e^{-jqz} A(q) \frac{J_1(\kappa r)}{\kappa J_0(\kappa b)} + E_r^{(s)}(r, z; \omega), \qquad (24)$$

$$Z_0 H_\theta(r, z; \omega) = jk \int_{-\infty}^{\infty} dq e^{-jqz} A(q) \frac{J_1(\kappa r)}{\kappa J_0(\kappa b)} + Z_0 H_\theta^{(s)}(r, z; \omega), \qquad (25)$$

where $\kappa^2 = k^2 - q^2$ and the source field components are given in Eq. (11). We now use the following definition of the longitudinal coupling impedance:

$$Z_\|(\omega) = -\frac{1}{Q} \int_0^g dz e^{-j\omega z/v} E_z(0, z; \omega). \qquad (26)$$

Using Eq. (23) and $\sigma = \omega/v\gamma$, we obtain

$$Z_\|(\omega) = -\frac{2\pi}{Q} \frac{A(\omega/v)}{I_0(\sigma b)}, \qquad (27)$$

which becomes

$$Z_\|(\omega) = -\frac{1}{Q I_0(\sigma b)} \int_0^g dz e^{j\omega z/v} E_z(b, 0, z; \omega). \qquad (28)$$

The fields within the cavity are expanded into an orthonormal set of modes which are defined with a metallic boundary condition at $r = b$ in the cavity region [13]. The magnetic field at $r = b$ can then be written as

$$Z_0 H_\theta(b, z; \omega) = \frac{jkb}{2\pi} \int_0^g dz' f(z') K_c(z, z'), \qquad (29)$$

where $f(z) = E_z(b, z; \omega)$, and where the cavity kernel is defined as

$$K_c(z', z) = 4\pi^2 \sum_\ell \frac{h_\ell(z) h_\ell(z')}{k^2 - k_\ell^2}. \qquad (30)$$

Here $h_\ell(z, r)$ is the $\theta$ components of the cavity magnetic field for the $\ell^{\text{th}}$ cavity mode.

Since $E_z$ is continuous across the boundary at $r = b$, we can take the Fourier transform of Eq. (23) to write

$$A(q) = \frac{1}{2\pi} \int_{-\infty}^{\infty} dz e^{jqz} f(z), \qquad (31)$$

leading to the pipe magnetic field component

$$Z_0 H_\theta(b, z; \omega) = -\frac{jkb}{2\pi} \int_0^g dz' f(z') K_p(z - z') + \frac{Q e^{-j\omega z/v}}{2\pi \epsilon_0 bc I_0(\sigma b)}. \qquad (32)$$

Here the pipe kernel is found to be [13]:

$$K_p(u) \equiv \int_{-\infty}^{\infty} dq e^{jqu} \frac{J_0'(\kappa b)}{\kappa b J_0(\kappa b)} = \frac{2\pi j}{b} \sum_{s=1}^{\infty} \frac{e^{-jb_s|u|/b}}{b_s}, \qquad (33)$$

with $u = z - z'$. As in Eq. (17), $b_s = (k^2 b^2 - p_s^2)^{1/2}$ for $p_s < kb$ and $b_s = -j(p_s^2 - k^2 b^2)^{1/2}$ for $p_s > kb$, where $p_s$ are the roots of $J_0(p_s) = 0$. Equating the magnetic fields in Eqs. (29), (32) then leads to the integral equation for $f(z)$:

$$\int_0^g dz' f(z')[K_p(|z'-z|) + K_c(z',z)] = -\frac{jQZ_0}{kb^2 I_0(\sigma b)} e^{-j\omega z/v}. \qquad (34)$$

If we renormalize $f(z)$ to include the factors on the right side of Eq. (34), we need to solve the integral equation

$$\int_0^g dz' F(z')[K_p(|z'-z|) + K_c(z',z)] = -je^{-j\omega z/v} \qquad (35)$$

for $F(z)$, and thus obtain from Eq. (28)

$$\frac{Z_\parallel(\omega)}{Z_0} = \frac{1}{kb^2[I_0(\sigma b)]^2} \int_0^g dz' e^{j\omega z'/v} F(z'). \qquad (36)$$

Equation (35) can be solved for a small obstacle leading to the result in Eq. (17) modified by the factor $[I_0(\sigma b)]^2$.

Equation (35) also serves as the starting point for the analysis of the smoothed high frequency behavior of a pillbox [13]. In particular, the kernels are averaged over rapid oscillations in frequency, and the resulting integral equation is solved for the averaged impedance.

One can also apply this approach to the problem of periodic cavities for both longitudinal and transverse impedance [14,15]. In this case there are separate integral equations obtained by matching the magnetic field in each cavity. The result is a set of integral equations involving the kernel of Eq. (33) supplemented by a coupling kernel. By invoking the Floquet solution for the fields in successive cavities one obtains a simple solution for the admittance as the sum of an imaginary term proportional to $k$, supplemented by the admittance of a single cavity. It seems that these results for the admittance, which appear to involve only magnetic effects for an azimuthally symmetric cavity, also require the factor $[I_0(\sigma b)]^2$ for finite $\gamma$.

## VI  DISCUSSION AND SUMMARY

We have reviewed some of the methods used to calculate the longitudinal impedance analytically, and have indicated the places where we believe some modification is needed for $v < c$. Similar results can be obtained for the transverse impedance. Although these have not been reviewed for changes needed when $v < c$, it seems that the factor $[I_0(\sigma b)]^2$ for the longitudinal impedance is replaced by the factor $[I_1(\sigma b)/(\sigma b/2)]^2$ for the transverse impedance. In any event, a combination of analytic and numerical approaches is advisable when applying these techniques to the design of actual machines.

## VII  ACKNOWLEDGEMENT

We wish to thank S. Kurennoy for helpful discussions.

## REFERENCES

1. A.W. Chao, *Physics of Collective Beam Instabilities in High Energy Accelerators*, Wiley (1993).
2. B.W. Zotter and S.A. Kheifets, *Impedances and Wakes in High-Energy Particle Accelerators*, World Scientific (1998).
3. A.W. Chao and M. Tigner, *Handbook of Accelerator Physics and Engineering*, World Scientific (1999).
4. S.S. Kurennoy, Phys. Rev. ST-AB**2**, 032001 (1999).
5. D. Davino, G. Miano, G. Panariello and L. Verolino, Phys. Rev. ST-AB**2**, 044401 (1999).
6. R.L. Gluckstern and F. Neri, IEEE Transactions of Nuclear Science, **32**, 2403 (1985).
7. L. Palumbo, V. Vaccaro and M. Zobov, *Wake Fields and Impedances*, Report 95-06, CERN-CAS (1995).
8. R.L. Gluckstern, R. Li and R.K. Cooper, IEEE Transactions on Microwave Theory and Techniques, **38**, 186 (1990); **38**, 1529 (1990).
9. R.L. Gluckstern and J.A. Diamond, IEEE Transactions on Microwave Theory and Techniques, **39**, 274 (1991).
10. S.S. Kurennoy, Phys. Rev. E**55**, 3529 (1997); R.L. Gluckstern and S.S. Kurennoy, Phys. Rev. E**55**, 3533 (1997).
11. R.L. Gluckstern and F. Neri, Proceedings of the Particle Accelerator Conference, Chicago, IL, p. 1271 (1989).
12. A.V. Fedotov and R.L. Gluckstern, Phys. Rev. ST-AB**1**, 024401 (1998).
13. R.L. Gluckstern, Phys. Rev. D**39**, 2773 (1989).
14. R.L. Gluckstern, Phys. Rev. D**39**, 2780 (1989).
15. A.V. Fedotov, R.L. Gluckstern and M. Venturini, Phys. Rev. ST-AB**2**, 064401 (1999).

# Nonlinear Features of the Longitudinal Instability For High-Current Machines

I. Hofmann and O. Boine-Frankenheim

*GSI, Planckstr.1, 64291 Darmstadt, Germany*

**Abstract.** We present results from experiments at the GSI machines as well as computer simulation for space charge dominated coasting beams (below transition). It is found that for the high-current machines presently under discussion the actual challenge lies in the nonlinear regime. Experiments are in good agreement with theory and simulation in the linear regime; for the nonlinear regime and long-time evolution rsp. saturation our experimental results show good agreement in some aspects, like wave steepening. To analyze the final momentum distribution we still depend on simulation, which shows that the behavior differs substantially, depending on whether the working point in the impedance plane lies close to the real (resistive dominated) or imaginary (space charge dominated) axis, or in between. For the space-charge-dominated regime ($Re\ Z << Im\ Z$) it is found by computer simulation that for currents far above the Keil-Schnell threshold self-stabilization occurs by formation of a momentum tail, hence linear instability criteria can be practically ignored. It is shown here that the global impedance distribution is of crucial importance.

## I INTRODUCTION

One of the challenging questions in connection with longitudinal instabilities is whether a high-current machine below transition must necessarily operate under conditions where the beam is linearly stable. This issue has been of particular interest in connection with driver accelerators for heavy ion inertial fusion, where the requirement of longitudinal stability cannot easily be reconciled with the small momentum spread needed for final focusing [1]. For spallation neutron sources and proton drivers for muon accelerators small momentum spread is desirable, and therefore similar questions can be raised. The theoretical threshold current in these applications is mainly defined by the relatively large space charge impedance, which causes a (real) frequency shift and thus reduces or even suppresses Landau damping; the actual instability is, of course, driven by the resistive impedance. It is well-known that for beams with significant space charge impedance the stability threshold is not just given by the Keil-Schnell criterion [2], which refers to an assumed circle in the impedance plane (here referred to as "circle criterion"). The real stability boundary depends on the distribution function as evaluated in detail in Ref. [3] and indicated in Fig. 1. It is recognized from Fig. 1 that the deviation from the circle criterion is largest in the "space charge regime", and relatively small in the "resistive regime". It turns out that the nonlinear behavior (saturation of the instability, overshoot, self-stabilization) has quite different features in these regimes. The physics in the "space charge regime" in particular is substantially different from the behavior in the resistive regime.

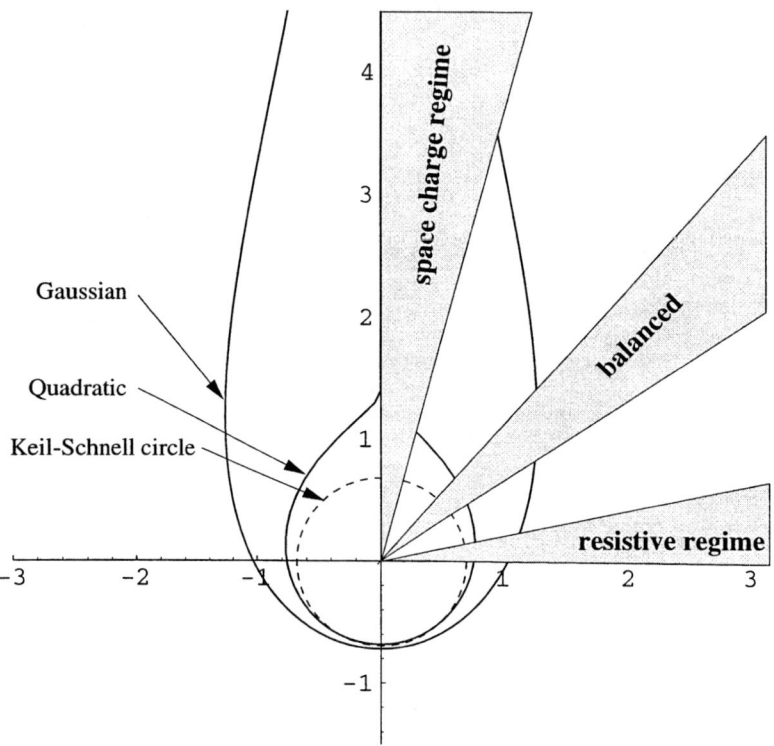

**FIGURE 1.** Stability boundaries in complex impedance plane for quadratic and Gaussian distribution functions. For the nonlinear behavior we find pronounced differences depending on whether the impedance is predominantly space charge dominated, resistive or balanced between both.

Experiments under such conditions are not easy to carry out in existing machines, though there is evidence in the ISIS neutron source that threshold currents (as given by the circle criterion) can be exceeded by an order of magnitude [4]. One therefore depends largely on predictions from computer simulation. Computer simulation has been found a necessary tool for interpreting results from observation of space-charge-dominated beams. At GSI a diversity of codes is available based on different methods and space charge calculation techniques. PATRIC(r,z) is the most versatile code containing RF options on different harmonics, an r-z Poisson solver, cavity impedances and Schottky noise analysis. More recently we have employed for longitudinal simulation a direct Vlasov integrator which avoids the noise inherent to PIC simulation [5]. This code is very useful to study long-term evolution and resolve fine structures in phase space. In particular it allows to determine the momentum distribution during the unstable phase, which so far we have not succeeded to measure in the experiment.

At the GSI machines several experiments with the RF cavity impedance as driving force for the instability have been carried out in past years in the "resistive" and "balanced impedance" regime and compared with computer simulation. We briefly summarize these experiments in section II (for details see Refs. [6,7]) before we present our simulation results on the influence of impedance in section III.

## II  EXPERIMENTAL RESULTS

### A  Linearized Regime

The linearized theory of the longitudinal instability goes back to the sixties [8,3], when the subject raised some concern as to whether the performance of the new proton machines would be affected. A special cavity was built for the ISR [9] to check if the theory predictions were right. The early measurements and accompanying simulations [10] were made above transition energy, where the space charge impedance is negligible.

In recent years the subject found new interest. A detailed study of it was undertaken in the time domain in a linear resistive electron channel at the University of Maryland [11]. The experiment confirmed the growth rates from linearized theory for fast and slow waves, which could be launched separately by appropriate initial conditions.

In the ESR and SIS we have investigated this instability using the electron cooling and a tunable RF cavity, which allows to carry out the experiments in the vicinity as well as far away from the stability boundary.

We have cooled a $C^{6+}$ beam at 250 MeV/u and 0.3 mA current in order to obtain a very small longitudinal momentum spread near $10^{-5}$. The frequency of the RF cavity was then shifted (within 15 ms) from an initially strongly de-tuned value towards the beam revolution frequency to obtain the expected unstable behavior. Different working points in the impedance plane could thus be achieved, depending on the de-tuning value (see Fig. 3). A small but finite gap voltage of 300 V was present to control the cavity impedance. The observed growth rates were found in excellent agreement with the results from computer simulation, which also included the finite gap voltage (see Ref. [7]). The finite gap voltage had the - not yet well-explained - effect of reducing the e-folding growth times below the values from linearized theory whenever the cavity frequency was matched (or nearly matched) to the revolution harmonic (see upper beam current trace in Fig. 3). For stronger de-tuning the gap voltage had no effect on the beam and the rise was exponential and in excellent agreement with both simulation *and* linearized theory [7]. (see Fig. 4). The more interesting question for the design of future machines is, of

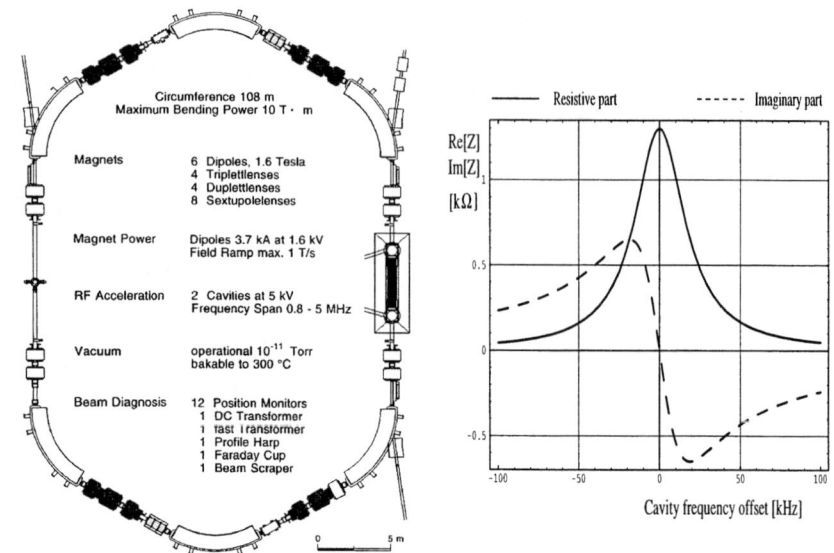

**FIGURE 2.** Frequency dependence of the RF cavity impedance in the ESR storage ring.

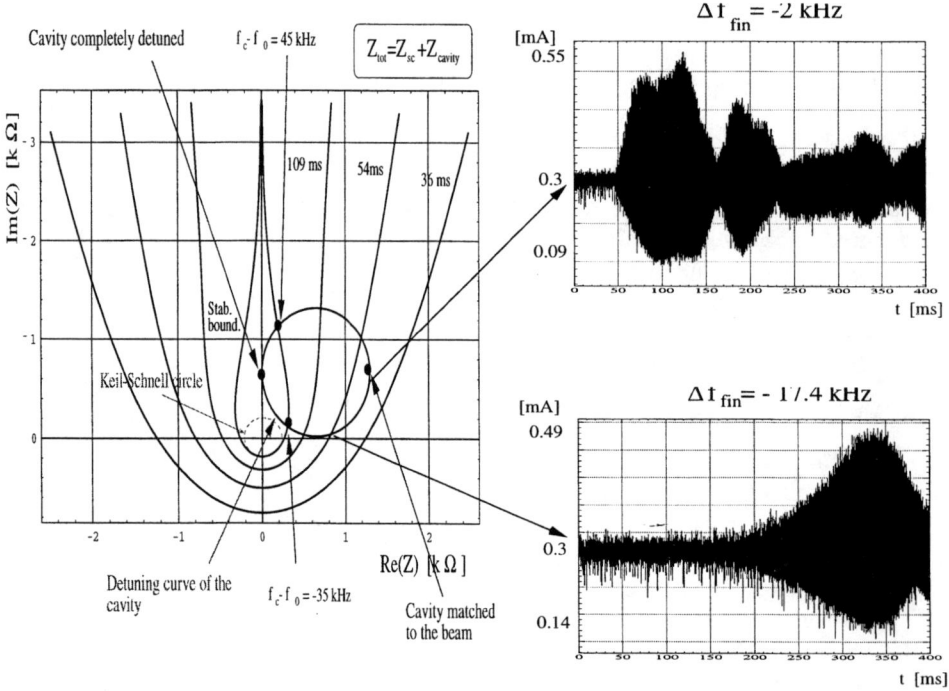

**FIGURE 3.** Measurements for different working points as function of de-tuning.

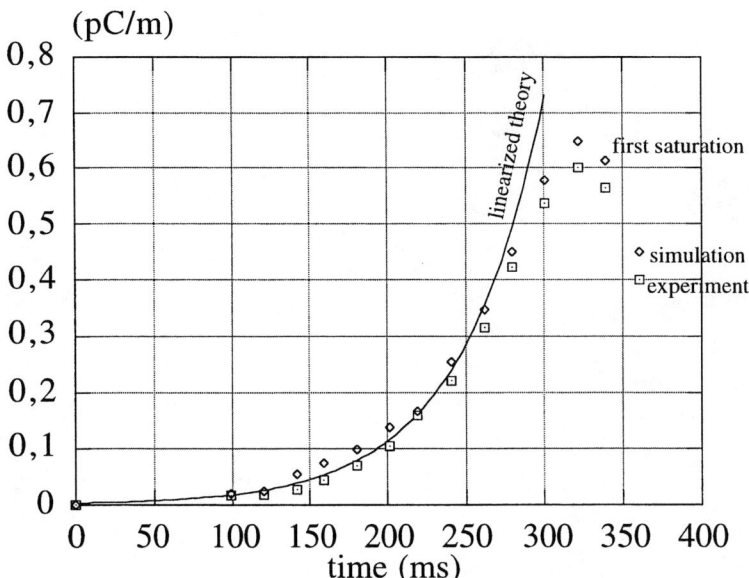

**FIGURE 4.** Comparison of initial growth (current modulation) for experiment, simulation and linearized theory for -17.4 kHz de-tuning.

course, whether linear stability is at all necessary. This emphasizes the need to study nonlinear behavior, where interesting beam physics appears as will be shown in the next section.

## B  Observation in the Nonlinear Regime

In order to evaluate the full history of the instability beyond the initial (exponential) growth we have recorded in the ESR experiments the line density synchronously with the revolution period (mountain range diagram of Fig. 5). An interesting nonlinear effect is that the strong coherent signals persist for at least as long as a second and are not Landau damped. This is confirmed by the longitudinal Vlasov code simulation. The working point in the example of Fig.5 (and likewise the simulation) was far in the unstable region pertaining to the Gaussian momentum distribution ("balanced impedance", corresponding to a detuning of 20 kHz in Fig. 3 due to the combined effect of the cavity and space charge impedances).

The diagram shows the initially exponential growth of the slow wave (moving to the right, with time increasing from bottom to top), nonlinear saturation and decay into additional fast waves moving to the left. The self-bunching effect is generally not exceeding 50% of the coasting beam current, which is in excellent agreement with our computer simulation. The appearance of significantly higher harmonic signals at some later time (0.2 seconds in Fig. 5) can be noted, which are not

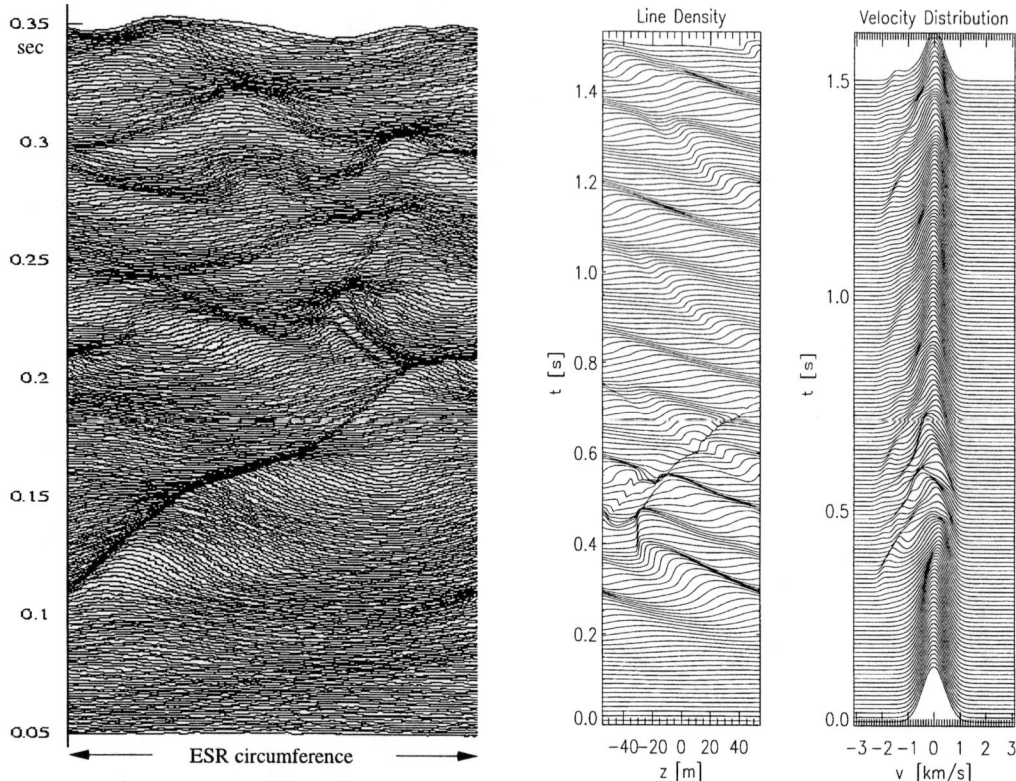

**FIGURE 5.** Exponential growth and nonlinear saturation phase of longitudinal resistive instability of cooled coasting beam in the ESR driven by the RF cavity on the first harmonic (left). The mountain range plot shows subsequent time traces from bottom to top over 350 ms (each trace is the current profile over one turn). For comparison we also show the line density and velocity distribution from a Vlasov simulation (right).

present if the instability on the fundamental harmonic is absent. We assume this results from a loss of Landau damping in the filamented phase space distribution of the saturated instability.

## III  INFLUENCE OF IMPEDANCE

### A  Long-Term Behavior for "Balanced Impedance")

The phase space plots of the Vlasov simulation for the case of Fig. 5 show that the long-term behavior in the "balanced impedance" regime is characterized by phase space holes or "bubbles" trapped by the nonlinear wave originating at the lower momentum edge (Fig. 6). These "bubbles" move upwards in momentum

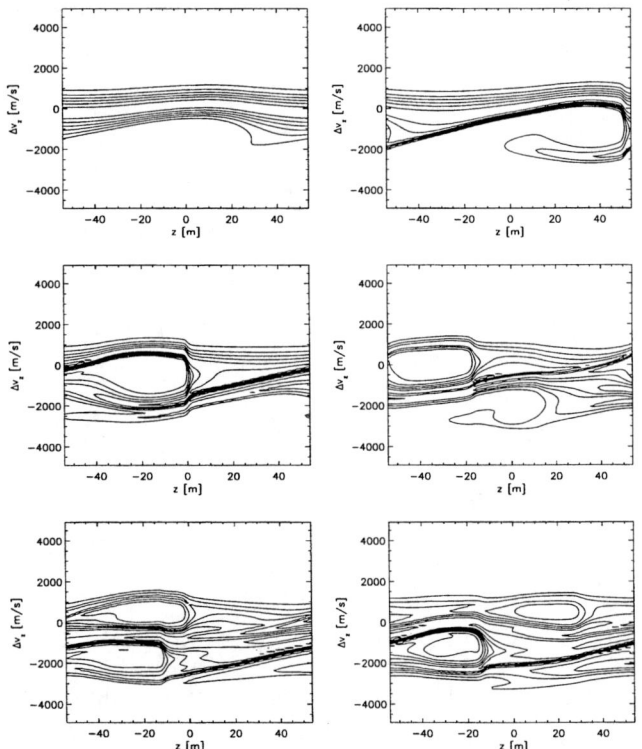

**FIGURE 6.** Computer simulation (Vlasov) of longitudinal instability at working point for parameters of the ESR experiment. Shown are longitudinal phase space density contours at equi-distant times starting from the linear phase into the nonlinear region in equidistant time steps until 850 ms.

space, while new waves are born (see fourth frame in Fig. 6), which explains the experimental observation of continuing oscillations of the line density. The initial e-folding time for this case is about 60 ms.

One of the interesting issues in this context is the question of overshoot. A very simplified picture of nonlinear saturation is that of a beam which self-stabilizes after it has gained enough momentum spread due to the instability. The overshoot theorem then states that the finally achieved momentum spread is actually larger than the threshold momentum spread necessary to avoid onset of the instability. The standard overshoot formula derived by Dory [12] in the absence of space charge

$$\Delta p/p_{final} \Delta p/p_{initial} \geq \Delta p/p_{thresh}^2, \qquad (1)$$

is inappropriate to explain our simulation results. It would suggest roughly a doubling of the momentum spread with respect to the threshold value for our example. Other overshoot criteria (based on simplifying theories) suggest weaker effects [13,14], but none of them is able to describe our simulation results for the "balanced impedance" regime as described in Fig. 7. This figure also contains mod-

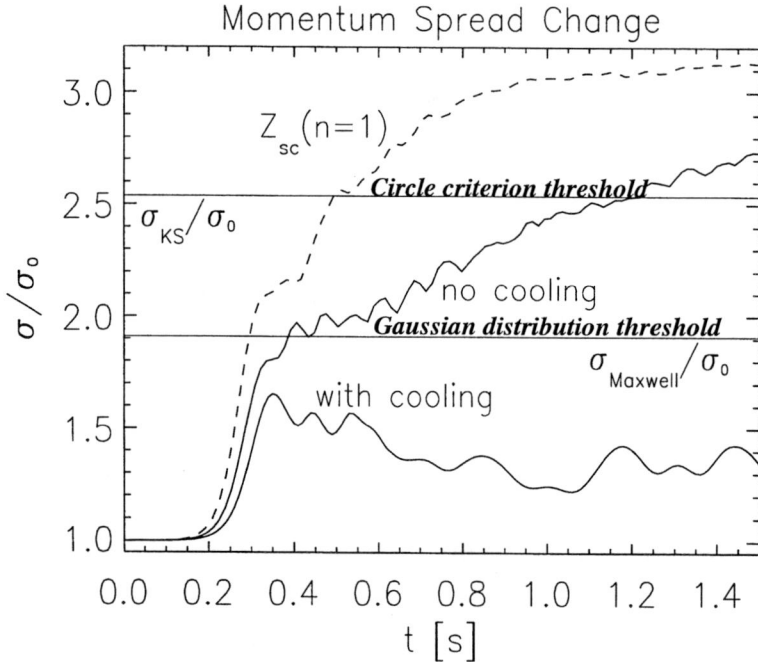

**FIGURE 7.** Evolution of the fwhm momentum spread showing the absence of overshoot at saturation of the instability. Note the curve "no cooling" continues to a slow, but steady growth at the threshold corresponding to a Gaussian distribution.

ifications due to electron cooling and an artificial impedance $Z_{sc}(n = 1)$ modeling the correct value of the space charge impedance at $n = 1$, but zero impedance at all other harmonics. The time evolution of the fwhm momentum spread ("no cooling") shows that the initially exponential growth levels off after roughly five e-foldings at a value which corresponds to the threshold momentum spread, hence no overshoot at this point. At later times there is slow growth of the momentum spread without evidence of saturation. The beam gradually looses energy, since the circulating energy is the only source of energy available for the resistive heating. The leveling-off at the threshold momentum spread permits the interpretation that the beam settles down near the stability margin, hence fluctuations (phase space "bubbles") are practically undamped and persist for a period of time. The case $Z_{sc}(n=1)$ shows more pronounced overshoot and saturation, hence the difference in behavior must be ascribed to the broad band nature of the space charge impedance. The effect of electron cooling in Fig. 7 on the other hand is to suppress the coherent fluctuations and lower the effective momentum spread threshold.

## B  Purely Resistive Impedance

Here we introduce an example which is linearly unstable due to a purely resistive impedance acting on *one* harmonic only. In Fig. 8 we have assumed an impedance (on harmonic $n=4$) corresponding to a working point $V=2$ in Fig.1, and no further impedance components. The beam develops a strong instability, which leads to a progressive filamentation of phase space and broadening (roughly doubling) of the momentum distribution. The result is well described by the overshoot theorem of Eq. 1 and resembles the behavior found in simulations for the ISR above transition energy, where the space charge impedance was negligible compared with the resistive component [10].

## C  Self-Stabilization and Momentum Distribution Tail

For the space-charge regime of Fig. 1 far up on the imaginary axis, and with a small resistive part of the impedance, it was found earlier by particle-in-cell simulation [15] that the longitudinal instability sets in, but levels off without causing harm to the effective momentum width. We have re-examined this in Fig. 9 using the longitudinal Vlasov code and assuming a current that exceeded the circle criterion threshold by a factor 25 ($U=25$, $V=2$ in Fig. 1).

Results from the latter code confirm the earlier conclusions of Ref. [15]. The instability sets in with no Landau damping initially, but leads to deceleration of a small fraction of particles to form a tail in the momentum distribution (shown as $f(v)$), which saturates and stops further growth. This case is of practical importance for the design of high current rings and operation far above the current threshold of the longitudinal resistive instability. The mechanism responsible for this self-stabilization of an originally unstable beam is believed to be "quasi-linear" diffusion of particles towards lower momenta by means of the nonlinear force of

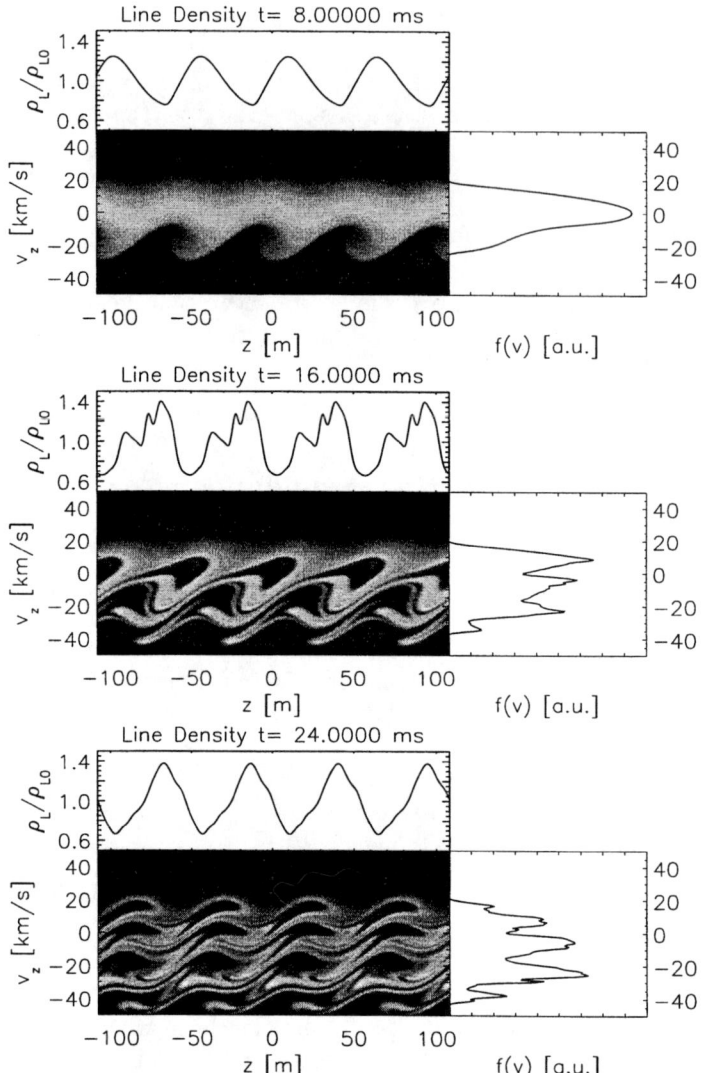

**FIGURE 8.** Vlasov simulation of phase space for purely resistive impedance, i.e. zero space charge case (negative image with brightness proportional to intensity).

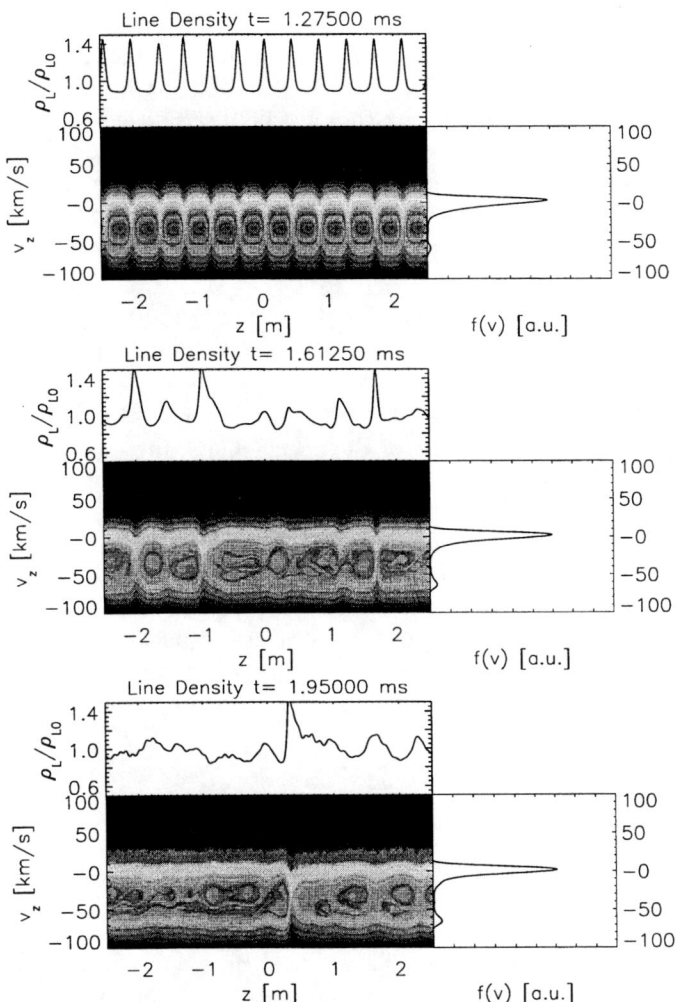

**FIGURE 9.** Vlasov simulation of self-stabilization by momentum tail for small resistive and dominant space charge impedance (20x above circle criterion threshold).

the steepened waves. Furthermore it must be noted that the resulting tails in momentum space open up an additional stable are in the stability diagram for large U as was suggested already in Ref. [15]. It is, however, not clear whether the final stabilization comes from Landau damping by tails in the distribution function, or it results from nonlinear effects in the excited mode spectrum, or a combination of both effects. This question subject requires futher study. We also mention that similar results were achieved for resistive impedances from a $Q = 1$ resonator to simulate the broad band impedance model assumed for the longitudinal microwave instability.

A further peculiarity of the nonlinear phase is the decay of the mode spectrum from the initially 12-th harmonic (with the driving resistive impedance) down to the first harmonic, where no resistive impedance is present. Such a nonlinear wave coupling is a typical nonlinear plasma phenomenon.

It is obvious that the broad-band nature of the space charge impedance plays a crucial role for this self-stabilization as was already pointed out in Ref. [15]. This is confirmed also by the Vlasov simulation of Fig. 10, where the impedance at harmonic $n = 12$ is the same as in the case of Fig. 9, namely $V = 2$ and $U = 25$, whereas the space charge impedance is canceled on all other harmonics. The beam is unstable, and significant momentum broadening results. In contrast with Fig. 8 we find, however, a very coherent and regular pattern of self-buckets propagating through the beam. We assume this is due to the large imaginary impedance at $n = 12$ and a resulting large coherent frequency shift suppressing filamentation.

## IV  SUMMARY

We have compared experimental observations on the longitudinal resistive instability with results from a new longitudinal Vlasov code and found good agreement in aspects where comparison was possible. The simulation was applied in particular to high-current behavior with large and dominant space charge impedance (typically an order of magnitude larger than the resistive part), where so far no measurements have been possible. Since the simulation result of self-stabilization far above threshold depends on the actual impedance distribution over the whole spectrum, experimental proof of this concept is highly desirable. This requires improved Schottky diagnostics after saturation of the instability, when the measurements are difficult due to the persistence of fluctuations from the preceding unstable phase.

*Acknowledgment*: The authors gratefully acknowledge discussions with G. Rumolo.

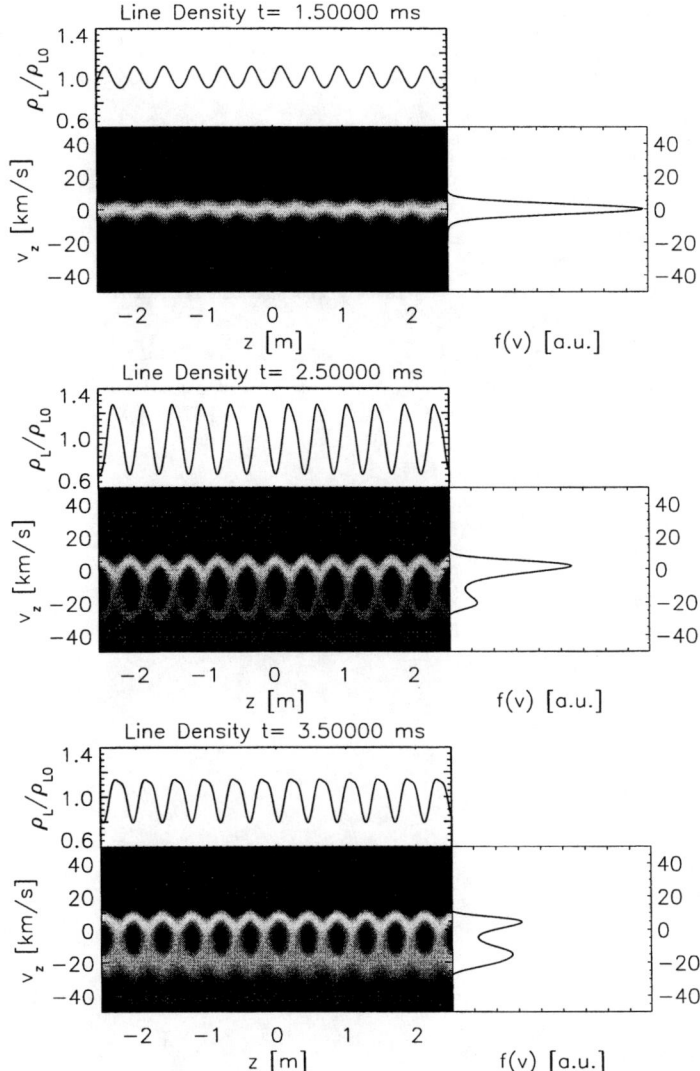

**FIGURE 10.** Artificial isolated impedance at n=12 showing absence of self-stabilization effect.

# REFERENCES

1. I. Hofmann, Proc. of the Linear Accelerator Conference, Chicago, Aug.23-28 1998, p. 1043
2. E. Keil and W. Schnell, CERN-Report No. ISR-TH-RF/69-48 (1968)
3. A.G. Ruggiero and V.C. Vaccaro, CERN-Report ISR-TH/68-33 (1968)
4. G. Rees, these proceedings
5. O. Boine-Frankenheim, I. Hofmann and G. Rumolo, Phys. Rev. Lett. **82**, 3256 (1999).
6. G. Rumolo et al., Nucl. Instr. Methods A **415**, 363 (1998)
7. I. Hofmann, G. Miano and G. Rumolo, Proc. European Particle Accelerator Conference, Stockholm, June 22–26, 1998, p. 2002 (1998)
8. L.J. Laslett, V.K. Neil and A.M. Sessler, Rev. Sci. Instr. **32**, 276 (1961)
9. B. Zotter and P. Bramham, IEEE Trans. Nucl. Sci **NS-20**, 830 (1973)
10. E. Keil and E. Messerschmidt, Nucl. Instr. Methods **128**, 203 (1975)
11. J.G. Wang, H. Suk and M. Reiser *Phys. Rev. Lett.,* **79**, 1042 (1997)
12. R.A. Dory, MURA-Report 654 (1962)
13. Y. Chin and K. Yokoya, Phys. Rev. D **28**, 2141 (1983)
14. S.A. Bogacz and K.-Y. Ng, Phys. Rev. D **36**, 1538 (1987)
15. I. Hofmann, Laser and Part. Beams **3**, 1 (1985)

# Impedances and Wakes in High-Energy Proton Accelerators

Bruno Zotter, CERN

## Abstract

The vacuum chamber and other elements surrounding particle beams in high-energy particle accelerators or storage rings constitute **impedances** [1]. They often lead to limitations of beam current by coherent instabilities or to energy loss which may overheat some of the elements. To control these limitations it is necessary to understand and minimize the impedances of the various structures surrounding the beam. For simple geometries analytic expressions can be derived which permit their estimation and scaling. For more complicated elements, powerful computer codes are nowadays available to obtain impedances or wake potentials. However, the results depend critically on the exact shape and dimensions of the structures and thus cannot be predicted in general. Therefore this presentation concentrates on three specific topics which were discussed recently: the resistive wall effect in non-circular chambers, the impedance of thin metal layers on ceramic beam pipes and that of wire cages used for rf shielding.

---

[1] **The longitudinal impedance** is defined as the ratio of the induced voltage to the beam current in the frequency domain and is often a rather complicated function of frequency with many resonances. **The longitudinal wake function** is defined similarly in the time domain as ratio of the induced voltage to the charge of the exciting point particle as function of the distance (or time delay). Hence impedance and wake function are Fourier transforms of each other. For the **transverse impedance** and **transverse wake function** the current and charge are replaced by the corresponding moments. The transverse wake function can be obtained from the transverse derivative of the longitudinal one by the **Panofsky-Wenzel theorem**. For bunches of finite length, the **effective impedances** and **wake potentials** are obtained by summation over all spectral frequencies.

# 1 Introduction

A basic requirement in the design of all new particle accelerators or storage rings is to identify and evaluate all major impedances which might limit the beam current of lead to overheating of some critical elements. Impedances are created by the vacuum chamber itself, as well as by all other structures surrounding the beam. Here we do not include the space charge impedance as it has been the subject of a special workshop last year in Shelter Island. Also we will not discuss the impedance due to synchrotron radiation as it is of little importance for proton machines. A number of excellent general reviews of the techniques used to calculate or compute structure impedances and wake functions have been published in the literature over last years[1, 2, 3, 4, 5], and nowadays it has become standard practice to prepare a complete "impedance budget" before any new particle accelerator is built.

The (longitudinal) **impedance** at any particular frequency $\omega$ is defined as the integral over the **synchronous** part of the electric field component $E_z(\omega)$ excited in a structure by a charged particle traveling in the $z$-direction with velocity $v = \beta c$, divided by the current $I_z(\omega)$. Usually a minus sign is attached to make the wake function positive in the decelerating field directly behind a particle:

$$Z_\|(\omega) = -\frac{1}{I_z(\omega)} \int_{-\infty}^{\infty} dz E_z(\vec{r}, z, \omega) \exp\frac{j\omega z}{v}, \tag{1}$$

where $\vec{r}$ is a transverse displacement from the axis which may often be assumed to be zero. The impedance describes the response of a structure to a continuous excitation at a single frequency. It is in general a complex function of frequency, the real part giving the in-phase and the imaginary one the out-of-phase components.

The (longitudinal) **wake function** is also called **delta-function wake potential** or **Green's function**, for which reason we abbreviate it with $G$ and reserve $W$ for the wake potential of a bunch of non-zero length. It is a real function of the distance $s = v\tau$ behind the particle of charge $Q$

$$G_\|(s) = -\frac{1}{Q} \int_{-\infty}^{\infty} dz E_z\left(\vec{r}, z, t = \frac{z+s}{v}\right). \tag{2}$$

The wake function describes the response of a structure to a short, pulsed excitation. It is the Fourier transform of the impedance.

Usually one assumes that the particle motion is ultra-relativistic, i.e. $v \to c$ or $\beta \to 1$, $\gamma \to \infty$. This has several simplifying consequences:

1) the EM fields cannot get ahead of the exciting particle, hence $G(s < 0) = 0$, which is often called **causality**[2]. For the impedance this assumption has the consequence that all poles are in the upper (lower) half of the complex frequency plane, depending on whether one uses a time factor $e^{j\omega t}$ or $e^{-i\omega t}$.

2) the real and imaginary parts of the impedance can be written as cosine and sine transforms of the wake function and yr therefore related by a **Hilbert transform**

$$ImZ_\|(\omega) = \frac{1}{\pi} PV \int_{-\infty}^{\infty} d\omega' \frac{ReZ_\|(\omega')}{\omega - \omega'} , \qquad (3)$$

where $PV$ stand for "principal value". This is usually known as the **Kramers-Kroning relation**.

3) In the integrals over the fields one assumes constant particle velocity and trajectory, i.e. a completely "stiff" beam. This is actually only correct in the limit $\gamma \to \infty$ when the relativistic particle mass $m = \gamma m_0$ also becomes infinite. For particle velocities $v \ll c$, calculation of the induced fields requires calculation of the changes of velocity and trajectory, and constant updating. This is done in some computer programs such as PIC or "particle-in-cell" codes.

Additional assumptions usually made to simplify the impedance calculations are

1) perfectly conducting walls, except for the resistive wall impedance which is usually treated separately;

2) the curvature of trajectories is usually neglected. Furthermore, for circular machines the frequency spectrum is not continuous but consists of discrete lines due to the periodicity of beam motion. Indeed, rf or kicker cavities and other large impedances are usually located in straight sections;

3) for simplicity of computation, structures are sometimes assumed to be axially symmetric while their cross sections are often oval or rectangular. The radius of the inscribed circle is used to get a pessimistic estimate. However, as will be discussed below, the assumed symmetry leads to the loss of quadrupole (and higher) modes which are important for transverse effects.

---

[2] actually it would be better to say "in agreement with causality"- causality is not violated when $v < c$ but the fields - and hence the wake function - can then get ahead of the exciting particle.

The **transverse impedance** has 2 components, usually chosen as $x$ and $y$ or $r$ and $\theta$. For a particle with offset $\xi$ the integral over the transverse Lorentz force component is divided by the dipole moment of the current in the chosen plane

$$Z_\perp(\omega) = \frac{j}{\xi I_\omega} \int_{-\infty}^{\infty} dz [E_\omega + v \times B_\omega]_\perp \exp\frac{j\omega z}{v} . \tag{4}$$

The factor $j$ has been introduced because of a 90° phase shift between source and test particle (an offset leads to a change of slope). The **transverse wake function** is found from

$$G_\perp(\omega) = \frac{j}{\xi Q} \int_{-\infty}^{\infty} dz [E_\perp + v \times B_\perp]_{t=(z+s)/v} . \tag{5}$$

The longitudinal and transverse field components are related by Maxwell's equations, which leads to the **Panofsky-Wenzel theorem**

$$\frac{\delta G_\perp}{\delta s} = \nabla_\perp G_\parallel(s) \tag{6}$$

and its equivalent in impedance

$$Z_\perp(\omega) = \frac{v}{\omega} \nabla_\perp Z_\parallel(\omega) . \tag{7}$$

The most important geometrical impedances are usually rf cavities or other elements constituting significant cross-section variations of the vacuum chamber. The (low-frequency) impedance as well as the (short-range) wake potential due to such localized elements can be readily estimated or computed. The latter can be used directly in simulation codes for single bunch instabilities working in time domain, which are dangerous for machines with high charge per bunch.

In proton accelerators, with typically many and rather long bunches compared to electron machines, more important are impedances with sharp resonances, corresponding to a long memory, as they may lead to multi-bunch instabilities. While such resonances can nowadays be calculated rather easily with several 2-D and 3-D computer programs, the exact resonant frequencies depend critically on all structure dimensions. They will thus often change considerably with temperature of an element or even with the momentary position of variable tuning loops. This leads to the problem that it is not possible to predict the total impedance of overlapping resonances of similar elements, in particular when they are present in large numbers in the machine - such as e.g. cavities or bellows. The "worst case assumption" - all elements exactly the same and all resonances overlap completely - is therefore far too pessimistic and may lead to exaggerated requirements. Fortunately the resulting instabilities can often be counteracted with rather simple feedback systems. It is thus only necessary to estimate the strength of the required feedback, then a pessimistic estimate is not too expensive.

In addition to the geometric impedances, the finite conductivity of the vacuum chamber and structures surrounding the beam is often even more important in particular for proton machines. This "resistive wall impedance" is essentially distributed over the whole machine circumference and is of particular concern in high-current proton accelerators such as spallation sources. The topics discussed here are mostly related to this aspect.

Since the values of the geometric impedances depend critically on the specific geometries of the structures investigated, they should be evaluated for every particular element and cannot be estimated in general. Therefore, in this talk I will concentrate on only a few subjects relevant to proton machines which have been under discussion in the last few years, and where at least some progress has been made in understanding: the resistive wall effect in non-circular chambers, the impedance of thin metal layers on ceramic walls, and the impedance of wire cages used for rf shielding.

## 2 The Resistive Wall Effect in Chambers with Non-circular Cross Section

In many accelerators the cross section of the vacuum chamber is non-circular, often elliptic or oval, sometimes nearly rectangular or square. Also circular chambers with holes or slots which break the rotational symmetry should be included in this category.

An interesting result for wakes in structures without rotational symmetry has been found as a by-product of 3-D simulation[6]. The transverse wake is no longer simply proportional to the offset of the leading particle ("dipole component"), but in addition has a part which is also proportional to the offset of the trailing (test) particle ("quadrupole component"). This can best be illustrated by the figures where we first plot the total transverse kick as function of the offsets of the test particle for 4 different positions of the beam, and then the difference between the kicks for an offset and those of a centered test particle. It can be seen that it is almost purely quadrupolar, higher multipole components become important only for test particles with much larger offsets[7]. Also structures with stronger asymmetries might have more pronounced multipolar wake components - however, many non-circular accelerator elements still have several symmetry planes such as the elliptic bellows and the separator plates presented here.

The long-range transverse resistive wall wake is generally known to be proportional to the inverse square root of the distance from the source particle[8]. The fields thus decay only slowly and in a circular machine the total effect should be obtained by summing over all previous turns. For a chamber with circular cross section, there is only a dipole component whose effect is proportional to the offset of the leading particle. Its betatron oscillations cause the terms to alternate their signs whenever the offset changes sign, and the infinite sum therefore converges. However, this is no longer the case for the quadrupole part of the wake in a non-symmetric chamber:- since both the leading and trailing particles perform the same betatron oscillations, their offsets change sign together. Since the effect is proportional to both signs, the terms all have the same sign. Thus the infinite sum over $1/\sqrt{s}$ would diverge, yielding an infinite tune shift!

Several attempts have been made to correct this unphysical result: E.g. it was proposed that the diffusion of fields through a metal wall of finite thickness would lead to a cutoff of the infinite sum when they reach the outer wall[9], or rather when they are reflected back to the beam position[10]. However, it should be realized that the square root decay is only an approximation, and changes to a much faster exponential decay at very large distances for finite values of the energy factor $\gamma$. This gives a much clearer explanation for the convergence as well as a better estimate of the sum[11].

A more direct approach can also be used to calculate the impedance of some non-circular vacuum chambers analytically. The first calculations of the resistive wall effect had indeed been for beams in chambers with rectangular cross section[8] and for beams between parallel plates[12, 13, 14]. The calculations were also performed for chambers with elliptic cross section[16, 17, 18], and the results were generalized to arbitrary cross sections a few years later[19, 20]. Also beams with finite offsets were analyzed[21], and an apparent difference between these results and those of an earlier paper[17] was explained as belonging to different integrals over the forces[24].

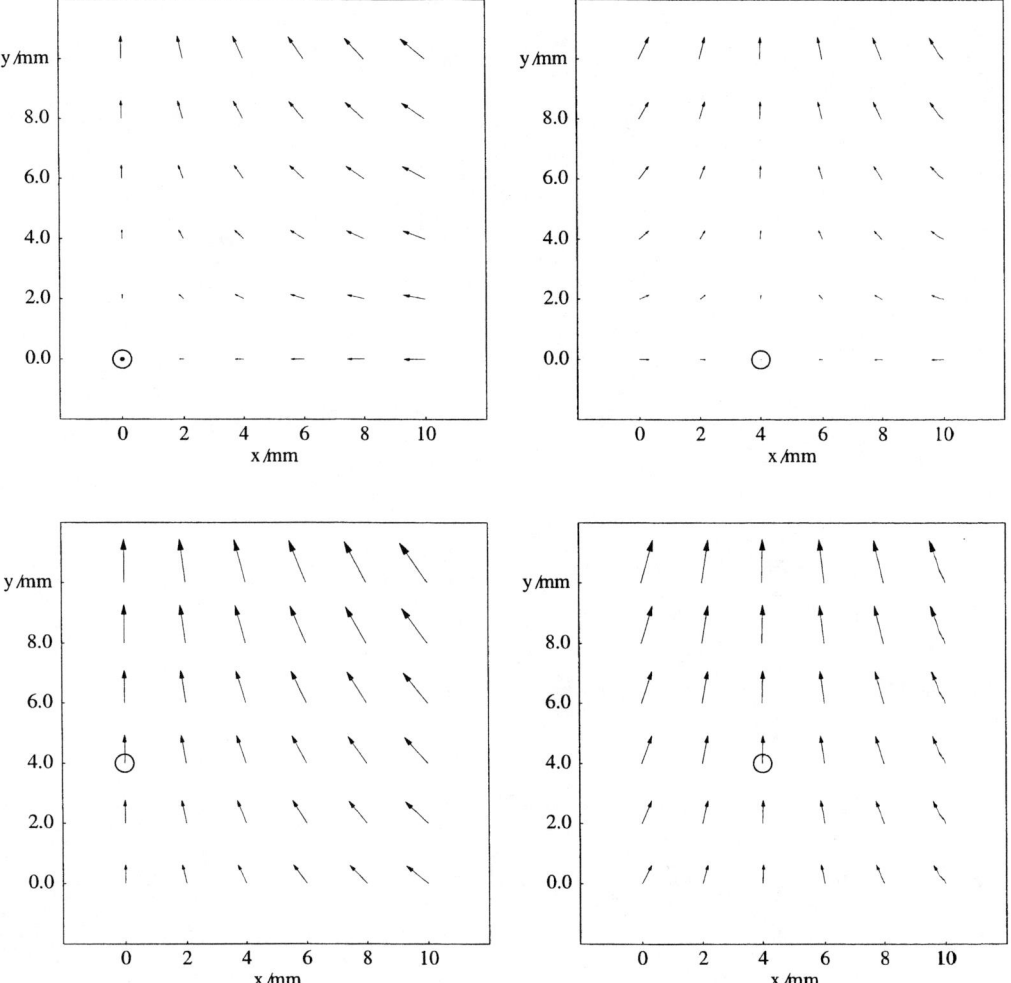

Figure 1: Transverse kicks as function of the position of the test particle in an elliptic structure (LEP shielded bellows) for different beam offsets (beam position small circle).

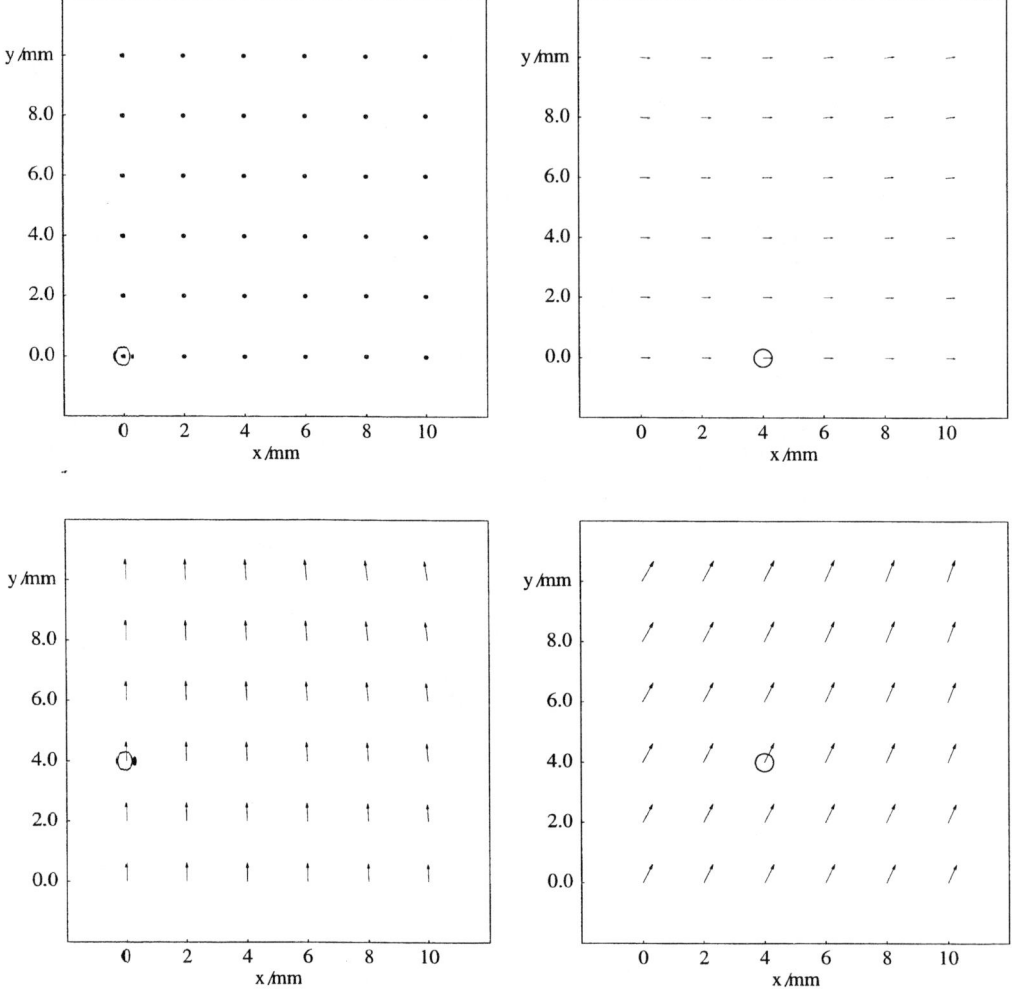

Figure 2: Transverse kicks reduced by the kicks due to an on-axis beam in an elliptic structure (LEP shielded bellows) for different beam offsets (beam position small circle).

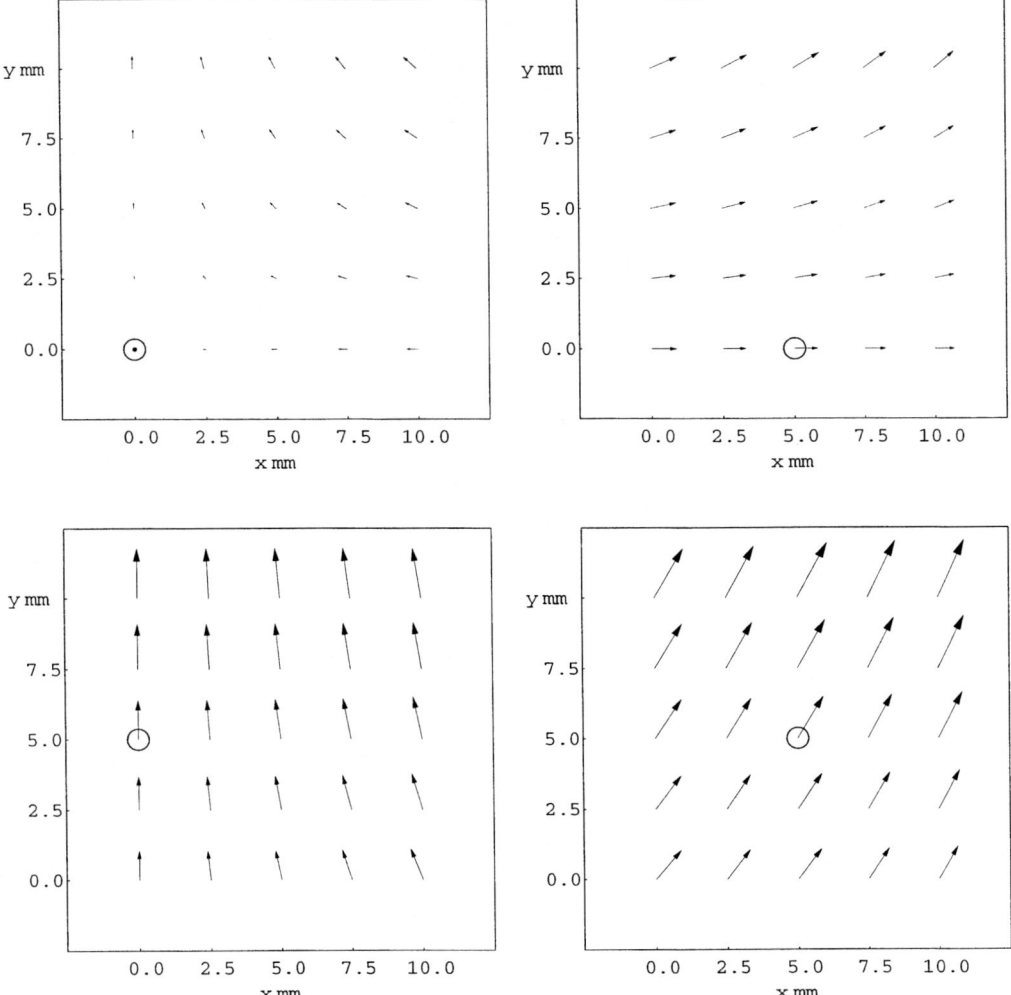

Figure 3: Transverse kicks as function of position of test charge in a LEP electrostatic separator for different beam offsets (beam position small circle).

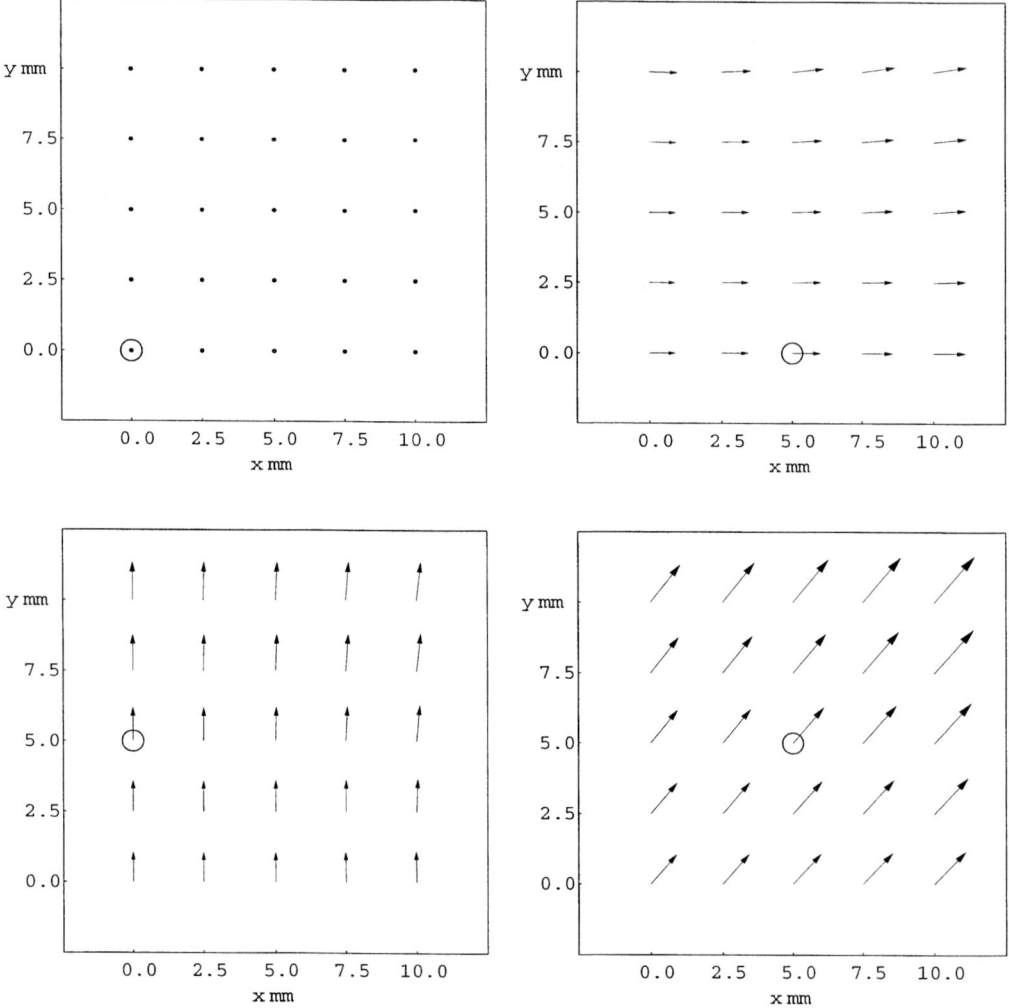

Figure 4: Transverse kicks reduced by those due to an on-axis beam as function of position in a LEP electrostatic separator for different source beam offsets (beam position small circle).

These calculation also gave finite results for the tune shift due to the resistive wall effect, in agreement with the results of the summation with the proper cutoff. A word of caution is on order here: the usual definition of the impedance in the limit $\gamma \to \infty$ is most convenient to avoid problems with wakes preceding the source (often called "causality"). However, in reality particle beams are always slower than light velocity, hence e.m. fields propagating with $v = c$ can get (arbitrarily far) ahead of them, depending on initial conditions. To solve the field equations for a point charge inside a resistive tube, it is important to take the limit $\gamma \to \infty$ last, otherwise the results show no penetration of fields into the wall at all[11].

For a finite wall thickness, the e.m. fields - and hence the impedances - depend on the structure outside the vacuum chamber proper. Often there are conducting metal surfaces or magnetic material of pole-shoes present, and the corresponding boundary conditions must be applied. Only in straight sections which are not surrounded by magnets the fields may extend quite far and it may then be assumed that they extend to infinity.

## 3 Metallized Ceramic Walls

Due to the fast change of the applied magnetic field in rapidly-cycling synchrotrons, rather strong eddy current would appear if their vacuum chambers were made of highly-conducting metal. They would counteract the applied field and would further heat the vacuum chamber. In order to reduce the losses, such chambers are therefore often made of ceramic material. Usually it is considered necessary to apply a thin metal layer to the inside of the chamber wall in order to avoid the accumulation of static charges. The layer should be very thin in order to limit eddy currents and thus to let the applied magnetic field penetrate without high losses. On the other hand, the layer should be thick enough to keep beam induced fields from penetrating to the outside and thus to avoid possible large impedances. An appealing variant of the continuous metallization is the application of metal stripes, usually parallel to the beam to carry the induced current. A small width of such stripes reduces the eddy currents further, and if the gap between them is kept narrow also the impedance remains small[26].

A rather emotional discussion has been centering over the last years on the question whether the thickness of a metal layer for shielding the e.m. fields generated by a particle beam from the outside needs to be of the order of the skin depth - which can reach several millimeters for low-frequency oscillations of beams in large accelerators - or whether a smaller thickness is sufficient[27, 28, 29]. Several authors agree on the second answer[25, 14], but there is still some disagreement on the exact design criteria. At present there are two different expressions for the "effective skin depth ":

$$\delta_{eff} = \frac{\delta^2}{d} \quad \text{or} \quad \delta_{eff} = \frac{\delta^2}{b} \tag{8}$$

where $d$ is the thickness of the ceramic behind the metal[14], while $b$ is the radius of the beam tube[11]. Since the radius $b$ is usually much larger than the thickness $d$, the required thickness for shielding is much smaller by the second expression, which seems to agree with recent measurements[27].

A related problem is the use of ceramic chambers of finite length to separate the beam vacuum from other regions such as experimental chambers or kicker tanks[22, 23]. Often the structure behind the ceramic wall then forms a resonant cavity, and field penetration should be limited at frequencies where the tank resonates[31]. Again this can be done by a thin metal layer on the ceramic. Recent measurements have shown that metal layers much thinner than the skin depth are sufficient to shield the e.m. fields[27].

A similar topic is the impedance of thin metal vacuum chamber walls. If there is dielectric material on its outside, the problem is the same as that of the metallized ceramic wall discussed above. However, if there is pure vacuum on the outside, one must specify its extent and the nature of the outer boundary. For a (very) far removed boundary one may assume that the vacuum extends to infinity and the analysis becomes simplified. The results are quite well known[2] but are probably of limited practical use. Indeed, already the small dielectric constant of air (1.0004) on the outside of the chamber wall is sufficient to slow down the light velocity in the medium to below the velocity of the beam whose energy factor exceeds $\gamma = 16$. Since at least some e.m. fields move with the beam, they travel faster than light in the medium and thus very weak Cerenkov radiation will occur.

If the outer boundary of the vacuum surrounding the chamber is another conductor - in particular a very good one - one can solve the problem by assuming that the vacuum extends to a perfect conductor, which reduces the impedance. However, if the outer boundary is (ferro-)magnetic material, such as is often the case for chambers inside bending or focusing magnets, the situation is worse. Assuming perfect magnetic material leads to very high impedances, in particular when the metal thickness is very small or tends to zero. Hence it is generally not recommended to have magnetic pole shoes penetrate into a vacuum chamber, as has been done in some early accelerator designs in order to minimize the magnet gap and hence the costs of magnets.

Finally, we mention also the calculations of the impedance of beam tubes made of two different metals. If the layers are thin enough that the e.m. fields can penetrate at least partially, the boundary conditions on the outside of the tube are again important and different results are obtained if the outer region is assumed to reach to infinity or to a perfect conductor or magnet.

# 4  Impedance of rf cages

The use of an rf cage, consisting of several metal wires or rods inside a ceramic vacuum chamber, was pioneered by the designers of the spallation source ISIS[32]. In addition to shielding the ceramic wall from the beam, this reduces the space charge impedance when its cross section of the cage follows the beam profile. When one keeps the ratio of the chamber radius $b$ to that of the beam $a$ as small as possible and approximately constant, both the longitudinal and the transverse impedance terms $\log(b/a)$ and $(1/a^2 - 1/b^2)$ become smaller. This reduces the space charge tune shift and thus permits more current to be stored in the beam before reaching limitations due to betatron resonances.

A practical difficulty of such a cage is the fact that ceramic chambers are usually produced only in short pieces of 1-2 m which are later connected to form a vacuum chamber. The cages must be held fixed in these sections and also connected to each other. It was found sufficient to make these connection for rf only by capacitive coupling, limiting the DC feed-throughs to a few locations to keep the cage at ground potential.

A similar design for the defunct K-factory in TRIUMF had planned to use metal stripes directly deposited on the ceramic sections which were shaped themselves to follow the beam profile. A special device was developed to deposit thick silver stripes on the inside aperture of a long ceramic tube[33]. The justification for developing this was the believe that the metal thickness must exceed the skin-depth at the lowest frequencies of concern, as the rods in the original cage for ISIS were also a few mm in diameter. The discussions reported above show that this might not have been necessary - a thin metallization of a few microns may have been sufficient for shielding, and much simpler to apply.

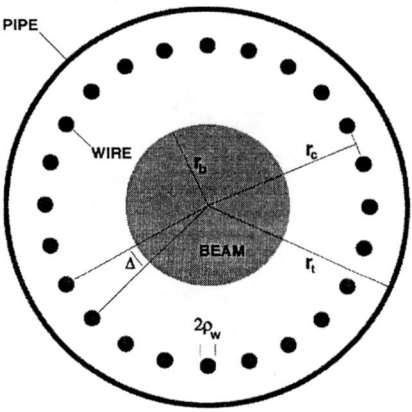

Figure 5: Schematic geometry of rf cage used in the impedance calculation.

Also in the design of the spallation source AUSTRON rf cages were included. For this purpose an analytic calculation of the impedance of such a complicated structure was made [34], although the geometry was somewhat simplified by assuming constant cross section. Usually the variation of the beam radius- and hence also the cage radius - are quite slow, so it should have only a small effect. This work was later revised[35] and extended to include an external beam pipe[36]. Here we only show the impedances of N equally spaced rods of thickness $d$ at the cage radius $r_c = \kappa b$ where $b = r_t$ is the tube radius. The resulting expressions can be written as:

$$Z_\parallel = \frac{ikLZ_0}{4\beta\gamma^2}\left[1 + 2\ln\frac{b}{a} + C_\parallel\right],$$
$$Z_\perp = \frac{iLZ_0}{2\pi\beta^2\gamma^2}\left[\frac{1}{a^2} - \frac{1}{b^2} + \frac{C_\perp}{b^2}\right] \quad (9)$$

In each of these expressions one can recognize the two space charge terms for a uniform beam of radius $a$ and the additional dimensionless contributions

$$C_{\parallel} = \frac{2N[\ln \kappa]^2}{N \ln \kappa + \ln \dfrac{Nd}{r_c} - \ln(1 - \kappa^{2N})},$$

$$C_{\perp} = [1 - \kappa^2] \frac{[\kappa^2 + \kappa^{-2}]\ln(1 - \kappa^{2N}) - 2\ln \dfrac{Nd}{r_c}}{N[1 - \kappa^2] - 2\ln \dfrac{Nd}{r_c} + [\kappa^2 + \kappa^{-2}]\ln(1 - \kappa^{2N})}. \quad (10)$$

The additional transverse impedance becomes quite small when the number of wires increases, while the longitudinal one approaches a constant (negative) value which depends mainly on the ratio of cage to tube radius $\kappa$ as can be seen in the following figures.

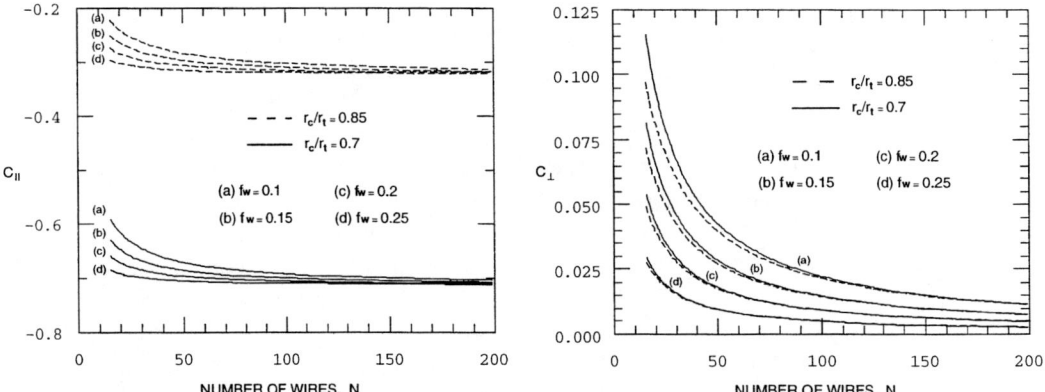

Figure 6: Longitudinal and transverse impedances of rf cage for various ratios of cage to tube radius as function of number of wires $N$ when the "filling factor" $Nd/\pi r_c$ is kept constant, i.e. the wire thickness decreases with increasing number $N$.

# 5 Conclusions

For each new particle accelerator - in particular for machines aiming at high beam currents - the careful establishment of an "impedance budget" is a prerequisite for reaching the desired performances. Many structures surrounding the vacuum chamber constitute "impedances" in which the beam current induces voltages which may lead to coherent instabilities or unwanted heating of sensitive elements. Sometimes it is possible to reduce the impedances by changing the design or by the choice of different materials. For other structures it may be necessary to shield them from the beam, e.g. with sliding contact fingers or rf cages. However, such measures are usually both expensive and may cause problems during operation if they are not designed to withstand long exposure to mechanical and thermal stresses, or if the machines are operated at higher beam currents than originally foreseen. A typical example for this were the shielded bellows in the ESRF, which were scaled down from the LEP design but operated at much higher bunch currents. When they overheated during operation they had to be replaced by a different design specially developed for such currents.

For proton machines - with typically many rather long bunches compared to electron machines - in particular sharp resonances with long memory are most important, as they may lead to multi-bunch instabilities. While such resonances can nowadays be calculated rather easily with several 2-D and 3-D computer programs, the exact resonant frequencies depend critically on all structure dimensions. They will thus often change considerably with temperature of an element or even with the momentary position of variable tuning loops. This leads to the problem that it is not possible to predict the total impedance of overlapping resonances of similar elements, in particular when these are present in large numbers - such as e.g. cavities or bellows. The so-called "worst case assumption" - that all elements are exactly the same and that all resonances overlap completely - is therefore usually far too pessimistic and may lead to exaggerated requirements.

The space charge and resistive wall impedances are usually more important for proton machines and can be evaluated in more detail. As the first topic has been treated in a special workshop last year, we concentrated here on aspects of the second one where recently some progress has been made in understanding. However, more work is needed also there to clear up some remaining discrepancies.

# References

[1] P. Wilson, AIP-184 (1987) p. 525 ff.
[2] A. Chao, Physics of collective Beam Instabilities, Wiley 1990.
[3] S. Heifets, S. Kheifets, Rev. Mod. Physics 63(1991) p.631 ff.
[4] L. Palumbo, V. Vaccaro, (1994)
[5] S. Kheifets, B. Zotter: Impedances and Wakes, World Sci. Publ. (1998)
[6] A. Wagner, thesis, CERN Report SL/AP 96- (1996)
[7] S. Heifets, A. Wagner, B. Zotter (1997) SLAC Report (1998)
[8] J. Laslett, K. Neal, A. Sessler, Rev. Sci. Instr. 36 (1965)
[9] A. Chao, private communication (1999)
[10] S. Heifets, SLAC-Pub-7985 (Oct.1998)
[11] B. Zotter, talk at the SLAC AP group (Feb.1999)
[12] A. Piwinski, Report DESY 84-097 (1984)
[13] H. Henke, O. Napoly, Proc. 2nd EPAC, Nice 1990, p.1046
[14] A. Piwinski, Report DESY-HERA 92-04 (1992)
[15] P. Morton, K. Neal, A. Sessler, J. Appl. Phys. 37 (1966), p.3875
[16] L. Palumbo, V. Vaccaro, Nuovo Cimento 89 A (1985), p. 243
[17] R. Gluckstern et al, CERN Report SL/AP 92-18 (1992)
[18] V. Balbekov, Proc. 3-rd EPAC, Berlin 1992, p.792
[19] R. Gluckstern J. van Zeijts, B. Zotter, Phys. Rev. E 47 (1993)
[20] F. Ruggiero, CERN Report SL/AP 95-95 (1995)
[21] A. Piwinski, Report DESY 94-068 (1994)
[22] B. Ng, SSC impedance workshop (1985) p.61
[23] S. Kurennoy, Proc. PAC (1993) p.3420
[24] K. Yokoya, KEK Preprint 92-296 (1993)
[25] B. Zotter, Part. Accel. 1 (1970) p.311.
[26] L. Walling et al, Nucl. Instr. Meth. A 281 (1989) p.433
[27] F. Caspers et al, PAC New York (1999)
[28] L. Vos, CERN/SL internal note (1999)
[29] G. Lambertson, LBL report42818 (1998)
[30] B. Zotter, CERN Report 69-16 (1969)
[31] G. Dome, Private comm.(1999)
[32] G. Rees et al, ISIS, Rutherford spallation source
[33] M. Craddock et al, TRIUMF K-factory proposal
[34] T.S. Wang, CERN Report PS 94-08 (1994)
[35] T.S. Wang, "The Space-Charge Impedance of RF-Shielding Wires,"
  in *Workshop on Space Charge Physics in High Intensity Hadron Rings,*
  edited by A. U. Luccio and W. T. Weng, AIP Conference Proceedings 448,
  New York, 1998, pp. 286-297.
[36] T.S. Wang, R. Gluckstern, PAC New York (1999)

# Collective Effects in the CERN-PS Beam for LHC

R. Cappi, R. Garoby, E. Métral

*CERN, PS Division, 1211 Geneva 23, Switzerland*

**Abstract.** This paper is an updated review of the collective effects observed and predicted in the CERN-PS machine for the LHC beam.

## INTRODUCTION

The PS machine as part of the LHC injector chain has to provide to the SPS a proton beam with specific characteristics [1]. To summarize, in the longitudinal plane the main problem is to generate a train of very short bunches (~3.8 ns) spaced by 25 ns, starting from very long bunches (~200 ns) coming from the PSB. While, in the transverse domain, the main issue is to provide a beam of high brightness (i.e. intensity to emittance ratio), with an intensity of $1.4 \times 10^{13}$ p/p (for the ultimate beam) and normalized rms transverse emittances of 3 µm.

The solution adopted is to accelerate in the PSB a beam with the right transverse emittance, but half the intensity, and inject two pulses into the PS machine at 1.4 GeV kinetic energy. The total circumference of the four PSB rings being equal to the PS circumference, a necessary condition, in order to fill only one half of the PS with a single PSB shot, is to use a $h=1$ RF system in the PSB. The second half of the PS is filled with a second shot, 1.2 s later (4+4=8 bunches). The 8 bunches are captured by the PS RF system on $h=8$ and then split into 16 with an adiabatic change of harmonic number from 8 to 16. They are subsequently accelerated to 26 GeV/c where the beam is debunched and rebunched at 40 MHz to provide the 25 ns spacing. Finally, the 84 bunches are compressed to 3.8 ns with a $2^{nd}$ harmonic RF system (80 MHz cavities, 300 kV each) and fast extracted to the SPS. This paper reports experiments performed to study and cure both longitudinal and transverse instabilities.

## LONGITUDINAL INSTABILITIES

### Coupled-Bunch Instabilities

To provide the required LHC beam characteristics many modifications have been recently made, in particular in the RF system [2]. New low-level beam controls to accelerate the beam on $h=8$ and $h=16$ (instead of the previous $h=20$) have been implemented and extra cavities have been installed in the ring (two 40 MHz and three 80 MHz). Accelerating the beam on $h=8$, the frequency spectrum of the beam and the tuning frequency of the ferrite cavities have changed. A new coupled-bunch instability has appeared in the vicinity of the 10$^{th}$ harmonic of the revolution frequency (i.e. coupled-bunch mode $n=2$ or 6) at about 3.5 GeV/c (see Fig. 1(a)). A damping system has been implemented by filtering the wall-current monitor signal at the instability frequency and reinjecting it into a pair of ferrite accelerating cavities (see Fig. 1(b)) [3].

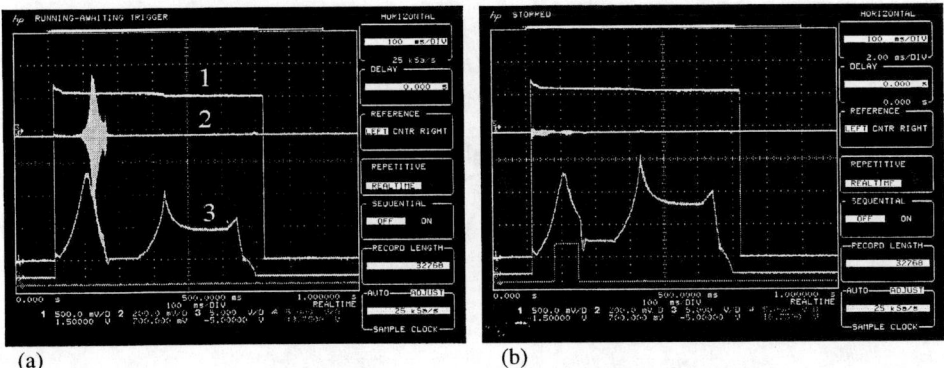

**FIGURE 1.** (a) Trace 1: Circulating beam current. Trace 2: Wall-current monitor signal filtered at $f=10f_0+f_s$. Trace 3: Detected wall-current monitor (i.e. bunch height). Time scale: 100 ms/div. (b) Same as Fig. 1(a), but with the longitudinal feedback in operation.

## Microwave Instabilities

For the LHC beam, the 8 bunches injected into the PS, split into 16 at 3.5 GeV/c, have to be "transformed" into a bunch train of 84 bunches (with a bunch spacing of 25 ns) before extraction to SPS. This implies a debunching-rebunching of the beam on $h=84$ (40 MHz). Moreover, because of longitudinal and transverse acceptances in the receiving machine their longitudinal emittances should not be larger than ~0.4 eVs/bunch. Unfortunately, microwave instabilities develop at the end of the debunching procedure, increasing the final longitudinal emittance by a factor 1.5 with respect to the desired value (see Figs. 2(a) and 2(b)).

The Keil-Schnell formula predicts a longitudinal wide-band impedance $Z/n$ of ~300 Ω, which is incompatible with the 20 Ω measured using other methods. HOMs in the 114 MHz pill-box cavities used for the lepton acceleration are suspected. Some UHF signals have been detected on the wideband mini-antennas inserted into these cavities. The definitive answer will be known after their removal in 2001, when LEP will be stopped. Moreover, a new scheme has been proposed, which does not use debunching-rebunching and ensures a "clean" extraction by preserving a gap in the train of bunches [4]. Experimental demonstration will take place in 1999 and 2000.

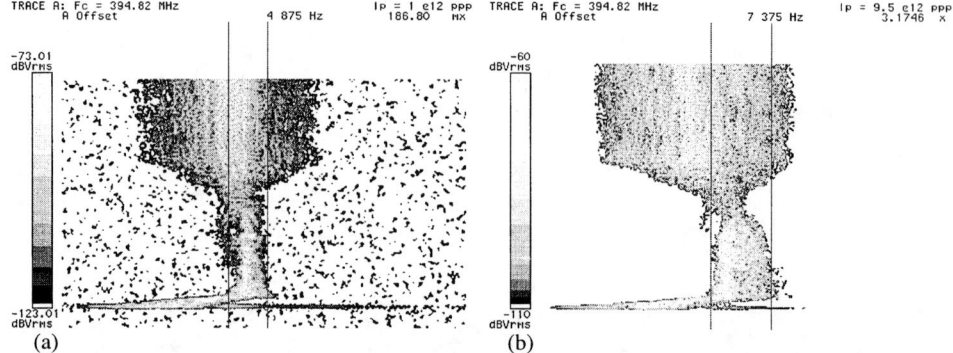

**FIGURE 2.** (a) Longitudinal Schottky scan spectrogram during the debunching of a low intensity beam ($10^{12}$ p/p). Time goes from top to bottom. Total time window is ~200 ms. In the first 100 ms the beam is still bunched by the RF voltage, which is adiabatically decreased and then switched OFF. During the following ~50 ms the beam is debunching with negligible momentum blow-up. The total (relative) momentum spread, indicated by the two line markers, is 0.5 $10^{-3}$. The last "transient" is produced by the fast extraction process. (b) Same as Fig. 2(a), but for a higher intensity beam ($10^{13}$ p/p). During the debunching there is a momentum blow-up. The final total (relative) momentum spread is ~0.8 $10^{-3}$.

# TRANSVERSE INSTABILITIES

## Theory

### Coherent Frequency Shifts of Bunched-Beam Modes

Sacherer's formula for the transverse coherent frequency shifts of bunched-beam modes is given by [5]

$$\Delta\omega_{c,m}^{x,y} = \left(|m|+1\right)^{-1} \frac{je\beta I_b}{2m_0\gamma Q_{x0,y0}\Omega_0 L} \frac{\sum_{k=-\infty}^{k=+\infty} Z_{x,y}\left(\omega_k^{x,y}\right) h_m\left(\omega_k^{x,y}-\omega_{\xi_{x,y}}\right)}{\sum_{k=-\infty}^{k=+\infty} h_m\left(\omega_k^{x,y}-\omega_{\xi_{x,y}}\right)}, \quad (1)$$

where $\omega_k^{x,y} = \left(k+Q_{x0,y0}\right)\Omega_0 + m\omega_s$. The transverse bunch spectra of mode $|m|$ are given by

$$h_m\left(\omega-\omega_{\xi_{x,y}}\right) = \left(|m|+1\right)^2 \frac{1+(-1)^{|m|}\cos\left[\left(\omega-\omega_{\xi_{x,y}}\right)\tau_b\right]}{\left\{\left[\left(\omega-\omega_{\xi_{x,y}}\right)\tau_b/\pi\right]^2 - \left(|m|+1\right)^2\right\}^2}, \quad (2)$$

and the PS transverse coupling impedances $Z_{x,y}(\omega)$ are approximated by the sum of the resistive wall ($Z_x^{RW}(\omega) \approx Z_c^{RW}(\omega) \times 0.45$ and $Z_y^{RW}(\omega) \approx Z_c^{RW}(\omega) \times 0.85$ [6]) and broad band impedances

$$Z_c^{RW}(\omega) = [Sgn(\omega) + j] \frac{R}{b_w^3} \sqrt{\frac{2\rho_w}{\varepsilon_0 |\omega|}}, \qquad (3)$$

$$Z_{x,y}^{BB}(\omega) = \frac{\omega_r}{\omega} R_{x,y} / \left[ 1 - jQ\left(\frac{\omega_r}{\omega} - \frac{\omega}{\omega_r}\right) \right]. \qquad (4)$$

Making the numerical computations for the single-bunch beam with nominal intensity (see Appendix), the following results, collected in Table 1, are obtained.

**TABLE 1.** Transverse instability growth rates and real frequency shifts of the nominal single-bunch beam for modes $m$=0 to 10.

| Head-tail mode $m$ | 0 | 1 | 2 | 3 | 4 | 5 |
|---|---|---|---|---|---|---|
| Horizontal growth rate [s$^{-1}$] | -95.5 | -49.1 | -34.4 | -28.6 | -26.9 | -11.3 |
| Vertical growth rate [s$^{-1}$] | -207.8 | -104.8 | -71.1 | -54.4 | -45.5 | -39.6 |
| Horizontal frequency shift [rad/s] | -1727.4 | -865.1 | -579.0 | -437.2 | -359.8 | -306.2 |
| Vertical frequency shift [rad/s] | -5050.4 | -2526.2 | -1685.4 | -1265.5 | -1014.5 | -848.1 |
| | 6 | 7 | 8 | 9 | 10 | |
| | 2.4 | 0.9 | 0.1 | -0.3 | -0.5 | |
| | -39.5 | -32.9 | -8.9 | -0.8 | -3.8 | |
| | -254.3 | -217.7 | -192.4 | -172.1 | -156.1 | |
| | -733.0 | -655.5 | -582.0 | -512.2 | -462.4 | |

The plot of the transverse instability growth rates as functions of the head-tail mode number is represented in Figure 3.

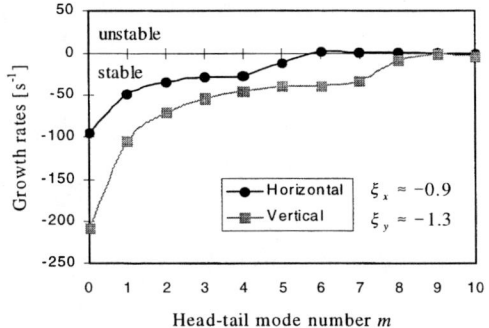

**FIGURE 3.** Transverse instability growth rates vs. head-tail mode number for the nominal single-bunch beam.

One concludes therefore that the theory, based on the above impedance model, predicts horizontal single-bunch instabilities with most critical head-tail mode number $m = 6$.

## Stabilization by Landau Damping

The transverse betatron frequency spreads (half widths at half height) are given analytically by [7]

$$\Delta\omega_{HWHH}^{x,y} \approx \frac{3 f_0 N_{oct}}{8} \left[ \left(\beta_{octx,y}^2 \varepsilon_{x,y}^{rms}\right)^2 + \left(2\beta_{octx}\beta_{octy}\varepsilon_{y,x}^{rms}\right)^2 \right]^{1/2} \int_0^{l_{mag}} K_{oct}\, dl. \quad (5)$$

A simplified stability criterion, which is drawn from dispersion relation analysis considering "elliptical" betatron frequency distributions, is given by [8]

$$\Delta\omega_{HWHH}^{x,y} \geq \sqrt{3}\, \left|\Delta\omega_{c,m}^{x,y}\right|. \quad (6)$$

From the numerical computations, the relations between the transverse betatron frequency spreads and the octupole current, for the nominal single-bunch beam, are given by

$$\Delta\omega_{HWHH}^{x}\,[\text{rad/s}] \approx 67\, I_{oct}\,[\text{A}], \qquad \Delta\omega_{HWHH}^{y}\,[\text{rad/s}] \approx 85\, I_{oct}\,[\text{A}]. \quad (7)$$

Therefore, the theory, based on the above impedance and frequency distribution models, predicts beam stability for $I_{oct} \approx 6.6$ A. Notice that the space-charge component of the impedance has not been taken into account in our calculations (as concerns both instability and damping [7]).

## Stabilization by Coupled Landau Damping

In the presence of linear coupling, but in the absence of external non-linearities, the necessary condition for the stability of the $m$th mode is that the sum of the transverse instability growth rates, in the absence of both coupling and Landau damping, is negative [6,9]

$$V_{eqx}^{m} + V_{eqy}^{m} \leq 0. \quad (8)$$

If Eq. (8) is true, then it is possible to stabilize this mode by increasing the skew gradient and/or by getting closer to the coupling resonance $Q_h - Q_v = 1$. The theoretical stabilizing values of the modulus of the $l$th Fourier coefficient of the skew gradient are given by

$$\left|\hat{\underline{K}}_0(l)\right| \geq \frac{2\left[-Q_{x0} Q_{y0} V_{eqx}^m V_{eqy}^m\right]^{1/2}}{R^2 \Omega_0} \times \frac{\left[\left(V_{eqx}^m + V_{eqy}^m\right)^2 + \Omega_0^2 \left(Q_h - Q_v - 1\right)^2\right]^{1/2}}{-\left(V_{eqx}^m + V_{eqy}^m\right)}, \quad (9)$$

where $Q_{h,v} = Q_{x0,y0} + U^m_{eqx,y}/\Omega_0$ are the horizontal and vertical coherent tunes in the presence of wake fields ($U^m_{eqx,y}$), but in the absence of coupling, and

$$U^m_{eqx,y} = \text{Re}(\Delta\omega^{x,y}_{c,m}), \qquad V^m_{eqx,y} = -\text{Im}(\Delta\omega^{x,y}_{c,m}), \qquad (10)$$

with $\text{Re}(\ )$ and $\text{Im}(\ )$ standing for real and imaginary parts. Furthermore, in the case of coupled-bunch instabilities, the mode numbers are related by $n_x = n_y - 1$.

In the presence of both linear coupling and external non-linearities, in addition to the exchange of energy (transfer of instability growth rates between the transverse planes), there can also be a partition of Landau damping for "optimum" coupling. In Refs. 6 and 9, the influence of linear coupling on Landau damping of coherent instabilities has been assessed using two typical frequency distributions (Lorentzian, $\rho(f) \propto 1/(1+u^2)$, and "elliptical", $\rho(f) \propto \sqrt{1-u^2}$ where $u = (f - f_0)/\Delta f$), knowing that they are limiting cases modeling spectra with and without important tails, and that realistic distributions are probably between them.

In the case of Lorentzian distributions, the necessary condition for the stability of the $m$th mode and the stability criterion are given by Eqs. (8) and (9), replacing $V^m_{eqx,y}$ by $V^m_{eqx,y} - \delta\omega_{x,y}$, where $\delta\omega_{x,y}$ are the half widths at half maximum of the spectra.

In the case of elliptical distributions, the situation is more involved due to the finite tails. Two cases appear depending on whether the transverse coherent tunes (in the absence of coupling) are "far" from or "near" each other ("near" means a tune separation smaller than the order of magnitude of the average of the transverse spreads). If $Q_h$ is "far" from $Q_v + 1$, then the necessary condition for the stability of the $m$th mode and the stability criterion are given by Eqs. (8) and (9). There is no transfer of Landau damping since the coherent tunes are too far from each other to share their stabilizing spreads. If $Q_h$ is "near" $Q_v + 1$, then in addition to the sharing of the instability growth rates, there is also a transfer of Landau damping for "optimum" coupling. The necessary condition for stability is

$$\text{Re}\left(\sqrt{\Delta\omega_x^2 - (2U^m_{eqx})^2} + \sqrt{\Delta\omega_y^2 - (2U^m_{eqy})^2}\right) \geq 2\left(V^m_{eqx} + V^m_{eqy}\right), \quad (11)$$

where $\Delta\omega_{x,y}$ are the half widths at the bottom of the spectra. If Eq. (11) is true then it is possible to stabilize the beam and a condition similar to Eq. (9) for the stabilizing values of the coupling coefficient may be approximated by

$$\left|\hat{K}_0(l)\right| \approx \frac{\left(-Q_{x0}Q_{y0}\right)^{1/2}}{R^2 \Omega_0} \times$$

$$\left\{ \begin{bmatrix} \text{Re}\left( \sqrt{\Delta\omega_x^2 - \left(2U_{eqx}^m\right)^2} \right) - 2V_{eqx}^m \end{bmatrix} \times \atop \begin{bmatrix} \text{Re}\left( \sqrt{\Delta\omega_y^2 - \left(2U_{eqy}^m\right)^2} \right) - 2V_{eqy}^m \end{bmatrix} \right\}^{1/2} \quad (12)$$

Notice that too strong coupling is detrimental here since it shifts the coherent tunes outside the spectra and thus prevents Landau damping.

Therefore, applying this theory, one sees from Figure 3 that the nominal single-bunch beam can be stabilized by linear coupling only (i.e. without octupoles), since for each mode, Eq. (8) is verified. Making the numerical computations, the stabilizing normalized skew gradient is given by $\left|\hat{K}_0(0)\right| = \left|\underline{K}_0\right| \approx 10^{-5}$ m$^{-2}$. Furthermore, one can notice that this result is still valid for "any" intensity (as concerns pure head-tail instability), since if the intensity is multiplied by a certain factor, the instability growth rates are both multiplied by the same factor and Eq. (8) remains then true. Notice also that this result is not modified by the transverse space-charge impedances (negative inductances), which have been neglected in this paper, since they do not affect the instability growth rates.

### *Stabilization by Chromaticity Tuning*

Changes in machine chromaticity shift the beam oscillation spectrum centered at the chromatic frequency. The beam spectrum-impedance spectrum interaction is therefore modified and leads to different oscillation modes. It has been shown in Ref. 10, that a small gap in the horizontal chromaticity values ($\sim 0 < \xi_x < \sim 0.05$), where all modes are stable, should exist according to Sacherer's theory.

## Observations

To insure the validity of Sacherer's one-dimensional theory, the skew quadrupole current must be set such as to have the minimum of linear coupling between the horizontal and vertical planes, i.e. $I_{skew} \approx 0.33$ A [11]. Setting the octupole current to zero, a head-tail instability appeared with the single-bunch beam.

### *Growth Rate Measurements and Determination of the Mode Number*

The instability was observed to be only in the horizontal plane. Figure 4(a) exhibits the first unstable betatron line, and Figure 4(b) shows that it is a head-tail instability with mode number $|m|=6$, which is in perfect agreement with theory.

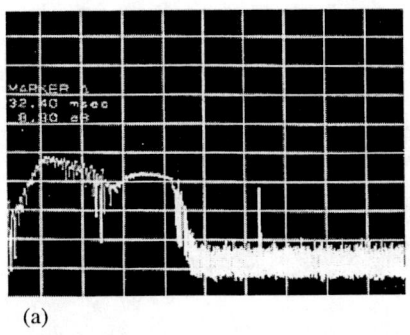 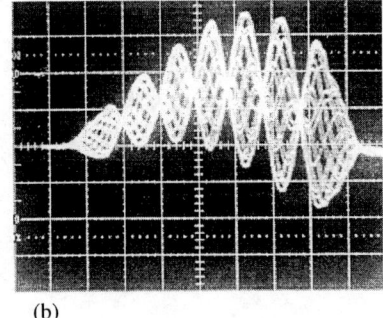

(a) (b)

**FIGURE 4.** (a) Measured horizontal instability growth rate on the first unstable betatron line (spectrum analyzer operating in zero frequency span) for the nominal single-bunch beam. Vertical scale: 10 dB/div. (b) $\Delta R$ signal from a radial beam-position monitor during 20 consecutive turns. Time scale: 20 ns/div.

## Stabilization by Landau Damping

Tuning the octupole current, the instability could be damped. The results of the measurements compared to theory are collected in Table 2, which shows the measured and theoretical stabilizing octupole currents, and the ratio between the two.

**TABLE 2.** Measured and theoretical stabilizing octupole currents for the nominal single-bunch beam.

| $I_{oct}^{exp}$ [A] | $I_{oct}^{theory}$ [A] | Ratio $= |I_{oct}^{exp} / I_{oct}^{theory}|$ |
|---|---|---|
| 8 | 6.6 | 1.2 |
| -10 | 6.6 | 1.5 |

No emittance blow-up has been observed during the first 500 ms, $\varepsilon_x^{norm,1\sigma} \approx \varepsilon_y^{norm,1\sigma} \approx 2\,\mu m$. The same analysis has been made for the ultimate single-bunch beam, i.e. with $N_b \approx 1.8 \times 10^{12}$ protons. The results of the stabilization by Landau damping are collected in Table 3. In this case also, no emittance blow-up has been observed during the first 500 ms, $\varepsilon_x^{norm,1\sigma} \approx \varepsilon_y^{norm,1\sigma} \approx 3.2\,\mu m$.

**TABLE 3.** Measured and theoretical stabilizing octupole currents for the ultimate single-bunch beam.

| $I_{oct}^{exp}$ [A] | $I_{oct}^{theory}$ [A] | Ratio $= |I_{oct}^{exp} / I_{oct}^{theory}|$ |
|---|---|---|
| 6 | 7.4 | 0.8 |
| -9 | 7.4 | 1.2 |

## Stabilization by Linear Coupling

By increasing the skew gradient instead of tuning the octupole current, the instability could also be damped, without emittance blow-up. The results of the measurements compared to theory are collected in Tables 4 and 5 for the nominal and ultimate beams. They both exhibit the measured stabilizing skew quadrupole current, its corresponding normalized skew gradient, the theoretical normalized skew gradient, and the ratio between the two.

**TABLE 4.** Measured and theoretical stabilizing normalized skew gradients for the nominal single-bunch beam.

| $I_{skew}$ [A] | $|\underline{K}_0|^{exp}(\times 10^{-5})$ [m$^{-2}$] | $|\underline{K}_0|^{theory}(\times 10^{-5})$ [m$^{-2}$] | Ratio = $|\underline{K}_0|^{exp} / |\underline{K}_0|^{theory}$ |
|---|---|---|---|
| 0.73 | 1.7 | 1 | 1.7 |
| -0.07 | 1.7 | 1 | 1.7 |

**TABLE 5.** Measured and theoretical stabilizing normalized skew gradients for the ultimate single-bunch beam.

| $I_{skew}$ [A] | $|\underline{K}_0|^{exp}(\times 10^{-5})$ [m$^{-2}$] | $|\underline{K}_0|^{theory}(\times 10^{-5})$ [m$^{-2}$] | Ratio = $|\underline{K}_0|^{exp} / |\underline{K}_0|^{theory}$ |
|---|---|---|---|
| 0.68 | 1.5 | 1 | 1.5 |
| -0.02 | 1.5 | 1 | 1.5 |

The relation between the skew quadrupole current and the modulus of the normalized skew gradient of the PS at 1 GeV kinetic energy is given in Figure 5(a). For the present 1.4 GeV kinetic energy, this measurement needs to be updated. However, a quick estimate has revealed that the minimum of linear coupling in the PS is obtained for the same skew quadrupole current, $I_{skew} \approx 0.33$ A. This result is in perfect agreement with those of Table 4, where it can be seen that 0.73 and –0.07 are symmetric with respect to 0.33, and thus correspond to the same skew gradient (as predicted by the stabilizing coupling theory). The future PS coupling measurement at 1.4 GeV should reveal this feature. Anyhow, the new curve should not deviate from the one at 1 GeV by more than 25%, and Figure 5(a) can therefore be used in a first approximation. Furthermore, as the new energy is greater than the previous one, for the same level of skew quadrupole current, the normalized skew gradient should be smaller, which means that with the updated curve the agreement between theory and experiment should be even better.

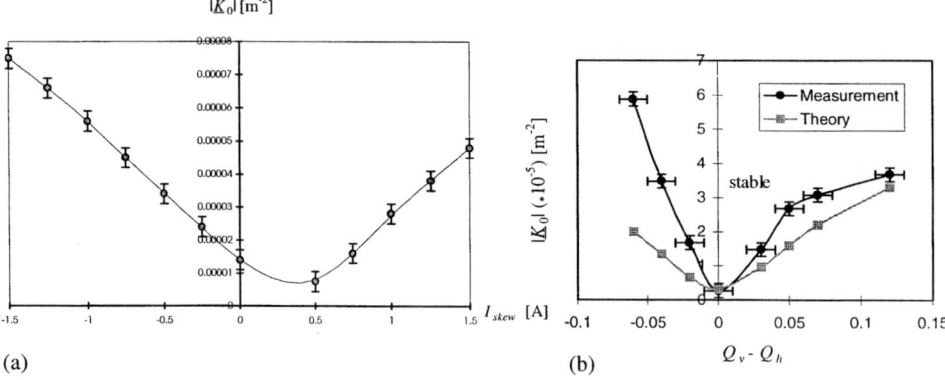

**FIGURE 5.** (a) Modulus of the normalized skew gradient vs. skew quadrupole current for the PS at 1 GeV kinetic energy. (b) Stability boundary in the plane $\left| \underline{K}_0 \right|$ vs. $Q_v - Q_h$ for the ultimate single-bunch beam.

As it can be seen from Eq. (9), the beam can be stabilized using the skew gradient and/or the tune separation. The results of damping measurements, made on the ultimate single-bunch beam, using both parameters are plotted in Figure 5(b).

## *Chromaticity Tuning*

Using the pole-face-windings and figure-of-eight-loop in addition to the normal quadrupoles, the chromaticity could be changed. Figure 6 exhibits different unstable modes ($m=4,5,7,8,10$) in the horizontal plane, in perfect agreement with Sacherer's theory, which have been obtained by tuning the chromaticity. However, one did not find stabilizing values of chromaticity, as could be predicted by the simplified Sacherer's theory.

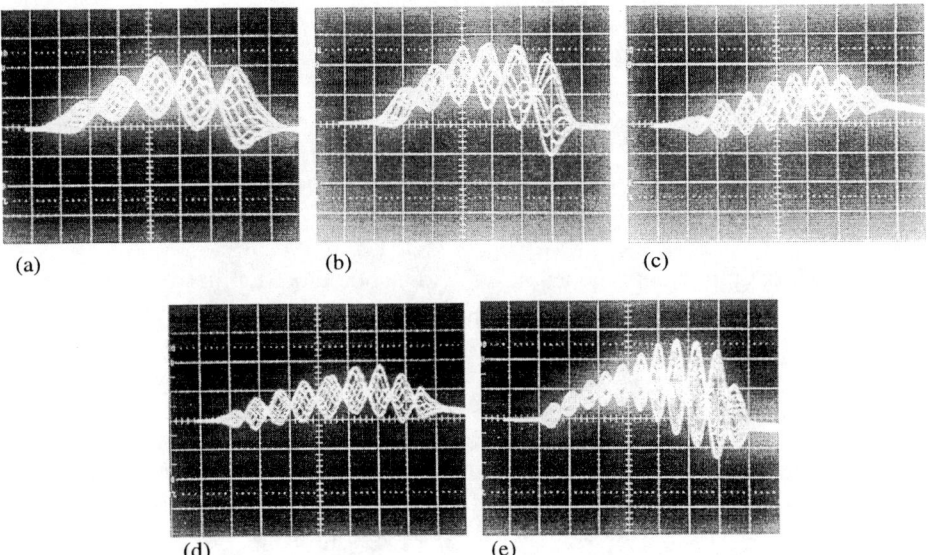

**FIGURE 6.** $\Delta R$ signal from a radial beam-position monitor during 20 consecutive turns. Time scale: 20 ns/div. (a) Nominal single-bunch beam with $Q_h = 6.08$, $Q_v = 6.32$, $\xi_x \approx -0.5$ and $\xi_y \approx -1.5$. (b) $Q_h = 6.18$, $Q_v = 6.21$, $\xi_x \approx -0.7$ and $\xi_y \approx -1.7$. (c) $Q_h = 6.21$, $Q_v = 6.18$, $\xi_x \approx -1.1$ and $\xi_y \approx -0.3$. (d) $Q_h = 6.21$, $Q_v = 6.16$, $\xi_x \approx -1.2$ and $\xi_y \approx 0.1$. (e) Ultimate single-bunch beam with $Q_h = 6.20$, $Q_v = 6.16$, $\xi_x \approx -1.3$ and $\xi_y \approx 0.1$.

## **Future Predictions**

Applying Sacherer's formula, coupled-bunch instabilities should appear with the final beam, which will be composed of 8 bunches. The first unstable transverse betatron lines are such that $n_{x,y} = 1$. The plot of the instability growth rates as functions of the head-tail mode number is represented in Figure 7 for the nominal beam.

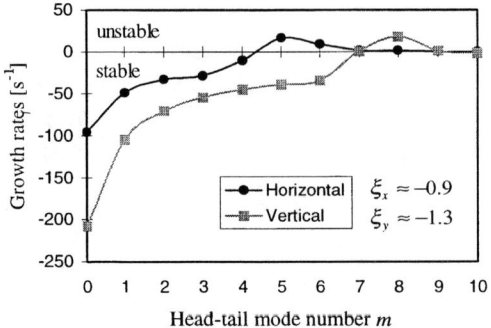

**FIGURE 7.** Transverse instability growth rates vs. head-tail mode number for the nominal 8 bunches beam.

One concludes therefore that the theory predicts transverse coupled-bunch instabilities ($n_{x,y} = 1$), with most critical head-tail mode number $m = 5$ for the horizontal plane.

Using linear coupling, both modes $m = 5$ and 6 should be stabilized, for $Q_h = 6.18$ and $Q_v = 6.21$, by putting $|\underline{K}_0| \approx 4.1 \times 10^{-5}$ m$^{-2}$, i.e. tuning the skew quadrupole current to $I_{skew} \leq -0.68$ A or $I_{skew} \geq 1.32$ A. However, under this condition, the modes $m = 7$, 8 and 9 should then become unstable, since both transverse instability growth rates are positive. One can perhaps imagine that these modes will be stabilized, remembering that Sacherer's theory is valid for the onset of coherent instabilities, and that the most critical modes will be damped by coupling. However, if a certain amount of octupole current is needed, it could be optimized using coupled Landau damping [12].

As concerns the final (8 bunches) ultimate beam, the transverse complex frequency shifts are multiplied by 1.8 (for the same coherent tunes), and the same results are obtained for the stabilization by linear coupling.

## Conclusions

The stability criterion for the damping of transverse head-tail instabilities in the presence of linear coupling has been verified experimentally and compared to theory, leading to a good agreement (to within a factor smaller than 2).

The high-order head-tail instabilities of the CERN-PS beam for LHC (single bunch with nominal or ultimate intensity) can be damped using coupling only (skew quadrupoles and/or tune separation). Furthermore, this result is predicted by theory for "any" intensity (as concerns pure head-tail instability).

The coupled-bunch instabilities should be damped also by coupling only (at least the most critical horizontal modes), or using coupled Landau damping (octupoles + coupling) to reduce the amount of external non-linearities.

## SPACE CHARGE

Space charge tune shifts can convey the beam onto non-linear resonances generating transverse emittance blow-up. The e.g. horizontal incoherent tune shift of the particle located in the center of a (transversally) Gaussian bunch, is given by (neglecting the wall effects) [13]

$$\Delta Q_{inc,x0} = -\frac{2 r_p I_p \beta_x R}{ec\beta^3 \gamma^3 a(a+b)}. \quad (13)$$

A similar equation is obtained for the vertical plane, replacing $x$ by $y$ and reversing the roles of $a$ and $b$ in Eq. (13). Making the numerical computations, one obtains for the single-bunch beam with nominal intensity, $\Delta Q_{inc,x0} = -0.18$ and $\Delta Q_{inc,y0} = -0.21$. For the single-bunch beam with ultimate intensity, it yields $\Delta Q_{inc,x0} = -0.22$ and $\Delta Q_{inc,y0} = -0.25$. The modulus of the previous values are below $0.3$, and therefore the absence of blow-up due to resonance crossings is in agreement with what was expected in Ref. 13.

As concerns head-tail instabilities, it has been shown before that the one-dimensional (horizontal) theory of Landau damping is in agreement with the observations if the space-charge impedance, given by $Z_x^{SC}(\omega) = -jRZ_0 \left( a_{round}^{-2} - b_{round}^{-2} \right)/\beta^2 \gamma^2$ for the simplified case of a round beam of radius $a_{round}$ circulating in a round pipe of radius $b_{round}$, is neglected (or at least the first incoherent term). Further work is needed to investigate this feature.

During the experiments, it has also been verified that the spread of the incoherent tune shift alone has no stabilizing effect on the high-order head-tail instabilities, as expected [7]. For the nearly round single-bunch beam with nominal intensity, a simple estimate for the e.g. horizontal space-charge tune spread, is given in Ref. 7, considering elliptical cross-section and parabolic density,

$$\Delta Q_x^{sc\,spread} \approx -\frac{\sqrt{13}}{8} \Delta Q_{inc,x0}. \quad (14)$$

From the numerical computations, $\Delta Q_x^{sc\,spread} \approx 0.08$, which should largely damp the head-tail instability $m = 6$, if the criterion of Eq. (6) could be used with the internal spread only. In practice, the instability is not damped in the absence of both octupoles and linear coupling, which shows that external non-linearities are required for Landau damping.

## CONCLUSION

Theoretical and experimental studies have been made on the longitudinal and transverse stability problems in the CERN-PS beam for the LHC. The longitudinal coupled-bunch instability can be damped by a longitudinal feedback. The longitudinal microwave instability will be avoided by adopting a new scheme, which is under study, to produce the LHC bunch train.

As concerns the transverse domain, until now experiments have been performed on a single-bunch beam with nominal and ultimate intensities. In both cases, linear coupling is sufficient to damp the high-order head-tail instabilities (in agreement with theory), without emittance blow-up. The next step consists in studying the final eight bunches beam.

## ACKNOWLEDGEMENTS

We wish to thank S. Hancock, G. Métral and R. Steerenberg for their help during the experiments.

## REFERENCES

1. K. Schindl, "The Injector Chain for the LHC", CERN/PS 99-018 (DI), 1999.

2. A. Blas et al., "Beams in the CERN PS Complex after the RF Upgrades for LHC", CERN/PS 98-022 (RF) or 6$^{th}$ European Particle Accelerator Conference (EPAC98), June 22-26, 1998, Stockholm, Sweden.

3. F. Pedersen and F. Sacherer, "Theory and Performance of the Longitudinal Active Damping System for the CERN PS Booster", CERN-PS-BR-77-8, 1977.

4. R. Garoby, "Bunch Merging and Splitting Techniques in the Injectors for High Energy Hadron Colliders", CERN/PS 98-048 (RF) or 17$^{th}$ International Conference on High Energy Accelerators (HEACC'98), September 7-12, 1998, Dubna, Russia.

5. F. Sacherer, "Transverse Bunched-Beam Instability-Theory", Proc. 9$^{th}$ Int. Conf. on High Energy Accelerators, Stanford 1974 (CONF 740522, US Atomic Energy Commission, Washington D.C., 1974), p. 347.

6. E. Métral, "Coupled Landau Damping of Transverse Coherent Instabilities in Particle Accelerators", Thesis, Joseph Fourier University of Grenoble (France), 1999.

7. D. Möhl, "On Landau Damping of Dipole Modes by Non-Linear Space Charge and Octupoles", CERN/PS 95-08 (DI), 1995.

8. A.W. Chao, *Physics of Collective Beam Instabilities in High Energy Accelerators*, New York: Wiley, 371 p, 1993.

9. E. Métral, "Theory of Coupled Landau Damping", Part. Accelerators, **62**(3-4), p. 259, 1999.

10. R. Cappi, "Observations of High-Order Head-Tail Instabilities at the CERN-PS", CERN/PS 95-02 (PA), 1995.

11. E. Métral, "Measurement of the PS Linear Coupling using FFT Analysis", CERN/PS/CA/Note 97-24, 1997.

12. E. Métral, "Measurements of Coupled Landau Damping of the PS Coherent Resistive Instability", CERN/PS/CA/Note 98-16, 1998.

13. R. Cappi, "The PS in the LHC Injector Chain", CERN/PS 97-16 (CA), 1997.

## APPENDIX: List of PS and beam parameters during the experiments

| | |
|---|---|
| $a = \sqrt{2}\left[\varepsilon_x^{rms}\beta_x + (D_x\sigma_p/p)^2\right]^{1/2}$ | $\sqrt{2}$ the rms horizontal beam dimension |
| $a_w = 7$ cm | Half major axis of the elliptical beam pipe |
| $b = \sqrt{2}\sqrt{\varepsilon_y^{rms}\beta_y}$ | $\sqrt{2}$ the rms vertical beam dimension |
| $b_w = 3.5$ cm | Half minor axis of the elliptical beam pipe |
| $c = 3\times 10^8$ ms$^{-1}$ | Velocity of light |
| $D_x \approx 2.5$ m | Average horizontal dispersion function |
| $D_y \approx 0$ m | Average vertical dispersion function |
| $e = 1.6\times 10^{-19}$ C | Elementary charge |
| $E_0 = 0.938$ GeV | Proton rest energy |
| $E_c = 1.4$ GeV | Proton kinetic energy |
| $I_b$ | Current in one bunch |
| $I_p = 3eN_b/2\tau_b$ | Bunch peak current considering a parabolic line density |
| $j = \sqrt{-1}$ | Imaginary unit |
| $k$ | $k = ...,-1,0,1,...$ for a single bunch or several bunches oscillating independently; $k = n_{x,y} + k'M$ with $k' = ...,-1,0,1,...$ for coupled motion of $M$ bunches |
| $L$ | Bunch length (in meters) |
| $m = ...,-1,0,1,...$ | Head-tail mode number |
| $m_0 = 1.67\times 10^{-27}$ kg | Proton rest mass |
| $M$ | Number of bunches of the beam |
| $n_{x,y} = 0,1,...,M-1$ | Transverse coupled-bunch mode numbers |
| $N_b$ | Bunch intensity. $N_b \approx 10^{12}$ protons for the nominal beam; $N_b \approx 1.8\times 10^{12}$ protons for the ultimate beam |
| $N_{oct} = 8$ | Number of octupoles |
| $Q \approx 1$ | Quality factor of the broadband impedances |

| | |
|---|---|
| $Q_h = 6.18$ | Horizontal coherent tune |
| $Q_v = 6.21$ | Vertical coherent tune |
| $Q_{x0,y0}$ | Transverse coherent tunes in the absence of wake fields |
| $r_p = 1.54 \times 10^{-18}$ m | Classical proton radius |
| $R = 100$ m | Average radius of the machine |
| $R_x \approx 1$ M$\Omega$/m | Shunt resistance of the horizontal broadband impedance |
| $R_y \approx 3$ M$\Omega$/m | Shunt resistance of the vertical broadband impedance |
| $Z_0 = 377 \Omega$ | Free space impedance |
| $\alpha_p = \gamma_{tr}^{-2} = 0.027$ | Momentum compaction factor |
| $\beta = 0.916$ | Relativistic velocity factor |
| $\beta_{octx} \approx 12.4$ m | Horizontal betatron function at the octupoles |
| $\beta_{octy} \approx 22.6$ m | Vertical betatron function at the octupoles |
| $\beta_x \approx R/Q_h = 16.2$ | Average horizontal betatron function |
| $\beta_y \approx R/Q_v = 16.1$ | Average vertical betatron function |
| $\gamma = 2.493$ | Relativistic mass factor |
| $\varepsilon_0 = 8.84 \times 10^{-12}$ Fm$^{-1}$ | Permittivity of free space |
| $\varepsilon_{x,y}^{norm,1\sigma} = \beta\gamma\varepsilon_{x,y}^{rms}$ | Normalized rms transverse emittances |
| $\varepsilon_{x,y}^{rms}$ | Rms transverse emittances. $\varepsilon_x^{rms} = 0.83 \times 10^{-6}$ m and $\varepsilon_y^{rms} = 0.87 \times 10^{-6}$ m for the nominal beam; $\varepsilon_x^{rms} = 1.36 \times 10^{-6}$ m and $\varepsilon_y^{rms} = 1.40 \times 10^{-6}$ m for the ultimate beam |
| $\eta = \alpha_p - \gamma^{-2} = -0.134$ | Slippage (or off-momentum) factor |
| $\xi_x \approx -0.9$ | Horizontal (relative) chromaticity |
| $\xi_y \approx -1.3$ | Vertical (relative) chromaticity |
| $\rho_w = 9 \times 10^{-7} \Omega$ m | Vacuum chamber resistivity |
| $\tau_b = 160$ ns | Total bunch length (in seconds) |
| $\omega_r = 2\pi f_r = 2\pi \times 1.4$ GHz | Vacuum chamber cut-off (angular) frequency |
| $\omega_s = 2\pi f_s = 2\pi \times 610$ Hz | Synchrotron (angular) frequency of the particles |
| $\omega_{\xi_{x,y}} = (\xi_{x,y}/\eta)Q_{x0,y0}\Omega_0$ | Transverse chromatic (angular) frequencies |
| $\Omega_0 = 2\pi f_0 = 2\pi \times 436.5$ kHz | Average revolution (angular) frequency of the particles |
| $B_y\rho_x = 7.14$ Tm | Beam rigidity |
| $\sigma_p/p \approx 10^{-3}$ | Rms relative momentum spread |
| $Sgn(\omega) = 1$ if $\omega > 0$, $-1$ if $\omega < 0$ | |
| $\int_0^{l_{oct}} K_{oct}\, dl\, [\text{m}^{-3}] = \dfrac{4.33}{6B_y\rho_x} \times I_{oct}\, [\text{A}]$ | Integrated octupole strength |

# Impedance Budget and Beam Instabilities of the JHF 50-GeV Proton Synchrotron

### Y. Mori and M. Yoshii
*KEK, Oho 1-1, Tsukuba-shi, Ibaraki-ken, Japan*

## 1 Inroduction

The planned 50-GeV proton synchrotron of the Japanese Hadron Facility (JHF) will consist of a 200MeV-proton linear accelerator (200-MeV linac) as an injector, a 3-GeV rapid cycling synchrotron as a booster and a 50-GeV proton synchrotron (main ring).[1] Original idea was that the accelerators would be constructed at the north site of KEK. Recently, this project has been reformed under the collaboration between KEK and JAERI(Japan Atomic Energy Research Institute) and a new site for the project will be Tokai campus of JAERI. Although some parameters have been changed, the basic design of the accelerator complex is not changed from the JHF one. In this paper, most of the discussions will be based on the design of the JHF 50-GeV proton synchrotron.

The main ring is to accelerate protons from 3 GeV to 50 GeV. The expected beam intensity in the main ring is $3.2 \times 10^{14}$ ppp and the repetition rate is about 0.3 Hz. The 50-GeV protons are extracted by slow and fast extraction schemes into two experimental areas: one is for experiments using secondary beams (K,antiproton, etc.) and primary beams by slow extraction, and the other is for the neutrino oscillation experiments by fast extraction. When operated in a slow extraction mode, the average current and duty factor, which is defined as the fraction of a cycle when a beam is available, are 16 µA and 0.20, respectively.

Protons are accelerated from 3 GeV to 50 GeV in the main ring. At the top energy of the 50-GeV main ring, $\gamma$ is 54.3. In a conventional way of designing a lattice using a regular FODO cell, the transition $\gamma$ approximately equals the horizontal betatron tune ($\nu_x$). In a machine of this scale, because $\nu_x$ is about 20-30, it is difficult to avoid the transition energy in the regular FODO lattice. Although techniques of the transition energy crossing have been developed in many operational proton synchrotrons, it is favorable to place the transition energy, where the phase focusing becomes zero, well above the top energy, avoiding the instabilities and associated beam losses. Thus, an imaginary transition $\gamma$ lattice, in which the momentum compaction factor is negative, is employed. To make the momentum compaction factor negative, either $\beta_x$ or $\rho$ should be modulated properly. In order to avoid a bigger beam size, $\rho$ modulation while invoking the missing bend sections in each arc of the ring is better than $\beta_x$ modulation, although the ring circumference becomes slightly large. In the $\rho$ modulation scheme, the momentum compaction factor and the dispersion function can be estimated analytically.[1] In the 50-GeV main ring, the superperiodicity is four and each arc section consists of six modules. Each module has three FODO cells starting from a defocusing quadrupole. In the center cell of the module, there is no bending magnets (missing bend cell). Thus, $\xi$ is about 1/3.

Table Main parameters of the 50-GeV ring

| | | |
|---|---|---|
| Injection energy, $E_i$ | 3 GeV | |
| Circumference, $C$ | 1445 m | |
| Harmonic number, $h$ | 17 | |
| Momentum compaction factor, $\alpha$ | -0.001 | |
| Natural chromaticity, $\xi Q_T$ | $\approx$-20 | |
| Horizontal tune, $Q_T$ | 21.80 | |
| Circulating current, $I_c$ | 6.86 A (at 3 GeV) | 7.06A (at 50 GeV) |
| Slippage factor, $\eta$ | -0.057 (at 3 GeV) | -0.00134 (at 50 GeV) |
| Momentum spread, $(\Delta p/p)_{FWHH}$ * | 0.42 % (at 3 GeV) | 0.23 % (at 50 GeV) |
| Rms bunch length in time, $\sigma_\tau$ * | 30 ns (at 3 GeV) | 4.2 ns (at 50 GeV) |
| Bunching factor, $B_f$ * | 0.273 (at 3 GeV) | 0.038 (at 50 GeV) |
| Synchrotron tune, $Q_s$ | 0.0034 (at 3 GeV) | 0.0003 (at 50 GeV) |
| Loaded shunt impedance of the cavity fundamental, $R_s$ | 14 k$\Omega$ | |
| Q-value of the cavity fundamental, $Q$ | 1 | |
| Resonant frequency of the cavity fundamental, $f_R$ | 3.43 MHz | |

* Values for the longitudinal emittance $\varepsilon_L = 3 eV \cdot s$ at injection

## 2 Beam Instabilities

Since the beam current in the main ring is very high (the circulating current of 10A and a peak current of 30A), beam instabilities and the impedance budget should be carefully examined. The beam instabilities are classified into two categories of single-bunch and multi-bunch phenomena. Both instabilities are induced by the impedance which is characterized as being either broadband or narrow-band.

### 2.1 Single-Bunch Instability
**Longitudinal**

The space charge, wall inductance and RF cavities are the main sources of the longitudinal broadband impedance. The threshold of the longitudinal impedance for the onset of the microwave instability and the negative-mass instability can be calculated using Keil-Schnell-Boussard criterion [4]:

$$\left|\frac{Z_L}{n}\right| \leq \frac{E\beta^2|\eta|}{eI_p}\left(\frac{\Delta p}{p}\right)^2_{FWHH}, \qquad (1)$$

### a. Space charge
A precise calculation of a geometrical factor using the actual lattice parameters shows $g_0=2.3$ on the average, resulting in a space-charge impedance of 25 $\Omega$ at 3 GeV.

### b. RF Cavity
Since a low-Q (Q≈1) MA(Magnetic Alloy) RF cavity is going to be used[3], the beam loading effect and the possibility of a microwave instability should be investigated. We found that there are no beam-dynamics problems associated with the single-bunch effects caused by the fundamental mode of the RF cavities, even without a feedback system for beam-loading compensation.

### c. Inductance
The threshold impedance at top energy becomes only $\left|\frac{Z_L}{n}\right| \leq 0.4\Omega$. At this energy, the inductive impedance is expected to be dominant contributor. To avoid bunch shortening due to the negative mass instability, we need to intentionally blow up the longitudinal emittance by a factor of, say, 3, to $\varepsilon_L = 3eV \cdot s$ after injection. Then, the stability condition at 50 GeV becomes

$$\left|\frac{Z_L}{n}\right| \leq 2\Omega \quad \text{(at 50 GeV: } \varepsilon_L = 3eV \cdot s \text{ at injection).} \qquad (2)$$

The dominant impedance source would be brought from the discontinuity of the chamber at flanges. About 500 such cavities can be accepted before the total inductance reaches a threshold value of 2 $\Omega$.

Regarding single-bunch effects, no instability is excited by the space-charge impedance and the fundamental mode of the RF cavity in the main ring. Although the wall inductance has a possibility to cause a negative-mass instability at 50 GeV, this can be avoided by increasing the longitudinal emittance to *3eV.s* after injection.

### Transverse
### a. Tansverse Mode-coupling
Although the transverse mode-coupling instability(TMCI) has been observed in many electron machines, the TMCI has been never observed in proton machines, presumably due to a long bunch length. It is not the case that TMCI is a major limiting factor.

## 2.2 Multi-Bunch Instability

## Longitudinal
### a. Cavity Fundamental and Parastic
The parasitic higher-order modes in the RF cavities are dominant sources of longitudinal narrow-band impedance. Neither a significant emittance distortion nor beam loss due to the potential-well distortion caused by the beam loading effect has been observed in the computer simulations. For multibunch effects, there are no significant problems associated with the longitudinal multibunch instabilities, provided by properly tuning the cavity fundamental, and damping the parasitic resonances below Q=5 where Q is the qualtiy factor of the cavity or pushing to a higher frequency above 9MHz.

## Transverse
the growth rate of the transverse coupled-bunch instability is proportional to the overlap integral between the transverse impedance and the spectrum of head-tail mode. The growth rate of the transverse coupled-bunch instability is roughly given by [8]

$$\tau^{-1} \approx \frac{1}{8\pi Q_T} \frac{1}{m+1} \frac{c}{E} I_c R_T, \qquad (3)$$

where $R_T$ is the transverse impedance evaluated at frequency $\omega = (n - Q_T)\omega_0$, and $n$ is the closest harmonic to the impedance. Like in the longitudinal case, the transverse instability is most serious at the injection energy of 3 GeV. The effect of Landau damping is not included in Eq.(3). Including the effect of Landau damping, we obtain the following criterion for the transverse impedance:

$$R_T \leq 1.1 M\Omega/m. \qquad (4)$$

The main sources of the transverse narrow-band impedance are the resistivity of the chamber wall, the kicker magnets, and the RF cavities. Since the natural chromaticity has a negative sign and sufficient amplitude, the head-tail instabilities are well damped up to very higher-order modes.

### a. Resistive Wall
The resistivity of the chamber material (or RF shieldings) is a major factor in determining the resistive-wall impedance. Stainless-steel 316 ($\rho \approx 100 \mu\Omega \cdot cm$) is assumed as the material of the vacuum chamber. In this case, the transverse resistive-wall impedance becomes 1.44 MΩ/m, which is more than the tolerance of 1.1 MΩ/m. A narrow-band transverse feedback system is essential to damp the dipole motion which causes the instability.

### b. Kicker Magnets
According to the reference, the total transverse impedance of kicker magnets is found to be 0.74 MΩ/m [6]. Although it is marginal, the instability caused by the kicker imped-

ance could be handled via the transverse feedback system.

## 3 Summary

The beam instabilities concerning to the 50-GeV proton synchrotron are summarized. Authors would like to make their sincere appreciation to Dr. Y. Chin for his cooperation.

References

[1] Y.Mori;"The Japanese Hadron Facility", Proc. of Particle Accelerator Conference,Vancouver, 1997.
[2] Y. Chin etal; "JHF Accelerator Design Study Report", KEK Reprot 97-16, 1998.
[3] A, Hofmann, "Theoretical Aspects of the Behavior of Beams in Accelerators and Storage Rings", CERN 77-13 (1977), p.139.
[4] K. Keil and W. Schnell, CERN-ISR-TH-RF/69-48 (1968).
[5] A. W. Chao, *Physics of Collective Beam Instabilities in High Energy Accelerators* (John Wiley & Sons, Inc. New York, 1993).
[6] T. Suzuki, Y. H. Chin and K. Satoh, Part. Accelerators **13**, 179 (1983).
[7] Y. H. Chin, CERN/SPS/85-2 (1985).
[8] R. Baartman, TRI-PP-87-12 (1987).

# Instability Issues at the SNS Storage Ring

## S.Y. Zhang

*Brookhaven National Laboratory, Upton, NY 11973*

**Abstract** - The impedance and beam instability issues of the SNS storage ring is reviewed, and the effort toward solutions at the BNL is reported. Some unsettled issues will be raised, indicating the direction of planned works.

## I. SNS Storage Ring

The parameters of the SNS storage ring are listed in Table 1. Several relevant issues in the machine design are as follows.

1. With the very high intensity of $2.08 \times 10^{14}$ protons per pulse, the space charge effect and the particle distribution in transverse phase space need attentions [1].

2. At 2 $MW$, the beam power of the SNS is about 12 times higher than the ISIS, and about 25 times higher than the PSR. In the 1996 survey, with 4.5 hours cooldown after the operations at about 56 $KW$ level, there are still 31 contact readings at the ends of sections show the activation higher than 2 $rem/hr$, 11 of them higher than 5 $rem/hr$ [2]. This activation affects operations at the PSR. For the SNS, therefore, the uncontrolled beam loss is not allowed to exceed 0.02% of the beam intensity. One approach to achieve this goal is to use large machine aperture, which then has impact on the impedance and instabilities.

| Parameter | | SNS | Unit |
|---|---|---|---|
| Beam Power | $P$ | 2 | $MW$ |
| Total Particle | $N$ | 2.08 | $10^{14}$ |
| Circumference | $C$ | 220 | $m$ |
| Kinetic Energy | $E_k$ | 1.0 | $GeV$ |
| Repetition Rate | | 60 | $Hz$ |
| Bunch Length | $t_B$ | 550 | $ns$ |
| Injection Turns | | 1200 | |
| RF Voltage, $h = 1/2$ | $V_{RF}$ | 40/20 | $KV$ |
| Beam Momemtum Spread | $\Delta p/p$ | 0.7 | % |
| Beam Current | $I_0/I_p$ | 40/80 | $A$ |
| Ave. Chamber Radius | $b$ | 10 | $cm$ |
| Uncontr. Beam Loss | | 0.02 | % |
| Beam Loss Power /m | | 1.8 | $W/m$ |

Table 1: SNS storage ring parameters

CP496, *Workshop on Instabilities of High Intensity Hadron Beams in Rings,*
edited by T. Roser and S. Y. Zhang
© 1999 American Institute of Physics 1-56396-910-6/99/$15.00

3. The SNS is intentionally designed with a relatively high RF voltage. The large beam momentum spread defined during the multiturn injection, therefore, may help the damping of the longitudinal microwave instabilities. The high RF voltage also brings along with the tolerance for the longitudinal space charge effect, which is defocusing and causes the bunch leakage into the gap.

4. The bunch in the SNS storage ring is very long, however, it only has to stay in the machine for about 1 $ms$. Immediately after the multiturn injection, the beam is extracted. This calls for better understanding of the impedance and instability issues during the beam stacking.

In this paper, selected impedance and instability issues will be discussed. Also, unsettled issues will be raised, indicating the directions of the planned works.

## II. Impedance

The impedances of the SNS storage ring, like other synchrotrons, fall into several categories as follows.

1. Space charge impedance is frequency independent. These are calculated for the SNS storage ring as $Z_\ell/n = -j196$ $\Omega$ and $Z_T = -j6.87$ $M\Omega/m$, for longitudinal and transverse, respectively. This impedance is a dominant one.

2. Resistive wall impedance is of interest at very low frequency range. The relation of $Z_T \approx (Z_\ell/n)\, 2R/\beta b^2$ exists. Sometimes, one estimates the resistive wall impedance by its value at the revolution frequency. At the SNS, the longitudinal and transverse resistive wall impedances at the revolution frequency are $Z_\ell(\omega_0) = 0.65(1+j)\Omega$ and $Z_T(\omega_0) = 5.22(1+j)K\Omega/m$, respectively.

3. Resonant frequency of broadband impedance is in a few $GHz$. Therefore, below the cut off frequency 1 $GHz$, this impedance can be seen as an inductance.

4. Resonant modes of low frequency impedance are in a few tens to several hundred $MHz$. In the relevant frequency range of up to 1 $GHz$, the imaginary part of the impedance cannot be seen as a pure inductance, and the real part may not be negligible. The contributors of the low frequency impedance include the BPM, the extraction and injection kickers. The impedance of the BPM of the SNS storage ring is handily calculated using conventional formulations [3,4], the real part of the impedance is peaked around 100 $MHz$, where the longitudinal and transverse impedances are $Z_\ell/n \approx 2$ $\Omega$ and $Z_T \approx 40$ $K\Omega/m$, respectively.

5. Narrow band impedances are mainly from the RF cavities. However, parasitic parameter may cause sharp resonance in the kicker impedance, and large steps and cavities of the vacuum chamber may also contribute to the narrow band impedance.

In the design of the SNS storage ring, some concerns with respect to the broadband and narrow band impedances, and also the extraction kickers have been raised. Therefore, these impedances will be discussed separately as follows.

## A. Broadband Impedance

The bellows, steps, ports, collimators, etc., contribute to the broadband impedance. The general relation between the longitudinal and transverse impedances, $Z_T \approx (Z_\ell/n) 2R/\beta b^2$, can be applied. In Table 2, the broadband impedances due to each component are shown for the SNS storage ring, using the analytical calculations [5].

|              | $Z_\ell/n$ | $Z_T$     |
|--------------|------------|-----------|
| Bellows      | 1.1        | 8.8       |
| Vacuum ports | 0.49       | 3.9       |
| Valves       | 0.28       | 2.2       |
| Steps        | 16.8       | 134.4     |
| Collimator   | 1.04       | 8.3       |
| Total        | 19.7       | 157.6     |
| **Unit**     | $j\Omega$  | $jK\Omega/m$ |

Table 2: Broadband impedance

The broadband impedance is dominated by large steps, typically 2 cm. At present, only moderate tapering and shielding are proposed. Further tapering and complete shielding can reduce the broadband impedance to just a few $\Omega s$. However, not only the cost is an issue, the machine aperture is also affected in further tapering and shielding.

On the other hand, the broadband impedance calculated for the SNS storage ring is comparable with $j14$ $\Omega$ for ISR [6], $j16$ $\Omega$ for SPS [7], $j30$ $\Omega$ for AGS [8], and $j17$ $\Omega$ for CPS [9]. Experience in the operations of these machines did not show that the broadband impedance posed the limit of beam intensity or caused additional beam loss.

## B. Narrow Band Impedance

The narrow band impedance comes mainly from the RF cavity high order modes. The large steps and other cavities in the chamber also contribute to the narrow band impedance in the range around $GHz$. In addition, it will be shown that the unmatched extraction kickers also contribute to the narrow band impedance.

Coupled bunch instability, caused by the narrow band impedance, is not relevant to the SNS. On the other hand, theoretically, narrow band impedance may cause instabilities for long bunches, for instance, the transverse Robinson instability [10].

For the low or medium energy hadron machines, however, no evidence has been shown that this impedance has posed serious limitations.

Better understanding of the impact of the narrow band impedance is needed, prior to take possible further steps beyond the moderate tapering and shielding.

## C. Extraction Kicker, Transverse

The extraction kicker of the SNS storage ring consists of 8 window frame magnet units, with the average length of $\bar{\ell} = 40\ cm$, the average width of $2\bar{b} = 14\ cm$, all have height $2a = 11.5\ cm$. The impedance of this kicker is not only relatively large, but also very sensitive to the terminations.

Consider the conventional calculation [11],

$$Z_T = \frac{c\omega\mu_0^2\ell^2}{4a^2 Z_k}\ \Omega/m \tag{1}$$

where $Z_k = j\omega L + Z_g$, with $L$ the magnet inductance, and $Z_g$ the termination impedance. For $Z_g = 200\ \Omega$, the extraction kicker transverse impedance is shown in Fig. 1, where we observe that the real part of the impedance, $Z_{T,\text{real}} \approx 20\ K\Omega/m$, is peaked at 30 $MHz$.

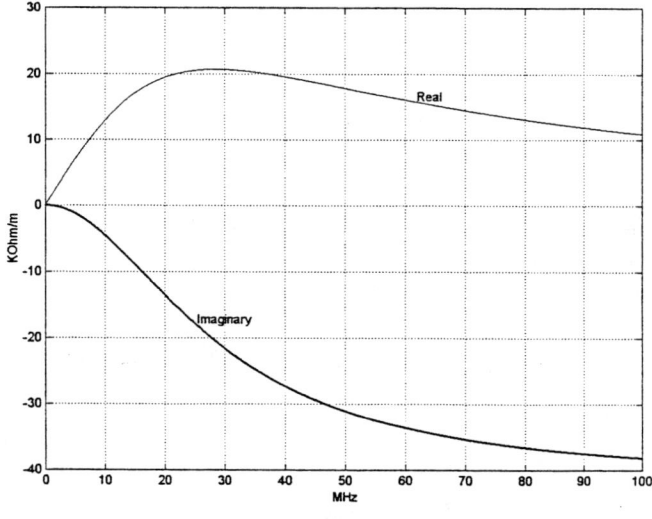

Fig. 1

Using a more realistic termination with 50 $pf$ capacitance presented around the stray inductance of 0.5 $\mu H$, the picture of the impedance is dramatically changed, as shown in Fig. 2. Two resonances with much narrower bandwidth are shown at 20 $MHz$, and 50 $MHz$, with the real parts of $Z_{T,\text{real}} \approx 100\ K\Omega/m$, and $Z_{T,\text{real}} \approx 200$

$K\Omega/m$, respectively. Sensitivity of the impedance with respect to the termination can also be shown if the charging resistance and the stray parameters are changed.

The AGS injection kicker and the AGS Booster extraction kicker have used the same structure of window frame magnets. The terminations of the magnet windings did not require special attentions. There are, however, some speculations that these kicker impedance might be responsible to certain instabilities. Therefore, the measurement of the extraction kicker impedance is undergoing in order to optimize the terminations.

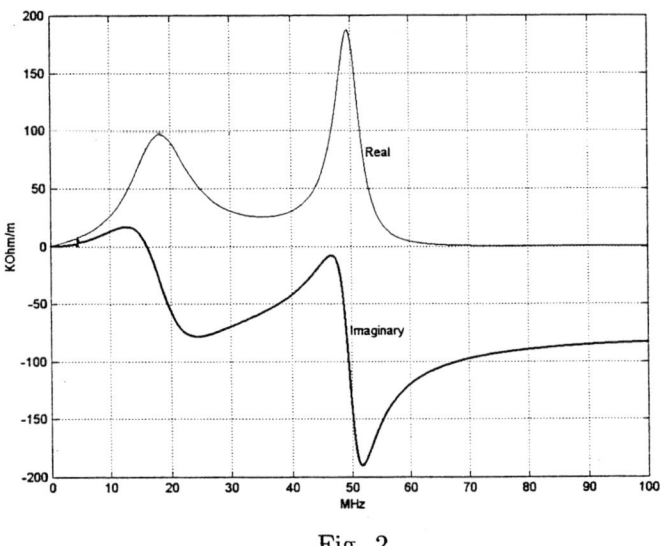

Fig. 2

## D. Extraction Kicker, Longitudinal

To reduce the massive ferrite loss, copper sheets are placed in the ferrite core of the window frame magnet as flux break. Taking the thickness of the copper sheet as 1 $mm$, the total leakage inductance around it is calculated as $L_{leak} \approx 6~\mu H$ [12]. This result is also verified by the simulation [13]. Using the formulation,

$$\frac{Z_\ell}{n} = j\omega_0 L_{leak} \qquad (2)$$

we get the equivalent longitudinal impedance $Z_\ell/n = j45~\Omega$.

It is interesting to note that the SNS longitudinal space charge impedance is $-j~196~\Omega$, which is negative inductive. Therefore, the magnet flux leakage may compensate a part of that impedance.

Note that the beam image current at the conductor, which is used as the magnet winding, will partly offset the field created by the beam. The present plan is, however, to use a single power supply to drive both conductors, thus, the image current effect will be negligible.

Compensation of the longitudinal space charge impedance has been studied by inserting ferrite rings in the beam pipe, such as that at PSR [14] and KEK PS [15]. This approach is also proposed for the muon-collider proton driver [16]. If handled carefully, the ferrite window frame used for the extraction kicker may be used for the same purpose. A thinner copper sheet might be used without causing a heating problem [17], which is associated with the real part of the impedance.

Finally, a complete model should include a differential flux leakage through the gap air. Assuming this leakage to be a half of the leakage around the copper sheet, and the parasitic capacitance to be 50 $pf$, the longitudinal impedance is shown in Fig. 3. It can be observed that the linear inductive part is extended up to 40 $MHz$, beyond that the real part of impedance will take place. The effect of this real part of the impedance on the longitudinal microwave instability probably needs attentions.

Termination is also very important in determining this impedance [18], which will be determined by the measurements.

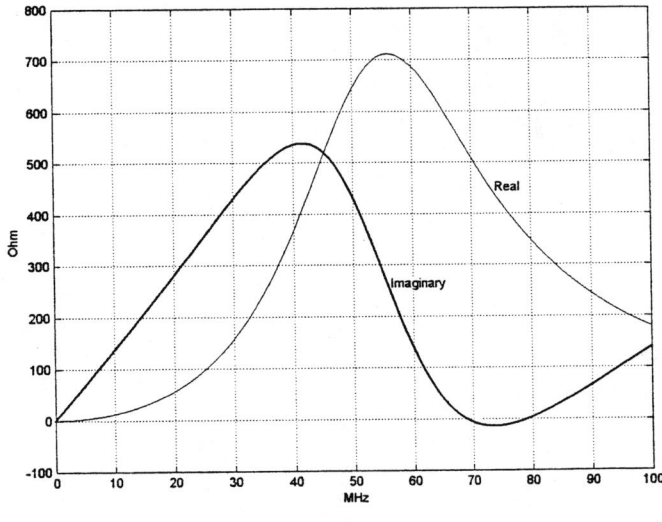

Fig. 3

## III. Instability

The most relevant instability issues at the SNS storage ring include the longitudinal microwave, the resistive wall, and the transverse microwave instabilities.

## A. Longitudinal Microwave Instability

Taking the beam peak current of 80 $A$, the Keil-Schnell criterion is satisfied for the SNS storage ring if the beam momentum spread $\Delta p/p \geq 0.65\%$. For the machines operated below transition, this criterion is believed to be too stringent. The SNS storage RF voltage of the fundamental harmonic is 40 $KV$, the beam momentum spread at the end of stacking is $\Delta p/p = \pm 0.7\%$. Therefore, the longitudinal microwave instability should not happen.

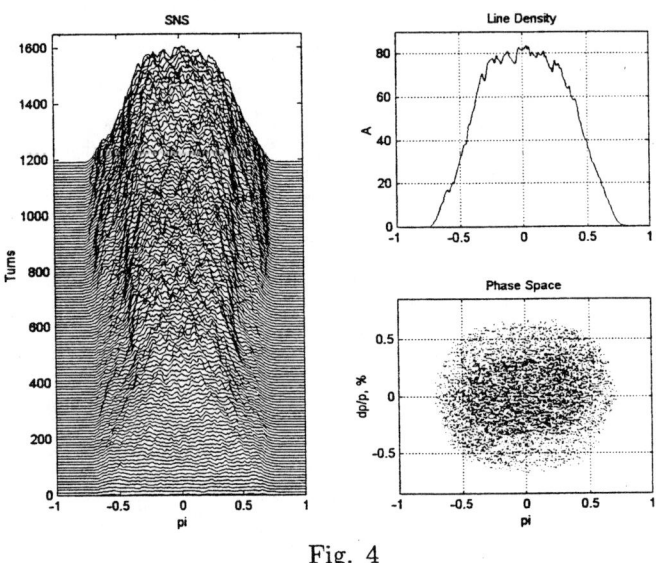

Fig. 4

One of the concerned issues in the beam stacking is the leakage of particles into the gap. The leaked proton beam is blamed for the survival of the electrons during the gap passage. The PSR study has shown that this leakage can lower the e-p instability threshold [19]. Also, for a storage ring, the leaked beam in the gap results in directly the beam loss at the extraction.

To reduce this leakage, the ramping of the RF voltage from 20 $KV$ at the beginning to 40 $KV$ at the end of stacking is proposed. The mountain range during the stacking, the line density and the particle distribution at the end of stacking are shown in Fig.4. It can be observed that the beam momentum spread has been reduced to $\Delta p/p = \pm 0.65\%$, which is acceptable from the longitudinal microwave instability point of view.

Using the typical Linac beam momentum spread at the end of the transport line, $\Delta p/p = \pm 0.1\%$, concerns have been raised on the beam momentum distribution in the ring. Simulations show that a large amount of particles have the momentum deviation less than $\pm 0.2\%$, during the stacking. The effectiveness of the longitudinal microwave instability damping mechanism, based on the beam momentum spread, is questioned.

To improve the beam momentum distribution in the ring, the Linac beam $\Delta p/p = \pm 0.3\%$ is requested, which is, in fact, used in the simulation that shown in Fig. 4. The beam momentum distribution in the ring has been improved, however, the beam loss associated with this large momentum spread probably needs attentions.

## B. Resistive Wall Instability

For the stainless steel vacuum chamber used for the SNS, at the tune of $\nu = 5.82$, the resistive wall instability growth rate at the end of stacking is calculated as $1\ ms$ for zero chromaticity. Also, the largest growth rate of mode $m = 1$ is about 1/5 of the $m = 0$ mode, i.e. $5\ ms$, which happens at $\xi \approx -0.04$. In this calculation, damping mechanism is not included.

Using the same approach, the AGS resistive wall instability at the injection porch is calculated with the growth rate of $0.37\ ms$, at the vertical tune of $\nu = 8.85$. The resistive wall instability observed at the AGS has a typical growth rate of $2\ ms$, which is slower than the calculated one by a factor of 5.

Also using the same approach, the AGS Booster resistive wall instability growth rate was calculated as $0.48\ ms$. In the operation, the instability has never been observed. The fast ramping at the Booster probably helps to further damp the instability.

We conclude that the resistive wall instability will not be strong enough to cause a serious problem at the SNS storage ring. Also, the choice of using stainless steel vacuum chamber is justifiable on this aspect.

## C. Transverse Microwave Instability

### 1. Transverse instability issues

The typical bunch length at the SNS storage ring is about $550\ ns$, giving rise to the full bandwidth of the bunch $3.6\ MHz$. In comparison, bunch lengths of the LHC and RHIC at the storage mode are about $1\ ns$, giving rise to the full bandwidth of the bunch $2\ GHz$.

The typical resonance of the broadband impedance is around $1\ GHz$. Therefore, the transverse mode coupling is of concern at both the LHC and RHIC [20]. Given the much smaller bandwidth of the bunch at the SNS storage ring, the mode coupling will not happen. Note that due to the difference between the spectrum of the modes, the mode crossing may take place, but the imaginary part will not rise.

As for the low frequency impedance, weak mode couplings may take place, but no strong mode coupling is expected. This is agreeable with the observation that no mode coupling has been causing problem in low and medium energy proton synchrotrons.

Since the entire beam life takes about a synchrotron period, conventional head-tail type instability will not be a serious problem.

The transverse microwave instability, however, may develop at a part of the long bunch, depending on the local peak current, associated impedance, and local coherent tune shift.

## 2. Microwave instability

The rule of thumb in the transverse microwave stability is that the coherent tune shift should not exceed the incoherent tune spread. The coherent tune shift comes mainly from the space charge image effect and the broadband impedance effect. This tune shift has been frequently measured in the operated machines.

The incoherent tune spread can be written as

$$\Delta\nu_{inc} = ((n+\nu_0)\eta - \xi\nu_0)\frac{\Delta p}{p} + \Delta\nu_{other} \tag{3}$$

where the slippage ($\eta$) and chromatic ($\xi$) effects are momentum dependent. Usually, $\Delta\nu_{other}$ includes the amplitude dependent octupolar tune spread and the synchrotron tune. A question often raised, is the space charge incoherent tune spread effective in damping the transverse microwave instability?

A review of the existing low and medium energy proton synchrotrons has shown that excluding the space charge incoherent tune spread, the coherent tune shifts consistently exceed the incoherent tune spread, which consists of chromatic, slippage, and synchrotron tune spread, but not the octupolar tune spread.

|  | ISIS | AGS B | PSR | AGS |  |
|---|---|---|---|---|---|
| $N$ | 4 | 2 | 3 | 6 | $10^{13}$ |
| $B_f$ | 1 | 0.4 | 0.4 | 0.3 |  |
| $E_k$ | 0.07 | 0.2 | 0.8 | 1.55 | $GeV$ |
| $\xi$ | -1.4 | -0.2 | -0.2 | -0.2 |  |
| $\Delta p/p$ | 0.2 | 0.7 | 0.34 | 0.4 | % |
| $\Delta\nu_{wall}$ | 10.9 | 5.63 | 0.77 | 1.77 | $10^{-2}$ |
| $\Delta\nu_{BB}$ | 0.73 | 0.74 | 0.35 | 1.95 | $10^{-2}$ |
| $\Delta\nu_{coh,total}$ | **11.63** | **6.37** | **1.12** | **3.84** | $10^{-2}$ |
| $|\xi|\nu_0\Delta p/p$ | 1.09 | 0.69 | 0.14 | 0.71 | $10^{-2}$ |
| $\eta\Delta p/p$ | 0.16 | 0.45 | 0.06 | 0.05 | $10^{-2}$ |
| $\Delta\nu_S$ | 0 | 0.3 | 0.04 | 0.27 | $10^{-2}$ |
| $\Delta\nu_{inc,total}$ | **1.25** | **1.44** | **0.24** | **1.03** | $10^{-2}$ |

Table 3. Coherent and Incoherent Tunes

Consider the normal operations of PSR, AGS and its Booster at the injection energies, and also a study performed at the ISIS for coasting beam at the injection energy. The broadband impedance is assumed to be $j30$ $\Omega$ for all the machines. For convenience, the mode $n = 1$ is used in the calculation of the slippage tune spread.

In Table 3, it is shown that for all these machines, the coherent tune shift is much larger than the incoherent tune spread. The gap is larger if the energy is lower.

The operations of these machines over the years, however, have not been hampered by the transverse microwave instability. It is of interest, therefore, to consider the role played by space charge tune spread in damping the transverse microwave instability.

## 3. Effect of space charge incoherent tune spread

With high intensities, the proton beam particle distribution is Gaussian in transverse. This distribution yields large tune spread, which is betatron amplitude dependent. In general, the particle with small betatron amplitude has large tune shift, and vise versa. For bunched beams, also the particle at the azimuthal center has larger tune shift, etc.

On the other hand, the coherent tune, caused by the image effect and the broadband impedance, is in general depressed into the same direction of the incoherent tune shifts.

The space charge incoherent tune spread, therefore, is likely damping the microwave instabilities, especially for the strong instabilities. Due to the averaging effect, the rigid dipole motion and probably higher order mode motion with slow growth rate are less likely to be damped by this mechanism.

Several complications relevant to this damping mechanism are presented as follows.

1. The coherent tune shift is not always depressed. The examples include the image effect for non-circular chamber, and the broadband impedance effect under certain chamber geometries.

2. The complex particle distribution in the tune diagram implies large uncertainties in the damping mechanism.

3. Since there is no tune spread for the uniform distribution of the particle in phase space, the simulations using uniform distributions may be qualitatively misleading.

## 4. AGS experience

The transverse microwave instability is only presented at the high energy end of the AGS cycle, well above transition, at around 20 $GeV$. In Fig. 5A, the measured bunch length for the AGS cycle is shown, which is lengthened by the VHF cavity dilution in several spots in the cycle. At the injection, it helps to reduce the slow loss. Immediately above the transition, it is for longitudinal reasons. Only well above transition, from 0.8 second, the VHF dilution is necessary to suppress the transverse microwave instabilities. The bunch length according to the phase damping is also shown for comparison.

In Fig. 5B, it is shown that at the injection energy, the required tune spread, which equals to the space charge image and broadband impedance coherent tune shift, is much larger than the combined tune spread of the slippage, chromaticity, and the synchrotron tune. Yet, the machine is very stable in transverse. This might be explained by the large space charge incoherent tune spread at the injection, also shown in Fig. 5B.

At the high energy end, the space charge incoherent tune spread is reduced rapidly to comparable with the required tune spread. The transverse microwave instability may, therefore, develop.

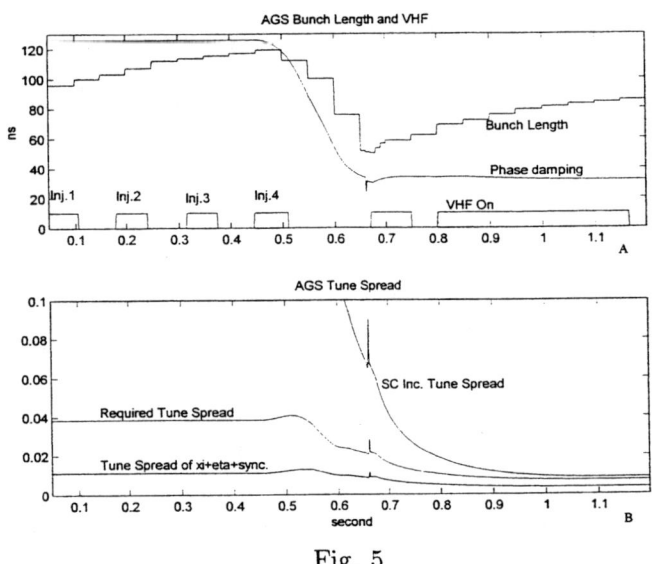

Fig. 5

## 5. Conclusion

Based on the observations of the low and medium energy proton synchrotrons, and the experiences at the AGS and the Booster, we may conclude that the transverse microwave instabilities will not take place at the SNS storage ring.

## IV. PSR Instability

The PSR type instability is quite different from the conventional transverse instabilities, and it is important to assure that the SNS storage ring will not be hampered by this type of instabilities. The study of the PSR instability at the BNL consists of several aspects. These include the analysis of the damping mechanism with respect to this type of instability, the issue of the gap clearing, and the secondary electron productions.

## A. The e-p instability

The high proton beam potential at the PSR is blamed to attract and probably to accelerate electrons. Therefore, it is one of the main culprits of the e-p type oscillation. It is straightforward to calculate the potential well for the PSR, the SNS storage ring for 1 $MW$ and 2 $MW$ operations, as shown in Table 4.

|  | PSR | SNS | SNS |  |
|---|---|---|---|---|
| $N$ | 4 | 10.4 | 20.8 | $10^{13}$ |
| $R$ | 14.35 | 35.1 | 35.1 | $m$ |
| $B_f$ | 0.4 | 0.4 | 0.4 |  |
| $b$ | 5 | 10 | 10 | $cm$ |
| $a = \sqrt{2}\sigma$ | 1.2 | 2.4 | 2.4 | $cm$ |
| $V_{pot.}$ | 6 | 6.5 | 13 | $KV$ |

Table 4. Potential well

It is shown that the potential well of the 1 $MW$ SNS storage ring is about the same as the PSR. For 2 $MW$ SNS, it is about twice as high.

One also notice that for e-p type instability, the space charge incoherent tune spread is not effective in damping. The reason is as follows.

1. The electron induced proton beam coherent tune shift is not depressed as the image effect or broadband impedance effect. Instead, with the electrons in the proton beam, the coherent tune tends to move out of the space charge incoherent tune spread, in the tune diagram.

2. If multipacting takes place, the neutralization factor could be large. At the SNS, the neutralization factor of $\eta_{neu} = 0.23$ could entirely offset the space charge effect, leading to zero tune spread.

These results have shown the existence of the mechanism of the e-p instability at the SNS storage ring. However, with the same importance, the damping mechanism needs also to be studied by comparing the two machines.

## B. Damping mechanism of momentum spread

Numerous evidences at the PSR have showed that the beam momentum spread is important in damping the e-p instability.

1. At the PSR, threshold is proportional to RF voltage $V_{RF}$. From the relation [21],
$$N \propto n\Delta p/p \propto \sqrt{N}\sqrt{V_{RF}} \qquad (4)$$
where $n$ is the electron bouncing mode number, we get
$$N \propto V_{RF} \qquad (5)$$
This implies also
$$N \propto (\Delta p/p)^2 \qquad (6)$$

2. Instability improved by inserting ferrite rings in the beam pipe, which cleared the gap, but also increased beam momentum spread.

3. Coasting beam threshold increases with the larger Linac beam momentum spread.

4. Significant chromatic effect has been recently shown in the study, which is related with the beam momentum spread.

5. Double RF study showed no change on the threshold. The peak current is reduced applying the second harmonic RF, but the beam momentum spread also reduced. Two effects may offset.

6. Increasing the bare tune by one unit improved the instability. It was explained by the effect of $\xi\nu_0\Delta p/p$.

The typical beam momentum spread at the SNS is about twice as much as the PSR. Also, for 2 $MW$ SNS, the electron bouncing frequency is about twice as high as the PSR. Given the similar slippage factor, the tune spread at the 2 $MW$ SNS is about 4 times as large as the PSR. These are shown in Table 5. Inserting the negative impact of the 2 $MW$ SNS high potential well, the factor of $\Delta\nu/V_{pot.}$ at the 2 $MW$ SNS is still about twice as high as the PSR. For the 1 $MW$ SNS, this factor is about 3 times in favor of the SNS.

|  | PSR | SNS, 1MW | SNS, 2MW |  |
|---|---|---|---|---|
| $\Delta p/p$ | 0.34 | 0.7 | 0.7 | % |
| $n$ | 60 | 86 | 120 |  |
| $\eta$ | -0.188 | -0.193 | -0.193 |  |
| $\Delta\nu = n\eta\Delta p/p$ | 3.8 | 11.6 | 16.2 | % |
| $\Delta\nu/V_{pot.}$ | 0.63 | 1.78 | 1.25 |  |

Table 5: Comparison of Tune Spread due to $\Delta p/p$

It is, therefore, possible that the e-p instability will not be seen at the SNS. Efforts at the SNS, however, go on for better understanding of the e-p type instability, damping mechanism, and other studies.

## C. Other studies

Other studies toward understanding and preventing the e-p type instability at the SNS are summarized as follows.

1. Gap clearing

    - As mentioned previously, in addition to the high RF voltage used in the stacking, a ramping of RF voltage from 20 $KV$ to 40 $KV$ at the end of stacking is proposed [22].
    - A gap cleaning kicker is proposed [23].
    - The effect of longitudinal impedance of the extraction kickers is under study.

2. Secondary electron (SE) production

    - The projectile scraping effect in SE production is verified in crashing the gold beam into the Booster injection septum. Translated equivalent SNS proton SE production rate was about 27 [24].
    - A systematic study of this effect at the BNL Tandem van de Graaff has been performed, using gold, oxygen, and proton ions. The scraping effect of $1/\cos\theta^{-1.15}$ has been obtained [25].
    - Also in the Tandem study, it has been shown that using the serrated surface for the collimator, the SE yield is reduced by more than a factor of 10 [25].
    - Further studies for different surface, coatings, conditioned surface are under planning.
    - Possible multipacting effect for the proton beams in the PSR and SNS is under studying [26].

## V. Research Bibliography

[1] W.T. Weng, AIP Conference Proceedings, 448, Shelter Island, New York, 1998.

[2] R. Macek, ibid.

[3] R.E. Shafer, IEEE Trans. NS, Vol.NS-32, No.5, p.1933, 1985.

[4] K.Y. Ng, Particle Accelerators, Vol.23, p.93, 1988.

[5] S.S. Kurennoy and G.V. Stupakov, Particle Accel., Vol.45, p.95, 1994.

[6] D. Boussard and J. Gareyte, SPS/AC, Impr. Report No.181, CERN 1980.

[7] T. Linnecar and E. Shaposhnikova, CERN-SL-96-45, RF, July, 1996.

[8] F. Pedersen and E. Raka, IEEE Trans. NS, Vol.NS-26, No.3, p.3592, 1979.

[9] R. Cappi, these proceedings.

[10] A. Chao, Physics of Collective Beam Instabilities in High Energy Accelerators, Wiley, New York, 1994.

[11] G. Nassibian and F. Sacherer, Nucl. Inst. Meth. Vol.159, p.21, 1979.

[12] G. Lambertson, Workshop on RHIC Performance, March, 1988.

[13] W.Z. Meng, private communication.

[14] J. Griffin et. al., FN-661, 1997.

[15] K. Koba, et. al., KEK Preprint 97-173, 1997.

[16] K.Y. Ng, Fermilab-FN-659, July 7, 1997.

[17] W.K. vanAsselt and Y.Y. Lee, Proc. of Part. Accel. Conf., p.881, 1991.

[18] F. Voelker and G. Lambertson, Proc. of Part. Accel. Conf., p.851, 1989.

[19] D. Neuffer et al, Nucl. Instr. and Meth. A321, p.1, 1992.

[20] J. Gareyte, Frontier of Particle Beams: Intensity Limitations, Springer-Verlag, New York, 1990.

[21] R. Macek, these proceedings.

[22] S.Y. Zhang and W.T. Weng, Proc. of Euro. Part. Accel. Conf., p.1139, 1998, Stockholm, Sweden.

[23] J. Wei, private communication.

[24] S.Y. Zhang, Proc. of Part. Accel. Conf., p.3297, 1999, New York, New York.

[25] P. Thieberger et. al., in preparation.

[26] M. Blaskiewicz, Proc. of Part. Accel. Conf., p.1611, 1999, New York, New York.

# Instability Issues for the ESS Linac and Rings

## G H Rees

*Rutherford Appleton Laboratory, CLRC, UK*

**Abstract.** Comments are made on beam instability issues in the ESS linac and rings. The topics of interest in the linac are halo generation in the absence and presence of machine imperfections, and also the stability of the momentum ramping of the output beam. In the case of the rings, the main concern is for fast coherent transverse instabilities due to the combined effect of coupled electron-proton oscillations and interaction with the wall impedances.

## INTRODUCTION

The reference designs for the ESS linac and rings have been changed since the initial proposal [1]. The main changes for the linac include an increase of the energy for chopping from 2 to 2.5 MeV and of the energy for funnelling from 5 to 20 MeV, with a new front end frequency of 280 MHz, and a frequency of 560 MHz after funnelling. The most difficult beam dynamics' areas are the chopping and funnelling lines [2], with a compromise required in the choice of the transverse beam emittances; a small value is desirable for the chopper, but a somewhat enhanced value to limit the tune depressions in the funnel and drift tube linac (DTL). These criteria set the specification for the transverse emittances to be provided by the ion sources and RFQ linacs. After the chopper all sections may be equipartitioned, but there is some flexibility in setting the equipartition ratios for the 560 MHz coupled cavity drift tube linac (CCDTL) and coupled cavity linacs (CCL1 and CCL2). Beam halo generation results from non-linearities and from mismatching due to machine imperfections, which cause coherent dipole and envelope motion. The envelope oscillations are a combination of three possible coupled transverse-longitudinal coherent modes [3]. Momentum ramping is achieved by phase modulation of a pair of CCL2 type cavities situated at an appropriate distance after CCL2. Errors in the amplitude and phase of the linac accelerating fields lead to errors in the range of the output momentum ramping, and field tolerances are set to limit the momentum range. The ramping cavities also act to reduce the instantaneous beam momentum spread.

For the accumulator rings, the circumference has been increased to reduce the number of injected turns and hence to reduce the number of proton foil traverses and foil temperature. Two new ring designs are under consideration, one in which the previous lattice is modified by including an additional straight section in each superperiod, and one in which the number of superperiods is reduced from three to two, but with the superperiod modified to allow the possibility of either foil stripping or laser stripping injection. The latter requires a straight section containing a wide aperture undulator magnet and involves a considerable change from the original superperiod. Instability issues are very similar for the different lattices, and the relevant topics have already been outlined in the report on ISIS [4] in these proceedings. Relevant parameters for the ESS rings are derived.

## LINAC DESIGN ISSUES

Individual sections of the linac have smooth changes of focusing and it is advantageous if the transitions between the sections have the same feature. Focusing patterns in the transverse planes may be defined in terms of $\beta\lambda$, where $\beta$ is the ion velocity relative to the velocity of light and $\lambda$ is the free space wavelength at the frequency of the CCL. In this notation for the ESS, the FODO focusing cells of the individual sections have lengths as follows: $2\beta\lambda$ for the RFQ, $4\beta\lambda$ for the DTL, $9\beta\lambda$ for the funnel, $10\beta\lambda$ for the CCDTL, $11\beta\lambda$ for CCL1 and $12\beta\lambda$ for CCL2. The choice of a $4\beta\lambda$ FODO for the DTL rather than an $8\beta\lambda$ FOFODODO is to allow smoother matching from the RFQ to the chopper and from the chopper to the DTL.

Chopper and funnel sections have been designed [2] with lengths reduced as far as is practical to restrict the bunch extents during the beam debunching and rebunching intervals. The funnel is as described in [2] but the choppers have been shortened and redesigned to spread the heat loads over an extended beam loss collector in each line, each capable of dissipating 2 kW of deflected beam power. The design philosophy of keeping equal beam aspect ratios, on average, in each cell has been retained. The reason why this restricts the development of beam halo may be seen from the form of the transverse space charge field for a non-linear beam distribution. The ratio of the linear to the non-linear field components varies as a function of the horizontal to vertical beam aspect ratio, a/b, while the linear effect on the beam is proportional to the $\beta_h$ - function and the non-linear effects to $\beta_h^2$ or $\beta_h\beta_v$. In the case of the 2-D parabolic transverse density distribution, for example, the transverse space charge electric field, $E_x$, at position (x,y) for a line charge density $(\div 4\pi\varepsilon_o)$ of $\lambda$ is:

$$E_x = \frac{8\lambda x}{a(a+b)} \left[1 - \frac{x^2(2a+b)}{3a^2(a+b)} - \frac{y^2}{b(a+b)}\right]$$

A schematic drawing of an ESS chopper line is shown in Figure 1, together with associated lattice functions under maximum space charge conditions (57 mA). The overall length is 40 $\beta\lambda$ (~ 1.56 m), and the central section has two doublet cells with regular lattice functions. In the straight section of the upstream cell, a push-pull travelling wave horizontal beam deflector is located and, in the downstream cell is a

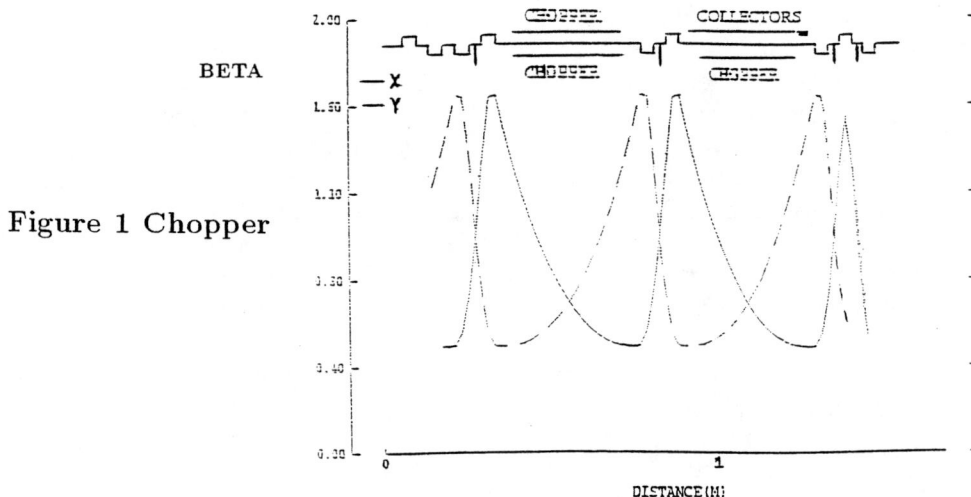

Figure 1 Chopper

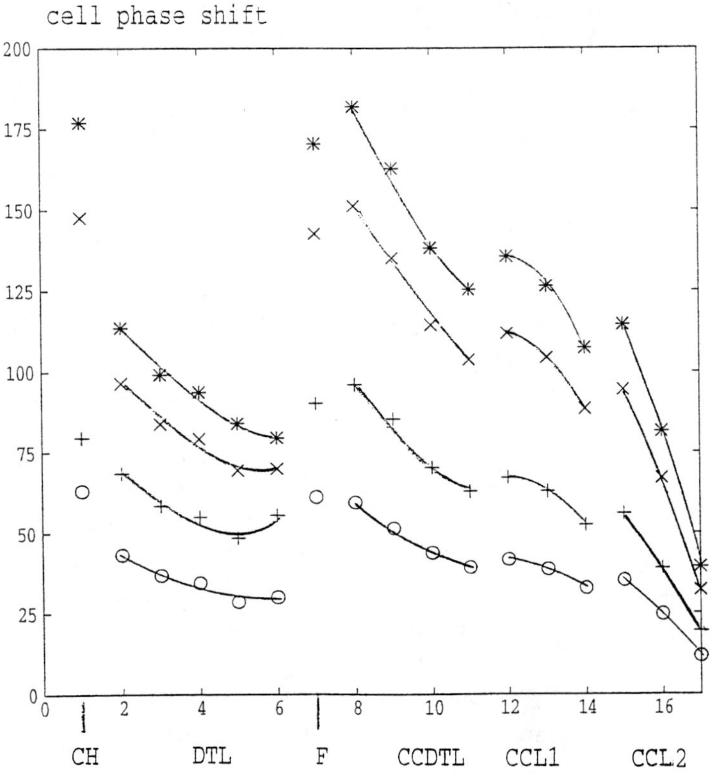

Figure 2 Coherent Modes

○ low mode
+ quadrupole mode
× high mode
∗ zero current quadrupole mode

single travelling wave deflector opposite to a horizontal beam loss collector. Deflection within the first unit is accommodated as there is a decreasing horizontal $\beta$-function in the direction of beam motion. Deflection within the second unit results in beam interception along the entire length of the loss collectors, thus distributing the heat load. Each chopper line has nine quadrupoles and four buncher cavities, as indicated in Figure 1, where the four vertical lines represent the locations of the cavities. Input matching from the RFQ is achieved using the initial four quadrupoles in the line and the initial buncher cavity, together with some adjustment of the RFQ output cell. Output matching to the following DTL is obtained using the last two quadrupoles and bunchers in the line together with the first two DTL quadrupoles.

The normalised transverse rms emittances specified for the RFQ are 0.225 ($\pi$) mm mr, and RFQ studies have indicated high transmission efficiency for these emittances when the related normalised longitudinal rms emittance is a factor of 1.83 larger. Design constraints in the chopper and funnel restrict the values that may be obtained for the transverse to longitudinal energy partitioning ratios. The ratios realised are 0.74 for the chopper and 1.10 for the funnel. A smooth transition to equipartitioning is then arranged for the DTL. After the funnel, the entire linac may be equipartitioned though there is flexibility that allows a smooth departure from this condition. Initial error studies have shown satisfactory results for equipartition, but comparisons are planned for other partitioning ratios. The ESS reference design assumes room temperature structures, though the opportunity exists of introducing superconducting cavities for the 250 to 1334 MeV CCL2 stage.

## COUPLED TRANSVERSE-LONGITUDINAL COHERENT MODES

Machine imperfections lead to coherent dipole, quadrupole and higher order mode motions, with the quadrupole oscillations one of the main sources of halo generation. Dipole motion results in the transverse plane for quadrupole misalignments and in the longitudinal plane for amplitude and phase errors in the radio frequency (rf) accelerating fields. Errors in quadrupole settings and rf focusing lead to a combination of three possible coupled transverse-longitudinal coherent modes; approximate expressions for the three modes are given in the reference [3] for the case of equal transverse tunes. Modified expressions may also be found for a limited range of unequal tunes, $\sigma_{yo} = \sigma_{xo}(1+\alpha)$, with $|\alpha| < 0.1$. The formulae overleaf may be used to find the trend of the coherent frequencies $\sigma_{env,q}$, $\sigma_{env,h}$ and $\sigma_{env,l}$ at increasing $|\alpha|$, with the associated nomenclature:

m =1 for equipartitioning, or m=2 for equal transverse emittances,
$\sigma_{yo}$ and $\sigma_{xo}$ the zero current transverse phase shifts per cell and
$\sigma_y$ and $\sigma_x$ the depressed tunes for linear space charge forces.

$$(\sigma_{env,q})^2 = (2\sigma_x)^2 (1+\alpha)^{1/m}$$

$$(\sigma_{env,h})^2 = A_1 + B_1$$

$$(\sigma_{env,l})^2 = A_1 - B_1$$

$$A_1 = (1+a)^{1/m}\left(\sigma_{xo}^2 + \sigma_x^2\right) + \frac{1}{2}\sigma_{lo}^2 + \frac{3}{2}\sigma_l^2$$

$$B_1^2 = \left[(1+\alpha)^{1/m}\left(\sigma_{xo}^2 + \sigma_x^2\right) - \frac{1}{2}\sigma_{lo}^2 - \frac{3}{2}\sigma_l^2\right]^2 +$$

$$(1+\alpha)^{1/m} 2\left(\sigma_{xo}^2 - \sigma_x^2\right)\left(\sigma_{lo}^2 - \sigma_l^2\right)$$

Values of the coherent frequencies are given in Figure 2 for the case of $\alpha = 0$ in the ESS at various stages in the chopper, funnel and in an equipartioned DTL, CCDTL, CCL1 and CCL2. The coherent shifts per cell (tunes) decrease smoothly through the individual linac sections but increase at each linac transition after the DTL. The relative mode spacing ratios do not change significantly and are set mainly by the ratio of the transverse and longitudinal rms beam emittances. The case of unequal transverse tunes remains to be investigated in tracking studies.

## HALO DEVELOPMENT

In the absence of machine imperfections, there is a gradual growth of the transverse and longitudinal rms emittances along the linac, even with nominally well matched transitions between the different stages. At the same time, there is a gradual growth of the beam halo. This emittance and halo growth appears inevitable because of some non-linear synchrotron motion at low energy in the DTL, together with a variation of the rf transverse defocusing forces along the extent of the bunches, and a change of the distributions as the beam tends towards a state of thermodynamic equilibrium. Growth in the emittance and halo is enhanced in the presence of machine errors, particularly those which cause mismatching and subsequent envelope oscillations. Different linac designs may be compared for their emittance and halo growth and also for their overall sensitivity to errors, including those which result in errors in output momentum and those which result in beam orbit distortions.

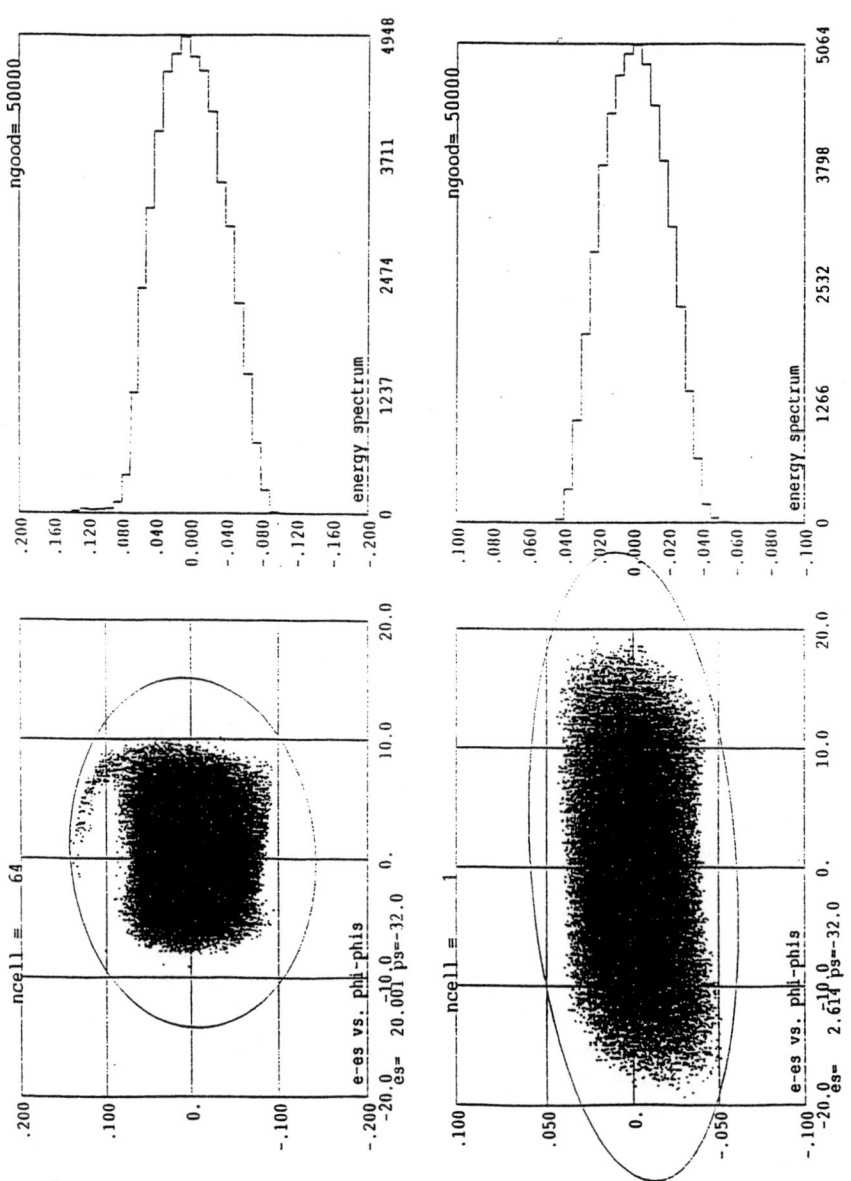

Figure 3 Longitudinal Halo Development in DTL

Various simulations have been made for the ESS linac and its associated external momentum reduction and ramping section. The simulations have involved the tracking of 50,000 or 100,000 particles through the chopper, DTL, funnel, CCDTL, CCL1, CCL2 and external line. Input beam distributions have been assumed which have the rms emittance values established in RFQ simulations, but with a 6-D waterbag distribution of uniform density for reference purposes instead of the actual RFQ output. Transmission efficiencies of 100% have been found in all cases, even for specified machine errors, suggesting that actual losses should be less than one part in $10^5$.

In passage through the chopper section, there is some redistribution of the particles in longitudinal phase space but very little observed halo and emittance growth. At entry into the DTL, the bunch phase extent is $\pm 18°$, resulting in non-linear longitudinal motion for the chosen synchronous phase angle, $\varphi$, of $-32°$. Phase space plots at the entry and exit of the DTL are given in Figure 3 and some distortion is apparent in the exit plot, with the development of a halo tail. The tail transmits through the remainder of the linac and is joined by other halo particles. Changing $\varphi$ to $-36°$ reduces the distortion of the phase space but does not eliminate the tail. The tail lies closer to the core, but at later stages of the linac some larger longitudinal but smaller transverse halo oscillation amplitudes result. The $\phi$ value of $-32°$ therefore appears adequate, particularly as it may prove possible to remove some of the halo tail in the funnel section, where there is an accessible region of finite dispersion.

Values found for the rms emittance growth and halo development are given in the following table, where halo oscillation amplitudes are given in terms of $\sigma$, the rms amplitude, and where rms emittance and halo growth are given for two positions, one for the linac output and one for the end of the momentum ramping line (significant growths continue in the external line). The first section of the Table gives the results for nominal six parameter matching, with the final line including the effect of additional rf system errors (a random distribution of 0.5% maximum in field amplitude and a random distribution of 0.5 degrees maximum in phase). These are chosen as the acceptable tolerances for the linac; the maximum effect found for twice these tolerances is shown by the energy and phase excursions of the dipole mode oscillations of Figure 4. The second part of the Table shows the additional effect of coupled envelope oscillations, and the case given is that for a single 10% quadrupole magnet error just ahead of the CCDTL. Significant emittance and halo growth occurs in he transverse planes for this case, most within the CCDTL, as may be seen in Figure 5. The mean beam current assumed is 2 x 57 mA, and at this level, no particle loss is found in the tracking simulations. The results are to be compared in the future with similar simulations of a non-equipartitioned linac.

## TABLE OF EMITTANCE GROWTHS FOR 2*57 mA

| a. | Case of nominal six parameter matching | | |
|---|---|---|---|
| | Longitudinal: | rms emittance growth | 10% , 12% |
| | | max halo amplitude | $7\sigma$ , $11\sigma$ |
| | Transverse: | rms emittance growth | 11.5% , 11.5% |
| | | max halo amplitude | $7\sigma$ , $7\sigma$ |
| | RF errors (.5% , .5°) | additional rms growth | 5% , 5% |
| b. | Case of a single 10% quad error just ahead of the CCDTL | | |
| | Longitudinal: | rms emittance growth | 15%, 33% |
| | | max halo amplitude | $6.5\sigma$ , $15\sigma$ |
| | Transverse: | rms emittance growth | 37% , 41% |
| | | max halo amplitude | $11\sigma$ , $13\sigma$ |

## MOMENTUM SPREAD MINIMISATION AND RAMPING

The section of beam line after the linac includes three CCL2 type cavities, one at the input end of the line and two more after an additional distance , $\ell$ , of approximately 70 m. The first external cavity is powered and phased to give some increase of the beam momentum spread and to hasten the debunching of the beam. In the following section of line, the beam spreads in phase, tilting in longitudinal phase space, and increasing in $\Delta p/p$ due to the effect of the longitudinal space charge forces. The second and third cavities are a distance , $\ell$ , downstream from the first, with the value of $\ell$ given by:

$$\ell * V = \beta^3 \gamma^3 (E_o/e) (c/2\pi f)$$

where   V   is the sum of the voltages on these second and third cavities,
           c   is the velocity of light,
           $f$   is the CCL2 cavity frequency of 560 MHz, and
           $e$, $E_o$ are the charge and rest mass of the ions.

The advantage of this arrangement is that any error in mean energy of the beam after the first external cavity is cancelled by the subsequent actions of the second and third cavities. If, however, there is an error in phase at the first cavity, this leads to a subsequent error in beam energy due to the effects of the phase error at the second and third cavities.

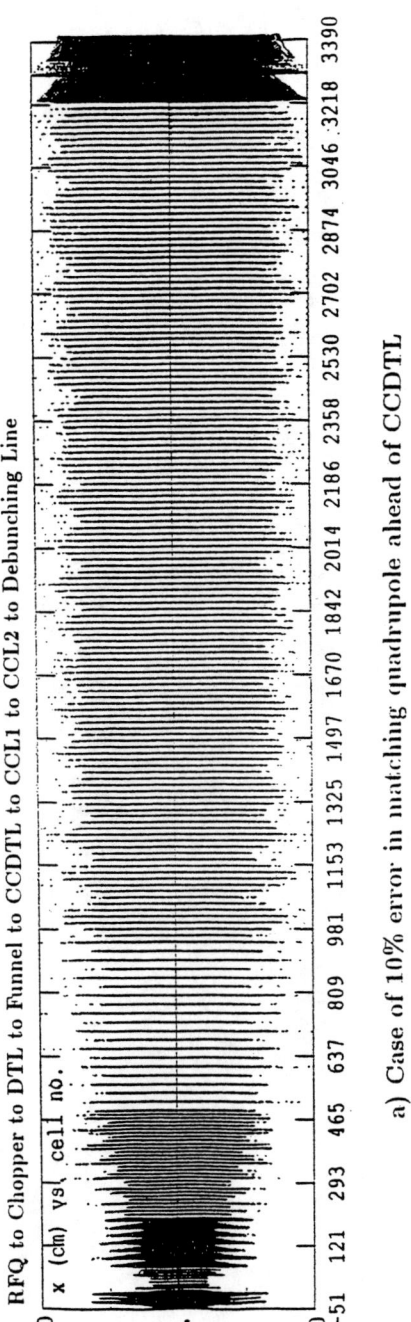

a) Case of 10% error in matching quadrupole ahead of CCDTL

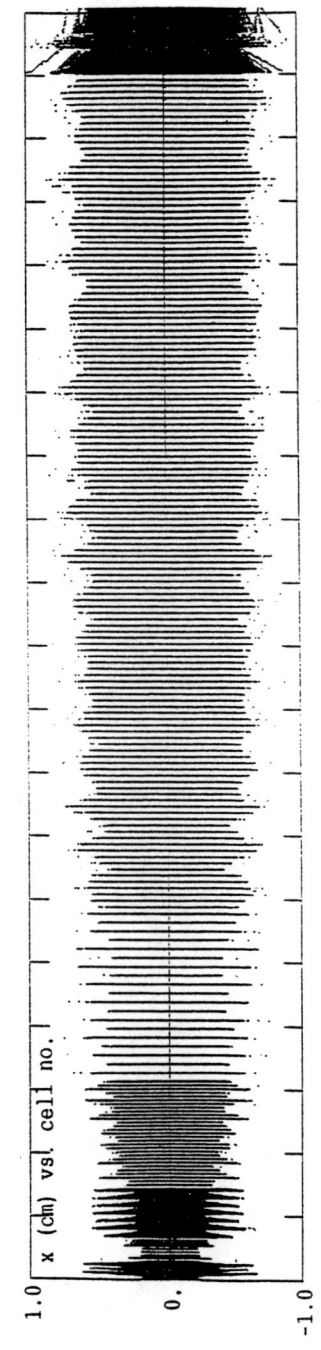

b) Case of nominal six parameter matching

Figure 5 Coherent Envelope Mode Motion

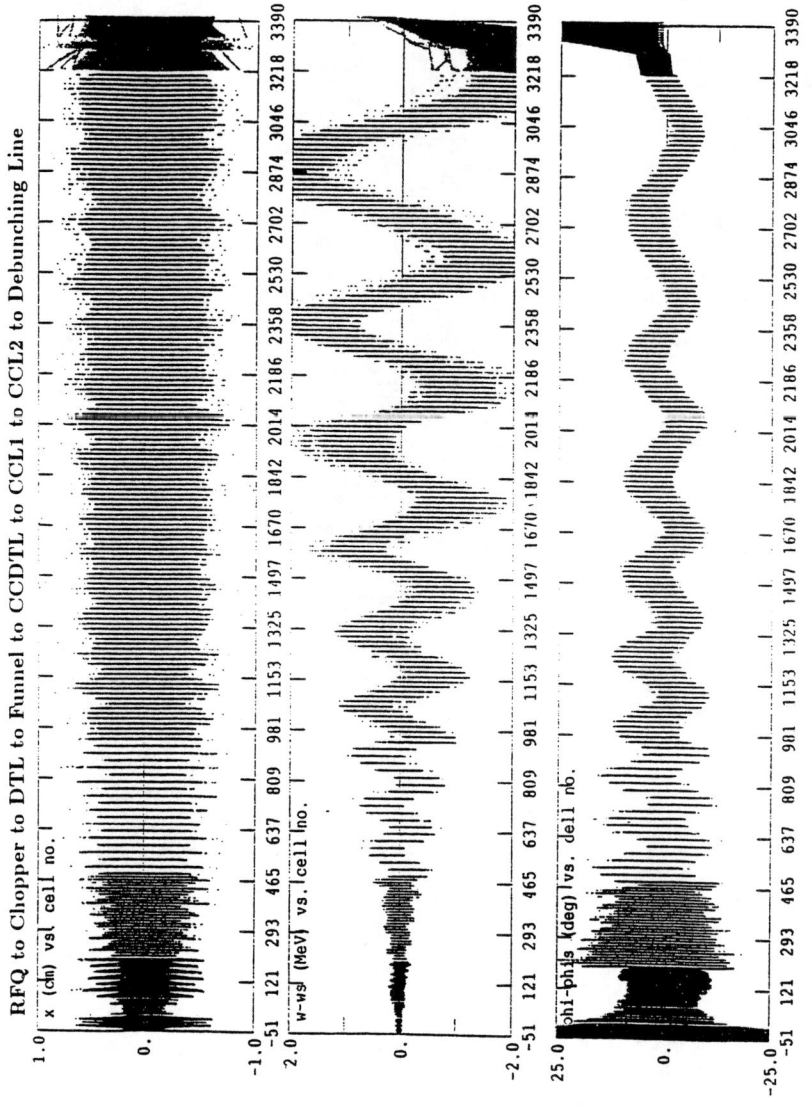

Figure 4 Randomly Distributed RF Field Errors, $\delta E < 1\%$, $\Delta\phi < 1°$

The value of V is selected so that the tilted ellipse in momentum phase space is rotated to minimise the final value of the beam momentum spread. A slow ramping in phase is also introduced in the two downstream cavities to provide the required ramping of momentum for the H⁻ injection painting in the accumulator rings. The range of momentum reduction and momentum ramping is shown in Figure 6, with the ellipses, drawn external to the main core of beam particles, representing the maximum longitudinal halo particles. The requirements for injection painting are also shown in Figure 6. The elongated momentum ellipses in longitudinal phase space are in the upright position so that there is little shearing of the beam in real space as it progresses down the following, nearly isochronous, high dispersion, achromatic collimation line which is ahead of the rings.

## INSTABILITY TOPICS FOR THE ESS RINGS

Five topics are of the most relevance for ESS ring instabilities:

1. Bunched beam longitudinal instability due to the space charge and wall impedance,
2. Coupled control loops and beam loading instability,
3. Fast transverse instability due to the space charge and wall impedance,
4. Head-tail instability, and
5. Electron-proton instability.

The longitudinal and transverse space charge impedances are large, though some reduction of both may be obtained by a smooth profiling of the vacuum chamber walls at a constant ratio to the beam dimensions. Practical values of momentum spread, $\Delta p/p$, may be found which give stability for the first item listed but not for the third. Stability for the second item may be achieved by appropriate use of beam feed-forward or by the application of rf feedback The fourth item, the head-tail instability, does not have time to develop as the injection interval is less than the synchrotron period and, for the planned natural values of the chromaticities, the possible head-tail modes are of very high order.

Thus the items of most concern are fast transverse instabilities due to the combined effect of transverse space charge forces, transverse wall impedances and coupled electron-proton motion. The space charge forces remove the Landau Damping, and the coherent transverse motion may become antidamped due to both the transverse wall impedance and the electron-proton interaction. The resistive components of the impedance and the delayed effects in the electron-proton motion provide the antidamping forces. Some stabilisation against these effects may prove to be necessary.

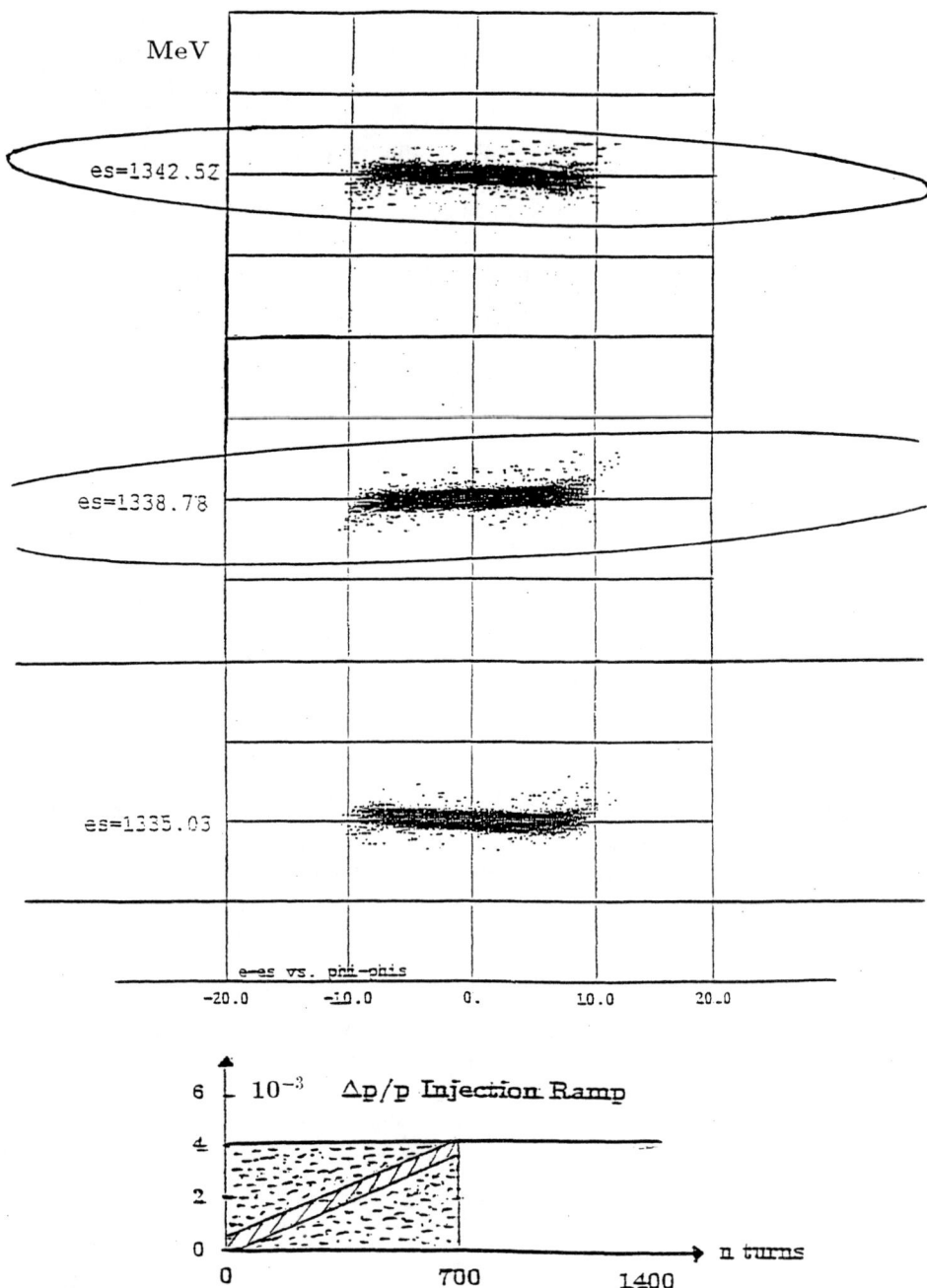

Figure 6 Debunched and Ramped Momentum Range

## LONGITUDINAL STABILITY

Bunched beam longitudinal stability may be achieved by appropriate choices of the rf containing fields, the rf control loops and the momentum painting of the injected beam. An approximate value for the required rf voltage per turn, V, may be found by assuming an elliptical 2-D longitudinal density distribution (the well known Hofmann-Pedersen distribution). The condition for stability for this distribution is:

$$V = gNe\, h^2 / \left[2\varepsilon_o R \gamma^2\, F\, F_1\, \eta_{sc}\right]$$

Where  $g$   is the longitudinal space charge g parameter,
$N$   is the number of particles per bunch,
$e$   is the electronic charge,
$h$   is the rf harmonic number,
$\varepsilon_o$   is the permittivity of free space,
$R$   is the mean ring radius,
$\gamma$   is the relativistic energy factor,
$F$   is the normalising factor of the elliptical distribution,
$F_1$   is the longitudinal instability from factor, and
$\eta_{sc}$   is the ratio of the longitudinal space charge to rf focusing forces.

Stability occurs for $\eta_{sc}$ values of less than 0.4, though it may be possible to exceed this value when below transition energy, as is the case for the ESS rings. An $\eta_{sc}$ value of 1.0 corresponds to the disappearance of the phase stability region; $\eta_{sc}$ values between 0.4 and 1.0 are of possible interest for bunch compression rings; for the ESS, however, $\eta_{sc}$ is kept less than 0.4.

Different rf voltage waveforms may be compared for the ESS rings using the stability formula. The normalising function is defined for each waveform:

| | | |
|---|---|---|
| Bunch extremities | $\varphi_2 = -\phi_1$ | $= 0.691\,\pi$ |
| Barrier extremities | $\varphi_3 =$ | $\pm\, 0.55\,\pi$ |
| Single rf harmonic | $F =$ | $2(\sin\varphi_2 - \varphi_2 \cos\varphi_2)$ |
| Dual rf harmonic | $FF =$ | $F - 0.25\,\sigma\, F(2\varphi_2)$ |
| Barrier bucket | $FB =$ | $\varphi_2^2 - \varphi_3^2$ |

Then, assuming $g = 2.0$, $h = 1$, $\sigma = 0.5$, $F_1 = 1$, T = 1.334 GeV and $\eta_{sc} = 0.34$:

|                          | Single Harmonic | Dual Harmonic | Barrier Bucket |
|--------------------------|-----------------|---------------|----------------|
| Normalising factor       | 4.1             | 3.94          | 1.727          |
| Bunching factor          | 0.417           | 0.512         | 0.621          |
| V for R = 26.0 m         | 19.8            | 20.6          | ± 47.0 kV      |
| V for R = 39.0 m         | 13.2            | 13.7          | ± 31.3 kV      |

The case of R = 26.0 m is for the earlier ESS ring, and R = 39.0 m for the later one. A dual harmonic rf system has been selected as the beam bunching factor is adequate and the extra complexity of the barrier bucket waveform does not appear justified. Injection painting is improved by the use of a dual rather than a single rf harmonic system, and tracking studies for the earlier ring gave a bunching factor of 0.47, after optimisation, close to that estimated for the elliptical distribution. For the parameters chosen the final beam momentum spread, $\Delta p/p$, is approximately $\pm .0.6\%$

Stability for the rf control loops under the heavy rf beam loading is envisioned with the aid of a one turn delay beam feed-forward system. A fixed tuning of the h = 1 cavities is proposed, corresponding to reactive beam loading compensation for a beam current of approximately 70% of the maximum level. At the same time, a correct programming of the voltages on the h = 1 cavities is provided for the case of zero beam current. The beam feed-forward system acts to cancel the effect of the fundamental beam loading as it is progressively increased. This arrangement limits the generator currents for the h = 1 system, and differs from that used on the ISIS synchrotron where the cavities are initially tuned to remain on resonance. The second harmonic system has a much reduced beam loading and is provided with continuous reactive beam loading compensation.

## TRANSVERSE COHERENT INSTABILITIES

The possibility of fast transverse instabilities exists in the ESS as the transverse space charge impedance is larger than the coasting beam threshold impedance at the peak circulating beam current, I, of approximately 100 A. This is so far many coasting beam modes, of $n$ values up to ~ 160, for the chosen $\Delta p/p$ value of $\pm$ 0.6%. A much larger $\Delta p/p$ value would be required to provide stability at the natural values of the ring chromomaticities:

$$\frac{RZ_o}{\beta^2\gamma^2}\left(\frac{1}{a^2}-\frac{1}{b^2}\right)|j| \nless \frac{\pi FQE_o\beta\gamma}{eIR}\left|(\xi Q-\eta(n-Q))\frac{\Delta p}{p}\right|$$

where  $Z_o$ is the characteristic impedance of free space,
a and b are the beam and aperture radii,
$\beta$ and $\gamma$ are the relativistic factors,

$\eta$ is equal to $\left(\gamma^{-2} - \gamma_t^{-2}\right)$,

$Q$ is the relevant betatron tune,

$F$ is the form factor of the transverse beam distribution,

$\xi$ is the ring chromomaticity in the relevant plane,

$\Delta p/p$ is the beam momentum spread, full width at half peak, and other parameters have been defined previously.

Concern exists if the predicted instability growth time is much less than either the beam occupancy time in the ring or the period of synchrotron oscillations, that is, if growth times are significantly less than a millisecond. Approximate growth times, neglecting Landau Damping, are:

$$\tau = Qe\gamma Z_o / (Ir_p R_\perp)$$

where  $r_p$ is the classical proton radius, and

$R_\perp$ is the resistive component of the transverse impedance

($\tau = 1$ ms for $R_\perp = 5k\Omega / m$)

The main contributions to $R_\perp$ are due to low frequency component of the resistive wall transverse impedance and higher frequency compnent of the fast extraction kicker magnets. This assumes a smooth vacuum chamber envelope with shielded bellows sections. The chambers are to be profiled at an approximately constant ratio to the beam radii, as at ISIS [4], with rectangular and not elliptical cross sections so a dedicated beam loss collection system may have a greater efficiency. The wall profiling reduces the reactive component of the transverse space charge impedance but increases the resistive part, which varies inversely as the cube of the chamber radius. The largest resistive wall impedance occurs for the lowest ($n - Q$) transverse mode with $n$ the lowest integer value greater than $Q$. Contributions to $R_\perp$ are kept less than 5 $k\Omega/m$ by using low resistivity material, copper or aluminium, for the inner vacuum chamber wall. Secondary electron emission coefficients are minimised by evaporating titanium nitride on the inner surfaces.

The transverse impedance presented by the ESS fast extraction kicker magnets depends on the design adopted. A similar design to that used in ISIS is proposed [1], where the magnet is sub-divided into a number of lumped, push-pull, window frame, ferrite modules, and with the two halves separated by a solid, conducting ground plane. The drive impedance is low, with each half module powered via six (ESS) or seven (ISIS) 50 $\Omega$ cables in parallel. In ISIS, each cable system is back terminated at its input end with the appropriate impedance of 50/7 $\Omega$, and the related pulse forming network is at half this impedance level. The disadvantage of this arrangement is that the thyratron switch has to provide twice the peak current (5000 A at 40 kV). An alternative is under consideration for the ESS with the back termination removed and a saturating inductor inserted between the cable output and the kicker input. In parallel, then, with each kicker magnet, a speed up network is added, a capacitance in parallel with a 5 $\Omega$ resistance. Simulations with the program PSPICE appear very promising, with smooth waveforms and a 99% rise time of 175 ns. The pulse forming network impedance is doubled, the thyratron peak current is halved and the presence of a speed up network ensures a low transverse impedance is presented to the beam. A prototype is required to prove the revised design.

Experience with ISIS suggests the value of $R_\perp$ for its three push-pull kicker magnets is less than 5 k$\Omega$/m. However, the impedance is difficult to measure as it arises due to the leakage inductance of the window frame magnets, as modified by the presence of the central ground plane. The reactive component of the impedance is a function of the beam distribution, and the resistive component depends on both the leakage fields and the external impedances. The central ground plane and low external impedances are required to reduce the levels of the peak beam induced fields.

Equivalent transverse coupling impedances may also be defined for coupled electron-proton interactions. The resistive components arise because of the delays involved in the interaction, with the electrons taking time to respond to the proton displacements, and vice versa. In the PSR at LANL [5], e-p type instabilities are observed with fast growth times, which infer corresponding values of $R_\perp$ of order 50 k$\Omega$/m. Nothing similar is observed on ISIS, either for a coasting or a bunched beam, so the differences between the two rings are of much interest.

Bunched beams only are created in the ESS rings, and the e-p interaction is then more complex than for the coasting beam case. The electrons involved in an e-p instability must have a continually changing oscillation frequency, with oscillation amplitudes greater than those for the protons. The electron frequencies are at a maximum when the peak of the bunch passes and are small or zero when the beam gap is adjacent. A large frequency range is involved, allowing the possibility of a range of $n$ values for $(n-Q)$ coherent proton modes. Values of $n$ up to 160 are potentially unstable for the ESS. Antidamping forces arise because the attractive forces on the protons, due to the electrons, are proportional to the proton coherent displacements of an earlier instant in time.

Longitudinal space charge forces cause a longitudinal drift of the electrons each turn, and electrons may escape transversely to the walls due to their transverse oscillations or due to their transverse velocities during the beam gap interval. Electrons are created continuously due to both ionisation of the residual gas and due to electron emission from the walls in multipactor type effects. This is the reason for the inner coating of titanium nitride on the chamber wall.

## PROFILING OF THE VACUUM CHAMBER WALL

There are disadvantages and advantages to profiling of the vacuum chamber wall. The disadvantages are the added complexity of construction, including that for shielding the bellows sections, and the increase in the resistive component of the transverse wall impedance, which scales inversely as the cube of the chamber radius. However, it is believed that these disadvantages are outweighed by the following advantages:

1. The reduced value for the transverse reactive space charge impedance.
2. The reduced value for the longitudinal space charge impedance, as characterised by the reduced longitudinal g-factor.
3. The reduced values required for the rf containment due to the reduced g-factor.
4. The reduction in the external betatron tune spread required to provide coherent transverse stability.
5. The reduced beam potential, due to the lower g values, for the trapping of the electrons involved in coherent e-p interactions.
6. The easier escape of the (e-p) electrons to the walls, due to their closer proximity.
7. The reduction in the number of secondary electrons produced by primary electrons due to their reduced velocity when striking the closer proximity walls.

The profiling of the vacuum chamber wall, including the shielding of the bellows, is one of the main differences between the PSR and ISIS rings and may be the reason why the latter is not troubled by the e-p instability.

[Note that the workshop impedance working group reported in its summary some different conclusions for profiling of the vacuum chamber]

## POSSIBLE STABILISATION MECHANISMS FOR TRANSVERSE INSTABILITIES

There is some partial stabilisation against fast transverse instabilities due to the finite beam momentum spread and the associated chromatic tune spreads at the natural chromaticity values in the ESS. Further partial stabilisation results due to the variation of the transverse image forces across the beam cross-sections for coherent transverse beam displacements. Estimates are given in [6] for such image force variations, for the case of a square vacuum chamber aperture within which is a beam of circular cross-section.

Storage time in the rings is insufficient for significant transverse head-tail motion to develop. Such motions are also inhibited because only very high order head-tail modes are possible, with mode numbers, m, of approximately 25 in value.

Instabilities grow from noise and the largest noise components are due to the residue of the linac bunch structure in the injected beam. However, the ESS linac bunch frequency is 560 MHz and the fundamental, higher harmonics, and the sidebands due to the chopping of the linac beam at the ring revolution frequency, are all of a sufficiently high frequency to lie in the stable $(n-Q)$ mode region for the ESS rings. It is of interest to note that this is not the case for the PSR ring, with $n \sim 70$ for the linac bunch frequency of $\sim 200$ MHz.

Additional methods for providing transverse stability include:

1. Inclusion of octupole magnets to enhance the beam tune spreads.
2. Application of fast, low delay, coherent feedback.
3. Fast variation of the betatron tunes, as used in ISIS [4].
4. Design of the linac beam chopper for an acceptable number of gap protons in the ring (e-p instability), and
5. Pulsed betatron excitation for fast clearing of protons from the beam gap (e-p instability).

Though space is available to include all these components, and some clearing electrodes may be added in the rings, it still appears prudent to design the ESS rings for lower beam densities $(N/(\overline{B} R \varepsilon))$ than in the PSR, with $\overline{B}$ the beam bunching factor and $\varepsilon$ the transverse emittance.

## REFERENCES

[1] ESS Volume III, The ESS Technical Study, ESS-96-53-M, 1996.

[2] C R Prior, Progress Report on ESS, Shelter Island Workshop, 1998, AIP Proceedings 448, pp 141-151.

[3] A Letchford, K Bongardt and M Pabst, Halo Formation of Bunched Beams in Periodic Focusing Systems. Proceedings of PAC 99, New York, 1999.

[4] G H Rees, Aspects of Beam Stability at ISIS, These Proceedings, 1999.

[5] R J Macek, PSR Experience with Beam Losses, Instabilities and Space Charge Effects. Shelter Island Workshop, 1998, AIP Proceedings 448, pp 116-127.

[6] G H Rees and C R Prior, Image Effects on Crossing an Integer Resonance, Particle Accelerators, 1995, Vol 48, pp 251-257.

# Longitudinal Impedance Tuner Using a New Material, FINEMET

## K. Koba

*KEK-Tanashi, 3-2-1, Midoricho Tanashi-shi, Japan*

**Abstract.** We have succeeded to cancel longitudinal space charge effects using an 'impedance tuner'[1][2][3] at the KEK-PS. The impedance tuner consists of the new material, 'FINEMET'[4][5][6][7], which has a large permeability over the beam spectrum region. The frequency shift of the coherent quadrupole mode is measured to infer a modified impedance.[8][9] It turns out that unavoidable nonlinearity in a longitudinal rf bucket has to be treated carefully in order to digest the observed beam signals. We will describe here how to measure the impedance from the beam signal, how to analyze the effect of the impedance tuner, and how to succeed in canceling any space charge effects.[10][11]

## INTRODUCTION

In a high intensity proton synchrotron, some of the emittance growth and beam instabilities are caused by space charge effects. In longitudinal phase space, the space charge forces weaken the rf focusing force. When a short bunch is required, for example in a proton driver of a muon collider, the effects are further enhanced and they limit the minimum bunch length.

An inductive device in a ring should be able to cancel the space charge force. Recently a very high permeability material, FINEMET, became available. It turns out this material has sufficient permeability at the beam frequency region, and possibly cancels the space charge impedance. We designed a device, and named it "impedance tuner". It consists of FINEMET cores and has been installed in the KEK PS main ring.

## ESTIMATE OF IMPEDANCE

The impedance tuner is installed in the KEK PS main ring. The main ring accelerates nine bunches of protons from 500 MeV to 12 GeV with 3 s. Each bunch is transferred from its booster ring, which is operated with a 20 Hz repetition. Therefore, the first bunch injected in the main ring circulates at the injection energy for around 400 ms by the time that the bucket is filled with the last bunch. One bunch has approximately $1 \times 10^{12}$ particles.

In order to observe space charge effects and their compensation with the impedance tuner, an experiment was carried out using beams at the injection energy. Let us first calculate the space charge impedance. The space charge impedance is a negative inductance, and is written as,

CP496, *Workshop on Instabilities of High Intensity Hadron Beams in Rings*,
edited by T. Roser and S. Y. Zhang
© 1999 American Institute of Physics 1-56396-910-6/99/$15.00

$$\frac{Z_{sc}}{n} = -j\frac{g_0 Z_0}{2\beta\gamma^2}, \tag{1}$$

where $Z_0$ is the free space impedance, $\beta$ and $\gamma$ are the Lorenz factors, and $g_0$ is a form factor defined by

$$g_0 = 1 + 2\ln\frac{a}{b}. \tag{2}$$

Here, $a$ is the radius of the beam pipe and $b$ is the transverse beam radius. Since the size of the beam pipe is about 150 mm and the average beam size is about 60 mm, the form factor ($g_0$) becomes 2.8. Together with the other parameters of a 500 MeV beam, the space charge impedance becomes $-j310\Omega$.

The FINEMET, on the other hand, has a positive inductive impedance. The inductive impedance is expressed as

$$\frac{Z_{ind}}{n} = j\omega_0 L, \tag{3}$$

where $\omega_0 = 2\pi f_0$ and $f_0$ is the revolution frequency (667 kHz); $L$ is the inductance at that frequency. We take the revolution frequency and characterize the impedance by it because we observe the effects of the impedance tuner on a single bunch whose fundamental frequency is the revolution. In order to cancel a space charge impedance of $-j310\Omega$ completely, the total inductance of 73.7 μH should be prepared with FINEMET.

The FINEMET is wound on a toroidal core with an outer diameters of 340 mm, an inner diameter of 140 mm, and a thickness of 25 mm, as shown in Fig.1. The inductance as a function of the frequency is measured while taking a bias current as a parameter, as shown in Fig.2. A measurement was carried out for a single core using a test cavity, which consisted of a conductor going through the core and a variable condenser. To estimate the inductance of the core, the resonance point was searched for frequencies between 0.3 and 10 MHz. Also a bias current was applied to the conductor from 0 to 48A. From this figure, the inductive impedance is about $j25.5\Omega$ when the bias current is 0 A. Therefore, the total of 12 pieces of FINEMET core are necessary to fully compensate for a space charge impedance.

 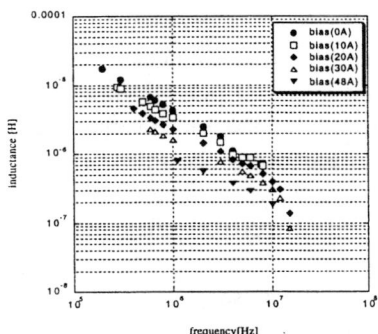

**FIGURE 1** FINEMET is a toroidal core with the outer diameters of 340 mm, the inner diameter of 140 mm, and the thickness of 25 mm.

**FIGURE 2** Inductance of FINEMET as a function of frequency is measured with a bias current from 0 to 48A*turn.

## EXPERIMENTAL SET UP

We have developed two different types of the impedance tuners so far. The reactance as a function of the frequency from 300 kHz to 30 MHz was measured for each type of impedance tuner by a network analyzer, and the results are shown in Fig.3.

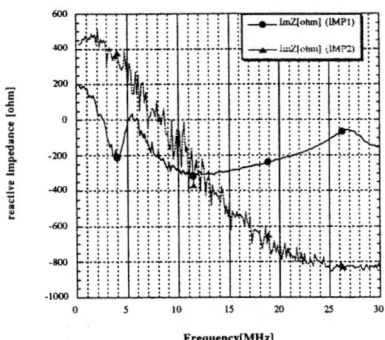

**FIGURE 3** Reactive impedance of the impedance tuner as a function of frequency. IMP1: first type, IMP2: second type.

The first impedance tuner consists of eight cores of FINEMET and a 6 turn bias coil is wound around the cores to control the inductance. All of the cores are installed in a single cylindrical vacuum chamber. The total length is 0.6m. The bias current is varied from 0 to 30 A (0 to 180 A*Turn).

The problem of this type was a large capacitance caused by the bias coil coupled with the inductance of FINEMET cores. It made a resonant circuit whose resonance frequency was about 2 MHz. That resonant frequency was relatively low compared to the fundamental frequency of the beam spectrum, about 667 kHz. To avoid a low frequency resonance, the second type of impedance tuner was developed.

This impedance tuner consists of three identical units, each unit has 4 pieces of FINEMET core, ceramic gaps and cooper shields, as shown in Fig.4. All of the FINEMET cores are placed outside of the ceramic gaps. There are short bars as shown in Fig.5. The space charge canceling effect could be obtained by observing the difference between with and without the short bars. The total length is 1.2m. With this type of the impedance tuner, the resonant frequency is raised to about 8 MHz. Because of the relatively flat frequency response of this device, this tuner is used for the experiment.

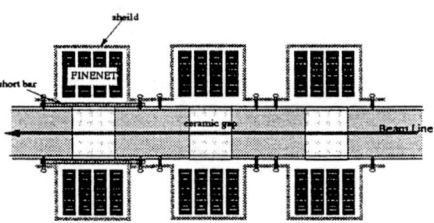

**FIGURE 5** Structure of the second type of the impedance tuner.

**FIGURE 4** Second type of the impedance tuner.

## MEASUREMENT

The frequency ($f_{s0}$) of synchrotron oscillations is perturbed with the potential of the space charge and the inductive impedance. We measure the shift of synchrotron frequency ($\Delta f_s$) as a function of beam intensity and obtain the total impedance as a coefficient.

Assuming that the bunch shape has a parabolic distribution, the perturbed incoherent frequency ($f_s$) by the total impedance is

$$f_s^2 = f_{s0}^2 \left( 1 - \frac{3ef_{rev}N}{\pi^2 hV_{RF}\cos\phi_s} \left(\frac{2\pi R}{\ell}\right)^3 \left[\frac{g_0 Z_0}{2\beta\gamma^2} - \frac{Z_{ind}}{n}\right] \right). \quad (4)$$

All symbols and nominal values in the KEK PS main ring are listed in Table 1. We consider only the space charge impedance and the inductive impedance (no resistive impedance) in a ring. When the shift is small ($\Delta f_s \ll f_{s0}$), equation (4) can be approximated by

$$\frac{\Delta f_s}{f_{s0}} = -\frac{3ef_{rev}N}{2\pi^2 hV_{RF}\cos\phi_s}\left(\frac{2\pi R}{\ell}\right)^3\left[\frac{g_0 Z_0}{2\beta\gamma^2} - \frac{Z_{ind}}{n}\right]. \quad (5)$$

If we can measure the linear relation between the frequency shift and beam intensity, the coefficient gives the total impedance.

**TABLE 1.** Nominal values in the KEK PS main ring.

| symbol | nominal value |
|---|---|
| $N$; number of particles per bunch | $1\times 10^{12}$ ppb |
| $R$; machine average radius | $54m$ |
| $\ell$; full bunch length | $80n\sec$ |
| $g_0$; form factor | 2.8 |
| $V_{RF}$; RF voltage | $120kV$ |
| $h$; harmonic number | 9 |
| $\phi_s$; synchronous phase | 0 deg |
| $f_{rev}$; revolution frequency | $667kHz$ |

In fact, the shift of incoherent synchrotron oscillations can not be measured directly. The frequency of coherent dipole oscillations does not move as a function of the beam intensity, either. However, an incoherent frequency shift can be inferred from the coherent quadrupole oscillations. If we take $\Delta f_{s2}$ as the frequency shift of the quadrupole oscillations, there is the following relation with the incoherent shift:

$$\frac{\Delta f_{s2}}{f_{s20}} = \frac{1}{4}\frac{\Delta f_s}{f_{s0}}, \quad (6)$$

where $f_{s20}$ is the coherent quadrupole frequency at zero beam intensity.

The frequency of the quadrupole oscillations is measured as the envelope oscillations of a bunch signal from a wall current monitor. The quadrupole oscillations are caused by a mismatch between a rf bucket and the bunch shape in the longitudinal phase space at the injection. An array of data indicating the peak height of a bunch as a function of time is fitted with the following function:

$$f(t) = p1 + p2\cdot\cos(2\pi\cdot p3\cdot t + p4)\exp(p5\cdot t). \quad (7)$$

Thus, the fitted value of $p3$ should be the frequency of the quadrupole oscillations.

The frequency of the quadrupole oscillations was measured as a function of the beam intensity using the second type of the impedance tuner (shown in Fig.6). The solid line is a fitted line of the data when all of the ceramic gaps were shorted. The dashed line is that when one ceramic gap was shorted and the remaining gaps were opened. The long dashed line is that when all of gaps were opened.

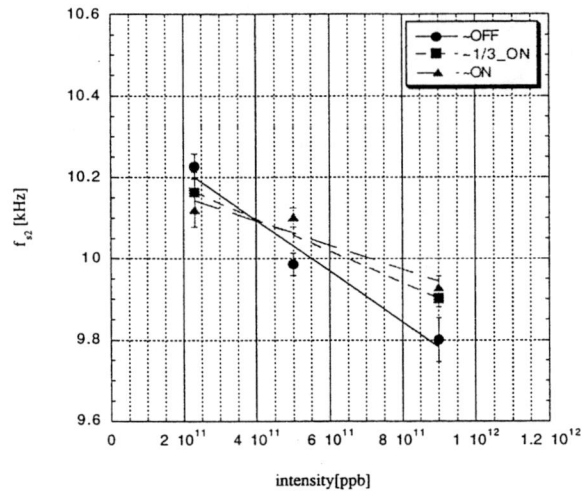

**FIGURE 6** Frequency of quadrupole oscillations as a function of intensity were measured. OFF: all of the gaps were shorted, , 1/3_ON: the one ceramic gap was opened and the rest of the gaps were shorted, ON: all of the ceramic gap were opened.

We estimate the non-perturbed synchrotron frequency ($f_{s20}$) by extrapolating the measured frequency to that of zero beam intensity; the results are listed in the Table2. The average synchrotron frequency is 10.27 kHz. The reactive impedance on each condition is derived from the slope using the average synchrotron frequency and listed in Table3. As expected, the more inductance that is installed, the lower is the negative of impedance is obtained. This demonstrates the space charge impedance is compensated by the impedance tuner.

**TABLE 2.** Non-perturbed synchrotron frequency ($f_{s20}$) by extrapolating the measured frequency to that of zero beam intensity.

| condition | $f_0$ [kHz] |
|---|---|
| without cores | 10.34 |
| with 1/3 of total cores | 10.25 |
| with total cores | 10.21 |

**TABLE 3.** Reactive impedance on each condition derived from the slope.

| condition | reactive impedance [$\Omega$] |
|---|---|
| without cores | $-j2475$ |
| with 1/3 of total cores | $-j1554$ |
| with total cores | $-j1182$ |

# EFFECT OF THE RF NONLINEAR FIELD

We have observed that the space charge impedance was reduced from $-j2475\Omega$ to $-j1182\Omega$ by the impedance tuner, which composed 12 pieces of the FINEMET cores. However, the measured impedance of $-j2475\Omega$ was 8 times as large as the impedance of $-j310\Omega$ obtained from the calculation. The quadrupole frequency at the zero beam intensity, given in Table2, was also lower than the frequency of 12.44kHz, estimated from the following equation using the parameters listed in Table1:

$$f_{2s0} = 2 \times f_{s0} = 2 \times f_{rev} \sqrt{\frac{-\eta h}{2\pi\beta^2 \varepsilon_s} eV_0 \cos\phi_s} \ . \tag{8}$$

## Simulation for the synchrotron frequency in an rf non-linear field

In fact, the synchrotron oscillation frequency could also be measured by observing the dipole oscillation; it was 5.2kHz. This result agrees with a measurement of the quadruple oscillation, but does not agree with the calculated value.

The typical bunch shapes at the injection and after a quarter of the synchrotron oscillation period are shown in Figs. 7 and 8, respectively. The particle distribution in the phase space was reconstructed from these two figures, as shown in Fig.9.

Including the non-linear effect of the rf field, the synchrotron frequency of single particle motion decreases when its amplitude increases. Each particle in the bunch has each incoherent synchrotron frequency as a function of the amplitude. However, it is not clear how the envelope oscillation is affected by the multi particle motion in the rf non-linear field.

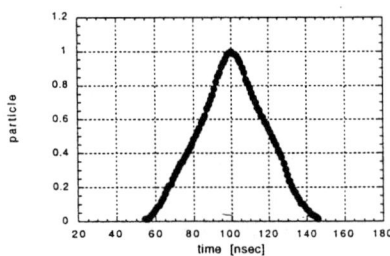 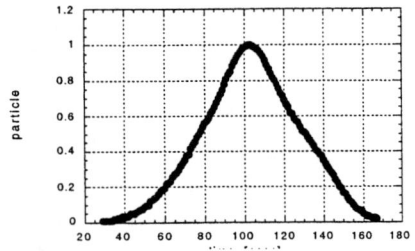

**FIGURE 7** The typical bunch shapes at the injection.  **FIGURE 8** The typical after a quarter of the synchrotron oscillation period.

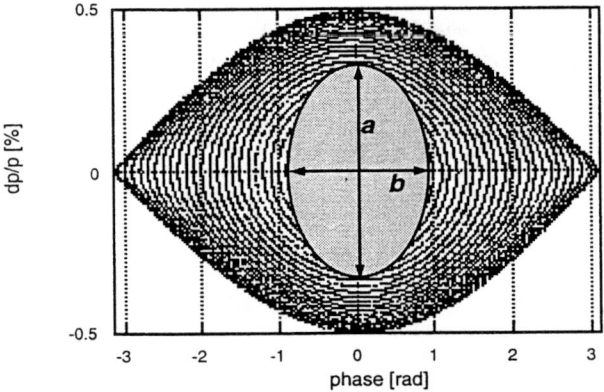

**FIGURE 9** The particle distribution in the phase space reconstructed from Figs. 7 and 8.

In order to clarify this, first, a multi particle simulation including the rf non-linear field without any space charge effects was carried out. In this simulations the bunch shape was matched to the rf bucket while changing the amplitude of the distribution. From the simulation result, we examined which frequency of the position in the bunch distribution is equal to the envelope oscillation frequency. We picked typical synchrotron frequencies for each different point of the distribution, center, peak, average and maximum point, as shown in Fig.10. The envelope oscillation obtained by the multi particle simulation was compared with four different kinds of frequencies listed in Fig.11. We found that the envelope oscillation frequency is close to the frequency of the average point.

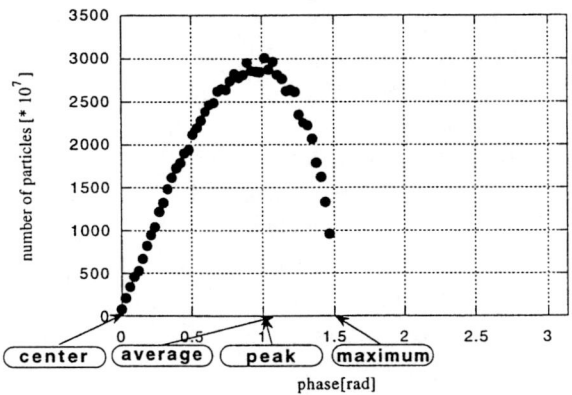

**FIGURE 10** Four different kinds of the position in the bunch distribution.

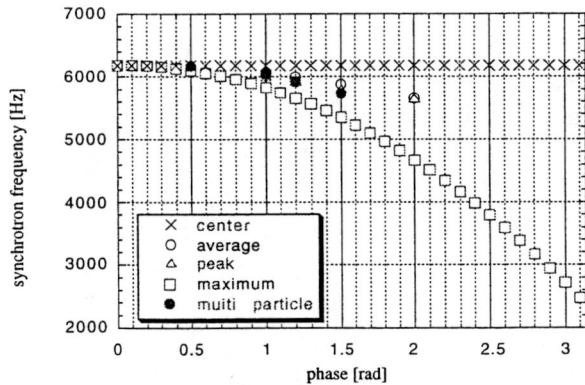

**FIGURE 11** The envelope oscillation obtained by the multi particle simulation was compared with four different kinds of frequencies.

Secondly, we performed a the simulation using the real distribution shown in Fig.9.

Assuming a linear field without any space charge effect, the frequency of the envelope oscillation was twice as large of the synchrotron frequency of the single particle motion, which was 12.46kHz.

On the other hand, assuming a non-linear field, the frequency obtained by the envelope oscillation was very close to the average frequency, which was 10.60kHz. It was consistent with the result of the measurement.

## Simulation of the quadrupole frequency shift in an rf non-linear field

Although the space charge impedance was $-j310\Omega$, which was estimated from the eq.(1), however, the experimental result is about $-j2400\Omega$, which is eight-times as large. In this estimation, it was assumed that the rf field is linear and that the bunch distribution is matched to the rf bucket. Thus, eq.(6) can be used to estimate the relation between the coherent and incoherent frequency shifts. However, as we pointed out previously, the rf field is non-linear and the bunch is not matched to the rf bucket.

In order to estimate the coherent quadrupole oscillation shift in a non-linear rf field, we have conducted milti particle simulations while changing the bunch length. Figure 12 shows the simulation result of the quadruple oscillation frequency as a function of the amplitude of the distribution for either case with and without any space charge effects. The simulation was carried out under the condition, $\frac{N}{\ell^3} = const.$ to keep that the space charge effects constant. As can be seen in Fig.12, the quadrupole oscillation shift increases when increasing amplitude. The quadruple oscillation frequency shift is affected by the rf non-linearity.

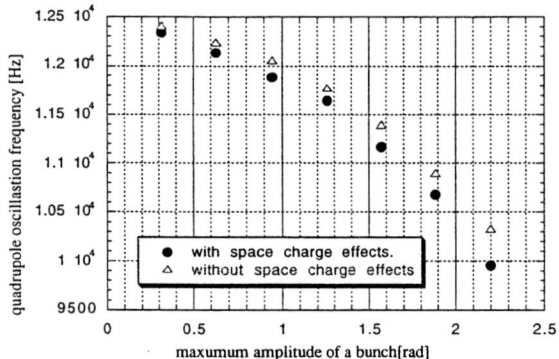

**FIGURE 12** The simulation result of the quadruple oscillation frequency as a function of the amplitude of the distribution for either case with and without any space charge effects.

Next, a simulation with real distribution (Fig.9) was tried in order to estimate the coherent frequency shift in an rf non-linear field. By assuming a non-linear field with space charge, the frequency shift was written as eq.(9) rather than eq.(6) under the condition of our measurement,

$$\frac{-\Delta f_{s2}}{f_{s20}} \approx \frac{1}{2}\frac{\Delta f_s}{f_{s0}}. \tag{9}$$

As a result of these examinations, it was found that if the bunch was influenced by the rf nonlinear field, the impedance obtained from the measurement was overestimated.

## Result of a new analysis by a simulation with a non-linear field

In order to analyze the impedance, we must consider the rf nonlinearity and the particle distribution in phase space. Moreover, the voltage induced by the space charge impedance is time-varying as a function of the bunch length and intensity, as shown in Fig.13 and Fig.14, respectively. In order to analyze the impedance exactly, we must consider these four effects.

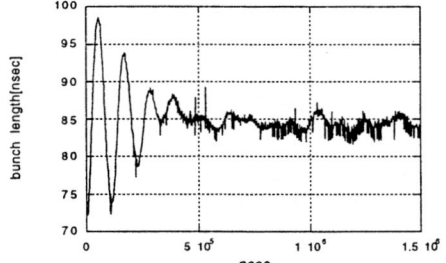
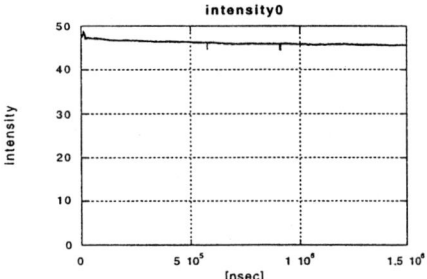

**FIGURE 13** The variation of the bunch length.   **FIGURE 14** The variation of the beam intensity.

The procedure to estimate the space charge impedance from a measurement is given as follows. Fist of all, a multi particle simulation was conducted using a real beam distribution. In this simulation, the voltage induced by the space charge effects was changed turn by turn, as shown in Fig.13 and Fig.14. The impedance was introduced as a parameter. It was compared with measurements from 0 turn (0 sec) to 250 turns (0.000375 sec) by making a least-square fit using the impedance as a parameter, which is shown in Fig.15.

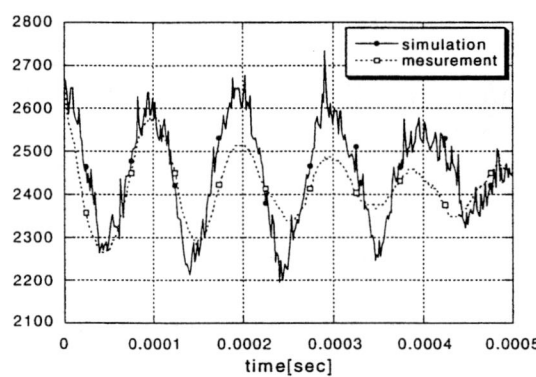

**FIGURE 15** The envelope oscillations obtained from measurement and simulation.

Figure 16 shows the least-square result as a function of the impedance. The sum of the least squares become minimum at an impedance of $-j660.5\Omega$ when there is no impedance tuner. When there is an impedance tuner the impedance is $+j442.4\Omega$, as shown in Table4.

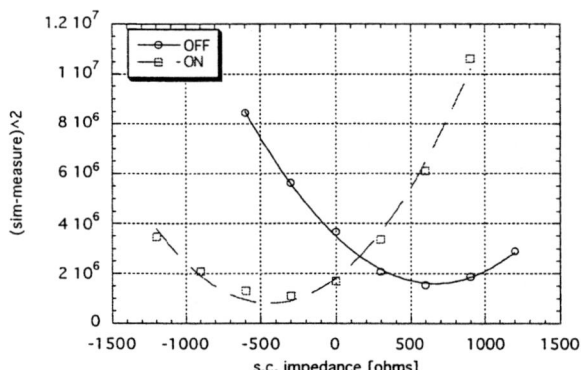

**FIGURE 16** Analyzed impedance including rf non-linearity.

**TABLE 4.** Reactive impedance on each condition obtained by the new analysis.

| condition | reactive impedance [$\Omega$] |
|---|---|
| without cores | $-j660.5$ |
| with total cores | $+j442.4$ |

# SUMMARY

We designed an impedance tuner consisting of the inductive material, FINEMET, to cancel the space charge impedance in the longitudinal direction. It was installed in the KEK PS main ring.

We observed the frequency shift of the coherent quadrupole oscillations and inferred the shift of the incoherent synchrotron oscillation. The total reactive impedance can be estimated as the coefficient between the shift and beam intensity. The measured impedance is reduced from $-j2475\Omega$ to $-j1182\Omega$ by the impedance tuner, which consists of 12 pieces of FINEMET cores. We demonstrated that the space charge impedance is compensated by the impedance tuner.

We analyzed the rf non-linear effect in detail by a calculation and a simulation. Thus, we have established an analysis method of the impedance with an rf non-linear field using the data in only a 1.5 synchrotron oscillation period.

## REFERENCE

1. K.Koba et al., *Proceedings of the 11th Symposium on Accelerator Technology and Science*, 1997, pp.409-411.
2. K.Koba et al, *Proceedings of the APAC*,1998.
3. K.Koba et al, RSI, vol.70, No.7, 1999.
4. T.Uesugi et al., *JHF internal report*, JHP-31, 1997.
5. Y.Yoshizawa et al., *Journal of Applied Physics*, vol.64, 1988, pp.6044-6046.
6. Y.Yoshizawa et al., *IEEE*, vol.25, No5, 1989, pp.3324-3326.
7. Y.Mori et al., *Proceedings of the EPAC*,1998, pp.299-301.
8. S.Hansen, *IEEE* vol.NS-22, No.3, 1975.
9. F.Sacherer, CERN/SI-BR/72-5, 1972.
10. K.Koba et al., *Proceedings of the PAC*,1999.
11. K.Koba, doctor thesis, 1999.

# Head-tail Instability and Microwave Instability in the KEK-PS

T. Toyama, D. Arakawa, S. Igarashi, J. Kishiro, K. Koba,
E. Nakamura, K. Takayama and M. Yoshii

*KEK, High Energy Accelerator Research Organization*
*1-1 Oho, Tsukuba-shi, Ibaraki-ken 305-0801, Japan*

**Abstract.** Head-tail instability and microwave instability has been studied and cured in the course of the intensity up-grade program of the KEK-PS. Experimental results, its interpretations and cures are given on the head-tail instabilities, emerging just after the acceleration start, and the microwave instability, arising after the transition energy in the 12GeV PS main ring.

## INTRODUCTION

Since 1976, the KEK 12-GeV Proton Synchrotron has delivered proton beams to a wide variety of physics experiments. The accelerator comprises two 750-keV Cockcroft-Walton accelerators, a 40-MeV drifttube linac, a 500-MeV fast-cycling synchrotron (Booster), and a 12-GeV synchrotron (Main Ring). The Main Ring stacks nine beam pulses from the Booster during the injection flat bottom, and delivers 12-GeV proton beams to two experimental halls as a DC beam spill in slow-extraction.

In 1994, the long-baseline neutrino-oscillation experiment using the KEK 12-GeV PS and Super KAMIOKANDE (K2K) [1] has been approved by the program advisory committee. The experiment demands a total number of protons of $10^{20}$ on target in three years, which corresponds to $6 \times 10^{12}$ protons per one Main Ring cycle ( 2 sec ). Just after the approval, the intensity up-grade program officially started. At the time the intensity was limited in the Main Ring due to beam losses during injection flat bottom, at the beginning of acceleration, at the transition energy and during slow-extraction. Fast-extraction ( one-turn extraction ) by kicker magnets was planned for the K2K experiment. Therefore the beam loss during extraction was considered not to be harmful at least for the K2K experiment.

In the following we briefly review the intensity up-grade proccess sequentially in time [2].

The first and long attack have been made on the beam losses during injection flat bottom and at the transition energy beam loss.

Among fast, medium and slow losses, the slow loss of a few 10 ms during the injection flat bottom was considered to be caused by a space charge induced fourth order resonance [3]. According to the assessment the operating point was moved from $\nu_y \sim 7.2$ to $\sim 5.2$. Nevertheless, a sizable beam loss remains during the injection flat bottom. We are now operationary choosing the best tune around $(\nu_x, \nu_y)=(7.1, 5.2)$, although we never have conclusive explanation of the loss yet.

Just after the transition the clear signal of a microwave instability has been observed. The wall-current type beam-position monitors (BPM) and cavity-like vacuum chambers (CVC) with an evacuation port, placed near 56 lattice quadrupoles, were identified as the dominant source of the instability. The phenomena was successfully explaned with the "proton klystron" model [4], where cumulative wake's effects due to highly resonant impedances as BPM's and CVC's play a key role. During the summer shutdown of 1996, two-thirds of those devices were replaced by newly designed ones; meanwhile the other one-third were left because of high residual radiation level. This replacement notably reduced the longitudinal emittance blow-up at the transition, although the instability itself is still observed to evolve at some level [5]. In addition the power supply of the $\gamma_T$ jump quadrupole magnets were replaced by new power supply with twice the current.

The next attack have been made at the beam loss at the beginning of acceleration. The beam loss was identified as a horizontal head-tail instability, emerging with the longitudinal mode number $\ell = 0, 1, 2$ during acceleration [6] [7] [8]. Especially the mode $\ell = 0$ is fatal. The horizontal chromaticity was shifting from a negative to a slightly positive value. The zero crosssing point was around 80 ms after the beginning of acceleration. The large chromaticity variation more than controlled by the sextupole magnets was attributed to an eddy current induced sextupole fields of the vacuum pipe in the bending magnets.

The injection kickers and resistive wall were identified as a main impedance source based on the calculation. Installation of new fast extraction kickers for the K2K experiment seemed to force the impedance larger. We decided to damp the head-tail instability by using rearranged octupoles of four-fold symmetry in addition to further careful chromaticity control.

In the commissioning from January 1999, the beam intensity had never reached at $6 \times 10^{12}$ protons per pulse (ppp), in spite of all our efforts.

Finally bunch shaping has been made using the fundamental rf cavities: uniform bunch formation by rf voltage modulation with a band-limited white signal [9]. We applied this method successfully to the moving rf buckets in the BR and MR. It affects on collective longitudinal instabilities at the injection and at the transition energy in the Main Ring. Although a horizontal head-tail instability occured during the injection flat bottom of the Main Ring with this bunch shaping, it is also suppressed by Landau damping with the octupole magnets. Consequently we have achieved the intensity goal of $6 \times 10^{12}$ ppp and have almost kept more than this intensity since May, 1999. The intensity up-grade is summarized in Fig. 1, where the maximum intensity records and recent average intensities are plotted with epoch-making improvements. The beam intensity evolves through the cycle as shown in Fig. 2.

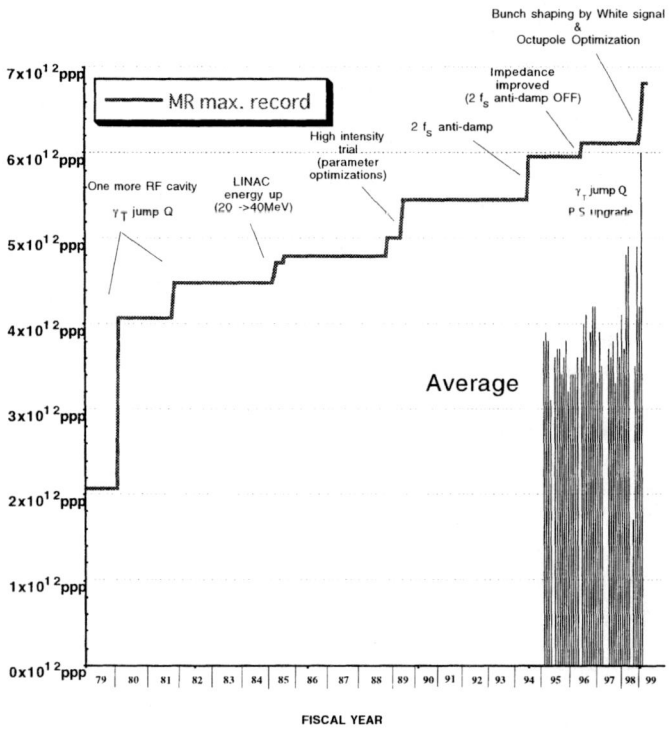

**FIGURE 1.** Beam intensity upgrade in the Main Ring.

In this paper we concentrate on two beam instabilities, the microwave instability and the head-tail instability studied in the Main Ring of the KEK-PS [2].

## MICROWAVE INSTABILITY

The 12 GeV PS has been subject to serious damage by beam loss at the transition energy ($\gamma_t$) since the beginning of its operation. As indicated in Fig. 1 two pairs of pulsed quadrupole magnets for $\gamma_t$ jump were installed at the beginning stage of PS operation. Trial of increasing intensity more, however, have been prevented by the loss.

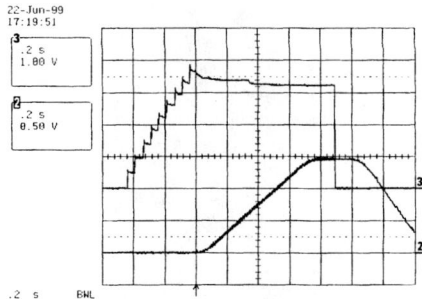

**FIGURE 2.** Beam intensity during acceleration in the Main Ring. Upper trace: beam intensity ($2\times10^{12}$ppp/div.), lower trace: fields of the bending magnet

## Experimental Results

Among many candidates which may cause the beam loss, microwave instabiity have been the most probable one. A systematic experimental study was performed with one bunch acceleration, the beam intensity range of $1 - 7\times10^{11}$ protons per bunch and the $\gamma_t$ jump range of 0 - 0.25 for 0.5 ms. The observed features were: (1) it never evolves to an observable level before transition crossing, (2) it grows from the bunch-tail portion, (3) just after transition crossing it dramitically starts to grow within 1 ms near 1 GHz, (4) then a substantial fraction goes away from the bunch-center. Figure 3 shows a typical evolution of the bunch shape projected on the time axis.

The threshold current for the microwave instability was obtained by measuring the size of the emittance blow-up ratio for different beam currents. It was defined as a ratio of the 10 ms down-stream emittance to 10 ms up-stream one.

Figure 4 shows the emittance blow-up ratio as a function of the particle number per bunch ($N_0$) at a $\gamma_t$-jump size ($\Delta\gamma_t$) of 0.15. Beyond some critical value of particle number the blow-up becomes obvious. The dependence of the emittance blow-up ratio on a size of $\gamma_t$-jump is depicted in Fig. 5 for a fixed beam current of $5 \times 10^{11}$/bunch.

## Impedance calculation and measurements

Resonant structures capable of exciting the observed microwave instability have been identified as a result of MAFIA calculations and impedance measurements at the test bench.

As shown in Table 1, the BPM and CVC which are placed at each side of the vacuum chambers in 56 lattice quadrupole magnets have a large shunt impedance $R_{shunt}$ and a high quality factor $Q$. In the numerical simulations, a magnitude of

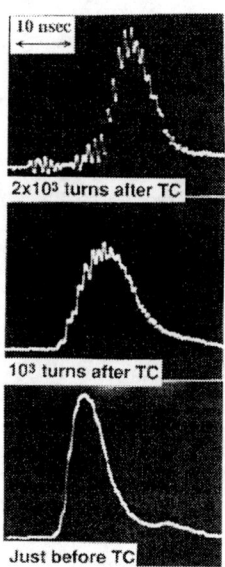

**FIGURE 3.** Bunch profiles from -1 msec to +2 msec (t=0 msec: transition).

$R_{shunt}/Q$ obtained by the MAFIA calculations and measurements has been used, which are in agreement to each other in order of magnitude; while the quality factor was treated as a kind of free parameter ($Q < 100$ is used, according to the observed fact that a downstream bunch is not affected by a upstream bunch) because a loaded $Q$ in the actual ring is not satisfactorily estimated.

## Proton-klystron model and simulation results

In the model highly resonant structures are periodically distributed along the beam-path and wake fields excited by the bunch-head can affect the bunch-tail. Member change in a global-scale between the bunch-head and -tail scarcely occurs

**TABLE 1.** Resonant impedance ( measured / calculated ).

|  | $\omega_\lambda/(2\pi)$ (GHz) | Q | $R_{shunt}$ ($\Omega$) | R/Q ($\Omega$) |
|---|---|---|---|---|
| BPM | 0.636/0.667 | 77/2650 | $1\times10^3/2.6\times10^4$ | 19.4/9.8 |
|  | /1.13 | /3769 | /$6.2\times10^4$ | /16.3 |
|  | 1.498/1.377 | 230/8222 | $5.3\times10^3/3.3\times10^3$ | 23/40 |
| CVC | /1.44 | /4846 | /$1.4\times10^5$ | /28.8 |
|  | /1.84 | /4048 | /$1.9\times10^5$ | /46.6 |

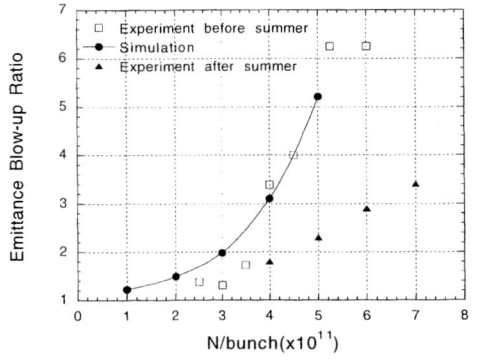 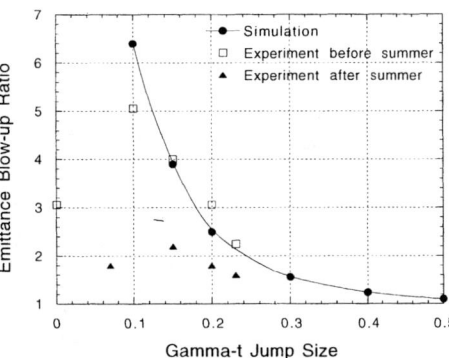

**FIGURE 4.** Emittance blow-up ratio vs the number of particles.

**FIGURE 5.** Emittance blow-up ratio vs the size of $\gamma_t$ jump ($\Delta\gamma_t$).

near transition energy. All of excited resonant structures can be regarded as idling cavities of a klystron. Build-up of the wakes in each resonant structure is treated in a formulation of the forced excitation of a damped harmonic-oscillator [4].

Longitudinal particle motion is tracked turn by turn using a set of discrete kinetic equations for the energy $E$ and the phase $\phi$ to the accelerating rf, taking into account normal rf acceleration including nonlinear kinematics effects, longitudinal space-charge forces, and interactions with the resonant scructures.

The temporal-evolution of a bunch was numerically obtained as seen in Fig. 6. The simulation reproduces the essential aspects of the experimental result. General features of the microwave instability are (a) efficient energy transfer from beam to microwaves is resulted from the deceleration of the micro-structure's core, yielding the drift of the cores in the negative direction of the momentum space, (b) the interaction between wakes and the beam tends to occur in the bunch-tail owing to the finite $Q$-value, (c) the convection of the microwave-structure formation toward the bunch-head is crucial from the same reason as that of (b). According to (b) and (c), the particles located in the region of $E > E_s$ and $\phi > \phi_s$ below the transition energy and $E < E_s$ and $\phi > \phi_s$ above the transition energy can contribute to the evolution of the microwave instability, where the suffix "s" indicates a synchronous particle. Since the wake's phase coincides with the motion of micro-structure's core, the particles located in the other side of momentum-space are affected by the wake in counter phase, suffering a rapid modulation in their motion. Below the transition energy a fraction of decceleated micro-bunch core neccesarily falls into the region of $E < E_s$ where the fraction is forced to move in the opposite direction of the phase $\phi$; the microbunch formation is likely to be wiped out. Thus the microwave instability is not able to evolve to a level comparable to that above the transition energy.

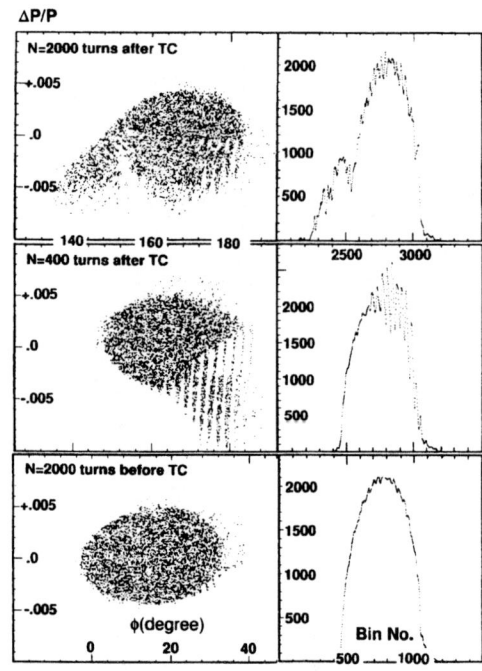

**FIGURE 6.** Typical simulation results of transition crossing.

In the early-stage (< 400 turns) bunching and amplification proceed like that in a microwave amplifier such as free-electron lasers; the repeated synchtorton-rotation in the micro-bucket doesn't occur but particles almost drift in the momentum-space. A fraction of the micro structure placed in the deccelerating phase is rapidly decelerated, amplifying the microwaves. The microwaves' phase follows the motion of the decelerated fraction because electro-magnetic waves are amplified at the expense of the kinetic energy of protons. Consequently the other fraction having been placed in the accelerating phase falls into the decelerating phase to turn to contribute to further microstructure formation. This looks like eruption in the bunch-tail (see Fig. 6).

Simulation results for the original impedance are plotted in Figs 4 and 5. It indicated that for the fixed value of $R_{shunt}/Q = 10$ Ω and $\gamma_t = 0.15$ the bunch eruption due to the microwave instability is remarkable beyond $N_0 = 3 \times 10^{11}$. Simulation showed larger suppression of the microwave instability as the $\gamma_t$ jump size ($\Delta\gamma_t$) increases and are in good agreement with the experimental results. The fast phase-mixing in a case of large $\Delta\gamma_t$ is likely to prevent the microwave instability from growing up.

## Cures

Based on the above analysis two improvement have been done: (1) the two-thirds of highly resonant impedance device pairs (BPM and CVC) have been replaced by newly designed ones (summer shutdown in FY1996), (2) the power supply of the $\gamma_t$ jump quadrupole magnets have been replaced by the more powerful one of twice the current (winter shutdown in FY 1997).

The new BPM is an electrostatic type. Its $R_{shunt}/Q$ is reduced by a factor of 2-3 according to the MAFIA calculation. The resonant frequency of the dominant mode moves to 342 MHz. In the new CVC the original cavity-like structure is discarded and the evacuation port is shielded by RF slits. As a result, a magnitude of resonant impedance is negligible small. The emittance blow-up ratio has been measured in the same way. The results are put in Figs 4 and 5. Certainly, the emittance blow-up ratio has notably decreased as expected, although the microwave instability itself is still observed to evolve at some level.

Consequently a new record of the maximum intensity, $6.2\times10^{12}$, was established and the stable operation with a beam intensity of $\sim 5\times 10^{12}$ ppp was realized without a significant beam loss at the transition energy.

## HEAD-TAIL INSTABILITY

The operating point, $(\nu_x, \nu_y)$, has been changed from (7.1, 7.2) to (7.1, 5.2) since January 1996 by the reason mentioned in the introduction. After the modification, the beam loss emerges around 80 $ms$ after the beginning of acceleration. The observation using the position monitor revealed that the instability is caused by a horizontal head-tail instability. The typical profile of the horizontal coherent dipole oscillation is shown in Fig.7.

## Experimental results

The systematic experimental study has been performed using single bunch. In Fig.8 the growth of the oscillation is shown in the lower trace and the degradation of a beam intensity in the upper trace. The signals are sampled at the timing of the bunch center, 5 k samples per division. The start point of the trace is 50 $ms$ after P2 (the beginning of acceleration). The instabilities occured repeatedly even decreasing beam intensity. Around 80 $ms$ from P2 the mode $\ell = 0$ was mostly observed and fatal. On the other hand the mode $\ell = 1, 2$ or some mixtures were observed at the neighbour of the above time region.

Near injection energy, there are some contributions to the chromaticity other than natural chromaticity [6]: an excitation of two families of correction sextupoles, a remnant of the correction sextupoles, a sextupole component of the dipole and quadrupole magnets, and an eddy current in the dipole vacuum pipe. Taking the energy dependence of the chromaticity into account, we obtain the time evolution

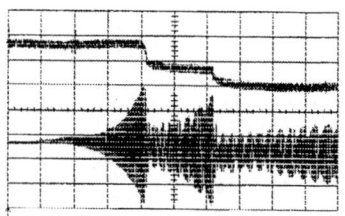

**FIGURE 7.** Horizontal coherent oscillation.

**FIGURE 8.** Growth of the coherent oscillation. Upper trace: beam intensity ($10^{11} protons/div$), lower trace: coherent oscillation, abscissa: 5 k samples/div.

of the chromaticity, which is shown with the measured chromaticity in Fig.9. The solid line and the filled circles are the calculated and measured results, respectively, for the operation in which the beam loss occurs due to the uncorrected chromaticity at (7.12, 5.21). The calculation including the eddy current effect well agrees with the measured result and well explains the beam loss around 80 $ms$ from P2. After

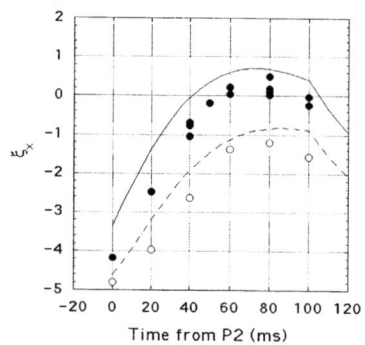

**FIGURE 9.** Evolution of the horizontal chromaticity during a beginning part of acceleration.

reducing the vertical tune from 7.2 to 5.2, the horizontal natural chromaticity increased by $\sim$ 1.3 and became closer to zero. This caused zero crossing of the horizontal chromaticity around 80 $ms$ after P2 and forced the horizontal coherent motion, mainly, of mode 0 unstable and resulted in the beam loss.

The cure was achieved by the sextupole family in the primary stage. Corresponding chromaticities are plotted in Fig.9 by the dashed line (calculated result) and the open circles (measured result).

To confirm further the eddy current induced sextupole fields, horizontal chromaticity was measured with two different ramping rates: $dB/dt = 2.3 [T/s] \times$

$t/(100[ms])$ for $0 < t < 100$ ms ( 100 ms smoothing pattern ), and $dB/dt = 2.3[T/s] \times t/(200[ms])$ for $0 < t < 200$ ms ( 200 ms smoothing pattern ). The calculated and measured chromaticities during a beginning part of acceleration are plotted in Fig.10 and Fig.11, respectively. Solid line and dashed line in Fig.10 indicate the case of the 100 ms and 200 ms smoothing pattern, respectively. Filled circles and open circles in Fig.11 indicate the case of the 100 ms and the 200 ms smoothing pattern, respectively. Global feature in the measured temporal variation of $\xi_x$ is qualitatively similar to the calculated one except an unknown $\xi_x$ shift of 1-2, which has been observed for these years. Direct field measurements

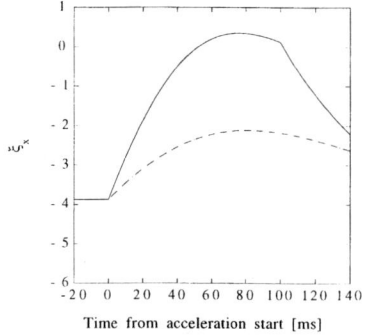 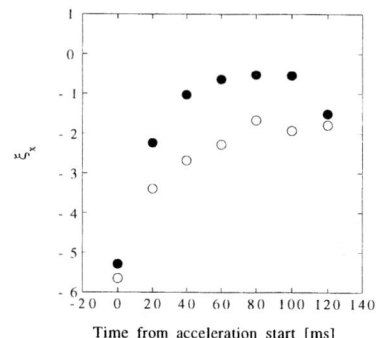

**FIGURE 10.** Calculated horizontal chromaticity

**FIGURE 11.** Measured horizontal chromaticity

of a model vacuum pipe of SUS316L in the monitor bending magnet gives a consistent results. Consequently, the change in the chromaticity is attributed to the eddy-current-induced sextupole fields.

## Analysis

The qualitative feature of the observed head-tail instability was well explained by a multi-particle simulation code, although a constant wake was used [6].

In order to obtain more quantitative feature, the growth rate is calculated using the Sacherer's formula [10] [11] [12].

For coupling impedances a resistive wall impedance, a kicker impedance and a broad band impedance are assumed. The resistive wall is modeled as rectangular beam pipes [13] in the bends and quads and cylindrical beam pipes in straight sections. There are five identical kickers for injection in the 12GeV PS [14]. Its impedance (real part) is calculated using the impedance formula in [15]. The real part of the total impedance which contributes the growth rate is depicted in Fig.12. The effective impedance is reduced using the form factor for a parabolic distribution.

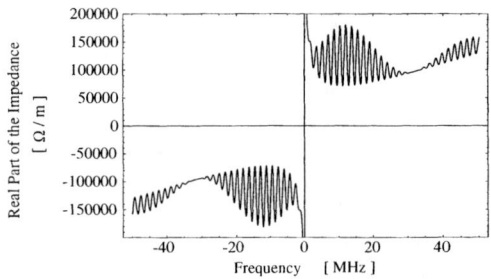

**FIGURE 12.** Real part of the total impedance ( calculated ).

The calculated results of the growth rate are depicted with the measured data for the mode $\ell = 0$ in Fig.13 and Fig.14. In Fig.13 the growth rate for the mode $\ell = 0, 1$ and 2 are plotted by a solid line, a dashed line and a dotted line, respectively. These results agree with the measured values very well. It should be noted that

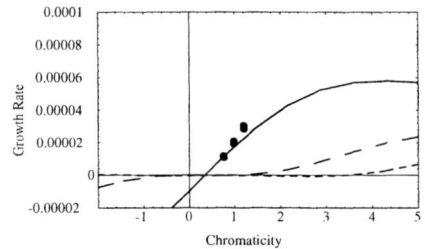

**FIGURE 13.** Growth rate as a function of horizontal chromaticity.

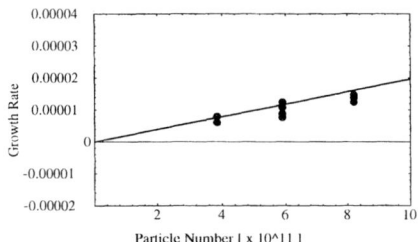

**FIGURE 14.** Growth rate as a function of bunch intensity.

three impedances contribute almost equally to the growth rate.

The growth rate measurement has suggested that the main sources of the coupling impedances are kickers and resistive wall. It was urgent to evaluate the coupling impedance of newly installed FX kickers [16] for the K2K experiment.

To evaluate the kicker coupling impedance, both of theoretical and experimental approach have been employed. The real part of the transverse coupling impedances of the injection and FX kickers were calculated by using the standard theory of travelling wave kickers [17] for the C-type kicker and for the "Twin c-type" kicker [16].

The measurement was performed with a copper wire of 1 mm in diameter which is coaxially stretched in the chamber. At first $S_{21}$ in "S" matrix was measured with a network analyzer ( hp8753E ) by displacing the wire in the horizontal direction. The $S_{21}$ component was translated to the transverse impedance by using the formula for distributed impedance [18].

The results are plotted in Fig.15 for the injection kicker and in Fig.16 for seven FX kickers. In the figures dots indicate the measured value and solid line the calculated one. Due to the limitation of the experimental setup, the terminations of the kickers were not same as the practical setup. In the present measurement, the matched loads were connected with the injection kicker at both ends. The FX kicker is conneted with a matched load at one end and short-circuited at the other end. The disagreement between the calculated and measured impedance may come from incomplete error correction or ferrite loss which were not included in the calculations. They are subjects to be solved in future.

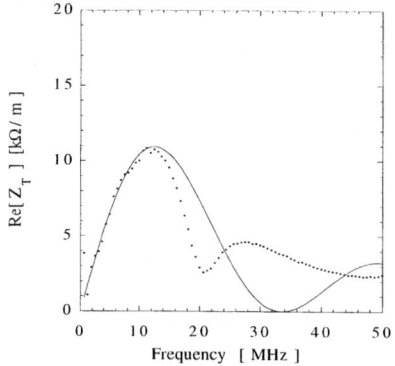

**FIGURE 15.** Coupling impedance of one injection kicker

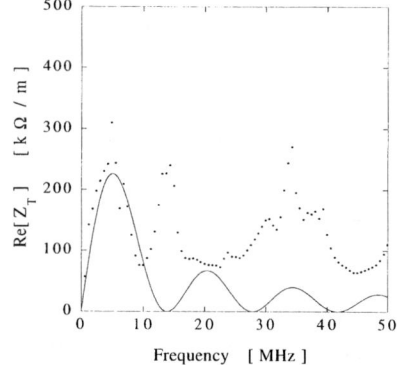

**FIGURE 16.** Coupling impedance of seven FX kickers

The additional impedance newly introduced by the seven FX kickers seemed to be much larger than that of the five injection kickers for the frequency range of our interest. The cure was decided to be performed by Landau damping with rearranged octupoles of four-fold symmetry in addition to further careful chromaticity control.

In addition, the longitudinal impedance measurement of the FX kickers embedded in the large rectangular chamber implied sharp resonances in a few hundred MHz region. It is potentially dangerous for the microwave instability serious at transition crossing. It seems to be caused by the gaps between kicker modules. Then the gaps between kickers were filled up by the short copper pipes of the similar inner size. It reduced the impedance peak height by a factor of two to four.

## BUNCH SHAPING BY RF VOLTAGE MODULATION

One of practical methods used to increase the beam-intensity limit is to reduce the peak current by rf manipulation [19], [20].

In the Main Ring a method so called "$2f_s$ anti-damping" was used [21], in which the quadrupole oscillation mode in the longitudinal motion is enhanced by feeding the bunch signal onto the accelerating voltage. This method had effects in some extent as indicated in Fig. 1. However the performance was not very reliable because it resorts to positive feedback.

As another possibility to manipulate the longitudinal distributon, fundamental rf voltage modulation with band-limited white signal was proposed [9], which is potentially able to manipulate the bunch without longitudinal emittance blow-up.

This method relies on the synchrotron frequency spread in the sinusoidal rf potential. The parametric resonance by rf voltage modulation with a single frequency of $2f_s - \Delta f$ has been well analyzed [22]. For a small perturbation a resonance island occurs and a synchrotron oscillation is perturbed. RF voltage modulation with a small number of adjacent resonances produces chaotic diffusion of particles [23], where islands overlap each other, satisfy the Chirikov criterion and make particles diffuse. Band-limited white signal may be a limiting case where the number of adjacent resonances is infinite. Particles diffuse in a bounded region and form a uniform distribution in the longitudinal phase space.

The temporal evolution of the distribution was derived analytically using a diffusion equation. The numerical simulation was also performed using one hundred of adjacent parametric resonances. The time constants evaluated with typical parameters at the Main Ring injection energy are shown in Fig. 17. The time constants are plotted as a function of the rms degree of modulation ($\xi_{RMS}$). Solid line indicates an analytical calculation and dots are simulation results. The results of both calculations agree very well and suggest time constants short enough for practical application.

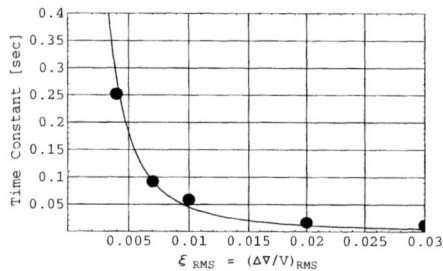

**FIGURE 17.** Time constant as a function of the degree of modulation.

The test was performed during the injection flat bottom of the main ring on June 11, 1998 and in successive experiments. A pseudo-uniform distribution was successfully obtained as shown in Figs 18 and 19.

A pseudo-uniform distribution has been also obtained by modulating the moving buckets in both the BR and the MR. For the moving buckets we have to track the synchrotron frequency. Figures 20 and 21 shows the line density without/with band-limited white signals that were observed ∼15 ms before transition energy. The discrepancy between both cases is clear.

In consequence an beam intensity increased about 24% comparing with the pre-

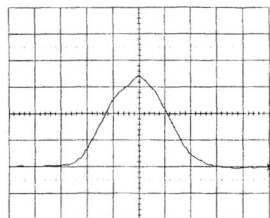 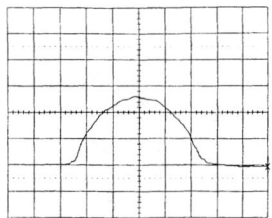

**FIGURE 18.** Bunch profiles without rf modulation, abscissa: 20 ns/div.

**FIGURE 19.** Bunch profiles with rf modulation. $\xi_{RMS} \approx 3.6\%$, abscissa: 20 ns/div.

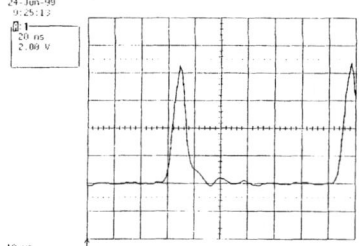 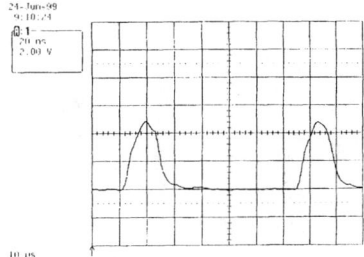

**FIGURE 20.** Bunch profiles without rf modulation, abscissa: 20 ns/div.

**FIGURE 21.** Bunch profiles with rf modulation. $\xi_{RMS} \approx 4\%$, abscissa: 20 ns/div.

vious method [24]. Proton beam of more than $6 \times 10^{12}$ ppp has been achieved at the main ring flat top and kept scince May, 1999, although a litttle increase in the longitudinal emittance was inevitable for stability at the transition energy. In this achievement additional machine tunings, especially a fine tuning of the octupole current to compromize Landau damping and dynamic aperture, have been indispensable.

## CONCLUSION

The critical issues for the KEK PS beam-intensity upgrade program: the microwave instability and the head-tail instability have been systematically studied and those instabilities have been practically overcome. The original intensity goal of $6 \times 10^{12}$ protons per cycle has been achieved. More intensity will be attained with removing the beam loss during injection flat-bottom, which is the next subject scheduled from this Fall.

# ACKNOWLEDGMENTS

The authors thank I. Yamane, H. Sato and other KEK-PS members for their interest and continual support through this work. Assistance of S. Ninomiya and H. Someya in preparing rf voltage modulation is also greatly acknowledged.

# REFERENCES

1. K. Nishikawa et al., K2K ( KEK to Kamioka ) Long Baseline Neutrino Oscillation Experiment, E362.
2. A concise summary of the KEK PS beam-intensity upgrade program including other subjects is available: K. Takayama, KEK Preprint 99-57. The head-tail instability was studied in the Booster about two decates ago: Y. Miyahara et al., Part. Accel. **10**, 125 (1980).
3. S. Machida and Y. Shoji, AIP Conf. Proc. 377 (1995) 160.
4. K. Takayama et al., Phys. Rev. Lett. 78(1997) 871.
5. K. Takayama et al., Proc. of the 1997 Part. Accel. Conf. (1997) 1548.
6. T. Toyama et al., Proc. of the 1997 Part. Accel. Conf. (1997) 1599.
7. T. Toyama et al., Proc. of the first Asian Part. Accel. Conf. (1998), 471.
8. T. Toyama et al., Proc. of the 1999 Part. Accel. Conf (1999).
9. T. Toyama, KEK Preprint-99-38.
10. B. Zotter et al., CERN 77-13, p.175.
11. A. Chao, "Physics of collective beam instabilities in high energy accelerators", John Wiley & Sons, Inc., 1993 and the references there in.
12. R. Cappi, Part. Accel. 50 (1995) 117.
13. K. Y. Ng, Part. Accel. 16 (1984) 63.
14. K. Takata et al., KEK Report, KEK-76-21 (1977).
15. G. Nassibian, CERN/PS 85-68 (BR), 1986.
16. T. Kawakubo et al., Proc. of the first Asian Part. Accel. Conf. (1998) 582.
17. H. Hahn et al., AD/RHIC/RD-111, 1997.
18. L. S. Walling et al., Nucl. Instru. Methods **A 281**, 433 (1989).
19. R. Cappi, et al., Proc. of Part. Accel. Conf., 1993, p.3570.
20. M. Blaskiewicz, et al., Proc. of Part. Accel. Conf., 1996, p.383.
21. S. Ninomiya et al., KEK Report 93-4 (1993). S. Ninomiya, private communication (1995).
22. D. Li et al., Nucl. Instru. Methods **A 364**, 205 (1995) and references therein.
23. W. Gabella et al., Part. Accel. **42**, 235 (1993).
24. T. Toyama et al., KEK Internal Report, SR-473, May 13, 1999 (Japanese).

# Intensity Dependent Effects in RHIC

Jie Wei

*Brookhaven National Laboratory, Upton, New York 11973, USA*[1]

**Abstract.** The Relativistic Heavy Ion Collider (RHIC) is currently under commissioning after a seven-year construction cycle. Unlike conventional hadron colliders, this machine accelerates, stores, and collides heavy ion beams of various combinations of species. The dominant intensity dependent effects are intra-beam scattering at both injection and storage, and complications caused by crossing transition at a slow ramp rate. In this paper, we present theoretical formalisms that have been used for our study, and discuss mechanisms, impacts, and compensation methods including beam cooling and transition jump schemes. Effects of space charge, beam-beam, and ring impedances are also summarized.

## 1. INTRODUCTION

The primary motivation for colliding heavy ions at ultra-relativistic energies is the belief that it is possible to create macroscopic volumes of nuclear matter at such extreme conditions of temperature and energy density that a phase transition will occur from hadronic matter to a confined plasma of quarks and gluons. The main goal of the Relativistic Heavy Ion Collider (RHIC) [1,2] is to provide head-on collisions at energies up to 100 GeV/u per beam for various species from proton to gold. Luminosity requirements for the heaviest ions are specified to be in the $10^{26-27}$ $cm^{-2}s^{-1}$ range. The higher gold-gold total cross-section results in interaction rates comparable to proton-proton colliders although this luminosity is several orders of magnitude lower than those machines. Based on these general requirements, RHIC machine parameters were derived and are outlined in Table 1.

Comparing with conventional proton colliders like the Tevatron and the Large Hadron Collider where limitation on beam lifetime and luminosity performance is due to beam-beam effects, performance of RHIC is primarily limited by intra-beam Coulomb scattering, which becomes increasingly important for high charge state ions. The scattering cross-section is proportional to $Z^4/A^2$ for particles of charge state $Z$ and atomic number $A$. For fully stripped gold ions, intra-beam scattering causes transverse emittance growth, increasing the emittance from 10 to more than 40 mm·mr in 10 hours, even in the optimal scenario when the transverse

---

[1] Work performed under the auspices of the US Department of Energy.

**TABLE 1.** Major parameters for the Relativistic Heavy Ion Collider.

| | | |
|---|---|---|
| Kinetic Energy, Injection – Top | | |
| gold | 10.8-100 | GeV/u |
| proton | 28.3-250 | GeV |
| Circumference, $2\pi R$ | 3833.845 | m |
| Number of bunches per ring | 60 | |
| Number of crossing points | 6 | |
| $\beta^*$, injection | 10 | m |
| $\beta^*$, low-beta insertion | 1 | m |
| Transition energy, $\gamma_T$ | 22.89 | |
| Betatron tunes | | |
| horizontal | 28.19 | |
| vertical | 29.18 | |
| Magnetic rigidity, $B_0\rho$ | | |
| injection | 97.5 | T·m |
| top energy | 839.5 | T·m |
| Dipole field at top energy | 3.45 | T |
| Quadrupole gradient at top energy | 71.2 | T/m |
| Accelerating RF system: | | |
| harmonic number, $h$ | 360 | |
| maximum voltage, $V$ | 0.6 | MV |
| Storage RF system: | | |
| harmonic number, $h$ | 2520 | |
| maximum voltage, $V$ | 6 | MV |
| Intensity per bunch, $N$ | | |
| gold | $10^9$ | |
| proton | $10^{11}$ | |
| Transverse emittance (95% normalized), $\epsilon_N$ | | |
| gold | 10 – 40 | mm·mr |
| proton | 20 – 27 | mm·mr |
| Longitudinal bunch area (95%), $S$ | | |
| gold | 0.3 – 1.2 | eV·s/u |
| proton | 0.3 – 1 | eV·s |
| Average luminosity: | | |
| gold | $10^{26}$ | $cm^{-2}s^{-1}$ |
| proton | $10^{31}$ | $cm^{-2}s^{-1}$ |

motion is fully coupled [3]. This growth results in a large beam dimension in the interaction region focusing quadrupole triplets. Consequently, the impact from the field errors of these superconducting magnets becomes significant, limiting the lowest achievable $\beta^*$ to about 1 meter at the interaction point and hence limiting the highest achievable luminosity. Longitudinally, intra-beam scattering causes bunch area growth, resulting in intensity loss of as much as 40% in 10 hours as particles escape the confining buckets of the radio-frequency (RF) system. In order to reduce beam loss and to improve luminosity lifetime, beam cooling has been investigated [4,18].

RHIC is the first superconducting machine where the beams have to be accelerated across the transition energy. Comparing with normal-conducting magnets, the superconducting magnets require a slow ramp rate for beam acceleration. Both chromatic nonlinear effects and beam self-field effects are strong during transition crossing [5]. A "matched first order" transition jump scheme is designed [6] to effectively increase the crossing rate by a factor of 8 during the 60 ms time around transition [7]. With such a scheme, the longitudinal emittance growth can be limited to less than 20% at transition with minimum disruption to the transverse particle motion.

This paper summarizes the dominant intensity dependent effects in RHIC. In Section 2, we review the scaling laws of intra-beam scattering, and discuss the Fokker-Planck formalism that describes the evolution of beam distribution and beam lifetime. In Section 3, we review formalisms for the non-adiabatic regime of transition crossing, emphasizing self-field mismatching and microwave instability. Space charge and beam-beam effects are briefly discussed in Section 4. Impedances and more conventional instabilities [8] are discussed in Section 5. The conclusion is given in Section 6.

## 2. INTRA-BEAM SCATTERING

Beam growth caused by intra-beam scattering (IBS) is of primary concern during both injection and storage of the heavy ion beams in RHIC. At injection, the IBS growth time for the momentum spread is about 3 minutes. Alternate filling of the two rings, each with 60 bunches, needs to be done within about 1 minute to prevent difficulty in transition crossing and top-energy RF recapture. At storage, emittance growth occurs in both the transverse and longitudinal dimension. Collimation systems are designed to remove particles escaped from the RF buckets.

### 2.1 Beam Rest Frame Hamiltonian

Intra-beam scattering mechanism can be described using the rest frame $(x, y, z, t)$ of the circulating synchronous particle. Measure dimensions in units of the characteristic distance $\xi_0$ with $\xi_0^3 = r_0 \rho^2 / \beta^2 \gamma^2$, time in units of $\rho/\beta\gamma c$, and energy in units of $\beta^2 \gamma^2 Z^2 e^2 / 4\pi\epsilon_0 \xi_0$, where $r_0 = Z^2 e^2 / 4\pi\epsilon_0 m_0 c^2$ is the classical radius, $\beta c$ and

$\gamma m_0 c^2$ are the velocity and energy of the synchronous particle, and $\rho$ is the radius of curvature in bending regions of magnetic field $B_0$. The Hamiltonian for particles in a simple system with bending dipoles and focusing quadrupoles of strength $n_1 = -(\rho/B_0)(\partial B_y/\partial x)$ is [9,10]

$$H = \begin{cases} \frac{1}{2}\left(P_x^2 + P_y^2 + P_z^2\right) + \frac{1}{2}x^2 - \gamma x P_z + V_C & \text{(bending section)} \\ \frac{1}{2}\left(P_x^2 + P_y^2 + P_z^2\right) - \frac{n_1}{2}(x^2 - y^2) + V_C + U_s & \text{(straight section)} \end{cases} \quad (1)$$

where $U_z$ is the potential provided by the RF system. The Coulomb potential is non-relativistic in the rest frame:

$$V_C = \sum_j \frac{1}{\sqrt{(x_j - x)^2 + (y_j - y)^2 + (z_j - z)^2}}. \quad (2)$$

In terms of dispersion function $D$ and betatron displacements $\beta_{x,y}$, this Hamiltonian is transformed to

$$\bar{H} = \frac{1}{2}\left(P_{\beta_x}^2 + P_{\beta_y}^2\right) + \frac{1 - \gamma^2 F_z}{2} P_z^2 + V_C + U_z, \quad (3)$$

where

$$F_z = \begin{cases} D + DD'' + (D')^2 & \text{(bending section)} \\ DD'' + (D')^2 & \text{(straight section)} \end{cases} \quad (4)$$

and

$$\langle F_z \rangle = \frac{1}{\gamma_T^2}. \quad (5)$$

Below transition energy, $\gamma < \gamma_T$, particles are in a positive-mass regime. In an idealized case that the machine lattice is uniform along the ring circumference, the Hamiltonian in the rest frame is time-independent. The particle system is thus conserved, so does the total temperature of the beam in the rest frame. The heat can be transfered from the high temperature to the low temperature direction. The system eventually reaches an equilibrium state when the temperature (i.e. rest-frame velocity) is the same in all directions.

In an actual alternating-gradient focusing ring, the beam sees a time dependent potential modulated by the ring lattice frequency. The beam structure absorbs "phonons" and heats up [11]. Intra-beam multiple scattering manifests as a mixture of thermal equalization and temperature growth asymptotically approaching equal temperature in all directions in the rest frame.

Above transition energy, $\gamma > \gamma_T$, the beam is in a negative-mass regime. The Hamiltonian (Eq. 3) indicates that even in the case of a uniform machine lattice, beam temperature can grow simultaneously in the longitudinal and transverse directions.

## 2.2 IBS Growth Scaling Laws

Energy exchange and temperature increase in the beam rest frame manifest as variation of beam emittance and momentum spread in the laboratory frame. In the laboratory frame, the rate of emittance and momentum growth is usually obtained [12,13] assuming multiple small-angle scattering among Gaussian-distributed beams. In the case that $D/\beta_x^{1/2}$ is nearly constant (e.g. for FODO lattice), the growth rate formula can be simplified into the following expression [14,3],

$$\begin{bmatrix} \frac{1}{\sigma_p}\frac{d\sigma_p}{dt} \\ \frac{1}{\sigma_x}\frac{d\sigma_x}{dt} \\ \frac{1}{\sigma_y}\frac{d\sigma_y}{dt} \end{bmatrix} = \frac{Z^4 N}{A^2} \frac{r_0^2 m_0 c^2 L_c}{8\gamma\epsilon_x\epsilon_y S_{rms}} F(\chi) \begin{bmatrix} n_b(1-d^2) \\ -a^2/2 + d^2 \\ -b^2/2 \end{bmatrix} \quad (6)$$

where $L_C \approx 20$ is the Coulomb logarithm, $\epsilon_{x,y} = \beta\gamma\sigma_{x,y}^2/\beta_{x,y} = \epsilon_N/6$ is the normalized rms transverse emittance, $S_{rms} = \pi m_0 c^2 \beta\gamma\sigma_s\sigma_p/cA = S/6$ is the rms longitudinal bunch area in phase space, $\chi = (a^2+b^2)/2$, $d = \frac{D\sigma_p}{(\sigma_x^2 + D^2\sigma_p^2)^{1/2}}$, $a = \frac{\beta_x d}{D\gamma}$, $b = \frac{\beta_y\sigma_x}{\beta_x\sigma_y}a$, $n_b$ is equal to 1 if the beam is azimuthally bunched, and is equal to 2 if it is not. For azimuthally bunched beams, $\sigma_s$ is the rms bunch length and $N$ is the number of particles per bunch; for unbunched beams, $N$ is the total number of particles and $\sigma_s = \sqrt{\pi}R$. In Eq. 6, $F(\chi)$ is an analytic function given by

$$F(\chi) = \frac{-3 + (1+2\chi)I(\chi)}{1-\chi} \quad (7)$$

where

$$I(\chi) = \begin{cases} \frac{1}{\sqrt{\chi(\chi-1)}} \text{Arth}\sqrt{\frac{\chi-1}{\chi}} & \chi \geq 1; \\ \frac{1}{\sqrt{\chi(1-\chi)}} \arctan\sqrt{\frac{1-\chi}{\chi}} & \chi < 1 \end{cases} \quad (8)$$

The growth rates are linearly proportional to the number of the particle $N$ in the beam, and are strongly dependent ($\sim Z^4/A^2$) on the charge state of the particle. Except for the form factors $\chi$, $d$, $a$, and $b$ that depend on the ratio of the beam amplitudes in different dimension, the rates are inversely proportional to the six dimensional phase space area. Below transition energy, the asymptotic distribution corresponds to the condition

$$\langle\frac{\sigma_x}{\beta_x}\rangle \approx \langle\frac{\sigma_y}{\beta_y}\rangle \approx \frac{\sigma_p}{\gamma}, \quad \gamma \ll \gamma_T.$$

Above transition energy, the asymptotic distribution corresponds to the condition

$$\sqrt{n_b n_c} \langle \sigma_x \rangle \approx \langle D \rangle \sigma_p, \quad \gamma \gg \gamma_T$$

where $n_c$ is equal to 1 if the horizontal and vertical motion are uncoupled, and is equal to 2 if they are fully coupled. In order to confine the horizontal emittance growth, we intensionally couple the horizontal and vertical motion. The growth rates at high energy become

$$\begin{bmatrix} \dfrac{1}{\sigma_p} \dfrac{d\sigma_p}{dt} \\ \\ \dfrac{1}{\sigma_x} \dfrac{d\sigma_x}{dt} \end{bmatrix} = \dfrac{Z^4 N}{A^2} \dfrac{\pi r_0^2 m_0 c^2 L_c}{16 \gamma_T \epsilon_x \epsilon_y S_{rms}} \begin{bmatrix} n_b(1-d^2)/d \\ \\ d/n_c \end{bmatrix}, \qquad (\gamma \gg \gamma_T) \quad (9)$$

which is to the first order independent of the beam energy.

## 2.3 Fokker-Planck Approach and IBS Beam Loss

In order to evaluate the beam intensity lifetime, we use the Fokker-Planck equation to describe the evolution of particle distribution in the phase space. The general 6 dimensional (6-D) equation can be greatly simplified by the fact that the IBS growth time is typically much longer than the synchrotron-oscillation period, which is again much longer than the multiple collision relaxation time. Since the leading source of beam loss is expected to be in the longitudinal direction due to the limited voltage of the RF system, we further assume in the transverse directions a time-evolving Gaussian distribution. After averaging over the machine circumference and the synchrotron phase for all the particles involved in the collision [15], we obtain a 1-D Fokker-Planck equation of the density function $\Psi(J)$ in the longitudinal direction in terms of the action variable $J$,

$$\dfrac{\partial \Psi}{\partial t} = -\dfrac{\partial}{\partial J}(F\Psi) + \dfrac{1}{2}\dfrac{\partial}{\partial J}\left(D\dfrac{\partial \Psi}{\partial J}\right), \text{ with } \begin{cases} J = 0: & -F\Psi + \dfrac{D}{2}\dfrac{\partial \Psi}{\partial J} = 0, \\ J = J_{max}: & \Psi = 0. \end{cases}$$

(10)

Here, the drift coefficient is given by the expression

$$F(J) = \oint \dfrac{2ds}{\pi R} \int_0^{\frac{1}{4}} dQ \left.\dfrac{\partial W}{\partial J}\right|_\phi^{-1}(Q,J) \int_{J_{min}}^J \left.\dfrac{\partial W}{\partial J}\right|_\phi (Q',J') \left[A_F(\lambda_1) + A_F(\lambda_2)\right] \Psi(J') dJ'$$

(11)

and the diffusion coefficient is given by the expression

$$D(J) = \oint \frac{2ds}{\pi R} \int_0^{\frac{1}{4}} dQ \left[ \left. \frac{\partial W}{\partial J} \right|_\phi (Q, J) \right]^{-1} \int_{J_{min}}^{\hat{J}} \left. \frac{\partial W}{\partial J} \right|_\phi (Q', J') \left[ A_D(\lambda_1) + A_D(\lambda_2) \right] \Psi(J') dJ' \tag{12}$$

where

$$A_F(\lambda) = -\frac{2Z^4 r_0^2 L_c E}{A^2 \beta^2 \gamma^4} \frac{I_F(\lambda)}{\sigma_x \sigma_y}, \quad A_D(\lambda) = \frac{Z^4 r_0^2 L_c E^2}{A^2 \gamma^3 h \omega_s} \frac{I_D(\lambda)}{\sigma_x \sigma_y}, \tag{13}$$

$\omega_s$ is the revolution frequency,

$$\lambda_{1,2} = \frac{h \omega_s g}{\gamma \beta^2 E}(W \mp W'), \quad g = \frac{1}{2}\sqrt{\frac{\beta \gamma \beta_{x,y}}{\epsilon_{x,y}}}, \tag{14}$$

$$\left. \frac{\partial W}{\partial J} \right|_\phi^{-1} (Q, J) = 8k \, \mathrm{K}(k) \cos 2\pi Q \left[ 1 - 4\xi \sin^2 2\pi Q + O(\xi^2) \right], \tag{15}$$

and $\xi = \exp\left[-\pi \mathrm{K}'(k)/\mathrm{K}(k)\right]$ and $\mathrm{K}'(k) = \mathrm{K}(\sqrt{1-k^2})$. The first integrals in Eqs. 11 and 12 represent the average over the machine lattice; the second integral represents the average over synchrotron-oscillation phase; while the third integral describes particles of different action $J'$ involved in the collision. The integration over $J'$ is performed such that $k(J') \sin 2\pi Q' \approx \sin[\phi(Q,J)/2]$, extending from $J_{min}$ to the bunch edge $\hat{J}$, with $k(J_{min}) \approx [\sin \phi(Q,J)/2]$. For a round beam with near constant $D/\beta_x^{1/2}$, we have

$$\begin{aligned} I_F(\lambda) &= 2g^2 \mathrm{sgn}(\lambda) e^{-(D\gamma\lambda/2\sigma_x)^2} \left\{ 1 - \sqrt{\pi} |\lambda| e^{\lambda^2} [1 - \Phi(\lambda)] \right\}, \\ I_D(\lambda) &= g e^{-(D\gamma\lambda/2\sigma_x)^2} \left\{ \sqrt{\pi}(1 + 2\lambda^2) e^{\lambda^2} [1 - \Phi(\lambda)] - 2|\lambda| \right\}, \end{aligned} \tag{16}$$

where $\Phi$ is the error function, and $\mathrm{sgn}(\lambda)$ is 1 if $\lambda \geq 0$, and is $-1$ if otherwise.

Starting from an initial distribution, Eq. 10 can be iterated to yield the time evolution of the longitudinal particle distribution, as shown in Fig. 1. Based on this information, evolution of the transverse beam dimension is obtained from the growth rate formulae given in Section 2.2. Beam loss through the RF bucket boundary is evaluated from the reduction of the integrated density $\Psi(J)$ over $J$. Typically, the longitudinal distribution under intra-beam scattering is Gaussian-like with zero density at the edge of the RF bucket, as shown in Fig. 1.

## 2.4 Luminosity and Beam Cooling

Due to emittance growth and intensity loss caused by intra-beam scattering, the luminosity in RHIC is significantly reduced at the early stage of storage, as

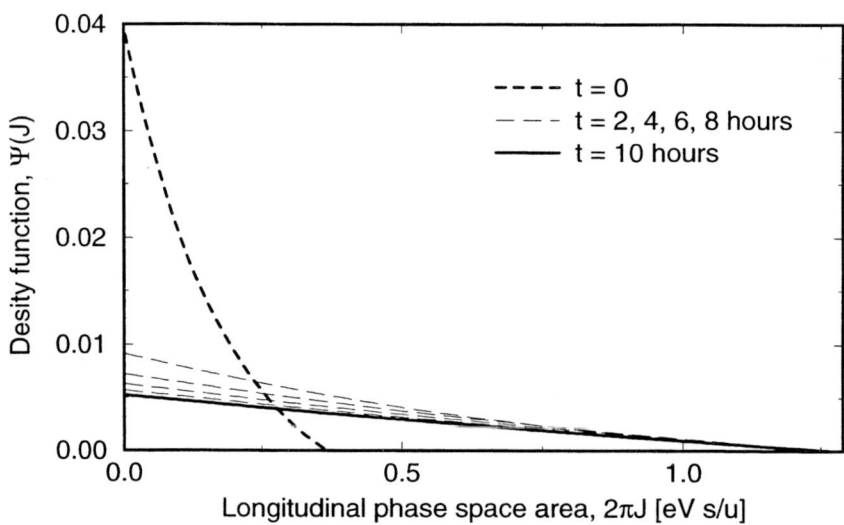

**FIGURE 1.** Evolution of the longitudinal density distribution under intra-beam scattering during a 10-hour storage in RHIC. The edge of RF bucket corresponds to 1.29 eV·s/u.

shown by the solid line in Fig. 2. Increasing peak RF voltage only improves the performance modestly, since transverse growth is strong and momentum acceptance can be a problem [7]. The ultimate improvement can be achieved if beam cooling methods are adopted. Fig. 2 shows the improvement of integrated luminosity over a 10-hour storage if transverse and longitudinal stochastic cooling [4] are employed. Cooling in both longitudinal and transverse planes provides an effective method to compensate for the beam growth, particle loss, and luminosity deterioration caused by IBS. With longitudinal and transverse cooling of bandwidth 4–8 GHz, the longitudinal beam loss resulted from the inadequacy of the RF voltage can be eliminated, and the transverse normalized beam emittance can be confined to about 30 mm·mr. With this scenario, the $\beta^*$ at the crossing point can be lowered under 1 meter without exceeding the transverse aperture limit at the focusing triplets. The integrated luminosity can be increased by at least a factor of 2 during the 10-hour storage period. Experimentally, however, stochastic cooling studies for bunched beam at the Tevatron indicates unexpected difficulty [16,17] due to large coherent signal saturation. Nevertheless, stochastic cooling in RHIC is expected to be much easier since the beam, instead of tightly bunched, occupies the entire RF bucket due to IBS. Also, the high charge state greatly improves signal-to-noise ratio. Recently, electron cooling has also been studied for RHIC employing separate electron storage rings for electron damping [18].

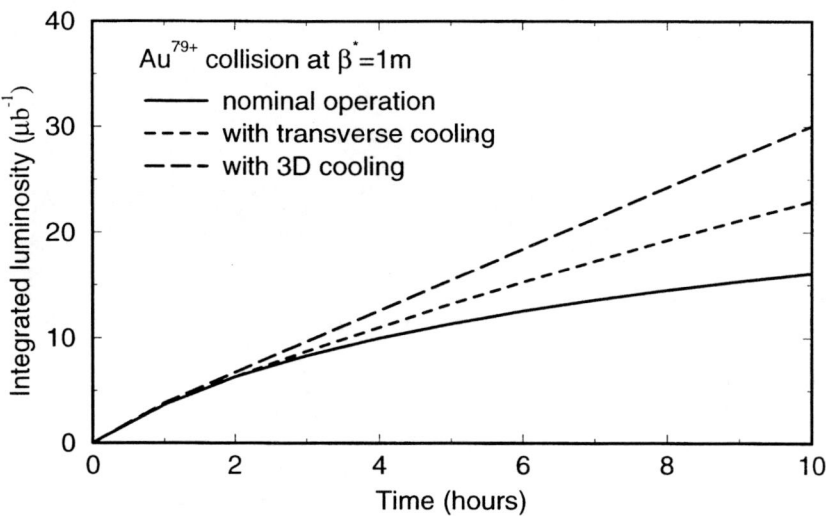

**FIGURE 2.** Integrated luminosities during a 10-hour storage for a) the nominal operation (without cooling), b) with transverse stochastic cooling, and c) with both transverse and longitudinal stochastic cooling. The bandwidth of the cooling system is assumed to be 4–8 GHz.

## 3. TRANSITION CROSSING

Transition energy crossing is likely to be the "bottle neck" for RHIC acceleration. In the presence of the beam self field, the bunch is mismatched to the RF bucket upon the shift of synchronous phase at crossing. The mismatch increases the momentum spread, enhancing the chromatic nonlinear effect and leading to particle loss. In the absence of a transition jump, this mechanism alone accounts for an intensity loss of about 70% at transition.

### 3.1 Non-Adiabatic Regime Formalism

During acceleration, the longitudinal particle motion is non-adiabatic within a characteristic time $\pm T_c$ near transition energy $\gamma_T$

$$T_c = \left( \frac{\pi E \beta^2 \gamma_T^3}{ZeV|\cos\phi_s|\dot\gamma h\omega_s^2} \right)^{\frac{1}{3}}, \tag{17}$$

where $E$ is the total energy of the particle, $\dot\gamma$ is the ramp rate, $\phi_s$ is the synchronous phase, and $h$ and $V$ are harmonic number and voltage of the RF system. The dynamics can be best described by the longitudinal amplitude function $\beta_L$ given approximately by [10]

$$\frac{\beta_L}{kT_c} = \frac{\pi}{3} x \left[ J^2_{-\frac{1}{3}}(y) + N^2_{-\frac{1}{3}}(y) \right] \approx 1.58 - 1.15x \tag{18}$$

where $y = 2x^{3/2}/3$, $x = |\Delta t|/T_c$, $k = ZeV|\cos\phi_s|/2\pi h$, and $\Delta t$ is the time delay from $\gamma_T$. With a normalized time $d\tau = k dt$, the longitudinal motion is described by a Hamiltonian expressed in terms of the action-angle variables $\varphi$ and $J$ as

$$H(\varphi, J; \tau) = \pm J/\beta_L. \tag{19}$$

The synchrotron frequency is

$$\Omega_s = k\beta_L^{-1}, \tag{20}$$

approaching a minimum but non-zero value of $\Omega_s(0) = 0.63\, T_c^{-1}$ at transition. The maximum excursions in the RF phase $\phi$ and energy deviation $W = \Delta E/h\omega_s$ are

$$\hat{\phi} = \sqrt{2\gamma_L J}, \text{ and } \hat{W} = \sqrt{2\beta_L J}, \tag{21}$$

where $1 + \alpha_L^2 = \beta_L \gamma_L$. For a bunch of phase space area $S = 2\pi J$ at transition, the RF phase spread of the bunch reaches the minimum value

$$\hat{\phi} = 0.52 \left( S/kT_c \right)^{1/2} \tag{22}$$

while the momentum spread reaches the maximum value

$$(\widehat{\Delta p/p}) = 0.71 h\omega_s \left( kT_c S \right)^{1/2} / E\beta^2. \tag{23}$$

## 3.2 Chromatic Nonlinearity and Self-Field Mismatch

The chromatic nonlinear effect [19,10] originates from the intrinsic mistiming among particles of different momentum deviation crossing transition, as shown in Fig. 3. Using the perturbation formalism for the non-adiabatic regime [20,21,5], the Hamiltonian system including the nonlinearity can be solved to obtain the longitudinal emittance growth during transition

$$\frac{\Delta S}{S} \approx \begin{cases} 0.76 \dfrac{T_{nl}}{T_c}, & \text{for } T_{nl} \ll T_c; \\ e^{\frac{4}{3}\left(\frac{T_{nl}}{T_c}\right)^{3/2}} - 1, & \text{for } T_{nl} \geq T_c, \end{cases} \tag{24}$$

where the total nonlinear time $\pm T_{nl}$ is given by

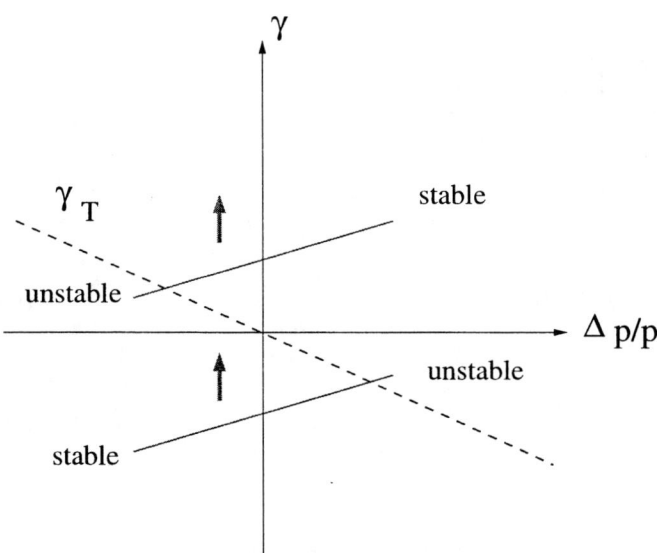

**FIGURE 3.** Mechanism of chromatic nonlinear effect at transition crossing. Part of the beam is unstable below transition before the RF phase switch, while the other part is unstable above transition.

$$T_{nl} = \left|(\alpha_1 + \frac{3\beta^2}{2})\right| \frac{\sqrt{6}\hat{\sigma}_{\Delta p/p}\gamma_T}{\dot{\gamma}} \qquad (25)$$

where $\alpha_1$ is the nonlinear momentum compaction factor [10]. In the absence of a transition jump, the chromatic effect alone (low intensity case) accounts for about 20% beam loss at transition in RHIC.

Both reactive and resistive impedances cause mismatch in the nominal bunch shape at the time the synchronous phase is jumped at transition. A reactive impedance changes the focusing force of the RF system differently below and above transition. The amount of mismatch is then proportional to the ratio of the self field to the RF field provided by the accelerating cavities. Again, we use the perturbation approach [22,5] to solve the Hamiltonian system in the non-adiabatic regime that includes the self field potential. For a parabolic distribution, the effective increase in the bunch area due to the mismatch, induced by a coupling impedance $|Z_L/n|$ around the bunch frequency, is

$$\frac{\Delta S}{S} = \frac{2h\hat{I}\,|Z_L/n|}{\hat{V}|\cos\phi_s|\,\hat{\phi}^2}, \qquad (26)$$

where

$$\hat{I} = \frac{3hNZe\omega_s}{4\hat{\phi}} \qquad (27)$$

is the peak current. The effect is usually a quadrupole-mode bunch tumbling and growth. For RHIC, the space charge impedance is about $|Z_L/n| \approx 1.2\ \Omega$. The induced force stretches the bunch momentum around transition. This momentum increase enhances the chromatic nonlinear effect, together resulting in a beam loss of about 70% in the absence of a transition jump.

## 3.3 Microwave Instability at Transition

As synchrotron frequency approaches a minimum value (Eq. 20) at transition, microwave instability is likely to occur. For a bunched beam, the instability threshold is inversely proportional to the ratio between the self field and the focusing field, i.e. inversely proportional to the bunch length (Eq. 21) to the cubic power. Quantitatively, we can solve the Vlasov equation in terms of the action variable using the non-adiabatic regime formalism [22]. An exact solution for the instability threshold can be obtained for the parabolic beam distribution as

$$D_{\parallel} \approx \frac{4h\hat{I}\left|Z_{\parallel}/n\right|}{9V|\cos\phi_s|\,\hat{\sigma}_\phi^2} \geq 1. \tag{28}$$

A capacitive (or inductive) longitudinal coupling impedance $Z_{\parallel}$ at a broad-band frequency will cause a microwave instability during a time

$$T_{mw} \approx 1.37\ (D_{\parallel} - 1)\ T_c$$

after (or before) transition. In the absence of the transition jump in RHIC, the beam at the nominal intensity is near the instability threshold due to the space charge force alone.

## 3.4 Compensation with a Transition Jump

The cause of strong chromatic nonlinear effect and self-field effect at RHIC is the slow ramp rate of the superconducting magnets. The most efficient way to reduce these effects is to effectively increase the crossing rate by a transition energy jump, which is achieved by pulsing two families of quadrupole correction magnets for about 60 ms. The change of $\gamma_T$ of the lattice is about $\pm 0.4$ units, effectively increasing the crossing rate by about a factor of 10.

A key issue in the design of transition jump schemes is to minimize the enhancement of machine lattice distortion. As experimentally demonstrated on the AGS [23,24] (Fig. 4), the jump could significantly enhance the nonlinear factor $\alpha_1$, and sextupoles were arranged to reduce the nonlinearity. With RHIC, we designed a "matched, first order" transition jump scheme [6] with which the longitudinal emittance growth can be limited to less than 20% at transition with minimum disruption to the transverse particle motion. Fig. 5 shows the residual beam growth at transition in the presence of the design transition jump.

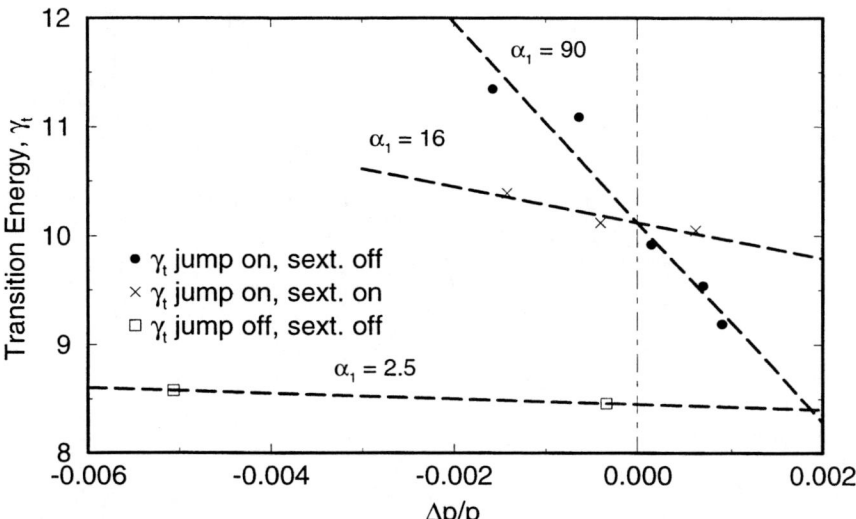

**FIGURE 4.** Enhancement of lattice nonlinearity due to transition jump and improvement from chromaticity sextupoles. The experimental study was done at the AGS with gold beam crossing transition. A "second order" transition jump scheme was used.

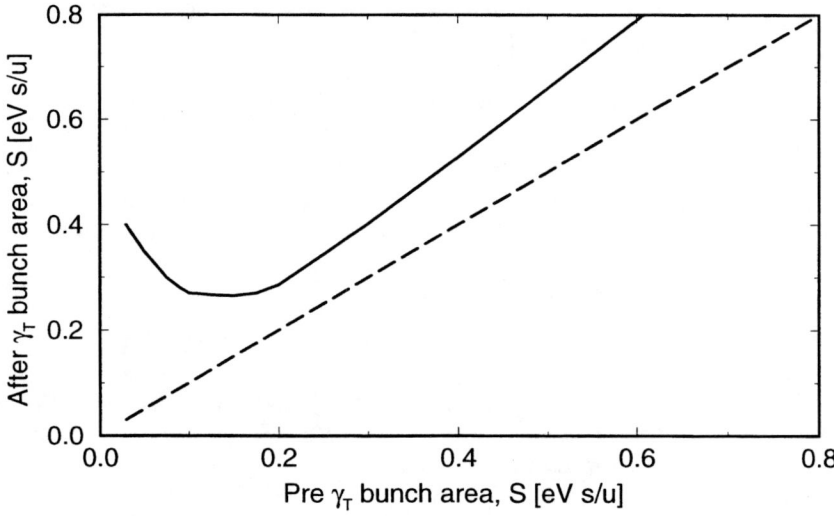

**FIGURE 5.** Longitudinal beam growth at RHIC transition energy. The growth for beams with smaller bunch area is mainly due to self field mismatch and microwave instability, while the growth for beams with larger bunch area is mainly due to the nonlinearity in momentum. The design bunch area is 0.3 eV·s/u.

## 4. SPACE CHARGE AND BEAM-BEAM EFFECTS

The mean field of the beam space charge modulates the transverse tunes of the particles and produces a tune spread that may compromise the dynamic aperture of the machine. The tune spread $|\Delta Q_{sc}|$ is the largest at the injection energy,

$$|\Delta Q_{sc}| = \frac{3Z^2 N r_0}{2\pi A \beta \gamma^2 \epsilon_N B_f} \approx \begin{cases} 0.030 \dfrac{N}{10^9} \dfrac{10[\mu m]}{\epsilon_N} & \text{for gold} \\ \\ 0.015 \dfrac{N}{10^{11}} \dfrac{20[\mu m]}{\epsilon_N} & \text{for proton} \end{cases} \qquad (29)$$

where $\epsilon_N$ is the normalized 95% emittance, and

$$B_f = \frac{\sigma_s}{\sqrt{2\pi} R}$$

is the bunching factor.

The mean field of the space charge between the colliding beams produces a tune spread at collision. With zero crossing angle, the head-on beam-beam tune spread per crossing is [25]

$$|\Delta Q_{bb}| = \frac{3Z^2 N r_0}{2 A \epsilon_N} \approx \begin{cases} 0.0012 \dfrac{N}{10^9} \dfrac{20[\mu m]}{\epsilon_N} & \text{for gold} \\ \\ 0.0037 \dfrac{N}{10^{11}} \dfrac{20[\mu m]}{\epsilon_N} & \text{for proton} \end{cases} \qquad (30)$$

which are independent of the energy and $\beta$ functions. Since the beams are separated immediately around the interaction point before entering into the triplet region, there are no parasitic collisions for the nominal storage scenario of 60 bunches per ring.

## 5. IMPEDANCES AND INSTABILITIES

In RHIC, the narrow-band impedance spectrum is dominated by RF cavities, while the broadband impedance spectrum is dominated by the resistive wall effect at very low frequencies and by kicker magnets in the intermediate frequency range up to 3 GHz [8].

The threshold for longitudinal microwave instabilities is lowest for heavy ions during transition crossing, and for protons during re-bucketing [8]. Transverse and longitudinal dampers are planned to compensate for coupled bunch instabilities [8]. The chromaticities of the machine are designed to shift from negative to positive for about 5 units in 30 ms at transition to prevent potential head-tail instability. Electron cloud effect has been [26] found to be tolerable during nominal operation with 60 bunches per ring.

# 6. CONCLUSIONS

Intra-beam scattering is the leading mechanism of luminosity degradation, emittance growth, and beam loss for RHIC. The beam is intentionally coupled in the transverse directions to reduce horizontal beam growth. The effects can be compensated by increasing the RF voltage, by a quicker re-filling, and ultimately by beam cooling methods. Transition crossing in RHIC is complicated by chromatic nonlinear effect, beam self-field mismatch, and microwave instability. The effects will be compensated by a first-order transition jump.

In year 2000, RHIC will enter its first year of operation. The machine will provide a test bed to verify the predictions discussed in this paper.

# REFERENCES

1. *RHIC Design Manual,* Brookhaven National Laboratory, Upton, New York (1998).
2. J. Wei, M. Harrison, *The RHIC Project — Design, Status, Challenges, and Perspectives,* XVI RCNP Osaka Inter. Sym. on Multi-GeV High-Performance Accelerators and Related Technology (Osaka, 1997).
3. J. Wei, *Evolution of Hadron Beams under Intra-beam Scattering,* Partical Accelerator Conference (Washington, D.C., 1993), p.3653.
4. J. Wei, *Stochastic Cooling and Intra-Beam Scattering in RHIC,* Workshop on Beam Cooling and Related Topics (Montreux, 1994), p. 132 (CERN 94-03).
5. J. Wei, *Longitudinal Dynamics of the Non-Adiabatic Regime on Alternating-Gradient Synchrotrons,* Ph. D. dissertation, Stony Brook, New York (1990); revised 1994.
6. S. Peggs, S. Tepikian, D. Trbojevic, *A First Order Transition Jump at RHIC,* Particle Accelerator Conference (Washington, D.C., 1993), p. 168.
7. J. Wei, J. Kewisch, V. Ptitsin, J. Rose, *RHIC Longitudinal Parameter Revision,* European Particle Accelerator Conference (Stockholm, 1998) p. 377.
8. M. Blaskiewicz, D.P. Deng, W.W. MacKay, V. Mane, S. Peggs, A. Ratti, J. Rose, T. Shea, J. Wei, *Collective instabilities in RHIC, Rev. 1.3,* RHIC/AP/36, Brookhaven National Laboratory (1994).
9. J. Wei, X-P. Li, A.M. Sessler, *The Low-Energy State of Circulating Stored Ion Beam: Crystalline Beams,* Physical Review Letters, **73** (1994) 3089.
10. *Handbook of Accelerator Physics and Engineering,* edited by A. Chao and M. Tigner, World Scientific (Singapore, 1999).
11. X-P. Li, A.M. Sessler, J.Wei, *Crystalline Beam in a Storage Ring: How Long Can It Last?* European Accelerator Conference (London, 1994), p. 1379.
12. A. Piwinsky, *Intra-Beam-Scattering,* CERN 92-01, CERN Accelerator School (Gifsur-Yvette, Paris, 1984), p. 405.
13. J. Bjoken, S.Mtingwa, *Intrabeam Scattering,* Particle Accelerators, **13** (1983) 115.
14. G. Parzen, *Intrabeam Scattering at High Energies,* Nuclear Instruments & Methods, **A256** (1987) 231.
15. J.Wei and A.G.Ruggiero, *Beam Life-Time with Intra-Beam Scattering and Stochastic Cooling,* Particle Accelerator Conference (San Francisco, 1991), p. 1869.

16. J. Marriner, *Theory, Technology, and Technique of Stochastic Cooling*, Workshop on Beam Cooling and Related Topics (Montreux, 1994), p. 14 (CERN 94-03).
17. D. Möhl, *The Status of Stochastic Cooling*, Nuclear Instruments & Methods, **A 391** (1997) 164.
18. A. Burov, V. Danilov, P. Colestock, Ya. Derbenev, *Electron Cooling for RHIC*, Fermilab-TM-2058 (1998).
19. K. Johnsen, *Effects of Non-Linearities on the Phase Transition*, Proceedings of CERN Symposium on High-Energy Accelerators and Pion Physics (Geneva, 1956), p. 106.
20. K. Takayama, *Phase Dynamics near Transition Energy in the Fermilab Main Ring*, Particle Accelerators, **14** (1984) 201.
21. S.Y. Lee and J. Wei, *Nonlinear Synchrotron Motion Near Transition Energy in RHIC*, European Particle Accelerator Conference (Rome, 1988), p.764.
22. J. Wei, S.Y.Lee, *Microwave Instability near Transition Energy*, Particle Accelerators, **28** (1990) 77.
23. J. Wei, A. Warner, L. Ahrens, J.M. Brennan, W.W. MacKay, S. Peggs, A. Ratti, K. Reece, T. Roser, W.A. Ryan, C. Saltmarsh, T. Satogata, D. Trbojevic, W. Van Asselt, *Experimental Study of Slow-Rate Transition Crossing in AGS*, European Accelerator Conference (London, 1994), p. 976.
24. J. Wei, J.M. Brennan, L. Ahrens, M. Blaskiewicz, D-P. Deng, W.W. MacKay, S. Peggs, T. Satogata, D. Trbojevic, A. Warner, and W. Van Asselt, *Effects of Enhanced Chromatic Nonlinearity during the AGS $\gamma_T$ Jump*, Proc. 1995 Particle Accelerator Conference and International Conference on High-Energy Accelerators (Dallas, 1995), p. 3334.
25. S. Peggs, *Beam-beam collisions and crossing angles in RHIC*, RHIC/AP/169, Brookhaven National Laboratory (1999).
26. K.A. Drees, *Beam Induced Electron Clouds at RHIC*, RHIC/AP/150, Brookhaven National Laboratory (1998).

# Short-Bunch Production and Microwave Instability Near Transition

K.Y. Ng[1] and J. Norem[2]

[1] *Fermi National Accelerator Laboratory,*[3] *P.O. Box 500, Batavia, IL 60510*
[2] *HEP Division, Argonne National Laboratory, Argonne, IL 60439*

**Abstract.** Some methods of making short bunches are reviewed. The experiment performed at the Brookhaven AGS for bunching near transition is reported. Microwave instability for coasting beam and bunched beam near transition is discussed and simulations are presented.

## I   INTRODUCTION

For the proton driver of the muon collider, bunching of intense proton bunches to rms length $\sigma_\tau \leq 2$ ns at extraction is desirable. There are two primary reasons. First, the proton bunch length is the only piece of information transmitted to the pions produced in the target and muons resulting in the decay of the pions. The shorter the length of the proton bunches, the less cooling of the muons will be necessary. Second, it will be easier to separate the muons polarized in the two helicities. The shorter the proton bunch length will result in higher muon polarization. The following are some ways to achieve narrow bunches using rf gymnastics:

**(1) Lowering and increasing rf voltage**

The rf is reduced adiabatically until the bunch spreads out and fills the bucket. The rf voltage is raised again suddenly. In a quarter synchrotron oscillation, a narrow bunch is obtained. The adiabatic process may take very long in order to allow the bunch to follow the change in the bucket. However, for a high-intensity bunch to stay at low momentum spread for a long time, it is likely that the microwave instability will develop. In order to avoid instability, we can snap the rf voltage down suddenly so that the rf bucket changes from Fig. 1(a) to 1(b). The bunch will be lengthened after a quarter synchrotron oscillation. The rf voltage is then snapped up again as in Fig. 1(c) and finally the lengthened bunch rotates into a narrow bunch. Of course, the rf nonlinearity will show up during bunch rotations. A second or third-order harmonic cavity will help in cancelling the rf nonlinearity. In practice, this method can shorten the bunch by a factor of at most 3 to 4.

**(2) Debunching at unstable fixed point**

The rf phase is suddenly shifted by 180° so that the bunch originally centered at the stable fixed point in Fig. 2(a) finds itself centered at the unstable fixed point in Fig. 2(b). The bunch will therefore spread out along the separatrices. After a while, the rf phase is shifted back by 180° as in Fig. 2(c). Synchrotron oscillation between $\frac{1}{4}$ and $\frac{1}{2}$ period will rotate the bunch into a narrow one. Again nonlinearity of the rf will show up in the bunch shape and a higher-order harmonic cavity will

---

[3)] Operated by the Universities Research Association, Inc., under contract with the U.S. Department of Energy.

help. Also, this process may be slow because movement along the separatrices is slow. In practice, this method can shorten the bunch by a factor of at most 3 to 4.

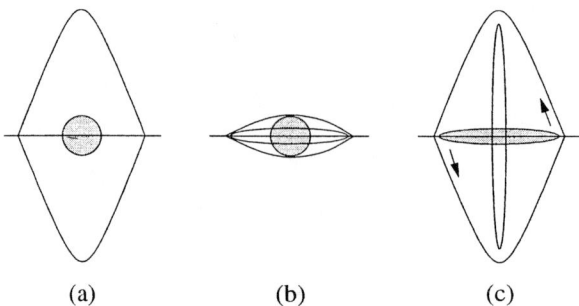

**FIGURE 1.** Bunch shortening is performed by snapping down the rf voltage $V_{rf}$, rotating for $\frac{1}{4}$ synchrotron oscillation, snapping up $V_{rf}$, and rotating for another $\frac{1}{4}$ synchrotron oscillation.

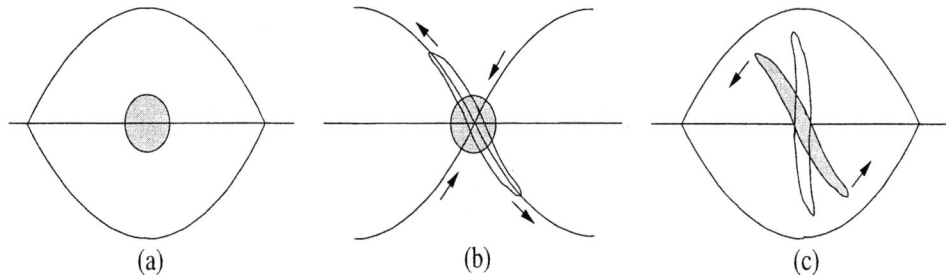

**FIGURE 2.** Bunch shortening is performed by shifting the rf phase by 180°, allowing the bunch to spread along the separatrices, shifting the rf phase by $-180°$, and rotating for $\frac{1}{4}$ to $\frac{1}{2}$ synchrotron oscillation.

### (3) Rebunching at higher frequency

During the end of the ramping, the frequency of the rf system is jumped to the next higher multiple of the circulating frequency. This process continues and the bunch will gradually be shortened by following the change in width of the bucket. In practice, the rf frequency of a rf system cannot be changed by very much. Therefore, there must be several rf systems with frequencies one above the other, so that the lower-frequency system will be replaced by the next higher one, etc. Thus, this method involve several high-frequency and high-voltage rf systems and will be expensive. Also the whole procedure will be slow.

### (4) Bunch-shortening near transition

At or near transition, there is little or no phase motion of the bunch particles. Thus, the particles continue to gain or lose energy according to the rf voltage they see, as is illustrated in Fig. 3(a), where the phase axis represents the rf phase of the particle when crossing the rf cavity gap. The bunch will shear in the momentum spread direction. A partial rotation will produce a narrow bunch, as depicted in Fig. 3(b). It is desirable to make the final bunching as fast as possible because of the large instantaneous currents produced in the ring, which can drive a variety of instabilities. The final bunching can be made quite fast if the transition energy can be moved farther away from the beam energy and/or the rf voltage can be raised during the final rotation, thus raising the synchrotron frequency just before extraction. One of the merits of this method is that no additional hardware, such as higher-harmonic cavities, is required. Also this method uses only the linear part of the rf wave and a small synchrotron phase rotation. Thus the rotation can be made quite linear. We do not need to operate the ring at or near transition all the time. With the flexible momentum-compaction (FMC) lattice, the transition gamma can be varied to a large extent by varying the gradients of a pair of quadrupoles [1]. An example is shown in Fig. 4, where each of the two F-quadrupoles at a distance about $\frac{1}{3}$ from the entrance and exit of the FMC module has been split into a pair denoted by QFS and QF2. By varying the gradients of the QFS and QF2, the large variation of transition gamma is shown in the left plot of Fig. 5, and the corresponding values of the momentum-compaction factor $\alpha$ are listed on the right. The betatron tunes have been kept nearly unchanged during the variation. As a

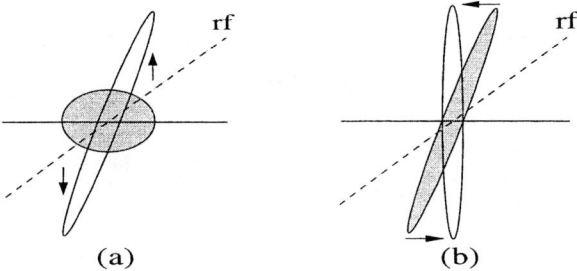

**FIGURE 3.** Bunch shortening is performed by allowing the bunch to shear very near to transition in the momentum direction, raising the rf, and rotating for $< \frac{1}{4}$ synchrotron oscillation.

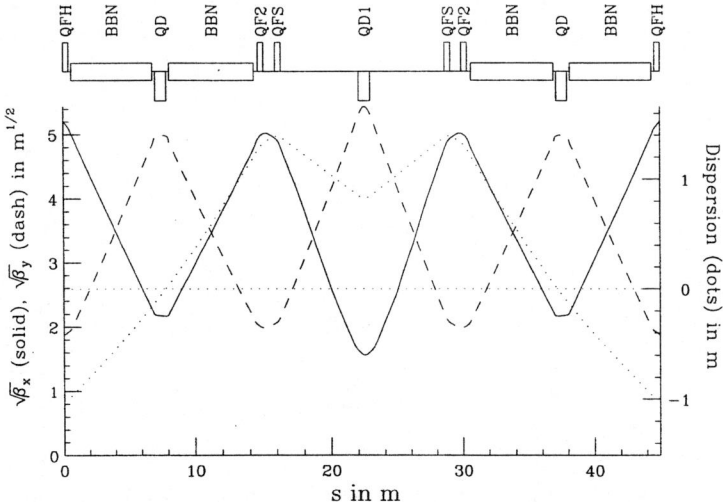

**FIGURE 4.** Two identical pairs of F-quadrupoles QFS and QF2 are installed at about $\frac{1}{3}$ from the entrance and exit of the FMC module. $\gamma_t$ can be varied by varying their gradients.

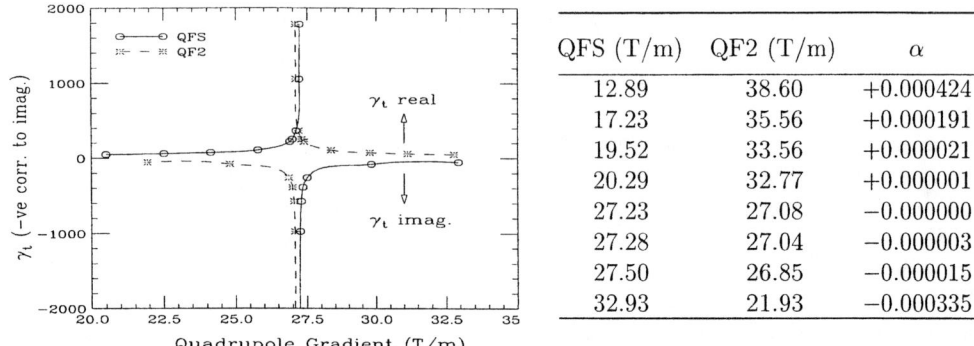

**FIGURE 5.** When the gradients of the quadrupole pairs are varied, the transition gamma (left) and the corresponding momentum compaction factor $\alpha$ (right) change by a wide range.

result, we can move close to or right at transition only at the moment when we want to make short bunches.

## II  EXPERIMENT ON BUNCH-SHORTENING NEAR TRANSITION

An experiment was performed at the Brookhaven AGS to demonstrate bunch-shortening near transition [2]. The operating mode of the AGS is shown in the left plot of Fig. 6. The maximum beam energy was reduced by flat-topping at 7 GeV, which shortened the acceleration period. The $\gamma_t$-jump system was modified to give a short flat-top period before the transition energy dropped. The beam was flat-

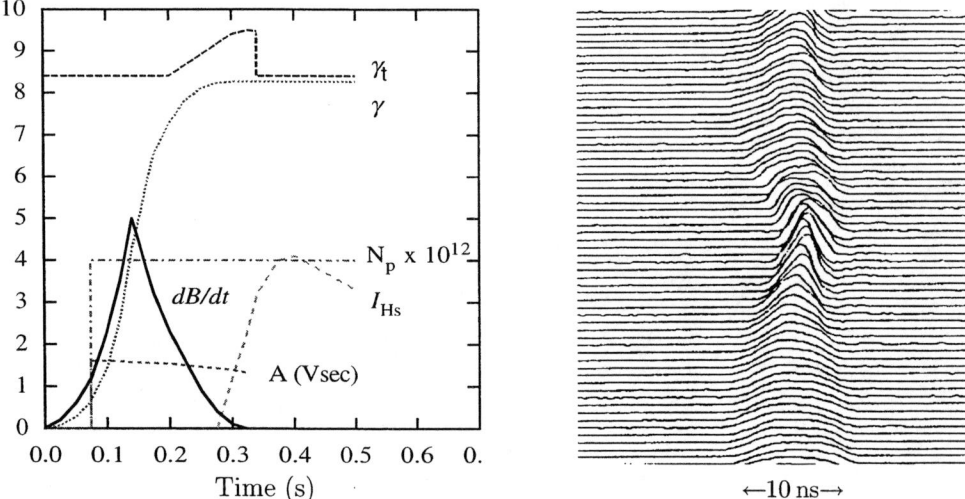

**FIGURE 6.** Left: The operating mode of the AGS, showing the magnet ramp $dB/dt$ in $10^{-1}$ T/s, beam gamma $\gamma$, transition gamma $\gamma_t$, sextupole current $I_{Hs}$ in 10 A, and longitudinal bunch area $A$ in eV-s. Right: Mountain range plots of the bunch for about 50 ms starting 10 ms before the transition energy was dropped to near the beam energy.

216

topped for 300 ms before the magnet guide field was raised slightly and then ramped down. Only one bunch was injected, which has a bunch area of $1.5\pm0.05$ eV-s and an intensity of 3 to $5\times10^{12}$ protons.

Because of the AGS $\gamma_t$-jump mechanism, the energy of the beam was kept at more than one unit of $\gamma$ below $\gamma_t$. At about 0.35 s after injection, $\gamma_t$ was dropped so that the beam was close to transition, with $|\gamma_t-\gamma|<0.05$. The beam started to shear and at the same time rotate slowly. Here, no special hardware was available to move the beam away from transition and no other higher rf voltage was provided to perform the final partial rotation depicted in Fig. 3(b). For this reason, the beam cannot be too close to $\gamma_t$, otherwise the partial rotation will take very long. Likewise, it cannot be too far from $\gamma_t$, otherwise the bunch will shear not only in the momentum direction as required in Fig. 3(a), but also in the phase direction, so that a tall and narrow beam will not be produced. During this run, a new measurement of $\gamma_t=8.34\pm0.05$ was made, which was best determined by measuring the synchrotron frequency.

A sample result is shown in the right plot of Fig. 6, which consists of mountain-range plots of the peak beam current and instantaneous current versus the machine phase sampled by the wall-gap monitor from 10 ms before the $\gamma_t$ dropped to 40 ms after. We can see obviously that the bunch became narrow after the transition energy was dropped to near the beam energy.

The beam current versus time during the final bunch rotation is shown in the left plot of Fig. 7. The bunch shape corresponding to the situation when it is shortest was shown in the right plot of Fig. 7, together with its shape before the transition energy was lowered. The shortest rms bunch length recorded was $\sigma_\tau=2.0$ ns and had been reduced 4 times.

Some important comments follow:
(1) During the whole experiment, no collective beam instability has been observed. However, the intensity has been 5 to 8 times below the required intensity of the $2.5\times10^{13}$ bunch for the proton driver of the muon collider. The proton driver ramps a batch of 4 such bunches at the cycling rate of 15 Hz. It is unclear whether collective instability will occur or not at such a high intensity.

(2) The slip factor $\eta$ is a function of momentum spread $\delta$:

$$\eta(\delta)=\eta_0+\eta_1\delta+\mathcal{O}(\delta^2)\,,\tag{2.1}$$

where

$$\eta_0=\frac{1}{\gamma_t^2}-\frac{1}{\gamma^2}\quad\text{and}\quad\eta_1=\frac{1}{\gamma^2}\left[\alpha_1+\frac{3\beta^2}{2}\right]+\eta_0\left[\alpha_1-\frac{1}{\gamma^2}\right]\,.\tag{2.2}$$

In above, $\alpha_0=\gamma_t^{-2}$ is the lowest-order momentum-compaction factor and $\alpha_1$ is the next higher order. They are defined as

$$C(\delta)=C_0\left\{1+\alpha_0\delta\left[1+\alpha_1\delta+\mathcal{O}(\delta^2)\right]\right\}\,,\tag{2.3}$$

with $C_0$ being the length of the closed orbit for the on-momentum particle and

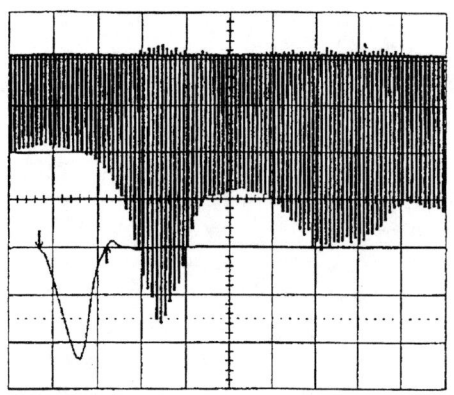 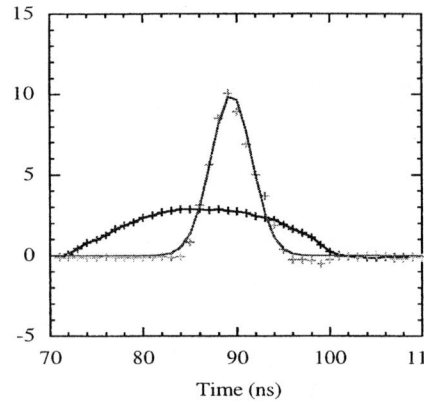

**FIGURE 7.** Left: Beam peak current versus time after $\gamma_t$ was dropped. The bunch was seen executing synchrotron oscillations slowly giving a maximum peak current as a result of the shearing in momentum. The bunch shape at its narrowest instant is shown at the corner. Right: Bunch shapes before and after bunching near transition, showing a final narrow 1.5 eV-s bunch in crosses of rms width $\sigma_\tau = 2.0$ ns together with the best fitted Gaussian in solid.

$C(\delta)$ the closed orbit length at fractional momentum spread $\delta$. Thus, even when $\gamma = \gamma_t$, $\eta$ is linear in $\delta$ and the time-slip $\Delta T$ per revolution period $T_0$, given by

$$\frac{\Delta T}{T_0} = \eta\delta = \eta_0\delta + \eta_1\delta^2 + \mathcal{O}(\delta^3) \;, \tag{2.4}$$

becomes quadratic in $\delta$. The drift in longitudinal phase will also be quadratic in $\delta$. Therefore, instead of shearing linearly in the momentum direction as illustrated in Fig. 3(a), the bunch will shear nonlinearly as in left plot of Fig. 8. The result is that the final bunch will be wider. This nonlinear phase-slip can be corrected, for example, by deploying sextupoles. It is clear that when $\alpha_1 = -\frac{3}{2}$ and $|\eta_0| \ll 1$, the first-order nonlinear drift will be eliminated.

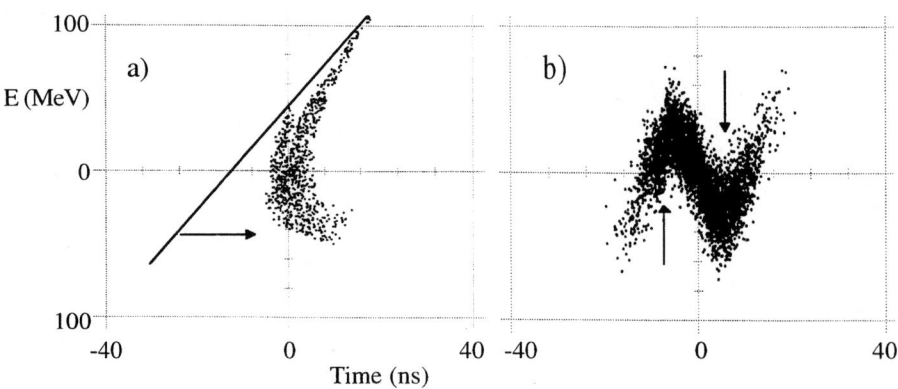

**FIGURE 8.** Effects of nonlinear $\eta$ and space charge on the final phase space distribution. Left plot shows the effect of a quadratic horizontal shear which occurs when $\alpha_1 \neq -3/2$. Right plot shows the effect of vertical shear from strong space-charge effects.

(3) For an intense proton beam, space-charge effect cannot be ignored. The wake potential is essentially proportional to the slope of the bunch linear distribution. Thus, staying near transition for too long, the bunch will shear into the shape of the tilted capital letter 'N', as shown by simulation in the right plot of Fig. 8. This is, in fact, a potential-well distortion of the rf wave, and can be cured, to a certain extent, by having rf systems of high frequencies, (see Sec. III B below).

# III  MICROWAVE INSTABILITY NEAR TRANSITION

## A  Analytic Solutions

In an operation near the transition energy ($\eta_0 \approx 0$), at least the next order, $\eta_1$ in Eq. (2.1), must be included for a meaningful discussion of the beam dynamics. Bogacz analyzed the stability of a coasting beam right at transition, $\eta_0 = 0$ [3], by including the $\eta_1$ term but neglecting other higher-order terms. For a Gaussian distribution with rms energy spread $\sigma_E$, he obtained an analytic expression for the growth rate at the revolution harmonic $n$:

$$\frac{1}{\tau_n} = -2\alpha_1 n \omega_0 \left(\frac{\sigma_E}{E_0}\right)^2 \phi_n \quad \text{with} \quad \tan\phi_n = \left[\frac{\mathcal{I}m\, Z_0^{\|}}{\mathcal{R}e\, Z_0^{\|}}\right]_n, \quad (3.5)$$

where $\mathcal{I}m\, Z_0^{\|} > 0$ implies capacitive and $\omega_0/(2\pi)$ is the revolution frequency of the on-energy particle which has energy $E_0$. He drew the conclusion that the beam will be completely stable. However, when he made this conclusion, he had in mind the assumption of $\alpha_1 > 0$ and $\phi_n > 0$, which is not always true. As a result, there will be microwave growth in general.

Holt and Colestock studied the same problem with coasting beam and Gaussian energy distribution, but allowing $\eta_0 \neq 0$ [4]. The dispersion relation is expressed in terms of the complex error function. Their conclusion is that there is no unstable region in the complex $Z_0^{\|}$-plane below transition. On the other hand, there are both stable and unstable regions above transition. They also claimed that their conclusion was supported by simulations. However, they did not specify the values of $\eta_0$ and $\eta_1$ in the simulations they presented or in their stability plots in the complex $Z_0^{\|}$-plane. It is hard to understand at least the situation below transition. It is clear that when $|\eta_0|$ is not too small, the contribution of $\eta_1$ is irrelevant. Thus their claim as stated can be interpreted as *no microwave instability below transition*, no matter how far away it is from transition. For this reason, this claim is quite questionable.

When we look into the stability plots of Holt and Colestock, we can see something that resembles a stability curve below transition, although the stability plots have been poorly drawn and are almost illegible. The presence of a stability curve implies the existence of both stable and unstable regions, in contradiction to their conclusion. We performed some simulations and have different results. We consider a coasting beam at 100 GeV in a hypothetoc ring of circumference 50 m, with a rms parabolic fractional momentum spread of 0.001, interacting with a broadband impedance of $Z_0^{\|}/n = 3.00\ \Omega$ at the resonance frequency of 600 MHz and quality factor $Q = 1$. This small size of ring is chosen because we want to limit the number of longitudinal bins around the ring so that not so many macro-particle will be necessary. The Keil-Schnell circle-approximated criterion gives a limit of $|Z_0^{\|}/n| = 1.00\ \Omega$ [5]. The results are shown in Fig. 9: the top 4 plots for $\eta = -0.005$ (below transition) and the lower 4 plots for $\eta = +0.005$ (above transition) at 0, 1200, 2400, and 3600 turns. We see that below transition irregularities develop at the low-momentum edge and the momentum spread broadens at the low-momentum side until the total spread is about 1100 MeV, about 2.75 times from the original total spread of about 400 MeV. This definitely confirms the occurrence of microwave instability below transition, and the eventual self-stabilization by overshooting. Above transition, irregularities also develop at the low-momentum edge and the momentum spread also broadens at the low-momentum edge. The total spread appears to be broader than the situation below transition. In addition, we see small bomb-like droplets launched at the low-momentum side, which is not observed below transition. We will come back to the simulations of coasting beam near transition later in Sec. III C.

## B  Bunched Beam Simulations

In this section, we study the stability of a bunched beam very close to transition. As an example, take a muon bunch in the proposed $50 \times 50$ GeV muon collider, which has a slip factor of $|\eta| = 1 \times 10^{-6}$. Everything we discuss here will apply to a proton bunch also, with the exception that the muons decay while the protons are stable. We will first discuss the situation with the decay of the muons taken into consideration, and later push the lifetime to infinity. We assume that sextupoles and octupoles are installed and adjusted so that the contributions of $\eta_1$ and $\eta_2$ become insignificant compared with $\eta_0$. The muon bunch we consider has an intensity of $N_b = 4 \times 10^{12}$ particles, rms width $\sigma_\ell = 13$ cm and rms fractional momentum spread $\sigma_\delta = 3 \times 10^{-5}$ or $\sigma_E = 1.5$ MeV. The impedance is assumed to be broadband with $Z_0^{\|}/n = 0.5\ \Omega$ at the angular resonant frequency of $\omega_r = 50$ GHz with quality factor $Q = 1$. The muons have an $e$-folding lifetime of 891 turns at 50 GeV in this collider ring. During the muon lifetime, there is negligible phase motion. Thus a bunching rf frequency system is not necessary. However, as will be explained below, rf systems are needed for the cancellation of potential-well distortion.

For bunched beams, there is the issue of potential-well distortion which must not be mixed up with the collective microwave instability. Potential-well distortion will change the shape of the bunch to something that looks like the right plot of Fig. 8, with the difference that the distortion of the beam does not come from the space-charge force, but mainly from the inductive part of the broadband impedance. The wake potential seen by a particle inside a Gaussian bunch at a distance $z$ behind the bunch center is shown in the left plot of Fig. 10 and is given by

$$V(z) = e \int_{-\infty}^{z} dz' \rho(z') W_0(z-z') = -\frac{eN\omega_r R_\|}{2Q \cos\phi_0} \mathcal{R}e\, e^{j\phi_0 - z^2/(2\sigma_\ell^2)} w\left[\frac{\sigma_\ell \omega_r e^{j\phi_0}}{c\sqrt{2}} - \frac{jz}{\sqrt{2}\sigma_\ell}\right],$$

where $\rho(z)$ is the bunch distribution, $W_0(z)$ the longitudinal wake function, $\sin\phi_0 =$

$1/(2Q)$, and $w$ is the complex error function. This distortion can be cancelled up to $\pm 3\sigma_\ell$ by 2 rf systems [6], which at injection are at frequencies $\omega_1/(2\pi) = 0.3854$ GHz and $\omega_2/(2\pi) = 0.7966$ GHz, with voltages $V_1 = 65.40$ kV and $V_2 = 24.74$ kV, and phases $\varphi_1 = 177.20°$ and $\varphi_2 = 174.28°$. This compensation is shown in the left plot of Fig. 10. Since only 2 sinusoidal rf's are used, the cancellation is not complete; however, the error is less than 1% of the original wake potential and is not important. Because of the lifetime of the muons, we first performed tracking for only 1000 turns in the time domain using the broadband wake function $W_0(z)$. The initial and final bunch distributions are shown in Fig. 11. During the simulation the

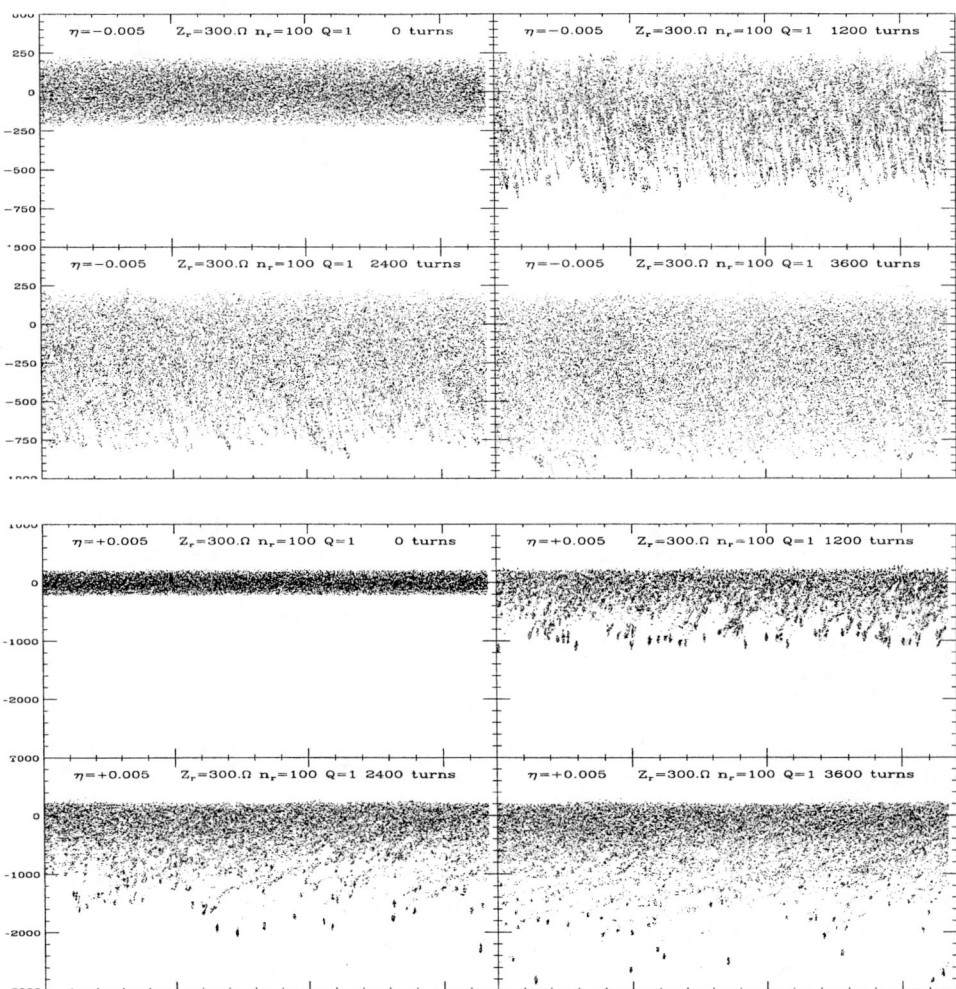

**FIGURE 9.** The top 4 plots and lower 4 plots are for $\eta = -0.005$ (below transition) and $\eta = +0.005$ (above transition), respectively, at 0, 12000, 24000, and 36000 turns. The impedance is a broadband with $Q = 1$, $Z_0^\parallel/n = 3.0\ \Omega$ at the resonant frequency of 600 MHz.

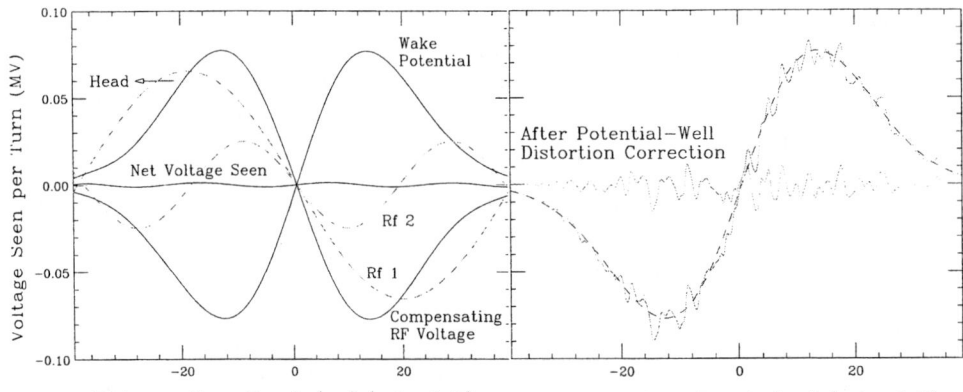

**FIGURE 10.** Left: Wake potential, compensating rf voltages, and net voltage seen by particles in the 13-cm bunch at injection. The compensating rf is the sum of two rf's represented by dashes. Right: Wake potential seen by the simulated bunch shown as dots is interlaced with the wake potential of an ideal smooth Gaussian bunch shown in dashes. The difference (center curve) represents the random fluctuation of the finite number of macro-particles.

compensating rf voltages were lowered turn by turn to conform with the diminishing bunch intensity due to the decay of the muons.

We see from the right plot of Fig. 11 that the bunch distribution has been very much distorted after 1000 turns. This comes mostly from the fact that the original distribution of the bunch in the left plot is not exactly Gaussian. It consists of $2 \times 10^6$ macro-particles randomly distributed according to a bi-Gaussian distribution. As a result, the wake potential of the actual bunch shown as a dotted curve in the right plot of Fig. 10 deviates slightly from and wiggles around the ideal wake potential curve of a smooth Gaussian bunch shown in dashes. The difference is the dotted jitter curve in the center of the plot. The fluctuation seen in the right plot of Fig. 11 is the result of the accumulation of this dotted jitter curve in 1000 turns with muon decay taken into account. Although this tiny fluctuation leads to a small potential-well distortion in one turn ($\leq 0.02$ MeV), it is unfortunate that it will accumulate turn after turn and will never reach a steady state, since the beam is so close to transition. This accumulated distortion can be computed exactly from the the dotted jitter curve. Any growth in excess will come from collective microwave instability. Note that the uncompensated potential-well distortion is quite different from the growth due to microwave instability. For the former, the growth in energy fluctuations every turn will be exactly by the *same* amount as given by the dotted jitter curve in the right Fig. 10 (if muon decay is neglected). This is because the wake potential of particles along the bunch does not depend on the energy distribution of the bunch, but only on its linear density and the latter is essentially unchanged since the particles do not drift much during the first 1000 turns. On the other hand, the initial growth due to microwave instability at a particular turn is proportional to the actual energy fluctuation at that turn

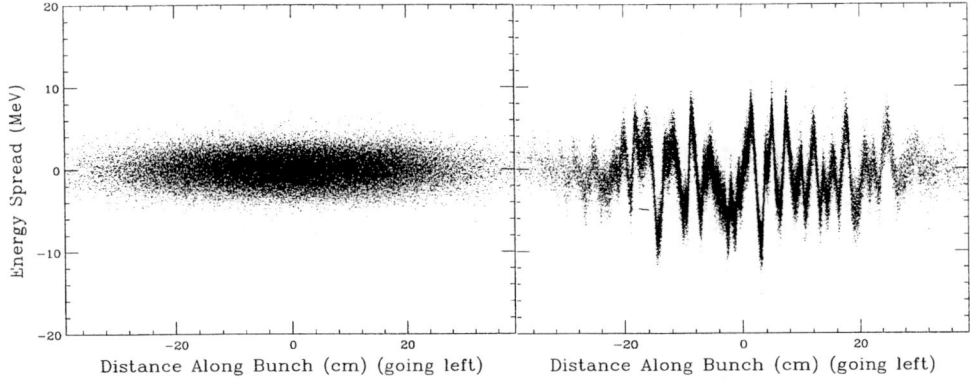

**FIGURE 11.** Simulation of the 13-cm bunch of $4 \times 10^{12}$ muons subject to a broad-band impedance with quality factor $Q = 1$ and $Z_\parallel/n = 0.5\ \Omega$ at the resonant angular frequency $\omega_r = 50$ GHz. The half-triangular bin width is 15 ps (0.45 cm) and $2 \times 10^6$ macro-particles are used. Left plot shows initial distribution with $\sigma_E = 1.5$ MeV and $\sigma_\ell = 13$ cm. Right plot shows distribution after 1000 turns with compensating rf's depicted in Fig. 10.

and the evolution of the growth is exponential. Thus, although the growth due to microwave instability is small at the beginning, it will be much faster later on when the accumulated energy fluctuations become larger. It is worth mentioning that even if the wake potential of the initial bunch with statistical fluctuations has been compensated exactly by the rf's, the bunch can still be unstable against microwave instability. An infinitesimal deviation from the bunch distribution can excite the collective modes of instability corresponding to some eigenfrequencies. In other words, the accumulated growth due to potential-well distortion is a static solution and this static solution converges very slowly close to transition until the momentum spread is large enough for the small $|\eta|$ to smooth the distribution. Microwave instability, on the other hand, is a time dependent solution.

In Fig. 12, the 3 plots on the left are for a 4000-turn simulation of the same muon bunch using $2 \times 10^6$ macro-particles with the decay of the muons considered. The two compensating rf systems are turned on. The first plot is for $\eta = 0$ so that microwave instability cannot develop. All the fluctuations are due to the residual potential-well distortion or the accumulation of the uncompensated jitters. The second and third plots are for, respectively, $\eta = -1 \times 10^{-6}$ (below transition) and $\eta = +1 \times 10^{-6}$ (above transition). We see that they deviate from the first plot, showing that there are growths due to microwave instability although the effect is small. The 3 plots on the right are the same as on the left with the exception that the muons are considered stable, or, in other words, the particles can be protons. We see that the second and third plots differ from the first one by very much (note the change in energy scale), indicating that microwave instability does play an important role for proton bunches in a quasi-isochronous ring. We also see that microwave instability is more severe above transition than below transition even when the beam is so close to transition.

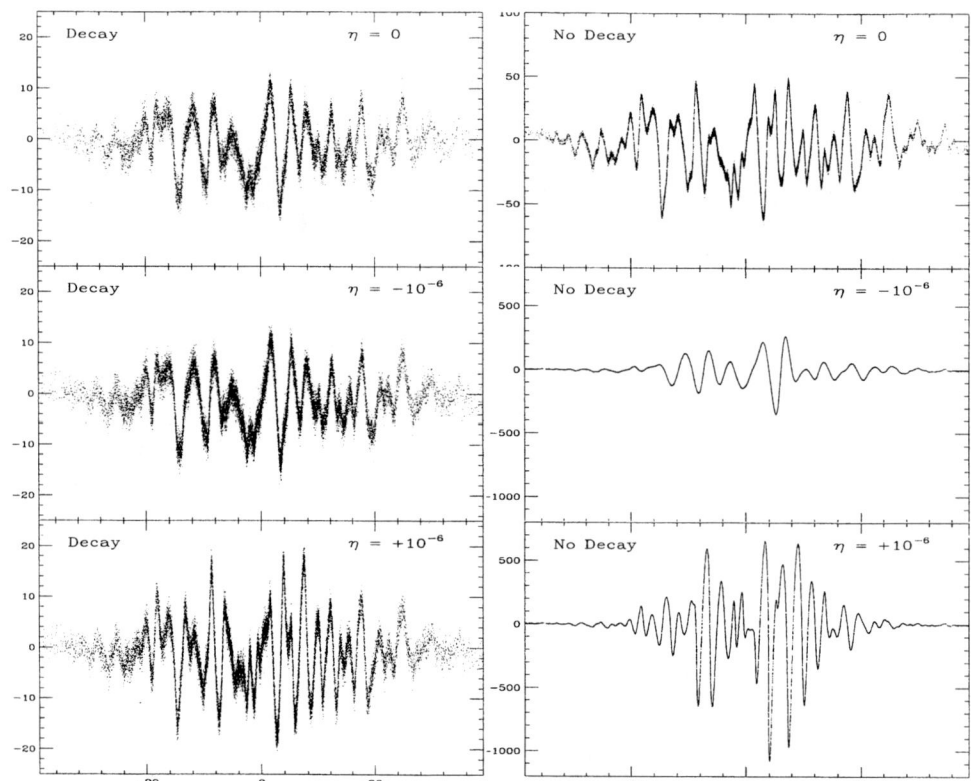

**FIGURE 12.** Phase-space plots of energy spread in MeV versus distance from bunch center in cm at the end of 4000 turns. All are simulating $4\times10^{14}$ micro-particles with $2\times10^6$ macro-particles. In the left 3 plots, the decay of the muons has been taken into account. The first left plot is for $\eta = 0$ so that it just gives the amount of potential-well distortion. The second and third plots are for, respectively, $\eta = -1 \times 10^{-6}$ and $+1 \times 10^{-6}$. The small deviations from the first plot are results of microwave instability. The right 3 plots are the same as the left, except that the muons are considered stable. Here, large microwave growths develop (note the change of energy scale).

## C  Coasting Beam Simulations

For coasting beams, we do not have the inverted tilted "N"-shape wake potential as in Fig. 10. Thus, no rf compensation will be required. However, the noise in the beam does result in a wake potential similar to the small residual wake-potential jitters in Fig. 10 after the rf compensation. Near transition where the phase motion is negligibly slow, these jitters will add up turn after turn without limit exactly in the same way as the bunched beam after having optimized the rf compensation. Thus, near transition, there is essentially no difference between a coasting beam and a bunched beam after the rf compensation. The only exception is that microwave instability develops most rapidly near the center of the bunch where the local intensity is highest, whereas in a coasting beam, microwave instability develops with equal probability along the bunch depending on the statistical fluctuations in the macro-particles.

In Fig. 13, we show some coasting beam simulations near transition by having $\eta_0 = 0$ or $\pm 5 \times 10^{-5}$ and $\eta_1 = 0$ or $\pm 0.05$. The coasting beam consists of $3.27 \times 10^{15}$ protons (or nondecaying muons) having an average energy of 100 GeV in a hypothetic ring with circumference 50 m. The initial momentum spread is Gaussian with rms fractional spread $\sigma_\delta = 0.001$ or $\sigma_E = 100$ MeV. Thus, at $1\sigma$, the contribution of $|\eta_1| = 0.05$ is the same as the contribution of $|\eta_0| = 5 \times 10^{-5}$. The simulations are performed with $8 \times 10^5$ macro-particles in 400 triangular bins. The impedance is a broadband with $Q = 1$ and $Z_0^\parallel/n = 2\,\Omega$ at the resonant frequency of $f_r = 300$ MHz.

All the plots in Fig 13 are illustrated with the same scale for easy comparison. The horizontal axes are longitudinal beam position from 0 to 166.7 ns, while the vertical axes are energy spread from $-4000$ to 3000 MeV. Plot (a) shows the initial particle distribution in the longitudinal phase space. All the other plots are simulation results at the end of 54,000 turns. Plot (b) is the result of having $\eta_0 = 0$ and $\eta_1 = 0$. It shows the accumulation of the wake-potential jitters over 54,000 turns. These jitters originate from the statistical fluctuation of the initial population of the macro-particles. Therefore, any deviation from Plot (b) implies microwave instability. Plots (c) and (d) are with $\eta_0 = 0$, but with $\eta_1 = +0.05$ and $-0.05$, respectively. We see the growths curl towards opposite phase directions nonlinearly as expected. This is due to the nonlinearity in $\delta$ in the time slip given by Eq. (2.4), similar to the simulations in Fig. 8(a). It appears that Plot (c) with $\eta_1 = -0.05$ gives a larger growth. Plots (e), (g), and (i) are for $\eta_0 = -5 \times 10^{-5}$ (below transition), but with $\eta_1 = +0.05, -0.05$, and 0, respectively. We see that the microwave instability is most severe when $\eta_1 = 0$, indicating that $\eta_1$ has the ability to curb instability. This is, in fact, easy to understand. The phase drift driven by $|\eta_1| = 0.05$ is much faster than that driven by $|\eta_0| = 5.0 \times 10^{-5}$ at larger momentum spread; for example, it will be 4 times faster at $2\sigma_\delta$, 9 times faster at $3\sigma_\delta$, etc. As a result, a nonvanishing $|\eta_1|$ tends to move particles away from the clumps, thus lessening the growth due to microwave instability.

Plots (f), (h), and (j) are for $\eta_0 = +5 \times 10^{-5}$ (above transition), but with $\eta_1 = +0.05, -0.05$, and 0, respectively. Again microwave instability is most severe when $\eta_1 = 0$, and $\eta_1$ does curb instability to a certain extent. Comparing Plots (e), (g), and (i) with Plots (f), (h), and (j), it is evident that the beam is more unstable against microwave instability above transition ($\eta_0 > 0$) than below transition ($\eta_0 < 0$) independent of the sign of $\eta_1$. For a fixed $\eta_0$, we also notice that negative $\eta_1$ is more unstable than positive $\eta_1$. The theoretical implications of these results are nontrivial and will be discussed in a future publication.

Now let us come back to the analytic investigations by Bogacz, Holt, and Colestock. Their results appear to contradict the simulations presented here. Analytic analysis often starts with the Vlasov equation. The time-dependent beam distri-

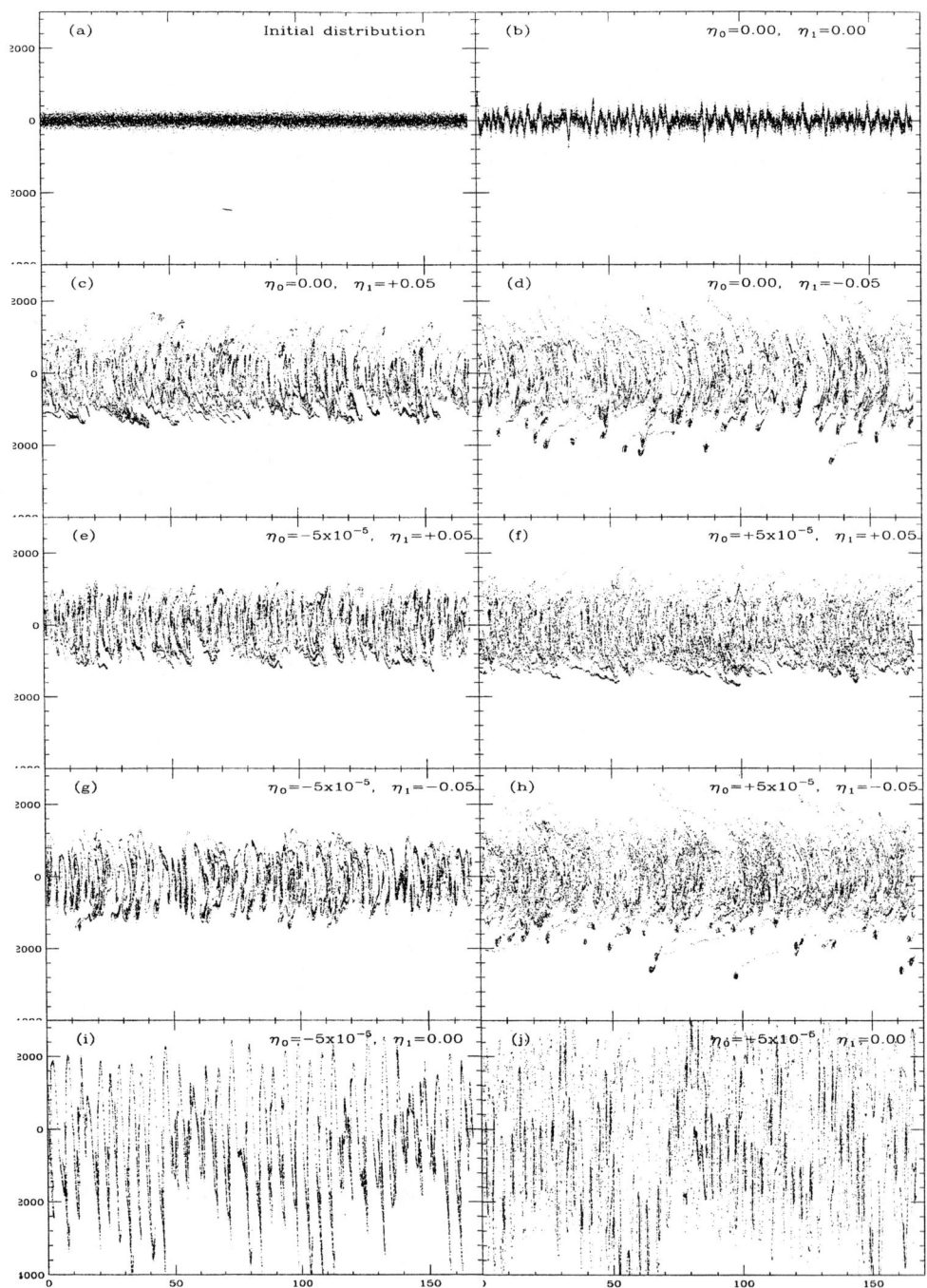

**FIGURE 13.** Energy spread (MeV) versus bunch position (ns) of coasting beam simulations. See text for explanation.

bution $\psi(\phi, \Delta E; t)$ can be separated into two parts:

$$\psi(\phi, \Delta E; t) = \psi_0(\phi, \Delta E) + \psi_1(\phi, \Delta E)e^{-i\Omega t} . \tag{3.6}$$

Here, $\psi_0$ is the *steady-state* solution of the Hamiltonian and $\psi_1$ describes the collective motion of the beam with the collective frequency $\Omega/(2\pi)$. After linearization, the Vlasov equation becomes an eigenequation with $\psi_1$ as the eigenfunction and $\Omega/(2\pi)$ the eigenfrequency. The equation also depends on $\psi_0$. Thus we must solve for the *steady-state* solution first before solving the eigenequation. The steady-state solution is the time-independent solution of the Hamiltonian which includes the contribution of the wake function. In other words, $\psi_0$ is the potential-well-distorted solution. Far away from transition, this distortion is mostly in the $\phi$ coordinate, for example, those brought about by the space-charge or inductive forces. Therefore, for a coasting beam, there will not be any potential-well distortion at all. The situation, however, is quite different close to transition. As was pointed out in above, the potential-well distortion is now in the $\Delta E$ coordinate. For this reason, not only bunched beams, even coasting beams will suffer from potential-well distortion as a result of the nonuniformity of the beam. In simulations, the nonuniformity arrives from the statistical fluctuation of the distribution of the macro-particles. This nonuniformity will accumulate turn by turn until the momentum spread is so large that the small $|\eta|$ is able to smooth out all nonuniformity. In other words, the steady-state distribution $\psi_0$ that goes into the Vlasov equation will be completely different from the original distribution in the absence of the wake. In the analysis of Bogacz, Holt, and Colestock, the ideal smooth Gaussian distribution in energy was substituted for $\psi_0$ in the Vlasov equation. However, this is a very unstable static distribution; even a small perturbation will accumulate turn by turn with extremely slow convergence. For this reason, it is hard to understand what their results really represent.

# REFERENCES

1. Trbojevic, D., Brennan, J.M., Courant, E.D., Roser, T., Peggs, S., Ng, K.Y., Johnstone, C., Popovic, M., Norem, J., *A Proton Driver for the Muon Collider Source with a Tunable Momentum Compaction lattice*, Proceedings of IEEE Particle Accelerator Conference, p.1030 (Vancouver, Canada, 1997).
2. Ankenbrandt, C, Ng, K.Y., Norem, J., Popovic, M., Qian, Z., Ahrens, L.A., Brennan, M., Mane, V., Roser, T., Trbojevic, D., and van Asselt, W., *Phys. Rev. ST Accel. Beams*, **1**, 030101 (1999).
3. Bogacz, A.S., *Microwave Instability at Transition—Stability Diagram Approach*, Proceedings of IEEE Particle Accelerator Conference, p.1815 (San Francisco, CA, 1991).
4. Holt, J.A, and Colestock, P.I., *Microwave Instability at Transition*, Proceedings of IEEE Particle Accelerator Conference, p.3067 (Dallas, TX, 1995).
5. Keil, E., and Schnell, W., CERN Report TH-RF/69-48, 1969.
6. Kim, E.-S., Sessler, A.M., and Wurtele, J.S., *Phys. Rev. ST Accel. Beams*, **2**, 051001 (1999).
7. Ng, K.Y., *Phys. Rev. ST Accel. Beams*, **2**, 091001 (1999).

# CONTRIBUTED PAPERS

# Convergence of basis expansions

Michael Blaskiewicz*
AGS Department Brookhaven National Laboratory

### Abstract

An exactly solvable model of longitudinal bunched beam stability is used to test the convergence of the basis expansion formalism. It is found that basis expansions can predict instability for equations that have no unstable solutions.

## 1 Introduction

The basis expansion technique[1, 2, 3, 4, 5] has been used for many years to find approximate eigenmodes in bunched beam instability problems. In general terms, one reduces the linearized Vlasov equation to an eigenvalue problem in one spatial ($\tau$) and one momentum ($v$) variable,

$$\lambda f(\tau, v) = L_{op}[f(\tau, v)]. \qquad (1)$$

In equation (1) $\lambda$ is the eigenvalue, $f(\tau, v)$ is its eigenfunction, and $L_{op}$ is a linear integro-differential operator. For all but the simplest cases [6, 7, 8, 9] no exact solutions are known. The basis expansion technique involves choosing a complete set of basis functions $g_n(\tau, v)$ and a weighting function $W(\tau, v)$ which satisfy an orthogonality relation

$$\int g_n^*(\tau, v) g_m(\tau, v) W(\tau, v) d\tau dv = \delta_m^n. \qquad (2)$$

In equation (2) the integral is over the domain where $W \neq 0$, the * represents complex conjugate, the index $n$ represents an enumeration of the basis functions, and $\delta_m^n$ is the Kronecker delta. To proceed one uses completeness to write

$$f(\tau, v) = \sum_{m=1}^{\infty} a_m g_m(\tau, v), \qquad (3)$$

---

*Work supported by the United States Department of Energy

where the $a_m$s are unknown coefficients.

One proceeds by multiplying equation (1) by $g_n^*(\tau, v)W(\tau, v)d\tau dv$ and integrates to obtain

$$\lambda a_n = \sum_{m=1}^{\infty} T_{n,m} a_m, \qquad (4)$$

where the matrix elements $T_{n,m}$ depend on the basis chosen, the impedance, etc. If $L_{op}$ is sufficiently well behaved the technique seems to be exact to this point. In practical applications the infinite matrix equation is intractable, the sum is truncated at some value $N$, and eigenmodes are obtained using numerical techniques[3]. There is circumstantial evidence[9] that truncating the sum can lead to grossly incorrect eigenvalues even for large $N$.

The purpose of this paper is to examine the expansion technique using a very simple, exactly solvable model of bunched beam stability. The model is quite naive, but seems to incorporate the fundamental elements of a longitudinal instability calculation. It is found that one must be careful in drawing conclusions from the basis expansion results.

## 2 Longitudinal instability model

The model assumes a waterbag distribution in a square well longitudinal potential with an impedance $Z = R - i\omega L$. The particles undergo perfect reflection at the edges of the bunch which makes the rf restoring force a boundary condition[11, 12, 8, 9, 10].

Let $\theta$ denote machine azimuth, $\omega_0$ be the angular revolution frequency of a synchronous particle, and $\tau$ be arrival time relative to the head of the bunch. Using $\theta$ as the time-like variable the Vlasov equation away from the reflective boundaries is,

$$\frac{\partial f}{\partial \theta} + v\frac{\partial f}{\partial \tau} - \kappa\left(IR + L\frac{\partial I}{\partial \tau}\right)\frac{\partial f}{\partial v} = 0. \qquad (5)$$

In equation (5): $f = f(\theta, \tau, v)$, $I = I(\tau, \theta) = q\int dv f(\theta, \tau, v)$, $v = d\tau/d\theta$, and

$$\kappa = \frac{\eta}{2\pi\omega_0\beta^2(E_0/q)},$$

with slip factor $\eta$. The synchronous particle's energy, charge, and velocity are $E_0$, $q$, and $\beta c$, respectively. For a waterbag distribution below the wave-breaking threshold the solution to the Vlasov equation is of the form:

$$f(\theta, \tau, v) = f_0 H(v_+(\theta, \tau) - v)H(v - v_-(\theta, \tau))H(\tau)H(\tau_b - \tau) \qquad (6)$$

where $f_0$ is a constant and $H(x)$ is the heaviside function with

$$H(x) = \begin{cases} 1, & \text{if } x > 0; \\ 0, & \text{otherwise.} \end{cases} \quad (7)$$

Substituting equation (6) into (5) results in differential equations for $v_+$ and $v_-$,

$$\frac{\partial v_\pm}{\partial \theta} + v_\pm \frac{\partial v_\pm}{\partial \tau} = F,$$

with

$$F = -\kappa \left( IR + L\frac{\partial I}{\partial \tau} \right).$$

For perfect reflection at the $\tau$ boundaries $v_+(\theta, 0) + v_-(\theta, 0) = 0$ and $v_+(\theta, \tau_b) + v_-(\theta, \tau_b) = 0$. The current is proportional to $v_+ - v_-$ and the solution is exact to this point.

To obtain an exactly solvable model neglect the effect of $R$ on the unperturbed distribution setting

$$v_\pm = \pm \hat{v} + \delta v_\pm(\tau) \exp(-iQ\theta).$$

Setting $\delta v_+(\tau) + \delta v_-(\tau) = D(\tau)$ and keeping first order terms yields a single equation for $D$ with

$$Q^2 D = -\hat{v}(\hat{v} - 2V)\frac{d^2 D}{d\tau^2} + 2U\hat{v}\frac{dD}{d\tau}, \quad (8)$$

where $U - iV = -q\kappa\omega_0 f_0(R - i\omega_0 L)$. For perfect reflection at $\tau = 0$ and $\tau = \tau_b$ the boundary conditions are $D(0) = D(\tau_b) = 0$. To solve (8) notice that

$$D(\tau) = \exp(\lambda_+ \tau) - \exp(\lambda_- \tau).$$

Inserting this expression yields a quadratic equation for $\lambda_\pm$,

$$\lambda_\pm = \frac{U}{\hat{v} - 2V} \pm \sqrt{\frac{U^2}{(\hat{v} - 2V)^2} - \frac{Q^2}{\hat{v}(\hat{v} - 2V)}}.$$

The boundary condition at $\tau_b$ gives $\lambda_+ - \lambda_- = 2\pi i k / \tau_b$ with $k \neq 0$ an integer. The eigenvalue satisfies

$$Q_k^2 = \hat{v}(\hat{v} - 2V)\left\{ \frac{k^2 \pi^2}{\tau_b^2} + \frac{U^2}{(\hat{v} - 2V)^2} \right\}, \quad (9)$$

and the eigenvector is

$$D_k(\tau) = \sin(k\pi\tau/\tau_b) \exp\left( \tau \frac{U}{\hat{v} - 2V} \right). \quad (10)$$

The functions $D_k$ for $k = 1, 2, \ldots$ form a complete set on $(0, \pi)$, so no eigenmodes have been missed. The right hand side of (9) is positive as long as $\hat{v} > 2V$, so resistance alone cannot cause instability. Additionally, reactance alone cannot cause instability, since no unperturbed distribution exists unless $\hat{v} > 4V$. To prove this consider the single particle Hamilonian which is given by

$$H = v^2/2 + \kappa L I + \text{boundary terms}.$$

For a stable unperturbed distribution the net pressure from the boundaries must be confining or zero. In particular $H \geq 0$ for $v = \hat{v}$. For $H = 0$, $\hat{v}^2 = -2\kappa L I = 4\hat{v}V$, which implies $\hat{v} > 2V$ and a stable system. This is in agreement with [7] which considers the case of a pure inductance with a parabolic line density confined by a linear rf force. For such a system any self consistent unperturbed distribution is stable.

While the reader may disagree with the physical behavior, the fact that equations (9) and (10) represent the complete solution of (8) is inescapable. In fact, everything before equation (8) may be viewed as purely motivational without affecting the main results.

## 3 Testing basis expansions

Consider equation (8) with $\hat{v}(\hat{v} - 2V) > 0$. Let $x = \pi\tau/\tau_b$, $\nu^2 = Q^2\tau_b^2/\pi^2\hat{v}(\hat{v} - 2V)$, and $\tilde{R} = U\tau_b/\pi(\hat{v} - 2V)$. The equation becomes

$$\nu^2 D = -\frac{d^2 D}{dx^2} + 2\tilde{R}\frac{dD}{dx}, \tag{11}$$

with boundary conditions $D(0) = D(\pi) = 0$. In these variables the exact eigenvectors and eigenvalues are

$$D_k(x) = \sin(kx)\exp(\tilde{R}x), \tag{12}$$
$$\nu_k^2 = k^2 + \tilde{R}^2. \tag{13}$$

Let the expansion functions be $g_n(x) = \sin(nx)$ with $W = 2/\pi$ for $0 < x < \pi$. The eigenvalue in equation (4) is $\lambda = Q^2$ and the matrix element is

$$T_{n,m} = nm\delta_m^n + \frac{8\tilde{R}nm}{\pi(n^2 - m^2)}odd(n - m), \tag{14}$$

where $odd(n - m) = 1$ if $n - m$ is odd, and $odd(n - m) = 0$ otherwise. In particular, when $n = m$ the term proportional to the reisistance is zero and there is no frequency shift in the weak coupling limit. This is consistent with the exact solution since the tune shift is $O(\tilde{R}^2)$.

A computer code was used to find solutions to

$$Q^2 a_n = \sum_{m=1}^{N} T_{n,m} a_m, \qquad (15)$$

for various values of $N$ and $\tilde{R}$. The only sophisticated part of the code is the eigenvalue solver[13] which is a well tested standard routine. Figure 1 shows the values of $Q$ as a function $\tilde{R}$ for $N = 2$ and predicts instability for $\tilde{R} \gtrsim 1$. Setting $N = 9$ gives Figure 2. The lowest frequency mode in Figure 2 is stable, and appears to be the case for all odd $N$. Also notice that large values of $Q$ go unstable first. This also seems to be generic. Setting $\tilde{R} \to -\tilde{R}$ has no effect on the eigenvalue spectrum.

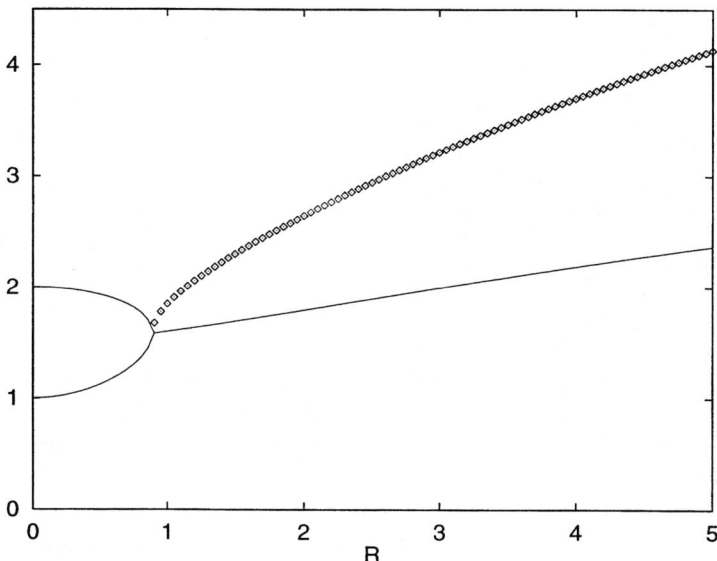

Figure 1: Real and imaginary parts of $Q$ versus $\tilde{R}$ for $N = 2$. The solid lines are the real part of $Q$ and the distance between the points and solid line is the imaginary part of $Q$ for the unstable mode.

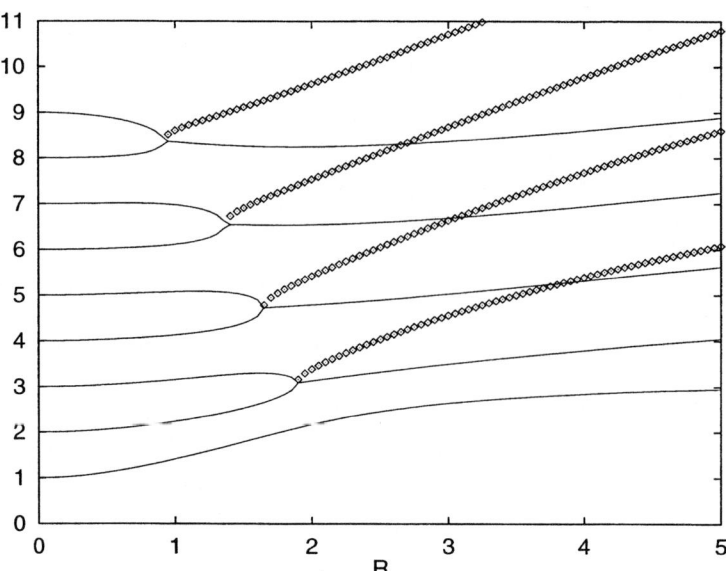

Figure 2: Real and imaginary parts of $Q$ versus $\tilde{R}$ for $N = 9$. The solid lines are the real part of $Q$ and the distance between the points and solid line is the imaginary part of $Q$ for the unstable mode.

As the number of modes increases the threshold value of $\tilde{R}$ for coupling between the lowest order modes continues to increase. Figure 3 shows the lowest 10 modes for $N = 100$. The expansion technique predicts that lowest order modes go unstable for $\tilde{R} < 5$, while the exact solution proves that the system is absolutely stable. Additionally, the threshold value of $\tilde{R}$ for the onset of unstable behavior without regard to mode varies only slightly with $N$. The threshold values of $\tilde{R}$ are 0.90, 0.95, and 0.95 for $N = 2$, 9, and 100, respectively. With basis expansions alone could one show that equation (11) has no unstable solutions?

At the workshop it was suggested that the lack of convergence in the basis expansion technique might be due to the wide bandwidth of the model impedance. To test this hypothesis equation (11) is modified to read

$$\nu^2 D = -\frac{d^2 D}{dx^2} + 2\tilde{R} \int_0^\pi S(x - x')dx' \frac{dD(x')}{dx'}, \qquad (16)$$

where $S(x)$ is a smoothing function. Since convolution commutes with differentiation this is completely equivalent to assuming a smoothed wake potential instead of a delta funtion for the resistive term. To proceed consider the Fourier series expansion of the smoothing function

$$S(x) = \sum_{n=0}^{\infty} S_n \cos(kx). \qquad (17)$$

This function is periodic with $S(x) = S(x + 2\pi)$ but since the domain of $D(x)$ is $[0, \pi]$, which is *half* the period of $S(x)$, the value of the convolution within $[0, \pi]$ will exactly agree with the value obtained using the bounded support smoothing function $S_b(x) = H(x + \pi)H(\pi - x)S(x)$.

Inserting equation (17) into equation (16) and proceeding as before one obtains,

$$T_{n,m} = nm\delta_m^n + \frac{4\tilde{R}nm}{(n^2 - m^2)}odd(n - m)(S_n + S_m). \qquad (18)$$

For a delta function $S_n = 1/\pi$ for $n \geq 1$ and equation (18) reduces to equation (14). To test the effect of smoothing set $S_n = \exp(-\sigma^2 n^2/2)/\pi$ for $n \geq 1$. This is the periodic extension of a Gaussian with a root mean square width $\sigma$. Note that since $S_0$ does not enter equation (18) one is free to assume that $S(\pi) = 0$. Figure 4 shows the lowest 10 modes for $N = 100$ with $\sigma = 0.1$. The threshold value of $\tilde{R}$ for the lowest order mode is reduced by smoothing the resistive term. However, the threshold value of $\tilde{R}$ without regard to mode is increased from 0.95 to 3.8. Reducing $\sigma$ to 0.05 increases this to 4.75. With $\sigma = 0.02$ the threshold value of $\tilde{R}$ is 4.55.

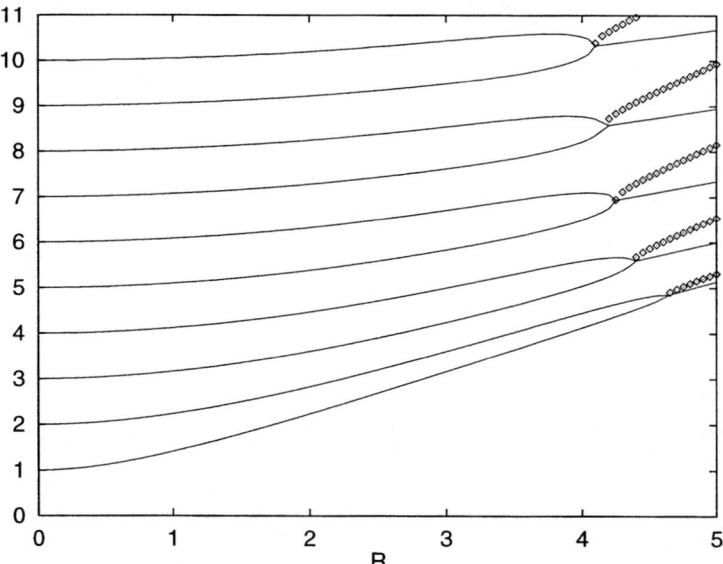

Figure 3: Lowest 10 eigenvalues versus $\tilde{R}$ for $N = 100$. The solid lines are the real part of $Q$ and the distance between the points and solid line is the imaginary part of $Q$ for the unstable mode.

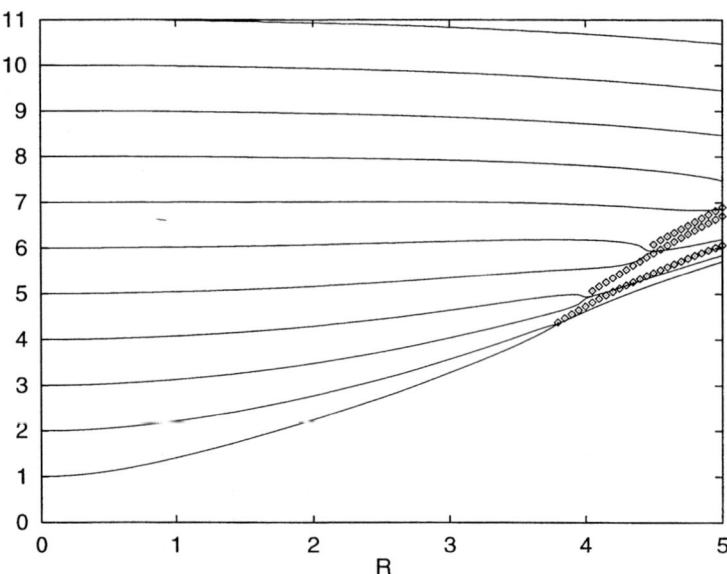

Figure 4: Lowest 11 eigenvalues versus $\tilde{R}$ using a Gaussian smoothing function for $N = 100$ and $\sigma = 0.1$. The solid lines are the real part of $Q$ and the distance between the points and solid line is the imaginary part of $Q$ for the unstable mode.

## 4 Conclusions

An exactly solvable model of longitudinal bunched beam stability including both resistance and inductance was presented and solved within the context of first order pertubation theory. It was found that the stability of the system depended only on the inductance. Including the effect of resistance on the unperturbed distribution (neglected here) may change this result. However, as discussed at the workshop, it may be that longitudinal microwave instability in bunched beams is an inherently nonlinear phenomena.

Regardless of the model's physical implications, the eigenvalue equation (11) seems quite reasonable and the basis expansion technique was applied to it. As is clear from Figures 1 through 3 the basis expansion technique predicts instability where none exists. Applying a spectral filter to the resistive term in the eigenvalue problem increased the stability threshold by as much as a factor of 4. The general application of this technique requires a better understanding of the mathematics.

# Acknowledgements

My thanks to Alex Chao, Ingo Hofmann, Bruno Zotter and several others for interesting questions and illuminating discussions.

# References

[1] K. Satoh, PEP Note 357 (1981).

[2] B. Zotter, CERN/SPS/81-20, (1981).

[3] Y.H. Chin, K. Satoh & K. Yokoya, *Particle Accelerators* **13**, p45 (1983).

[4] G. Besnier, D. Brandt, B. Zotter, *Particle Accelerators* **17**, 51-77, (1985).

[5] Y.H. Chin, CERN/SPS/85-9, (1985).

[6] F. Sacherer, CERN/SI-BR/72-5 (1972).

[7] D. Neuffer *Particle Accelerators* **11** p23 (1980).

[8] V.V. Danilov, E.A. Perevedentsev, 15th ICHEA p1163 (1992).

[9] M. Blaskiewicz, *Phys Rev ST Accel Beams* **1** 044201 (1998).

[10] M. Blaskiewicz, PAC99 TUP60 (1999).

[11] A.G. Ruggiero, BNL 51236, p 91. (1979)

[12] K.J. Kim, BNL 51236, p 100. (1979)

[13] The eispack routine cg, available through http://www.netlib.org

# Non-linear Mode Coupling and Saw-Tooth Instability

## S. Heifets

*Stanford Linear Accelerator Center, Stanford University, P.O. Box 4349, Stanford, CA 94309*

**Abstract.** Dynamics of the longitudinal relaxation oscillations of a single bunch above the threshold of microwave instability is discussed.

## I INTRODUCTION

Usually, beam dynamics in accelerators is linear by design and nonlinear effects are considered in context of long-run single particle stability. There are, however, a number of phenomena where nonlinear behavior is essential:

1. Beam-beam blow-up and O-type oscillations,
2. Beam-Ion instability
3. Saturation of single/multi-bunch instabilities,
4. Saw-tooth instability,
5. Relaxation oscillations in the beam interacting with high-Q resonator (SPEAR, J. Sebek, C. Limborg)
6. Transverse relaxation oscillations (K. Harkey, Argon).

Similar problems exist and were studied in hadron machines (P.Colstoke).

The 1D saw-tooth instability [1] provide a simple case for study of these phenomena. It was observed in many laboratories as periodic relaxation oscillations of the rms bunch length with corresponding excitation of the second- o third harmonics of synchrotron frequency in the bunch spectra. The instability manifests itself in seemingly simple situation of a single bunch under steady-state external conditions and can be considered as onset of the microwave instability. Hopefully, approach described below may be relevant to hadron machines.

## II HAISSINSKI SOLUTION

At low current, a bunch can be considered as $N_b$ uncorrelated individual particles oscillating in an one-dimensional time-independent potential of the rf bucket with the synchrotron frequency $\omega_{0s}/2\pi$. The particle motion is described in terms of dimensionless canonical variables

CP496, *Workshop on Instabilities of High Intensity Hadron Beams in Rings*,
edited by T. Roser and S. Y. Zhang
© 1999 American Institute of Physics 1-56396-910-6/99/$15.00

$$x = z/\sigma_0, \qquad p = -\delta/\delta_0, \qquad \{x,p\} = 1 \qquad (1)$$

where $z$ is position of a particle in respect with bunch centroid ($z > 0$ in the head of the bunch), $\delta = \Delta E/E$, $\sigma_0$ and $\delta_0$ are the rms bunch length and the rms energy spread in a bunch at zero current. It is convenient to use also dimensionless time $s = \omega_{0s} t$.

At large number of particles per bunch $N_b$, the nonlinearity of the potential is dominated by interaction of particles through a longitudinal wake-field excited by the beam. The wake field is characterized by the longitudinal beam impedance $Z(\omega)$. The Hamiltonian $H(x,p,s) = \frac{p^2}{2} + V(x,s)$ where the self-consistent potential depends on the distribution function $\rho$, $\int \rho \, dx \, dp = 1$,

$$V(x,s) = \frac{x^2}{2} + \lambda \int_x^\infty dx' dp' \rho(x',p',s) S(x' - x), \qquad (2)$$

and

$$S(x) = \frac{4\pi}{Z_0} \int \frac{d\omega}{2\pi i} \frac{Z(\omega)}{\omega} [1 - e^{-i(\omega \sigma_0/c_0)x}]. \qquad (3)$$

Here $Z_0 = 4\pi/c_0 = 120\pi$ Ohm is impedance of vacuum,

$$\lambda = \frac{N_b r_0}{2\pi R \gamma \alpha \delta_0^2}, \qquad (4)$$

$r_0$ is electron classical radius, $2\pi R$ is machine circumference, and $\alpha$ is momentum compaction.

In a simple case of a narrow-band impedance centered at $\omega_r$ with shunt impedance $R_r$ and quality factor $Q_r$,

$$S(x) = \frac{4\pi R_r}{Z_0 Q_r \zeta} \sin[\mu \zeta x] e^{-\mu x/2Q_r}, \qquad (5)$$

where $\mu = \omega_r \sigma_l/c$, and $\zeta = \sqrt{1 - 1/(2Q_r)^2}$. This impedance is used below.

Diffusion and radiation damping caused by synchrotron radiation are described by the Fokker-Plank equation for the distribution function $\rho$. In steady-state, there is Haiisinski solution [2] $\rho_H(x,p) = (1/Z_H) e^{-H_H(x,p)}$ where $H_H$ depends on $\rho_H$ in a self-consistent way, and temperature $T = 1$ in the chosen variables. Explicit expression for Haiisinski distribution and for self-consistent potential $U_H(x)$ can be obtained numerically.

In this note we consider relatively small $\lambda$ where $U_H(x)$ has only one minimum and coherent frequency shift is small. In this case, neither the well-known mode coupling instability nor the Baartman-Dyachkov mechanism [3] can explain the instability.

Because phase mixing is fast process, it is convenient to use action-angle variables $J$, $\phi$ introduced in such a way that, in the steady-state $\rho_H$, Hamiltonian

$H_H(p,x) = H_H(J)$. Then $\rho_H$ is independent on $\phi$ and normalized by $dJd\phi\rho_H(J) = 1$. Numerical integration $J = (1/\pi)\int dx\sqrt{2(H-U_H(x))}$ where integral is taken between turning points, gives $J(H)$ and defines frequency $\omega_H(J) = dH_H/dJ$.

In the time-dependent case,

$$\rho(J,\phi,s) = \rho_H(J) + \sum_n \rho_n(J)e^{in\phi}, \quad H = H_H(J) + \sum_n U_n(J,s)e^{in\phi}, \tag{6}$$

where

$$U_n(J,s) = \lambda \int dJ' d\phi' \rho_m(J,s) R_{m,m'}(J,J'). \tag{7}$$

Note that $U_n$ is defined only by amplitudes $\rho_m$ while, by definition of $J, \phi$ variables, all azimuthal harmonics $U_n$ generated by $\rho_H$ are cancelled by terms arising from $p^2/2 + x^2/2$ part of the Hamiltonian.

Coefficients $R_{m,m'}$ are related to the impedance

$$R_{m,m'}(J,J') = -\frac{4\pi}{Z_0}\int \frac{d\omega}{2\pi i}\frac{Z(\omega)}{\omega}C_m(J,\omega)C_{m'}^*(J',\omega), \tag{8}$$

$$C_m(J,\omega) = \int \frac{d\phi}{2\pi}e^{-im\phi}e^{ix(J,\phi)}. \tag{9}$$

where $x(J,\phi)$ is particle trajectory in the self-consistent Haiisinski potential. From the symmetry $t \to -t$ it follows that $x(J,-\phi) = x(J,\phi)$. Then $R_{m,m'}$ are real what follows from the symmetry of impedance $Z^*(-\omega) = Z(\omega^*)$.

## III FOKKER-PLANK EQUATION. LINEAR APPROXIMATION

The Fokker-Plank equation in $J, \phi$ variables gives for harmonics $\rho_n(J,s)$, $n = 0, 1, ..$

$$\frac{\partial \rho_n(J,s)}{\partial s} + in\omega_H(J)\rho_n - in\rho'_H(J)U_n + i\sum[\frac{\partial U_m}{\partial J}(n-m)\rho_{n-m} - \frac{\partial \rho_{n-m}}{\partial J}mU_m] = \gamma_0 F_n \tag{10}$$

Here $\rho'_H = \frac{\partial \rho_H(J)}{\partial J}$, $\gamma_0$ is synchrotron radiation damping, and $F_n$ describes diffusion and damping. Neglecting anharmonic terms of trajectory and terms of the order of $\gamma_0$, one gets

$$F_n = \frac{\partial}{\partial J}[\frac{J}{\omega_H}\frac{\partial \rho_n}{\partial J} + J\rho_n] - \frac{n^2}{4J\omega_H}\rho_n \tag{11}$$

For small $\lambda$ (low current), harmonics oscillate with time as $\rho_n \propto e^{-i\omega_H s}$. Hence, $U_n$ is given by the sum of terms oscillating proportional to $R_{m,k} e^{i(m-k)\omega_H s}$. Averaging of fast oscillating terms leaves only diagonal term

$$U_m(J) = \lambda \int dJ' d\phi' R_{mm}(J,J') \rho_m(J',s). \tag{12}$$

In the linear approximation, all azimuthal modes are independent and each is described by the superposition of radial modes $X_\nu$,

$$\rho_n(J,s) = \rho'_H(J) \sum_\nu b_\nu X_\nu(J) e^{-i(\omega_H - \bar\omega)s}. \tag{13}$$

$X_\nu$ are eigen vectors of the matrix

$$M_m(J,J') = 2\pi\lambda R_{mm}(J,J')\rho'_H(J') - \delta(J-J')(\omega_H(J) - \bar\omega), \tag{14}$$

$$\int dJ' M_m(J,J') X_\nu(J') = -\nu X_\nu(J). \tag{15}$$

The beam is linearly unstable if at least one of the eigen-values $\nu$ has positive imaginary part $\Gamma_\nu = Im[\nu] > \gamma_n$. In Eq. (14) we introduce constant $\bar\omega$ so that $\Omega = Re\nu$ is the coherent frequency shift. The diagonal approximation for $R_{mk}$ is valid if $Re\nu \ll \omega_H \simeq 1$.

Note, $U_m(J,s) = \sum_\nu (\omega_H - \bar\omega - \nu) b_\nu X_\nu(J)$.

If $\rho'_H(J)$ is monotonic function, the kernel of Eq.(15) can be written in a more symmetric form if vectors $\sqrt{\rho'_H} \tilde X_\nu(J)$ are used instead of $X_{nu}(J)$. In this form it is easy to see that, apart of Landau damping, the eigen values are real if $R_{mm}$ is symmetric, $R_{mm}(J,J') = R_{mm}(J',J)$. Hence, the beam stability depends on the anharmonicity of the trajectories and given by the asymmetric part of $R_{mm}$ [4].

Eq. (15)

$$[\omega_H(J) - \bar\omega - \nu] X_\nu(J) = 2\pi\lambda \int dJ' R_{mm}(J,J') X_\nu(J') \tag{16}$$

shows that the structure of the mode is

$$X_\nu(J) = \frac{r_m(J)}{\omega_H(J) - \bar\omega - \nu}, \tag{17}$$

where $r_m(J)$ is a smooth function of $J$. Hence, the mode is localized around the resonance value $J_r$, $\omega(J_r) = \bar\omega + \Omega$ with the width $\Gamma_m$.

# IV COMPARISON WITH A NONLINEAR OSCILLATOR

Well above the threshold, large number of unstable interacting modes leads to the turbulent regime of instability.

We are interested in the case where coherent tune shift is by an order of magnitude smaller than the dimensionless synchrotron frequency (which is of the order of one in our variables). In this case, it could be that only single mode is unstable, and interaction of this coherent mode with particles is important. These interaction can be described in the quasi-linear approximation [5].

The hint on how such a small perturbation can lead to non-perturbative effects can be obtained from the behavior of nonlinear oscillator in periodic external field $F = -\epsilon \cos(\Omega s + \psi_0)$. This case is quite analogous to interaction of a particle with an unstable dipole mode of a bunch which produces the periodic perturbation $U_1 \propto e^{-i\omega_H t}$. Effect of this perturbation on particles can be described by the Hamiltonian

$$H(J, \phi, s) = H_0(J) + \frac{\epsilon}{2}\sqrt{\frac{2J}{\omega_0}} \cos(\phi - \Omega s - \psi_0), \qquad (18)$$

where $H_0(J) = \omega_0 J + \kappa J^2/2$, $\kappa$ is parameter of nonlinearity $\omega(J) = \omega_0 + \kappa J$. For $\Omega = \omega_0 + \Delta$, $\Delta \ll \omega_0$. The resonance Hamiltonian in the new canonical variables $J, \alpha$, $\alpha = \phi - \Omega s - \psi_0$, is time-independent

$$H(J, \alpha) = -\Delta J + \frac{\kappa J^2}{2} + \frac{\epsilon}{2}\sqrt{\frac{2J}{\omega_0}} \cos \alpha. \qquad (19)$$

There are fixed points (FP) at $\alpha = 0$ or $\alpha = \pi$ given by equation

$$-\Delta + \kappa J \pm \frac{\eta}{\sqrt{J}} = 0, \qquad (20)$$

where $\eta = \epsilon/(2\sqrt{2J\omega_0})$.

If $\Delta/\kappa < 0$, there is one FP at $\alpha = \pi$ for any $\eta$. If $\Delta/\kappa > 0$, one FP is located at $\alpha = \pi$, $J_r \simeq \eta^{2/3}$ for $\eta \gg (\Delta/\kappa)^{3/2}$, and $J_r \simeq (\Delta/\kappa)$ for $\eta \ll (\Delta/\kappa)^{3/2}$). For small $\eta < \eta_{max} = 2(\Delta/3\kappa)^{3/2}$ there are two additional FPs at $\alpha = 0$, one of them is stable at $J_r \simeq (\kappa/\Delta)^2 \eta^2$, and one is unstable at $J_r \simeq \Delta/\kappa$. The stable FP is at the center of a narrow separatrix with the width $(\delta J) = 8\eta\sqrt{\Delta/\kappa}$, while trajectories of particles outside of the separatrix are circles centered at $J_r = (\kappa/\Delta)^2 \eta^2$ and $\alpha = \pi$. The width of the separatrix grows with the amplitude of perturbation $\epsilon$ and more and more particles are trapped inside of the separatrix. Trajectories of particles outside of the separatrix retain in the shrinking area which center shifts to large $J$ at $\alpha = \pi$. This area disappears eventually when amplitude $\eta$ exceeds $\eta_{max}$.

It is worth noting that the resonance exists at arbitrary small amplitudes $\epsilon$ and the width of the resonance depends on the ratio of $\epsilon$ to nonlinearity $\kappa$ and, hence, can be large even for small amplitudes for small nonlinearities.

The split of a beam into two beamlets corresponding trapping in the separatrix was observed recently in experiment [6].

Let us consider the same case of external force with Fokker-Plank equation. Harmonics $\rho_n$ satisfy Eq. (13) while $U_n$ has in this case only one harmonics $U_1 = U_{-1}^* = \eta\sqrt{J}e^{-i\Omega s}$, where $\eta$ may grow $\eta = \eta_0 e^{\Gamma s}$. Such an analysis, although not self-consistent, may clarify possible effects of the mode coupling.

Equation for $\rho_1$ in the linear approximation has a familiar solution $\rho_1(J) = \rho_{res}(J)$,

$$\rho_{res} = \frac{\eta \rho_H'(J)\sqrt{J}}{(J - J_r - iw)} e^{-i\Omega s}, \tag{21}$$

were $J_r = \Delta/\kappa$ and the width of the resonance $w = \Gamma/\kappa$. Hence, the linearized equation describes the resonance mode but not modification of the rest of the phase plane.

Let us include now coupling to harmonics $\rho_0$ and $\rho_{\pm 2}$

$$\dot{\rho}_1 + i\omega_H \rho_1 - i\rho_H' U_1 + 2i\frac{\partial U_1^*}{\partial J}\rho_2 - i\frac{\partial \rho_0}{\partial J}U_1 + i\frac{\partial \rho_2}{\partial J}U_1^* = 0. \tag{22}$$

Equation

$$\dot{\rho}_0 + \gamma_0 \rho_0 = i\frac{\partial}{\partial J}[U_1\rho_1^* - c.c] \tag{23}$$

can be solved explicitly giving

$$\rho_0 = i\frac{\partial}{\partial J}[\frac{U_1\rho_1^*}{2\Gamma} - c.c]. \tag{24}$$

Note, that the condition $\int dJ \rho_0(J) = 0$ gives the boundary condition $U_1\rho_1 = 0$ at $J = 0$.

Similarly, for $\rho_2 = \bar{\rho}_2 e^{-2i\omega_H s + 2\Gamma s}$ we get

$$\bar{\rho}_2 = \frac{\bar{U}_1^2}{\omega_H - \Omega - i\Gamma}\frac{\partial}{\partial J}[\frac{\bar{\rho}_1}{\bar{U}_1}], \tag{25}$$

where $\bar{\rho}_1 = \eta\sqrt{J}f$ is defined as $\rho_1 = \bar{\rho}_1 e^{-i\omega_H s + \Gamma s}$.

Function $f(J)$, satisfies equation

$$(J - J_r - iw)f = \rho_H'(J) + \frac{i\eta^2}{2w}\frac{\partial^2}{\partial J^2}[Jf^* - c.c] - \{\frac{\eta^2(\partial f/\partial J)}{J - J_r - iw} + \frac{\partial}{\partial J}[\frac{\eta^2 J(\partial f/\partial J)}{J - J_r - iw}]\}. \tag{26}$$

Here $J_r = \Delta/\kappa$, and $w = \Gamma/\kappa$. The first term in the RHS gives resonance solution of the linearized equation, the second term gives correction due to coupling to $\rho_0$, and the last term in curly brackets describes effect of the quadrupole mode.

For small $\eta$, equation can be solved by iterations substituting the resonance solution in the last two terms. Fig. 1 shows distribution function $\rho_H + \rho_0$ (left column) and $\rho_0$ (right column) vs $J$ perturbed by the dipole (upper row) and quadrupole modes (second row). Parameters $\eta = 0.03$, $w = 0.2$, $J_r = 0.75$, $\kappa = 0.1$ and $\omega_H = 0.7 + \kappa J$ were used in calculations.

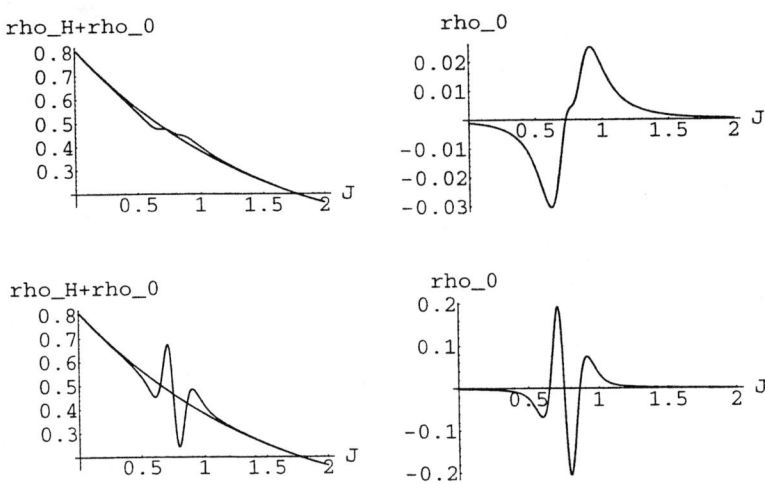

**FIGURE 1.** Distortion of the distribution function by the coherent modes: top row: by dipole mode, bottom row: quadrupole mode.

The second and the third terms in Eq. (26) are small compared to the first term if $(J - J_r)^2 < \eta\sqrt{J_r}$ what corresponds to the width of the separatrix of the nonlinear resonance discussed above.

The dipole mode affects particle distribution flattening it at the resonance, result which is well known in the quasi-linear approximation. Particles with low energy move to the right. As a result, energy of the system increases what leads, in a self-consistent Vlasov equation, to saturation of the unstable mode. It is interesting, that effect of the quadrupole mode is opposite. This suggest that interaction of two modes may lead to oscillatory regime.

To summarize, we note that linear approximation describes the resonance while the zero-harmonic $\rho_0$ may be required to describe the distortion of distribution function due to perturbation of the non-resonance particles by the coherent mode.

## V  NONLINEAR REGIME

Below certain threshold current, eigen-values of all azimuthal modes in linear approximation are real, the beam is stable, and is described by Haissinski distribution. Just above the threshold, there is a single unstable mode. Distribution function still is a series of eigen-functions Eq. (13) but with time-dependent coefficients $b_n(s)$. The non-trivial phenomena above the threshold corresponds to the non-linear interaction of the modes. Which one of the azimuthal modes become unstable first depends on the character of the impedance and the answer can be obtained from the linear analysis. Usually, it is a competition between dipole ($m = 1$) and quadrupole ($m = 2$) modes while sextupole and higher order modes come into play at higher currents. Considering instability close to the threshold, we can take into account only harmonics $\rho_m$ with $m = 1, 2$ expanding them in eigen vectors $V_\nu = (X_\nu, Y_\nu$ of the linear problem

$$\rho_1(J, s) = \rho'_H(J) \sum_\nu b_\nu X_\nu(J) e^{-i(\omega_H - \bar\omega)s} \qquad (27)$$

$$\rho_2(J, s) = \rho'_H(J) \sum_\mu a_\mu Y_\mu(J) e^{-2i(\omega_H - \bar\omega)s} \qquad (28)$$

where

$$M_m(J, J') = 2\pi\lambda R_{mm}(J, J')\rho'_H(J') - \delta(J - J')(\omega_H(J) - \bar\omega), \qquad (29)$$

with $m = 1, 2$, and

$$\int dJ' M_1(J, J') X_\nu(J') = -\nu X_\nu(J), \qquad (30)$$

$$\int dJ' M_2(J, J') Y_\mu(J') = -\mu Y_\mu(J). \qquad (31)$$

The radial amplitudes satisfy equations

$$\dot b_\nu + (i\nu + \gamma_d) b_\nu + i \sum_{\mu,\sigma} C^\nu_{\mu,\sigma} a_\mu b^*_\sigma + i \sum_\sigma d_{\nu,\sigma} b_\sigma = 0, \qquad (32)$$

$$\dot a_\mu + (2i\mu + \gamma_q) a_\mu + i \sum_{\sigma,\lambda} g^\mu_{\sigma,\lambda} b_\lambda b_\sigma + i \sum_\sigma f_{\mu,\sigma} a_\sigma = 0, \qquad (33)$$

Here we used orthogonality $\int dJ \bar X_\nu X_{\nu'} = \delta_{\nu,\nu'}$, and $\int dJ \bar Y_\mu Y_{\mu'} = \delta_{\mu,\mu'}$ where $\bar X_\nu$ and $\bar Y_\mu$ are eigen-vectors of transposed matrices $M_1^T$ and $M_2^T$. Coefficients are defined by the integrals

$$C^\nu_{\mu,\sigma} = 2 \int dJ \bar X_\nu Y_\mu \frac{\partial}{\partial J}[(\omega_H - \bar\omega - \sigma^*) X^*_\sigma] - \int dJ \bar X_\nu X^*_\sigma \frac{\partial}{\partial J}[(\omega_H - \bar\omega - \mu) Y_\mu] + \qquad (34)$$

$$\int dJ \bar X_\nu X^*_\sigma \frac{(\omega_H - \bar\omega - \sigma^*)}{\rho'_H} \frac{\partial [Y_\mu \rho'_H(J)]}{\partial J} - 2 \int dJ \bar X_\nu Y_\mu \frac{(\omega_H - \bar\omega - \mu)}{\rho'_H} \frac{\partial [X^*_\sigma \rho'_H]}{\partial J}, \qquad (35)$$

$$d_{\nu,\sigma} = 2\pi\lambda \int dJ \bar{X}_n u X_\sigma \int dJ' \frac{\partial R_{00}(J,J')}{\partial J} \rho_o(J') - \int dJ \frac{(\omega_H - \bar{\omega} - \sigma)}{\rho'_H} \frac{\partial \rho_0}{\partial J} \bar{X}_\nu X_\sigma, \tag{36}$$

$$g^\mu_{\sigma,\nu} = i \int dJ \bar{Y}_\mu X_\sigma \frac{\partial}{\partial J}[(\omega_H - \bar{\omega} - \nu)X_\nu] - i \int dJ \bar{Y}_\mu X_\sigma \frac{(\omega_H - \bar{\omega} - \sigma)}{\rho'_H} \frac{\partial}{\partial J}[X_\nu \rho'_H], \tag{37}$$

$$f_{\mu,\sigma} = 4\pi i\lambda \int dJ \bar{Y}_m u Y_\sigma \int dJ' \frac{\partial R_{00}(J,J')}{\partial J} \rho_o(J') - 2i \int dJ \frac{(\omega_H - \bar{\omega} - \sigma)}{\rho'_H} \frac{\partial \rho_0}{\partial J} \bar{Y}_\mu Y_\sigma. \tag{38}$$

Dipole and quadrupole modes interact with each other directly (terms proportional $C^\nu_{\mu,\sigma}$, $g^\mu_{\sigma,\lambda}$) and through perturbation of the ground state $\rho_0(J)$ (terms proportional to $d_{\nu,\sigma}$ and $f_{\mu,\sigma}$).

Substitute expansion Eqs. (27), (28) in Fokker-Plank equation for $\rho_0$

$$\frac{\partial \rho_0}{\partial s} + \gamma_0 \rho_0(J,s) = i\sum_m (\nu^* - \nu') b^*_\nu b_{\nu'} \frac{\partial}{\partial J}[\rho'_H X^*_\nu X_{\nu'}] + 2i \sum_m (\mu^* - \mu') a^*_\mu a_{\mu'} \frac{\partial}{\partial J}[\rho'_H Y^*_\mu Y_{\mu'}]. \tag{39}$$

This allows us to write another set of equations for $d_{\nu,\sigma}$ and $f_{\mu,\sigma}$:

$$\dot{d}_{\nu\sigma} + \gamma_0 d_{\nu\sigma} = -i(\nu'^* - \sigma') P^{\nu,\sigma}_{\nu',\sigma'} b^*_{\nu'} b_{\sigma'} - 2i(\mu'^* - \lambda') Q^{\nu,\sigma}_{\mu',\lambda'} a^*_{\mu'} a_{\lambda'}, \tag{40}$$

$$\dot{f}_{\mu\lambda} + \gamma_0 f_{\mu\lambda} = (\nu'^* - \sigma') F^{\mu,\lambda}_{\nu',\sigma'} b^*_{\mu'} b_{\sigma'} + 2(\mu'^* - \lambda') G^{\mu,\lambda}_{\mu',\lambda'} a^*_{\mu'} a_{\lambda'}. \tag{41}$$

Coefficients $P$, $Q$, $F$, and $G$ are given by convolution of eigen-vectors and $\rho'_H$. For example,

$$P^{\nu,\sigma}_{\nu'\sigma'} = -\int dJ \frac{\partial}{\partial J}[\frac{\omega_H - \bar{\omega} - \sigma}{\rho'_H} \bar{X}_\nu X_\sigma] \frac{\partial}{\partial J}[\rho'_H X^*_{\nu'} X_{\sigma'}]. \tag{42}$$

Coefficients $Q$, $F$, and $G$ have similar structure where some vectors $X$ are replaced by $Y$.

## VI  SINGLE MODE

Let us consider single unstable radial dipole mode $\nu" \equiv Im[\nu] > 0$ taking into account coupling to $\rho_0$. The system of equations in this case is

$$\dot{b} + (i\nu + \gamma_d)b + idb = 0, \quad \dot{d} + \gamma_0 d = -2\nu'' P|b|^2. \tag{43}$$

Here $P = P_{\nu,\nu}^{\nu,\nu}$. In the linear approximation the mode is unstable if $\nu'' > 0$.

In the nonlinear regime, $d$, the momentum of $\rho_0$, modifies the linear coherent frequency $\nu$ and can stop and even reverse the sign of the growth rate.

In terms of variables $y = 4\nu'' Im[P]|b|^2/\gamma_0^2$, $x = (2/\gamma_0)Im[d]$, and $s = \gamma_0 t$, equations

$$y' + \zeta y - xy; \quad x' + x + y = 0 \tag{44}$$

depend only on single parameter $\zeta = 2(\gamma_d - \nu'')/\gamma_0$. Additional to the FP $x = y = 0$, there is another FP $x = \zeta$, $y = -\zeta$. The first FP is stable provided $\zeta > 0$ and unstable otherwise. The second FP is stable for $\zeta < 0$ and unstable for $\zeta > 0$. Because $|b|^2 > 0$, the second FP exists if $(\nu'' - \gamma_d)\nu'' Im[P] > 0$. For $\nu'' > \gamma_d$ this requires $Im[P] > 0$. In the case of linearly stable system, $\gamma_d > \nu''$, the nontrivial FP exists if $Im[P] < 0$ assuming non-zero initial conditions or large fluctuations.

The non-trivial FP corresponds to a limiting cycle where $b$ changes periodically, $b = Be^{i\Omega s}$ with real $\Omega$, while $\dot{d} = 0$, and distribution function has a constant distortion $d = i\gamma_d - \nu$. For $\Omega$ and $|b|$ we get

$$\Omega = -Re(\nu) + (Im[\nu] - \gamma_d)\frac{Re[P]}{Im[P]}, \quad |b|^2 = \frac{\gamma_0}{2Im[P]}(1 - \frac{\gamma_d}{Im[\nu]}). \tag{45}$$

This is the main result of quasi-linear theory: $\rho_0$ is distorted by the unstable mode in such a way that the mode is stabilized. A constant distortion changes both rms bunch length and the energy spread.

## VII  TWO INTERACTING MODES

Analysis in more general cases is complicated. Let us consider two dipole modes, one unstable with eigen-values $\nu$, $Im[nu] = \nu'' > \gamma_d$, and another stable mode with eigen-value $\mu$. Effective radiation damping for a mode can be calculated convoluteeing eigen-vectors with the right-hand-side of Eq. (11). The system of equation in this case is

$$\dot{b}_\nu + (i\nu + \gamma_d + id_{\nu,\nu})b_\nu + id_{\nu,\mu}b_\mu = 0, \tag{46}$$

$$\dot{b}_\mu + (i\mu + \gamma_d + id_{\mu,\mu})b_\mu + id_{\mu,\nu}b_\nu = 0, \tag{47}$$

and four equations for $d$, for example

$$\dot{d}_{\nu\mu} + \gamma_0 d_{\nu\mu} = -i\sum_{\nu',\mu'}(\nu'^* - \mu')P_{\nu',\mu'}^{\nu,\mu}b_{\nu'}^*b_{\mu'}. \tag{48}$$

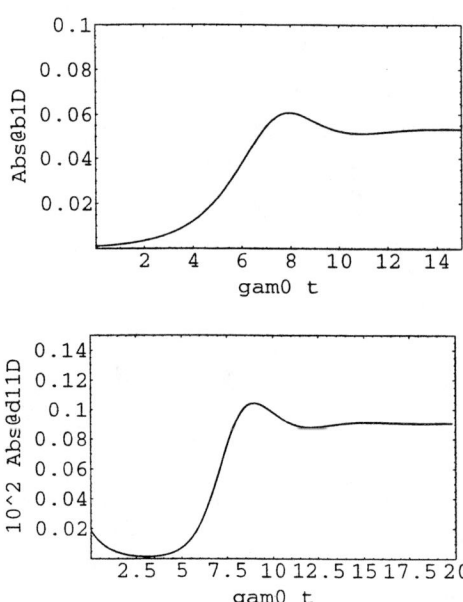

**FIGURE 2.** Nonlinear saturation of a single dipole mode. Parameters $\lambda = 6.8$, $\alpha = 0.4$, $\mu = 1.5$. Eqs.(45), (43) give $|b1| = 0.053$, $|d| = 0.0910^{-2}$.

Let us neglect for a while the non-diagonal terms in Eqs. (46), (47). The mode stability then depends on the dynamic increment $\Gamma_\nu = \gamma_d - Im[\nu + d_{\nu,\nu}]$. Let us take initial conditions where amplitudes $b[0]$, $[0]$ are small. Then $d$ in $\Gamma_\nu$ is initially negligible and mode is unstable. The unstable mode leads, first of all, to growth of the diagonal $d_{\nu,\nu}$ described by $\dot{d}_{\nu\nu} + \gamma_0 d_{\nu\nu} = -2\nu'' P_{\nu,\nu}^{\nu,\nu}|b_\nu|^2$. If $Im[P_{\nu,\nu}^{\nu,\nu}] > 0$, $d_{\nu\nu}$ is negative and can stabilize mode and reverse sign of $\Gamma_\nu$, see Fig. 3.

Due to the same mechanism, $\Gamma_\mu$ of stable mode is modified due to variation in time of $\dot{d}_{\mu\mu} + \gamma_0 d_{\mu\mu} = -2\nu'' P_{\nu,\nu}^{\mu,\mu}|b_\nu|^2$. If $Im[P_{\nu,\nu}^{\mu,\mu}] < 0$, the linearly stable mode can become unstable when linearly unstable mode saturates. After that, their roles interchange and the process can repeat itself. For small $b_\nu$ the dominant term defining growth of the mode $\mu$ is $d_{\mu\mu}$ given by $\dot{d}_{\mu\mu} + \gamma_0 d_{\mu\mu} = -2Im[\mu] P_{\mu,\mu}^{\mu,\mu}|b_\mu|^2$. If $Im[P_{\mu,\mu}^{\mu,\mu}] < 0$, the fastest growing mode is the mode with the minimum $Im[\mu]$, that is, the most stable mode in the linear approximation.

If we take from the beginning $\mu = [\nu]^*$, then Eq. (48) is simplified to

$$\dot{d}_{\nu\nu} + \gamma_0 d_{\nu\nu} = -2\nu''[P_{\nu,\nu}^{\nu,\nu}|b_\nu|^2 - P_{\mu,\mu}^{\nu,\nu}|b_\mu|^2]. \tag{49}$$

It depends only on $|b|^2$ terms. The coefficients $P$ have symmetry

$$P_{\nu,\nu}^{\nu,\nu} = P_{\mu,\mu}^{\nu,\nu} = [P_{\nu,\nu}^{\mu,\mu}]^* = [P_{\mu,\mu}^{\mu,\mu}]^*, \tag{50}$$

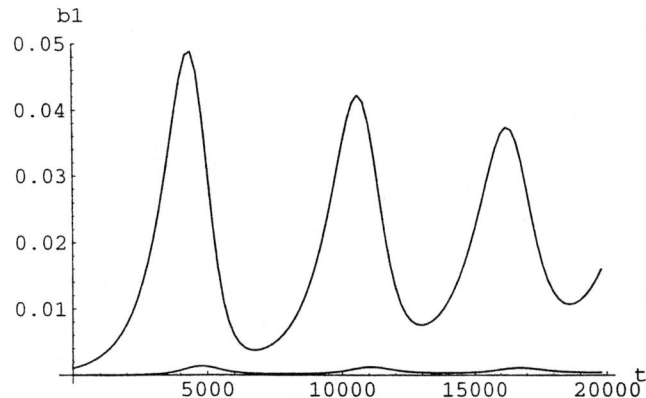

**FIGURE 3.** Amplitude of linearly unstable dipole mode (upper curve) and dynamic decrement. The growth changes sign when the dynamic increment is larger than linear growth rate, in this case, 0.002.

$$P^{\nu,\mu}_{\nu,\nu} = P^{\nu,\mu}_{\mu,\mu} = [P^{\mu,\nu}_{\nu,\nu}]^* = [P^{\mu,\nu}_{\mu,\mu}]^*. \tag{51}$$

Hence, $Im[P^{\mu,\mu}_{\mu,\mu}] < 0$ if $Im[P^{\nu,\nu}_{\nu,\nu}] > 0$. The last requirements is satisfied because $P^{\mu,\mu}_{\mu,\mu}$ is given by the second derivative of a function at the maximum, see Eq. (42).

This preliminary arguments show already that magnitude of the decrement $\gamma_d$ is related to the flip-flop dynamics. Let us study in a more systematic way fixed points (FPs) of the problem, considering interaction of the most stable and the most unstable modes. Allowing the limiting cycles, wee assume that $b_\nu \propto b_\mu \propto e^{-i\Omega s}$ were $\Omega$ is to be defined. Eqs. (46) and (47) give

$$b_\mu = -\frac{d_{\mu\nu} b_\nu}{\nu^* + d_{\mu\mu} - \Omega - i\gamma_d}, \quad (\nu - \Omega + d_{\nu\nu} - i\gamma_d)(\nu^* - \Omega + d_{\mu\mu} - i\gamma_d) = d_{\nu\mu} d_{\mu\nu}. \tag{52}$$

Eq. (49) defines $d$,

$$d_{\nu\mu} = -\frac{2}{\gamma_d}[P^{\nu\mu}_{\nu\nu}|b_\nu|^2 - P^{\nu\mu}_{\mu\mu}|b_\mu|^2]. \tag{53}$$

Symmetry Eq. (50),(51) gives

$$d_{\nu\nu} = d^*_{\mu\mu} = -\frac{2}{\gamma_d} P^{\nu\nu}_{\nu\nu}[|b_\nu|^2 - |b_\mu|^2]. \tag{54}$$

$$d_{\nu\mu} = d^*_{\mu\nu} = -\frac{2}{\gamma_d} P^{\nu\mu}_{\nu\nu}[|b_\nu|^2 - |b_\mu|^2]. \tag{55}$$

Hence, $d_{\nu\mu} = \zeta d_{\nu\nu}$ where $\zeta = P_{\nu\nu}^{\nu\mu}/P_{\nu\nu}^{\nu\nu}$.

The first of Eq. (52) gives for the amplitudes

$$|b_\mu|^2 = |b_\nu|^2 \frac{|d_{\nu\mu}|^2}{(Re[\nu + d_{\nu\nu}] - \Omega)^2 + (Im[\nu + d_{\nu\nu}] + \gamma_d)^2}. \tag{56}$$

The second of Eq. (52) is quadratic equation for $\Omega$,

$$\Omega = (Re[\nu + d_{\nu\nu}] - i\gamma_d) \pm \sqrt{|d_{\nu\mu}|^2 - (Im[\nu + d_{\nu\nu}])^2}. \tag{57}$$

If $|d_{\nu\mu}|^2 > (Im[\nu + d_{\nu\nu}])^2$, then $Im[\Omega] = -\gamma_d$ and the trivial FP $b_\nu = b_\mu = 0$ is stable. In the opposite case, $|d_{\nu\mu}|^2 < (Im[\nu + d_{\nu\nu}])^2$, FP exists if

$$(Im[\nu + d_{\nu\nu}])^2 = \gamma_d^2 + |\zeta d_{\nu\nu}|^2, \quad \Omega = Re[\nu + d_{\nu\mu}]. \tag{58}$$

In the last case,

$$|b_\mu|^2 = |b_\nu|^2 \frac{|\zeta d_{\nu\nu}|^2}{(Im[\nu + d_{\nu\nu}] + \gamma_d)^2}, \tag{59}$$

$$d_{\nu\nu} = -\frac{2}{\gamma_d} P_{\nu\nu}^{\nu\nu} |b_\nu|^2 [1 - \frac{|\zeta d_{\nu\nu}|^2}{(Im[\nu + d_{\nu\nu}] + \gamma_d)^2}]. \tag{60}$$

Equation $d_{\nu\mu} = \zeta d_{\nu\nu}$ relates $Red_{\nu\nu} = \eta Im d_{\nu\nu}$, where $\eta = Re[P_{\nu\nu}^{\nu\nu}]/Im[P_{\nu\nu}^{\nu\nu}]$. Then, finally,

$$Im d_{\nu\nu} = \frac{1}{|\zeta|^2(1+\eta^2) - 1}[\nu" \pm \sqrt{\gamma_d^2 + |\zeta|^2(1+\eta^2)[(\nu")^2 - \gamma_d^2]}]. \tag{61}$$

Eq. (59) defines $b_\nu^2$. It has to be positive what impose some constrain on parameters $P$ and $\gamma_d$.

Computer simulations based on Eqs. (46-48) confirmed these results. Fig. 4 depicts saw-tooth behavior in the system of two dipole modes interacting through perturbation of distribution function $\rho_0$. The phases of oscillations of two modes are shifted and the moment $d_{\nu\nu}$ of $\rho_0$ also oscillates in time. The oscillations are anharmonic and anharmonicity depends on the initial conditions and radiation damping.

In the same way interaction of other modes can be explored. Fig.5 depicts interaction of two linearly unstable modes, one quadrupole and another dipole. The nonlinear interaction leads to saturation of both modes.

Fig.6 shows interaction of three modes: linearly unstable and stable dipole modes and linearly unstable quadrupole mode. When current increases, the quadrupole mode become linearly unstable. At the same time, increasing nonlinearity of the potential well may suppress relaxation oscillations of the dipole modes. Interaction of these mode with dipole modes may lead to their stabilization while saw-tooth behavior may be preserved in quadrupole radial modes.

Certainly, more simulations should be carried out in a consistent way. This would allow direct comparison with the experimental data.

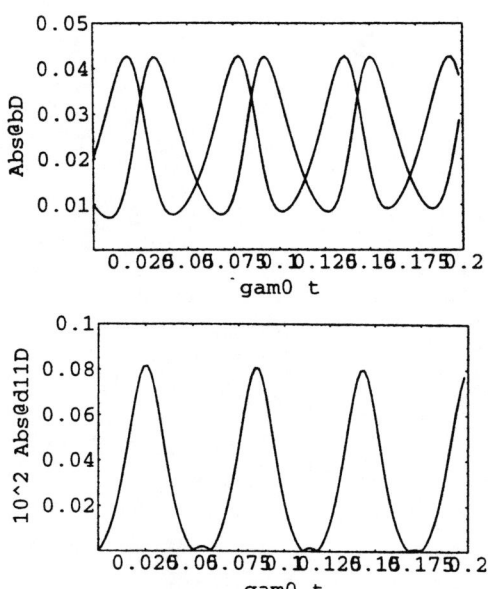

**FIGURE 4.** Coupled oscillations of two dipole modes. One is linearly unstable and another one is the most linearly stable. Parameters $\lambda = 6.88$, $\alpha = 0.4$, $\mu = 1.5$, $\gamma_0 = 10^{-5}$. Initial conditions are $b_1 = 2\,10^{-2}$, $b_2 = 10^{-2}$, all $d = 0$.

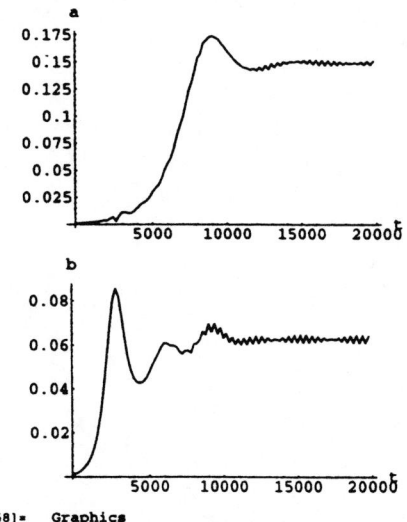

Out[58]= Graphics

lambda=6.88447 alpha=0.9 mu=1.5

gam0=0.001 gamdip=0.0001 gamquad=0.0001

Dipole + Quadrupole modes, both
linearly unstable. Stabilization is due to nonlinear interaction.

**FIGURE 5.**

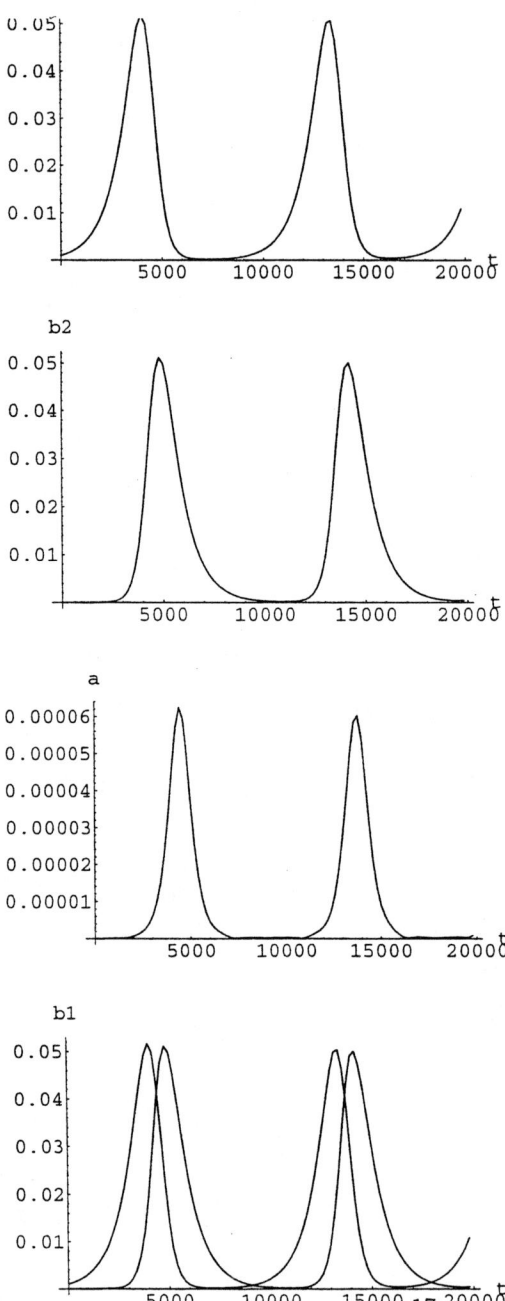

**FIGURE 6.** Interaction of two dipole modes and one quadrupole mode. One of the dipole modes is linearly unstable, another is the most linearly stable mode. Quadrupole mode is linearly stable.

## VIII  CONCLUSION

Dynamics of the system in the nonlinear regime above the threshold of instability may be quite complicated and substantially depends on the impedance, radiation damping, and beam current. Additional to already known mechanisms of linear mode coupling and Baartman-Dyachkov mechanism, both of which require high currents leading to large coherent tune shifts or appearance of the second minima in the self-consistent potential, there is another mechanism of nonlinear mode coupling. We explore the last mechanism and demonstrated that it may lead to the saw-tooth oscillations for beam current close and above the threshold of microwave instability. Numerical results are qualitatively similar to that obtained in reference [7] but, from our point of view, allows more systematic approach to the nonlinear collective phenomena. Application of this approach to other cases mentioned in the introduction will be reported elsewhere.

## REFERENCES

1. Krejcik, P. et al., Proc. IEEE Part. Accel. Conf., Washington D.C., 1993, p. 3240.
   B.V. Podobedov and R.H. Siemann, Signals from Microwave Unstable Beams in the SLC Damping Rings, PAC99, 1999, New York.
2. Haissinski, J. Nuovo Cimento, 1973, 18B, pp. 72.
3. Baartman, R., and Dyachkov, M. Proc. IEEE Part. Accel. Conf., Dallas, 1995.
4. Oide, K., "A New Mechanism of Longitudinal Single Bunch Instability in Storage Ring", KEK, 1995 (unpublished).
5. O'Neil, T. Phys. Fluids 1965, 8, pp. 2255.
   Chin Y., and Yokoya, K., Nucl Instr. and Methods, 1984, 226, 223-249.
   Morales,G.T. and O'Neil,T. Phys. Rev. 1965, 28, pp. 417.
   Schonfeld, J., Ann. Phys., 1985, 160, pp. 149.
   Meller, R.E., Ph. D. Thesis, "Statistical Method for Nonequilibrium Systems with Application to Accelerator Beam Dynamics", Cornell, 1986.
   Heifets, S.A., Phys. Rev. 1996, 54, p.2889.
6. J. Byrd, F. Zimmermann Private communication
7. Stupakov, G.V.,. Breizman, B.N, and Pekker, M.S., SLAC-PUB-7377, 1996

# Analysis of coupled bunch instability spectra

E. Shaposhnikova

*CERN, Geneva, Switzerland*

**Abstract**

Beam spectra observed during the development of a coupled bunch instability contain information about the coupled bunch mode n, which describes the phase shift between adjacent bunch oscillations. This number indicates the possible frequency of the guilty impedance with accuracy up to an integer multiple of the bunch spacing frequency. However when there are many possible candidates with imprecisely known frequencies this can be insufficient. In this paper we discuss what additional information about the frequency of the source of instability can be obtained from the analysis of the unstable beam spectra envelope. This is applied to the measurements in the CERN SPS.

## 1 Introduction

Longitudinal coupled bunch instabilities can present a serious limitation to the beam intensity in an accelerator. In the CERN SPS in normal operation the result of these instabilities is a longitudinal emittance blow-up of the proton beam by almost a factor 10 during the acceleration cycle, [1]. It is not harmful for the present fixed target beam operation, total intensity $4.7 \times 10^{13}$, but may become a problem for future roles of the SPS as LHC injector or as a high intensity machine for neutrino experiments.

There are several possible solutions to this problem. They are for example: increased passive damping of resonant impedances, increased Landau damping by applying a high harmonic RF system, active damping using a specially designated feedback system and others. At the moment the most efficient is to use the 4-th harmonic RF system in bunch shortening mode through the cycle. This can stabilize the beam almost up to the top energy (450 GeV), [2]. Nevertheless in all cases identification of the source of instability seems to be an important issue.

For the single bunch in the SPS it was possible to use measurements of the unstable spectrum to determine the sources of microwave instability. For multi-bunch operation, the development of coupled bunch instabilities after transition crossing towards higher energies, is associated with the growth of a broad-band spectrum, see examples in Fig.1.

The spectrum of the unstable bunched beam has components at frequencies

$$\omega = (n + lM)\omega_0 + m\omega_s,$$

where $\omega_0 = 2\pi f_0$ and $\omega_s = 2\pi f_s$ are revolution and synchrotron frequencies, M is the number of bunches in the ring, $n = 0, 1...M - 1$ is the coupled bunch mode number,

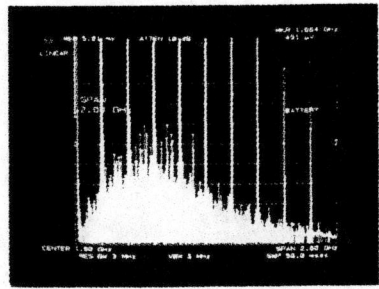 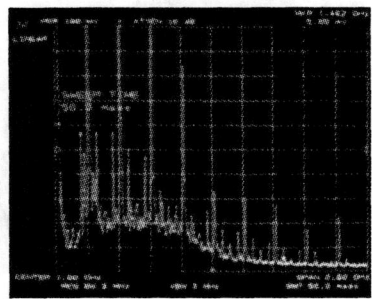

Figure 1: Beam spectrum from 0 to 2 GHz just after transition crossing (left) and at the end of fixed target proton cycle in the CERN SPS (right) for beam intensity $4.2 \times 10^{13}$ (left) and $1.6 \times 10^{13}$ (right). 200 MHz vertical lines are stable beam spectrum.

describing the phase shift $2\pi n/M$ between adjacent bunches, $m = 0, 1...$ is the multipole number (m = 1 - dipole, 2 - quadrupole and so on) and $l = 0, \pm 1, ...$ Often measurement of $n$ alone is sufficient to guess which HOM in the cavity drives this instability. However sometimes it is not obvious, as in the case of the SPS with its 5 different RF systems installed in the ring and many other cavity-like objects. Then any additional information about the frequency of the guilty impedance would be appreciated. Below we show that under certain conditions the envelope of the longitudinal coupled bunch mode spectrum contains this information. This analysis is then applied to the SPS.

## 2 Review of theory

### 2.1 Main equations

Let us consider an accelerator with $M$ identical and equally spaced bunches. A general equation describing stability of the system in the presence of coupling impedance $Z(\omega)$ can be obtained from the equations of motion and linearised to give the first order perturbation Vlasov equation, see [3], [4]. This equation can be written in the form:

$$j_k = \sum_{l'=-\infty}^{\infty} G_{kk'} \frac{Z_{k'}}{k'} j_{k'}, \qquad (1)$$

where $k = n + lM$, $k' = n + l'M$, $-\infty < l, l' < \infty$, $Z_k = Z(k\omega_0 + \Omega)$ and $j_k = j(k\omega_0 + \Omega)$ is the Fourier transform of the beam current perturbation

$$j(\theta, t) = e^{i\Omega t} \int_{-\infty}^{\infty} j(\omega) e^{-i\frac{\omega}{\omega_0}\theta} \frac{d\omega}{\omega_0}.$$

The elements of matrix $G_{kk'}$ are

$$G_{kk'} = A \sum_{m=-\infty}^{\infty} m\omega_{s0} \int_0^{\infty} \frac{d\mathcal{F}}{dr} \frac{I_{mk}(r) I^*_{mk'}(r)}{\Omega - m\omega_s(r)} dr, \qquad (2)$$

where
$$A = -i\frac{J_A}{V_0 h \cos(h\theta_s) S}.$$

Here $J_A$ is the average beam current, $V_0$ is the RF voltage amplitude, $h\theta_s$ is the synchronous phase, $h$ is the RF harmonic number, $\mathcal{F}(r)$ is the unperturbed distribution function and

$$I_{mk}(r) = \frac{1}{2\pi}\int_{-\pi}^{\pi} e^{ik\theta(r,\psi)-im\psi} d\psi. \tag{3}$$

The normalisation factor $S$ is defined as

$$S = \omega_{s0} \int_0^\infty \frac{\mathcal{F}(r) r\, dr}{\omega_s(r)}. \tag{4}$$

To describe the system two sets of variables are used: $(\theta, \dot{\theta})$, where $\theta$ is the azimuthal coordinate in the rotating coordinate system, and $(r, \psi)$ - the amplitude and phase of the synchrotron oscillations. For small oscillations with linear synchrotron frequency $\omega_{s0}$ and maximum oscillation amplitude $r_{max} \simeq \omega_0 \tau/2$, we have $r^2 = \frac{\dot{\theta}^2}{\omega_{s0}^2} + \theta^2$. Here $\tau$ is the bunch length in seconds.

Equation (1) can be used for instability threshold calculations in very general cases taking into account the nonlinearity of synchrotron motion (long bunches), an arbitrary RF waveform (double RF system) and so on, see for example [5]. Below we will restrict ourselves to linear synchrotron motion in the single RF system. Then the integral defined in (3) is $I_{mk} \simeq i^m J_m(kr)$, where $J_m(x)$ is the Bessel function of order $m$.

We will consider the excitation of a single multipole $m$ for given frequency $\Omega$ and will discuss this assumption later.

Assuming that $\Omega \ll \omega_0$ we neglect the dependence on $\Omega$ in $j_k$ and $Z_k$.[1] This allows us to simplify eq.(1) to the following eigenvalue problem, [6]:

$$\frac{\Omega - m\omega_{s0}}{m\omega_{s0}} j_k = A \sum_{l'} g_{kk'}^m \frac{Z_{k'}}{k'} j_{k'}, \tag{5}$$

where

$$g_{kk'}^m = \int_0^\infty \frac{d\mathcal{F}}{dr} J_m(kr) J_m(k'r) dr. \tag{6}$$

Eigenvalues of this equation give the frequency shifts of the coherent modes and the eigenfunctions - their spectrum. Below we will concentrate on the properties of these eigenfunctions.

## 2.2 Analysis of beam spectra

Let us consider the narrow band impedance with bandwidth $\Delta\omega_r \ll M\omega_0$. The most simple case for analysis is when the resonant frequency of the impedance $\omega_r = 2\pi f_r =$

---

[1] We exclude the Robinson instability from consideration.

$p_r\omega_0$ is far away from an integer or half integer multipole of the bunch spacing frequency $M\omega_0$. (More general case is considered in Appendix.) Then in the sum in equation (5) we can keep only one term with $l' = l_p$, where $p = n + l_p M \simeq \pm p_r$ and (5) becomes

$$\lambda j_k = A g_{kp}^m \frac{Z_p}{p} j_p, \qquad (7)$$

where $k = n + lM$ and $-\infty < l < \infty$.

The coherent frequency shift is

$$\lambda = \frac{\Omega - m\omega_{s0}}{m\omega_{s0}} = -i\frac{J_A}{V_0 h \cos(h\theta_s) S} g_{pp}^m \frac{Z_p}{p}. \qquad (8)$$

For instability one needs resistance. Since $\text{Re}Z$ is always positive and $g_{pp}^m$ - negative, $p$ is positive or negative depending on whether we are above or below transition.

As follows from (7) the unstable spectrum for coherent mode $(m, n)$ consists of lines at frequencies $\omega_k = 2\pi f_k = (n + lM)\omega_0 + m\omega_s$, $-\infty < l < \infty$. Negative frequencies appear on the spectrum analyzer as lower synchrotron sidebands at $(l+1)M\omega_0 - n\omega_0 - m\omega_s$, $0 < l < \infty$. The amplitudes of these lines are defined by an eigenfunction which can be written as [6]

$$j_p = 1, \quad j_k = g_{kp}^m / g_{pp}^m, \qquad (9)$$

where $g_{kp}^m$ is given by expression (6). For negative frequencies $j_{-k} = (-1)^m j_k$ and for odd $m$ lower sidebands will have a phase shift $\pi$, which of course will not be seen on the spectrum analyzer.

In measurements with low frequency resolution similar spectra will be seen due to the resonant impedance at frequencies $(lM + n)\omega_0$ and $[(l + 1)M - n]\omega_0$. However with high enough frequency resolution, the position of the synchrotron sidebands around the revolution lines $n$ and $M - n$ gives information about which $n$ is driven by the impedance: above transition internal sidebands indicate an impedance situated at a higher frequency and external - at a lower, see also [7]. The opposite is true below transition. However the value $lM$ is still unknown.

To proceed further let us consider the binomial family of distribution functions:

$$\mathcal{F}(r) = \mathcal{F}_0 (1 - \frac{r^2}{r_{max}^2})^\mu \qquad (10)$$

with $\mu \geq 1$. Then function $g_{kp}^m$ can be written as

$$g_{kp}^m = -2\mu \mathcal{F}_0 \int_0^1 (1 - x^2)^{\mu-1} x J_m(y_k x) J_m(y_r x) dx, \qquad (11)$$

where $y_k = k r_{max} \simeq \pi f_k \tau$ and $y_r = p_r r_{max} \simeq \pi f_r \tau$. In Fig.2 a few examples of beam spectrum envelope are shown for different sets of parameters $m$, $\mu$ and $f_r \tau$.

From these examples it is not obvious to derive a connection between the shape of the unstable beam spectrum envelope and the resonant frequency, although we notice that

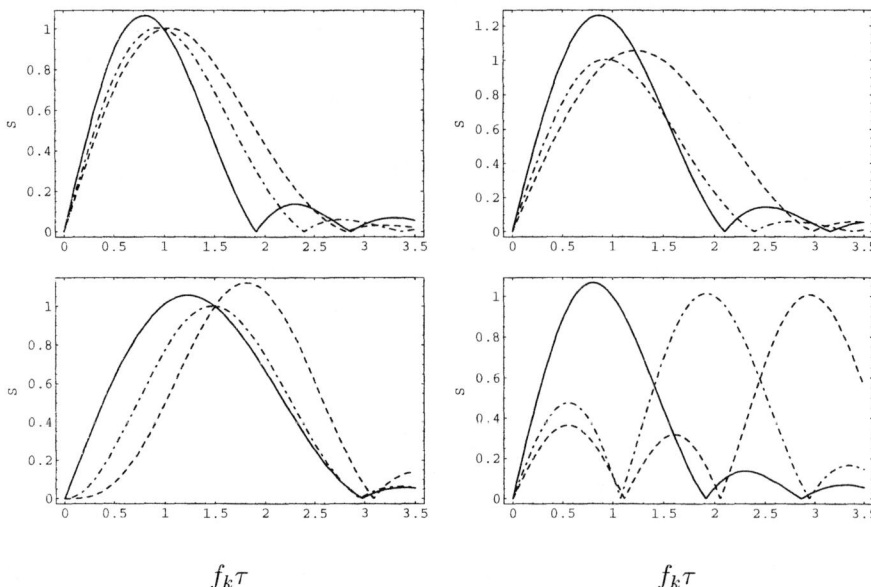

Figure 2: Beam spectrum envelope for: top, left - $m = 1$, $f_r\tau = 1$ and $\mu = 1.5$ (solid line), $\mu = 2$ (dash-dot) and $\mu = 2.5$ (dash); top, right - $m = 1$, $\mu = 2$ and and $f_r\tau = 0.5$ (solid), $f_r\tau = 1$ (dash-dot) and $f_r\tau = 1.5$ (dash); bottom, left - $f_r\tau = 1.5$, $\mu = 2$ and $m = 1$ (solid), $m = 2$ (dash-dot) and $m = 3$ (dash); bottom, right - $m = 1$, $\mu = 1$ and $f_r\tau = 1$ (solid), $f_r\tau = 2$ (dash-dot) and $f_r\tau = 3$ (dash).

for some parameters the maximum of the spectrum is close to the resonant frequency. If we now plot the position of the absolute maximum $f_{max}\tau$ as a function of the resonant frequency $f_r\tau$ (both normalised to bunch length) we get dependencies as shown in Fig.3 for the dipole mode $m = 1$ and the distribution function with $\mu = 1$ (left) and $\mu = 3$ (right).

One can clearly see two different regimes: one, when $f_{max}\tau \simeq const$ and another, when $f_{max} \simeq f_r$. The transition between them for $m = 1$ occurs around $f_r\tau \sim 1$. The *const* is in fact a function of $\mu$ and $m$. Similar behaviour for the maximum of the spectrum envelope can be observed for higher multipoles $m$. Examples for $m = 2, 3$ and $\mu = 1$ are presented in Fig.4.

The existence of two different regimes can formally be understood from the behaviour of the Bessel functions for small and large arguments. Using the first term in the expansion of $J_m(y_r x)$ for $y_r < 1$ we see that

$$j_k \propto \int_0^1 x(1-x^2)^{\mu-1} x^m J_m(y_k x) dx \qquad (12)$$

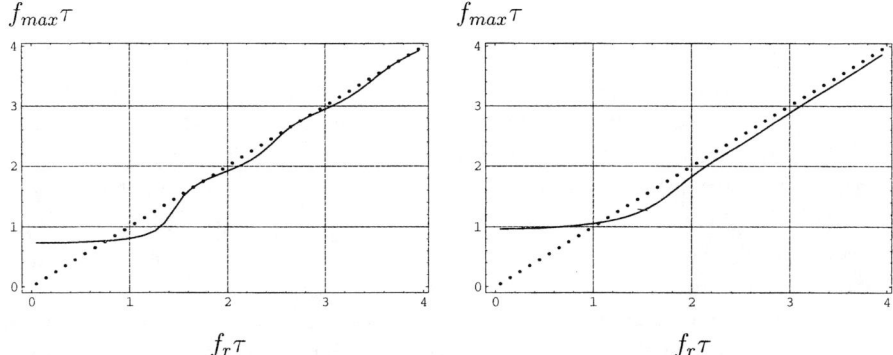

Figure 3: Position of the maximum in the beam spectrum envelope as a function of $f_r\tau$ for dipole mode $m = 1$ and two different distribution functions with $\mu = 1$ (left) and $\mu = 3$ (right).

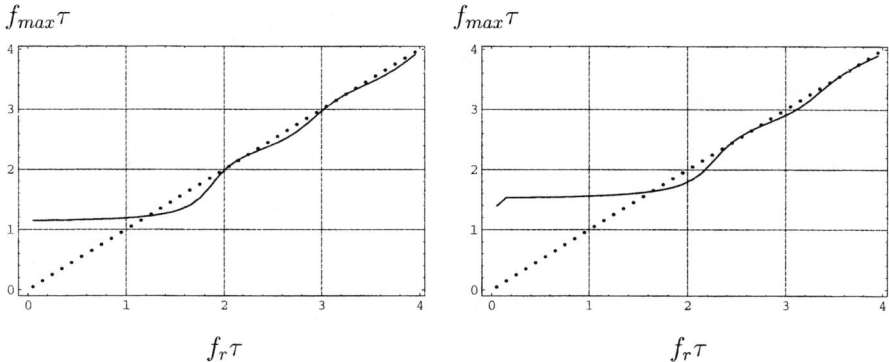

Figure 4: Position of the maximum in the beam spectrum envelope as a function of $f_r\tau$ for $\mu = 1$ and modes $m = 2$ (left) and $m = 3$ (right).

and the position of the maximum does not depend on $y_r$. For $\mu = 1$ the beam spectrum has amplitudes

$$j_k \propto \frac{J_{m+1}(y_k)}{y_k}. \tag{13}$$

The asymptotic behaviour of the Bessel functions for large arguments $x \gg 1$

$$J_m(x) \simeq (\frac{2}{\pi x})^{1/2} \cos(x - \frac{m\pi}{2} - \frac{\pi}{4})$$

explains why starting from some value of $f_r\tau$ the spectrum peaks at $f_{max} \sim f_r$. Indeed, the function under the integral in (11) is zero at $x = 0$ and the error using this asymptotic at small $x$ can be shown to be of the order $ln(y_r)/(y_r)$. Then for $y_r, y_k > 1$

$$j_k \sim \frac{1}{y_k^{1/2}} \int_0^1 (1 - x^2)^{\mu-1} \cos[(y_k - y_r)x]dx. \tag{14}$$

For $\mu = 1$ this gives

$$j_k \sim \frac{\sin(y_k - y_r)}{y_k^{1/2}(y_k - y_r)}. \tag{15}$$

These results have a clear physical meaning as well. If the wavelength of the wake field is longer than the bunch length, the spectrum of the perturbation does not contain any information about it. For higher frequencies or smaller wavelength the bunch length is already sufficiently long to "resolve" the frequency of the perturbation. The most efficiently excited is the perturbation with a maximum of the spectrum closest in frequency to the driving impedance.

From this analysis we can already conclude that for a fixed bunch length the resonant frequency of the impedance driving the instability is either below or very close (and slightly above) the maximum of the bunch spectrum envelope. However if one also takes into account values of $f_r\tau$ at which the given mode $m$ is most efficiently excited then, as we will see below, for all modes, except dipole, the maximum of the spectrum is close to the resonant frequency.

To understand under which conditions a given spectrum will be excited, one should consider simultaneously the thresholds or growth rates for different modes $m$. [2] Growth rates (8) for the distribution function (10) can be presented in the form:

$$\frac{\mathrm{Im}\,\Omega}{\omega_s} = \frac{4}{\pi^2} \frac{J_A \mathrm{Re} Z}{hV_0 \cos(h\theta_s)} \frac{F_m^*}{f_0\tau}, \quad F_m^* = \frac{m\mu(\mu+1)}{f_r\tau} \int_0^1 x(1-x^2)^{\mu-1} J_m^2(y_r x)\,dx. \tag{16}$$

In Fig.5 the dependence of the formfactor $F_m^*$ on $f_r\tau = y_r/\pi$ for modes $m = 1, 2, 3, 4$ is shown for distribution functions with $\mu = 1$ (left) and $\mu = 2$ (right).

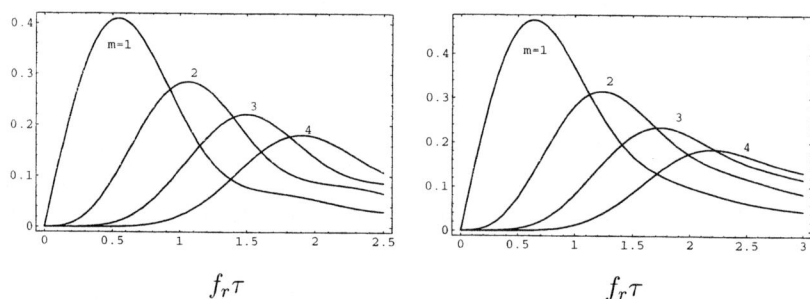

Figure 5: Formfactor $F_m^*$ from eq.(16) for modes $m = 1, 2, 3, 4$ as a function of $f_r\tau$ for distributions with $\mu = 1$ (left) and $\mu = 2$ (right).

From Fig.5 one can notice that for $\mu = 1$ mode $m = 1$ is dominant up to $f_r\tau \simeq 0.9$, $m = 2$ from this value up to 1.4, $m = 3$ from 1.4 to 1.85 and so on... At low values of $f_r\tau$ higher modes are strongly supressed in comparison with the dipole mode. Then we see that in Fig.4 in the first regime ($f_{max}\tau \simeq const$) only the part of the curve which is close to the diagonal has sense.

---

[2] In this analysis bunch length is assumed constant.

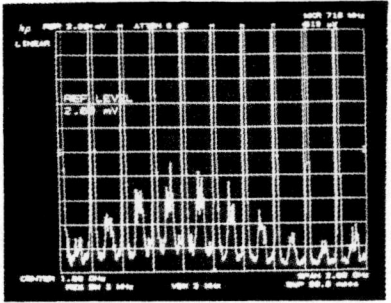 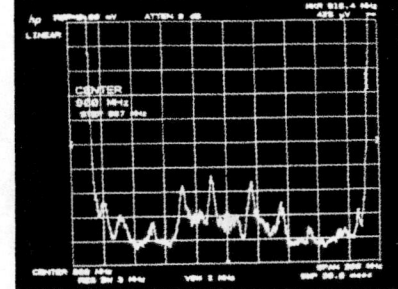

Figure 6: Beam spectrum from 0 to 2 GHz (left) and from 0.8 GHz to 1 GHz (right) at the end of the fixed target proton cycle in the CERN SPS for low beam intensity $4 \times 10^{12}$.

Above we considered independent excitation of the different modes m. This assumption is valid for bunches with small synchrotron frequency spread and resonant frequency which is not too high [5]. If one of these conditions is not satisfied, then many multipoles can be excited simultaneously. Then it is possible, [5], [8], to sum up terms with different $m$ in equation (1). This situation describes the so called microwave instability with a spectrum having a maximum at the frequency of the resonant impedance.

## 2.3 Coupled bunch instabilities in the CERN SPS

Calculations show [9] that after transition crossing during the fixed target cycle in the SPS the threshold for coupled bunch instabilities is continuously decreasing. It also has a frequency dependence so that the lowest threshold is for $f_r \sim 0.4/\tau$. As a result, different HOM can be successively excited during the cycle due to a bunch length change.

One other example of the bunch spectrum during the development of a coupled bunch instability towards the end of the cycle in the CERN SPS is shown in Fig.6. This beam had a total intensity 10 times lower than in the measurements presented in Fig.1 (left). As one can notice, at this intensity the signals are much clearer.

From the measured value of $nf_0 \sim 113$ MHz a few candidates were possible. These measurements were done with at low intensity. At the end of the cycle the bunch length was (1.5 - 2) ns, significantly shorter than in the measurements presented in Fig.1 (left) for the high intensity beam (2.5 - 3 ns). However both spectra have a maximum around 700 MHz. In both cases parameter $f_{max}\tau$ is more than 1. From this and also the fact that the maximum is the same, we expect the resonant frequency to be close to the maximum of the spectrum. This can be a HOM in the travelling wave 200 MHz RF system with resonant frequency 912 MHz.

In Fig.1 (right) we presented the signal obtained with quite long bunches, during the development of instability due to the fundamental frequency of the 352 MHz superconducting RF system, installed in the SPS for lepton acceleration, when non-optimal passive damping was accidentally used. In this case the spectrum had a maximum close to the resonant frequency.

## 3 Conclusions

As a conclusion we summarise which additional information about the resonant frequency of the guilty impedance can be obtained from the measured beam spectrum:
- The measured bunch profile and the stable beam spectrum can give some idea about the distribution function ($\mu$)
- Measurements of the bunch length $\tau$ and $f_{max}$ from the unstable bunch spectrum give $f_{max}\tau$
- Now there are two possibilities:
  - parameter $f_{max}\tau < 1$; then one only can say that $f_r\tau \leq 1.2$
  - parameter $f_{max}\tau > 1$, then the most probable $f_r \sim f_{max}$. The uncertainty can be estimated, for example, from the vertical displacement of the curves from the diagonal and in the region with efficient excitation (see Fig.5) does not exceed $\pm 0.2/\tau$. The smaller the bunch length - the larger the uncertainty.

## Acknowledgements

I am grateful to T.Linnecar and T.Bohl for stimulating discussions and support.

## Appendix

Now let us consider the possibility that the resonant frequency of the narrow-band impedance is close to an integer or half integer multipole of the bunch spacing frequency $M\omega_0$. Then we should keep two terms in the sum in equation (5) and it becomes

$$\lambda j_k = A g_{kp_1}^m \frac{Z_{p_1}}{p_1} j_{p_1} + A g_{kp_2}^m \frac{Z_{p_2}}{p_2} j_{p_2}, \qquad (17)$$

where $p_1 = p = n + l_1 M \simeq \pm p_r$ and $p_2 = n + l_2 M$. Here $p_1 p_2 < 0$ and for values of $l_2$ we have 2 possibilities, depending on where the resonant impedance is situated: $l_2 = -l_1$ or $l_2 = -(l_1 + 1)$.

Taking for $k$ values $p_1$ and $p_2$ we can find eigenvalues for this situation. For the narrow band impedance assumed above $\Delta\omega_r \ll M\omega_0 \ll 1/\tau$. In this case we can write

$$\begin{aligned} g_{p_1 p_1}^m &\simeq g_{p_2 p_2}^m = g_p^m \\ g_{p_1 p_2}^m &= g_{p_2 p_1}^m \simeq (-1)^m g_p^m \\ g_{kp_1}^m &\simeq (-1)^m g_{kp_2}^m. \end{aligned}$$

Then for the growing mode, the eigenvalue is

$$\lambda \simeq A g_p^m \left( \frac{Z_{p_1}}{p_1} + \frac{Z_{p_2}}{p_2} \right). \tag{18}$$

There is no instability if $p_1 = -p_2$, which can happen [4] when the $\omega_r$ is equal to an integer or half integer (for even $M$) multipole of the bunch spacing frequency $M\omega_0$.

For the beam spectrum we finally have an expression similar to (1):

$$j_k = A \left[ \frac{Z_{p_1}}{p_1} g_{kp_1}^m + (-1)^m \frac{Z_{p_2}}{p_2} g_{kp_2}^m \right] / \lambda \simeq g_{kp}^m / g_p^m. \tag{19}$$

# References

[1] T.Bohl, T.Linnecar, E.Shaposhnikova, Emittance control by the modification of the voltage programme, CERN SL-MD Note 246, 1997.

[2] T.Bohl, T.Linnecar, E.Shaposhnikova, J.Tückmantel, Study of different operating modes of the 4th RF harmonic Landau damping system in the CERN SPS, proceed. EPAC'98, 1998.

[3] A.N.Lebedev, Proc. VI Int. Conf. on H.E. Accel., CEAL-2000, p.284, 1967; Atomnaya Energia, v.25, N2, p.100, 1968.

[4] F.J.Sacherer, A longitudinal stability criterion for bunched beams, IEEE Trans. Nucl. Sci. NS-20, p.825, 1973.

[5] V.I.Balbekov, S.V.Ivanov, Longitudinal beam instabilities in the proton synchrotrons, Proc. XIII Int. Conf. H.E. Accel., Novosibirsk, v.2, p.124, 1987, in Russian.

[6] J.L.Laclare, Bunched beam coherent instabilities, CERN Acc. School 1985, CERN 87-03, p.264, 1987.

[7] F.Sacherer, F.Pedersen, Theory and performance of the longitudinal active damping system for the CERN PS Bosster, NS-24, N.3, p.1396, 1977.

[8] J.M.Wang, C.Pelegrini, On the condition for a single bunch high frequency fast blow-up, Proc. XI Int. Conf. on H.E. Accel., Geneva, p.554, 1980.

[9] E.Shaposhnikova, Longitudinal instabilities in the SPS, Proc. of the workshop on LEP-SPS performance, Chamonix IX, Chamonix, p.69, 1999.

# The Coupling Impedance of the RHIC Injection Kicker System

## H. Hahn

*Brookhaven National Laboratory, Upton, NY 11973*

**Abstract.** In this paper, results from impedance measurements on the RHIC injection kickers are reported. The kicker is configured as a "C" cross section magnet with interleaved ferrite and high-permittivity dielectric sections to achieve a travelling wave structure. The impedance was measured using the wire method in which a resistive match provides a smooth transition from the network analyzer to the reference line in the set-up. Accurate results are obtained by interpreting the forward scattering coefficient via the log-formula. The four kickers with their ceramic beam tubes contribute a $Z/n = 0.22$ $\Omega$/ring in the interesting frequency range from 0.1 to 1 GHz, and less above. At frequencies above $\sim$100 MHz, the impedance is ferrite dominated and not affected by the kicker terminations. Below 100 MHz, the Blumlein pulser with the $\sim$75 m feeding cables is visible in the impedance but makes no significant contribution to the results. The measurements show that the kicker coupling impedance is tolerable without the need for impedance reducing measures.

## INTRODUCTION

The RHIC injection kicker[1] is configured as "C" cross section magnet with interspersed ferrite[2] and dielectric[3] blocks as shown in Fig. 1 The deflection properties of the kicker are dominated by the magnetic field and thus by the ferrite properties and its geometrical configuration. The dielectric blocks provide the capacitance required to achieve a transmission line kicker and to reduce the charcteristic impedance so as to match the 25 $\Omega$ of the Blumlein pulser.

The longitudinal coupling impedance of a half-size kicker model was previously measured by Mane et al.[4] with the results raising some concern due to the presence of sharp resonances in the GHz region. A more extensive study identified the origin of these resonances as local resonances in the dielectric blocks exposed to the beam.[5] In the modified kicker configuration, the resonances were completely suppressed by removing the dielectric blocks at the sides and replacing them by ferrite. The two kicker versions are shown in Fig. 1.

---

Work performed under the auspices of the U. S. Department of Energy

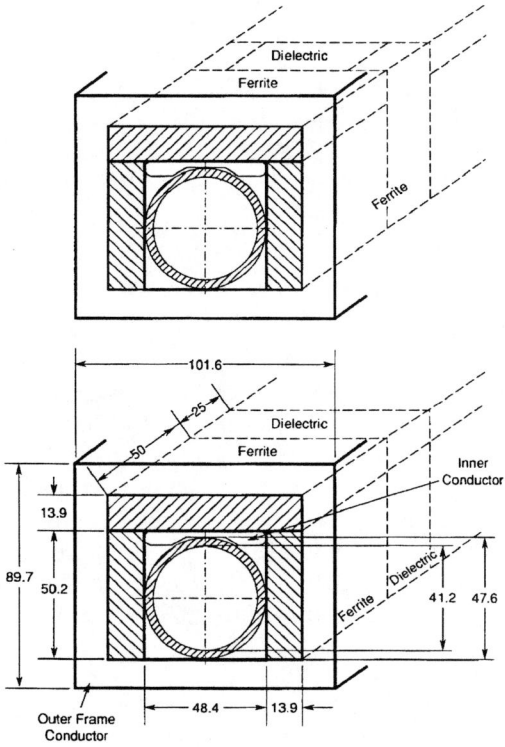

**FIGURE 1.** Geometry of original (bottom) and modified (top) RHIC injection kicker design.

In this paper, measured results for the longitudinal coupling impedance of a half-size injection kicker model, both free-standing as well as part of the kicker system with Blumlein pulser, are presented.

# WIRE MEASUREMENT OF THE COUPLING IMPEDANCE

The coupling impedance of the kicker system was measured with the existing bench setup for wire measurements previously used and described,[4] but now using the wire method in which a resistive match provides a smooth transition from the 50 Ω of the network analyzer to the characteristic impedance of the reference line in the setup.[6] The advantage of this method stems from the simplicity of its calibration procedure and the fact that a sufficient approximation of the results for the impedance in terms of real/imaginary or amplitude/phase is directly provided by the network analyzer and can be plotted without post processing. However the results quoted were obtained using the more accurate "log-formula" appropriate for distributed systems.

The measurements were performed with the Hewlett-Packard Network Analyzer 8753C connected to the S-Parameter Test Set 85078A. The characteristic impedance of the connecting reference waveguides was measured directly to be $Z_{ref} = 165$ Ω by shorting the output port of a waveguide section, for which the input impedance becomes

$$Z_{in} = Z_{ref} \tan\left(\frac{\pi}{2}\frac{f}{f_{\lambda/4}}\right)$$

Matching the 50 Ω of the network analyzer to the 165 Ω of the reference line is achieved by adding at the input port, a parallel resistor of ∼59.9 Ω and a series resistor of 137.7 Ω which provides forward and backward matching. At the output port, a series resistor of ∼ 115 Ω provides forward matching. The frequency dependence of the carbon resistors and stray inductances/capapcitances destroys the match at higher frequencies, but is corrected by the calibration procedure.

The coupling impedance is obtained from the change in the forward scattering coefficient $S_{21DUT}$ with respect to the reference system with $S_{21ref} = 1$ if properly calibrated. In the case of a single lumped disturbance, the coupling impedance is obtained by the well known formula

$$Z_{hp} = 2Z_{ref}\frac{(1 - S_{21DUT}/S_{21ref})}{(S_{21DUT}/S_{21ref})}$$

which as a first approximation can also be used in the case of a distributed impedance such as the kicker.[7] This formula is used by the $hp$ network analyzer and is quite useful in an exploratory study involving distributed impedances. However, if accurate results are required, at least the "log-formula"

$$Z_{\log} = -2Z_{ref}\ln\frac{S_{21DUT}}{S_{21ref}}$$

in which the scattering coefficients are complex quantities should be used; [8,9] although itself an approximation, it is extremely simple to use and sufficient in the present application.

## THE KICKER SYSTEM WITH CERAMIC BEAM TUBE

The RHIC injection kicker system consists of four separate units, each 1.12 m long with separate pulser. The impedance measurements were performed in a half-length model with pulser and then extrapolated to the total system.[10] The kicker will operate in air and thus requires a ceramic beam tube, the dimensions of which are 47.6 mm o.d. and 41.3 mm i.d. Insertion of the ceramic beam tube into the

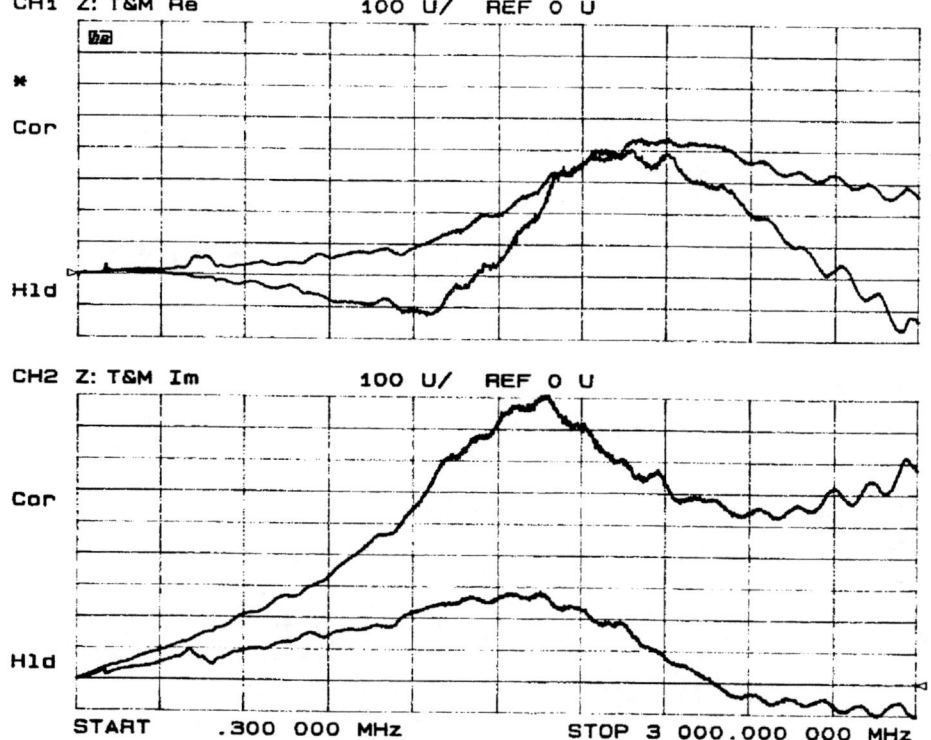

**FIGURE 2.** Comparison of impedance of kicker with and without beam tube.

kicker increases the capacity by ~2% and thus the impedance by only 1%. The coupling impedance of the kicker system, that is with the output port terminated and the input port connected to the pulser, has been measured with and without the ceramic beam tube. The hp impedance for both cases is compared in Fig. 2. Inspection of Fig. 2 shows an unphysical negative real part of the impedance. It has been verified, that this problem is caused by the use of the hp formula and can be corrected via the log-formula.

The results for the kicker system impedance are at frequencies above ~100 MHz independent of the port terminations and can be summarized by the impedance values with beam tube of

$$|Z/n| = 0.22 \ \Omega/\text{ring}$$

and without beam tube of $|Z/n| = 0.14 \ \Omega/\text{ring}$, in the range up to ~1 GHz, but lower at higher frequencies. Note that the impedance is resonance free in the measured region up to 3 GHz. The low-frequency end is discussed in the next section.

Overall, one finds that the impedance contribution by the injection kicker is tolerable without the need for special impedance reduction measures such as a low-impedance ($R$-square, $R_{sq} \sim 1 - 10 \ \Omega$) coating. A low-impedance coating can limit the rise time and cause arcing at higher fields. A high-resistance coating ($R_{sq} \sim 1\text{M} \ \Omega$) will not reduce the coupling impedance but is required to prevent electrostatic charging of the beam tube, and will be provided in the RHIC kicker.

## LOW FREQUENCY COUPLING IMPEDANCE

At frequencies below $\sim$100 MHz the ferrites have low losses and the kicker acts as a transmission line magnetically coupled to the beam. As a result, the kicker terminations and in particular the pulser with the $\sim$75 m long connecting cable become visible in the coupling impedance, as seen in Fig. 3.

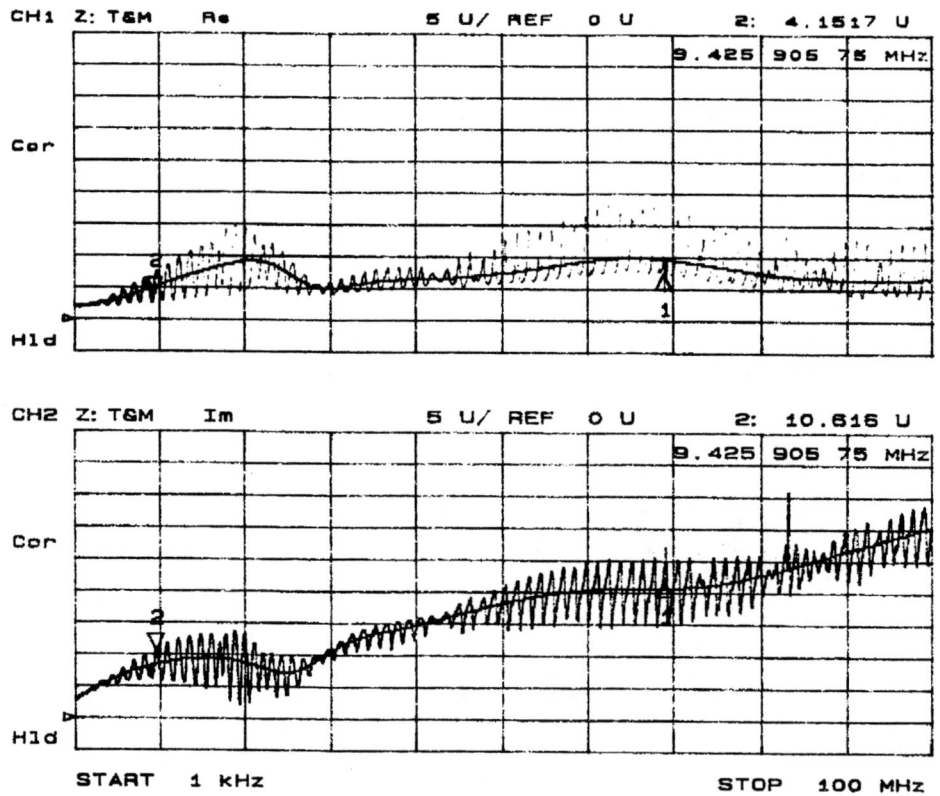

**FIGURE 3.** Low frequency coupling impedance of kicker system (rapid oscillation) and free standing kicker (smooth curve).

Nassibian and Sacherer[11] have theoretically analyzed this situation, but their results were limited to fully terminated kickers and a later report by Nassibian[12] gave only expressions for the real part of the impedance. A generalized derivation of the longitudinal coupling impedance valid at "low" frequencies, is presented in the appendix.[13] The kicker is still treated as a transformer-coupled uniform transmission line, but the constraints on the terminations are removed. The kicker is defined by its characteristic impedance $Z_K$ and the wave number $k_K$, both beam-independent parameters measurable directly on the magnet. The generalized theory of the coupling impedance has two additional free parameters: (1) the mutual inductance $M$ and (2) the "self"-inductance of the beam $L_B$. The values of these parameters are obtained from "wire"-measurements, using the theoretical expressions for their interpretation. As intuitively expected, the "self"-inductance of the beam follows from the results for the kicker open at both ports. The mutual inductance is expected to be

$$M = \frac{1}{2}L_K(1 + x/a)$$

with $L_K$ the kicker inductance, $a$ the kicker half-aperture and $x$ the displacement from the center towards the bus bar. For a centered beam, one has $M = L_K/2$.

The general expression for the coupling impedance is best handled via the computer program MACSYMA. The case of one port terminated, $R_o = Z_K$, and the input port represented by the general impedance $R_i = R + jX$ is of practical importance, as it reflects the typical situation of the kicker system. Here one finds

$$Z = \frac{1}{2}Z_K \left(\frac{M}{L_K}\right)^2 \{Z_K^2(1 - \cos 2\theta)$$

$$+ 4Z_K R(1 - \cos\theta)$$

$$+ (R^2 + X^2)(3 - 4\cos\theta + \cos 2\theta)$$

$$- 2Z_K X(2\sin\theta - \sin 2\theta)\}/D$$

$$+ jZ_K \frac{L_B}{L_K}\theta - j\frac{1}{2}Z_K \left(\frac{M}{L_K}\right)^2 \{Z_K^2(2\theta - \sin 2\theta)$$

$$+ 4Z_K R(\theta - \sin\theta)$$

$$+ (R^2 + X^2)(2\theta - 4\sin\theta + \sin 2\theta)$$

$$- 2Z_K X(1 - 2\cos\theta + \cos 2\theta\}/D$$

with the denominator $D = (Z_K + R)^2 + X^2$,

$$Z_K = \sqrt{<L>/<C>}$$

and $\theta = \omega l \sqrt{(<L><C>)} = 2\pi f/f_{2\pi}$ the electrical length of the kicker where $<L>$, $<C>$ are the quantitites per unit length. Theory and measured results for the coupling impedance are in reasonable agreement as long as the "electrical length" of the kicker is less than one free-space wavelength. At higher frequencies, the assumption of magnetic coupling is still valid, but the kicker is no longer the ideal lossless, uniform transmission line and a more general treatment, for example based on equivalent circuits, is required.[14]

## REFERENCES

1. H. Hahn, N. Tsoupas, and J. E. Tuozzolo, Proc. 1997 PAC, Vancouver, BC, Canada, p. 213.
2. CMD 5005 by Ceramic Magnetics, Inc., Fairfield, NJ
3. MCT-100 by Trans-Tech, Inc., Adamstown, MD.
4. V. Mane, S. Peggs. D. Trbojevic, and W. Zhang, Proc. 1995 PAC, Dallas, TX, p. 3134.
5. H. Hahn, M. Morvillo, A. Ratti, BNL Report AD/RHIC/RD-95 (1995).
6. A. Ratti, Fourth EPAC, London 1994, vol. 2, p. 1262.
7. H. Hahn and F. Pedersen, Report BNL #50870 (1978).
8. L. S. Walling, D. E. McMurry, D. V. Neuffer, and H. A. Thiessen, Nucl. Instr. Meth., A 281, p. 433 (1989).
9. V. G. Vaccaro, et al, Proc. 1993 PAC, Washington, D.C., p. 2154 (1993).
10. H. Hahn and A. Ratti, BNL Report AD/RHIC/RD-105 (1996).
11. G. Nassibian and F. Sacherer, Nucl. Instr. & Meth., vol. 159, p. 21 (1979).
12. G. Nassibian, Report CERN/PS 84-25(BR) (1984) and CERN/PS 85-68(BR) (1986).
13. H. Hahn and A. Ratti, BNL Report AD/RHIC/RD-111 (1997).
14. H. Hahn and A. Ratti, Proc. 1997 PAC, Vancouver, BC, Canada, p.216; BNL Report AD/RHIC/RD-112 (1997).

# APPENDIX: Theoretical Low-Frequency Coupling Impedance of Transmission Line Kickers

In contrast to the "good" agreement between experiment and theory reported by Nassibian, the results from the RHIC injection kicker impedance measurements showed significant differences with the above theories. Furthermore, the impedance measurements of the kicker system can only be done with one port resistor-terminated and the other changing with frequency from short to open due to the long feeding cables. The need to amend the intrinsic limitations of the Sacherer-Nassibian theories thus provided the impetus for the present study.

In the low- frequency range here considered, the kicker acts as a transmission line with uniform, albeit anisotropic properties. The kicker and the beam are treated as magnetically coupled transmission lines, for which the differential equations are well known. By limiting the considerations to the extreme relativistic case, where the space charge effect can be neglected, the beam can be represented by a "transmission line" in which the inductance per unit length, $L_B/l$, is determined by the coupling impedance of the un-terminated kicker, and the capacitance is negligible. One finds, with the harmonic time dependence $\exp(j\omega t)$ suppressed, the following set of differential equations in the position-dependent variables $i_K$, $u_K$, $u_B$ representing the kicker current, kicker voltage, and beam voltage respectively

$$\frac{du_K}{ds} = -jk_K Z_K i_K + j\frac{M}{L_K}k_K Z_K i_B$$

$$\frac{di_K}{ds} = -jk_K u_K/Z_K$$

$$\frac{du_B}{ds} = j\frac{M}{L_K}k_K Z_K i_K - j\frac{L_B}{L_K}k_K Z_K i_B$$

Assuming an extreme relativistic, filamentary beam current of unit strength

$$i_B = e^{-jks}$$

one obtains the coupling impedance from

$$Z = -\int_0^l \frac{du_B}{ds} e^{jks} ds$$

were $k = \omega/c$, $k_K = \omega\sqrt{<L><C>}$ and $Z_K = \sqrt{<L>/<C>}$.

The solution of the above differential equations are found without difficulty, for example by means of the MACSYMA program, together with the boundary conditions established by the kicker input and output terminations, $R_i$ and $R_o$

$$u_K(s=0) = -R_i i_K(s=0)$$
$$u_K(s=l) = R_o i_K(s=l)$$

The general expression for the coupling impedance is somewhat lengthy, but reduces at low frequencies, where $k \ll k_K$, to a manageable size. Several special cases must be expected in the field or are accessible to measurement and thus are of theoretical interest. The case of one port terminated with the characteristic impedance and the other with a general impedance representing the pulser and connecting cables is given in the main part of this paper. A few simpler cases relevant to bench measurements are discussed here:

1. **Case of $R_i = \infty$ and $R_o = \infty$**

$$Z = jZ_K \frac{L_B}{L_K}(1-\kappa^2)\theta$$
$$+j2Z_K \left(\frac{M}{L_K}\right)^2 \frac{1-\cos\theta}{\sin\theta}$$

with the magnetic coupling coefficient

$$\kappa^2 = \frac{M^2}{L_K L_B}$$

This case, i.e. with both kicker ports open, clearly shows the inadequacy of the original Nassibian-Sacherer theory which predicts a singularity in the limit of $\theta \to 0$, in contrast to the experimental and the present theoretical result

$$\lim_{\theta \to 0} Z = jZ_K \frac{L_B}{L_K}\theta = j\omega L_B$$

2. **Case of $R_i = Z_K$ and $R_o = Z_K$**

$$Z = Z_K \left(\frac{M}{L_K}\right)^2 (1-\cos\theta)$$
$$+jZ_K \frac{L_B}{L_K}\{(1-\kappa^2)\theta + \kappa^2 \sin\theta\}$$

It is to be noted, that the above expressions for the real part of the coupling impedance is identical to Nassibian's result, whereas the imaginary part differs. At very low frequencies and in the limit of $\theta \gg 1$ one finds

$$\lim_{\theta \to 0} Z = jZ_K \frac{L_B}{L_K}\theta = j\omega L_B$$

$$\lim_{\theta \to \infty} Z = jZ_K \frac{L_B}{L_K}(1 - \kappa^2)\theta = j\omega L_B(1 - \kappa^2)$$

comparing these limits with wire-measurements yields $L_B$ and $\kappa$ since the parameters $Z_K$ and $L_K$ are known from previous kicker measurements. The smooth curve in Fig 3 represents this case with input and output fully matched. The results can be fitted with $L_B = 132$ nH and $(1 - \kappa^2) = 0.325$, leading to

$$L_B/L_K = 0.18, \quad M/L_K = 0.34, \quad \text{and} \quad \kappa = 0.82.$$

The values thus obtained can then be checked by comparison with the following two cases.

### 3. Case of $R_i = Z_K$ and $R_o = 0$

$$Z = \frac{1}{2}Z_K \left(\frac{M}{L_K}\right)^2 (1 - \cos 2\theta)$$
$$+ jZ_K \frac{L_B}{L_K}\left\{(1 - \kappa^2)\theta + \frac{1}{2}\kappa^2 \sin 2\theta\right\}$$

### 4. Case of $R_i = Z_K$ and $R_o = \infty$

$$Z = \frac{1}{2}Z_K \left(\frac{M}{L_K}\right)^2 (3 - 4\cos\theta + \cos 2\theta)$$
$$+ jZ_K \frac{L_B}{L_K}\left\{(1 - \kappa^2)\theta + 2\kappa^2 \sin\theta - \frac{1}{2}\kappa^2 \sin 2\theta\right\}$$

Theory and measured results for the coupling impedance are in reasonable agreement at "low" frequencies. At higher frequencies, the assumption of magnetic coupling is still valid, but the kicker is no longer the ideal lossless, uniform transmission line and a more general treatment, for example based on equivalent circuits, is required.

# Longitudinal Space Charge Impedance

J. G. Wang

Oak Ridge National Laboratory, Oak Ridge, TN 37831
and Visiting at Brookhaven National Laboratory, Upton, NY 11973

**Abstract.** We review the definition and concepts of the longitudinal space charge impedance, and the formulas for its calculations. The conditions, limitations, and corrections in applying these formulas are discussed. Especially, we emphasize the various geometry factors associated with this parameter.

## Introduction

The longitudinal space charge impedance is an important physical parameter in the analysis of longitudinal instabilities in charged particle beams. It relates the perturbed longitudinal electrical field due to space charge force with the perturbed beam current. The effect of the transverse dimension on the longitudinal force appears as a geometry factor. In the literature, there are many different g-factors in carrying out calculations of the longitudinal space charge impedance. These g-factors result not only in different numerical values for a given beam, which may be significant for practical machine designs, but also in different physical pictures of the instability process in some cases. A review of the definition of the longitudinal space charge impedance and the formulas for its calculations are given first. We then discuss the conditions and limitations in applying these formulas, and corrections being made for specific situations. An interesting case to distinguish the so-called body wave from the surface wave is also presented.

Another useful beam parameter in accelerator design is the space charge potential. This parameter not only determines how fast a beam expands in both transverse and longitudinal dimensions, but also leads to the perturbed longitudinal electrical field associated with the longitudinal space charge impedance. The space charge potential can be calculated with a static charge distribution in the beam frame. There are also the geometry factors appearing in analysis. We review their calculations under various conditions.

Both space charge potential and longitudinal space charge impedance are measurable, physical quantities. The longitudinal space charge impedance can be determined experimentally by measuring the propagation speed of perturbations. The space charge potential can be measured by beam expansion in space. Some experimental examples concerning these parameters are presented.

CP496, *Workshop on Instabilities of High Intensity Hadron Beams in Rings,*
edited by T. Roser and S. Y. Zhang
© 1999 American Institute of Physics 1-56396-910-6/99/$15.00

## Definition and Formula

Let us consider a charged particle beam in a meta-equilibrium state with the beam parameters of current I, line charge density $\Lambda$, and velocity v. A longitudinal perturbation to the meta-equilibrium state at a time t and longitudinal position z produces a perturbed longitudinal electrical field $E_1(z,t)$ and other perturbed beam parameters such as $i_1(z,t)$, $\Lambda_1(z,t)$, and $v_1(z,t)$. The perturbed axial field $E_1(z,t)$ usually consists of two components: a contribution $E_{1s}(z,t)$ due to space charge force of beam particles and a contribution $E_{1w}(z,t)$ due to the perturbed beam image current on the surrounding pipe wall. Transferring these physical quantities from the real space-time domain (z,t) to the complex frequency domain (k,$\omega$), we define the longitudinal space charge complex wave impedance per unit length as [1,2]

$$Z_s^*(k,\omega) = -\frac{E_{1s}(k,\omega)}{i_1(k,\omega)} ,$$

(1)

which has a unit of $\Omega$/m.

For a coasting beam with a uniform circular transverse profile in a straight, round channel under the long wavelength perturbation limit, Eq. (1) becomes

$$Z_s^*(k,\omega) = i\frac{g}{4\pi}\left(-\frac{ck^2}{\omega} + \frac{\omega}{c}\right)Z_0 .$$

(2)

Here $Z_0$=377 $\Omega$ is the characteristic impedance of free space, c speed of light, and g a geometry factor being discussed later. In Eq. (2) the first term in the bracket is a capacitive component due to the axial electrical field produced by the line charge density perturbation, while the second term is an inductive component due to the magnetic field associated with the perturbed beam current. Under a sinusoidal perturbation of frequency $\omega$ with a linear wave satisfying $\omega \cong kv$, Eq. (2) can be approximated as

$$Z_s^*(\omega) = -i\frac{g\omega}{4\pi\beta\gamma^2 v}Z_0 .$$

(3)

The longitudinal space charge impedance per perturbation wavelength $\lambda$ is thus calculated as

$$Z_z = Z_s^* \lambda = -i\frac{g}{2\beta\gamma^2}Z_0 .$$

(4)

Equation (4) shows that the longitudinal space charge impedance per perturbation wavelength depends only on beam energy. It is negative imaginary, and thus, pure capacitive.

In a circular accelerator with coasting beams the perturbation wavelength $\lambda$ is simply related to the average main radius R and harmonics number n by $\lambda = 2\pi R/n$. Therefore, the longitudinal space charge impedance for circular accelerators is usually expressed as [3]

$$\frac{Z_{//}}{n} = -i\frac{g}{2\beta\gamma^2}Z_0 .$$

(5a)

Note that $Z_{//} = 2\pi R Z_s^*$ is in fact the total longitudinal space charge impedance along the accelerator circumference. The geometry factor g widely used in the high energy accelerator community is

$$g = 2\ln\left(\frac{b}{a}\right) + 1 ,$$

(5b)

where b is the pipe radius and a is the beam radius. For the SNS accumulator ring where the proton beam has an energy of 1 GeV ($\beta = 0.875$) and the ratio of b/a on average is about 2.7, Eq. (5b) yields a geometry factor g of about 3.0, and Eq. (5a) results in a longitudinal space charge impedance $Z_{//}/n$ of about 150 ohms.

## Conditions and Limitations

Equations (4) and (5) are derived under many assumptions. Thus, there are a number of limitations in applying them. Firstly, the equations are valid only under the long wavelength limit. At higher frequencies, the interaction between beams and surrounding walls becomes weak, or even screened off [4, 5]. Secondly, they require a straight channel consisting of smooth, conducting wall with a circular cross section. A particle moving in a curved trajectory is subject to the "pipe resonances" [6], thus, corrections may have to be made for rings. For non-smooth walls, an example is the effect of a uniform beam propagating inside a round, rf-screening wire cage with ceramic pipes [7]. Though a circular, transverse cross section is applied to the above formulas, other geometry, such as rectangular and elliptical, were investigated in the past [8,9].

As far as for the beam itself concerned, Equations (4) and (5) require coasting beams with a uniform particle density distribution and uniform, circular cross section. The betatron motion of particles and the beam envelope oscillation have been neglected. For bunched beams, there are variations of beam parameters in space and time along the bunch. It is common practice to define the longitudinal space charge impedance in the sense of average over the bunch length. The various particle distributions, rather than the uniform one, will be discussed in next section.

There is another subtle assumption in the derivation of the longitudinal space charge impedance, which has been largely overlooked. That is, the particle volume density varies with perturbations, while the beam radius remains unchanged. Strictly speaking, this is true only for a beam produced from an immersed gun and focused by an infinitely strong magnetic field. When both the particle volume density and beam radius change due to a perturbed longitudinal field, the longitudinal space charge impedance changes as well. This issue will be discussed later.

**Corrections**

In the original derivation of Equation (5), the geometry factor g has the form [8]:

$$g = 2\ln\left(\frac{b}{a}\right) + 1 - \frac{r^2}{a^2} .$$

(6)

Here r is the radial position of particles within the beam. It is clear that the longitudinal space charge impedance, as well as the perturbed longitudinal electrical field, is dependent on the radial position. Equation (5b) only shows the value on the axis, which is the maximum. It is argued that an average over the transverse dimension should be used instead of the value on axis. This results in

$$\bar{g} = 2\ln\left(\frac{b}{a}\right) + \frac{1}{2} .$$

(7)

For the SNS ring, Eq. (7) yields a geometry factor of about 2.5, leading to a longitudinal space charge impedance of 125 ohms, which is significantly smaller than that on the axis.

As mentioned above, Equation (5) was derived under the assumption that the beam radius remains constant, while the particle volume density varies with perturbations. In reality, both the particle volume density and the beam radius changes with perturbations since the external focusing magnetic field is finite. The relative change of these two parameters depends on specific beam and focusing parameters. An extreme case is that the particle volume density remains constant, while the beam radius varies with perturbations. The theory and experiment show that the geometry factor for this case is [10]

$$g = 2\ln\left(\frac{b}{a}\right) .$$

(8)

Equation (8) indicates that the longitudinal space charge impedance, as well as the perturbed longitudinal electrical field within the beam, is a constant.

The waves produced by perturbations are often classified as body waves if particle volume density varies with perturbations, or surface waves if particle volume density remains unchanged with perturbations [11]. Equations (4-7) describe the dynamics of the body wave, where the perturbed longitudinal field, as well as space charge impedance, peaks on axis and gradually decreases towards the beam edge. The word "surface wave", on the other hand, has the connotation that the electrical fields do not penetrate into the interior. It is indeed in general that the perturbed longitudinal electrical field, as well as space charge impedance, has their maximum on the beam surface, and gradually dies off towards the axis. Equation (8) describes a special case of the surface wave under the long wavelength limit where the reactive skin depth $c/\omega_p$ with c the speed of light and $\omega_p$ the plasma frequency is much larger than the beam transverse dimension. Thus, the perturbed longitudinal field, as well as space charge impedance is a constant across the beam. In Fig. 1 we compare different space charge impedance for the body wave, surface wave, and its long wavelength limit.

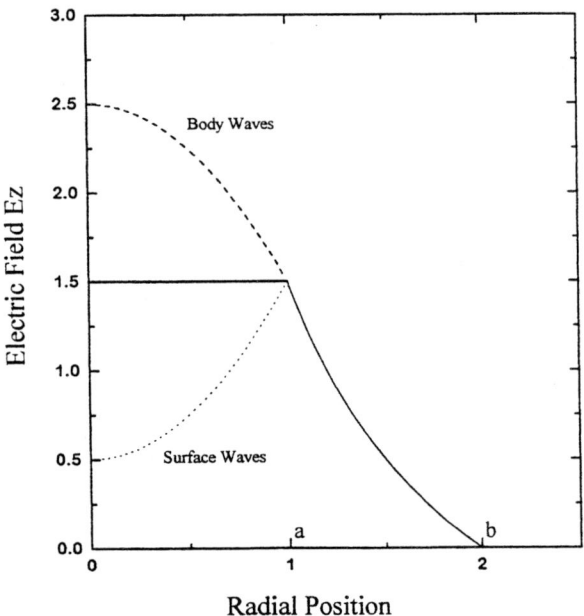

Fig. 1 Perturbed longitudinal electrical field vs. radial position for body wave, surface wave and its long wavelength limit.

The previous discussion is valid only for a uniform particle distribution in a round beam. For other particle distributions, a rigorous derivation and mathematical expression of the longitudinal space charge impedance is rather complicated. The common practice is to employ Equation (4) or (5) with the geometry factor g calculated from the concept of equivalent beams. For example, a round beam with a Gaussion particle distribution has an equivalent uniform beam of radius twice its rms radius. Thus, following Eq. (7), the geometry factor for the Gaussion beam is usually expressed by

$$\overline{g} = 2\ln\left(\frac{b}{2\tilde{x}}\right) + \frac{1}{2} .$$

(9)

## Space Charge Potential

1. Uniform beam

The perturbed longitudinal electrical field in the calculation of space charge impedance can be determined by the space charge potential in beams. For a uniform beam with circular cross section, the space charge potential is

$$\phi(r,z) = -\frac{g}{4\pi\varepsilon_0}\Lambda ,$$

(10a)

with the geometry factor g given by

$$g = \begin{cases} 2\ln\left(\frac{b}{a}\right) + 1 - \frac{r^2}{a^2} & 0 \leq r \leq a \\ 2\ln\left(\frac{b}{r}\right) & a \leq r \leq b . \end{cases}$$

(10b)

This is usually done in the beam frame for a static field, and a transformation to the lab frame will recover the particle energy dependence. The perturbed longitudinal electrical field is then calculated as

$$E_s(r) = -\frac{g}{4\pi\varepsilon_0}\frac{\partial \Lambda}{\partial z} .$$

(11)

2. Particle distribution

For particle distributions rather than a uniform beam, the calculation of the space charge potential is more complicated. R. Baartman has shown that for a binomial family of distributions [12]

$$\rho = (\mu+1)\frac{\Lambda}{\pi a^2}\left(1-\frac{r^2}{a^2}\right)^\mu,$$

(12)

the geometry factor for the calculation of the longitudinal space charge potential can be expressed as

$$\bar{g} = 2\ln\left(\frac{b}{2\tilde{x}}\right) + \frac{1}{2}.$$

(13)

Indeed, this expression is the same as we use the concept of equivalent beams for a Gaussion distribution, Eq. (9).

3. Parabolic bunch

For particle bunches the beam parameters, such as particle density, current, and radius, varies in time and space. And, the image fields reduce the axial defocusing space-charge force, while increase the radial defocusing space charge force. Therefore, a single geometry factor is only valid in the average sense. For ellipsoidal bunches with a uniform volume density distribution

$$\Lambda(z) = \Lambda_0\left(1-\frac{z^2}{z_m^2}\right),$$

(14)

where $z_m$ is the maximum bunch half length and $\Lambda_0$ is the peak line charge density, the geometry factor can be approximated as

$$\bar{g} \approx 2\ln\left(\frac{b}{a}\right) + \frac{1}{2}.$$

(15)

A more detailed discussion on this topic can be found in Reference [13].

## Measurements

1. Longitudinal space charge impedance of a uniform beam

The longitudinal space charge impedance can be measured in experiment by launching space charge waves and measuring their propagation speed on the beam [14]. An experiment was performed at the University of Maryland with space charge dominated electron beams [10]. Figure 2 shows the beam current signals modulated with localized perturbations at different locations of a transport channel. The data from these signals yield the wave speed and the geometry factor, as shown in Fig. 3. The longitudinal space charge impedance is then obtained from Eq. (4).

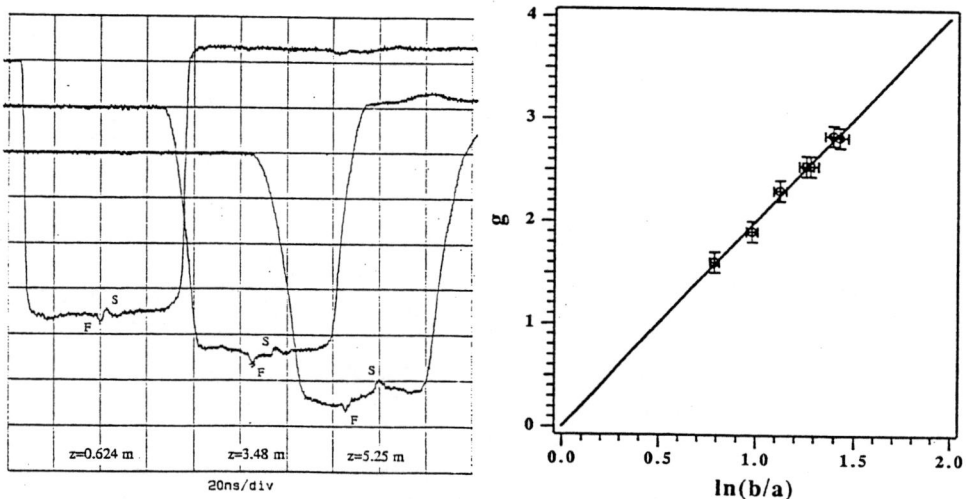

Fig. 2 Beam current with perturbations.    Fig. 3 Geometry factor from experiment.

2. Geometry factor of a parabolic bunch

The geometry factor of a parabolic bunch can be determined in experiments by measuring the free expansion of the bunch profile [15]. Figure 4 shows a parabolic bunch propagating in a transport channel. Its bunch length increases with distance due to space charge expansion. Figure 5 shows the experimental data with curve fitting to the longitudinal envelope equation. This results in the geometry factor g under two different focusing conditions.

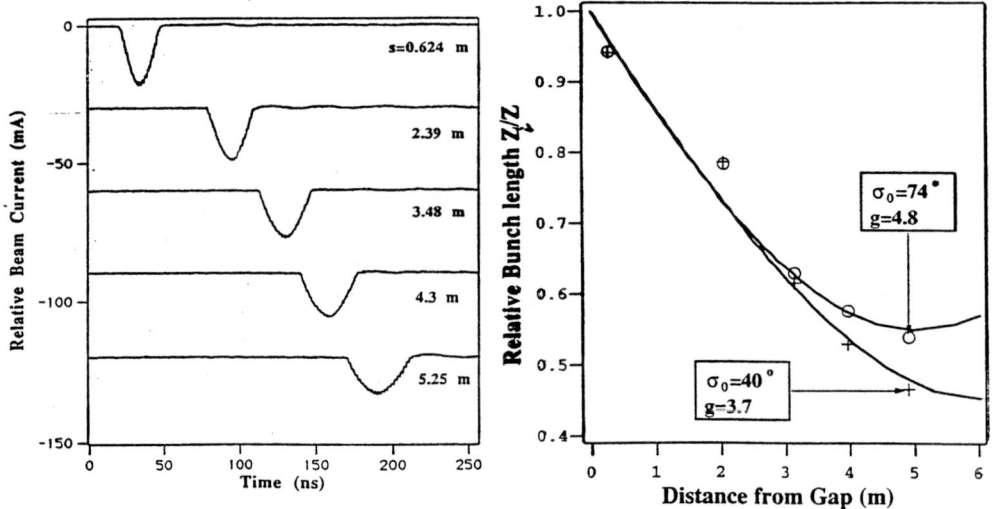

Fig. 4 Expansion of a parabolic bunch.   Fig. 5 Geometry factor of the bunch.

## References

[1] M. Reiser, "Theory and Design of Charged Particle Beams", John Wiley & Sons, Inc. New York, 1994.
[2] J. G. Wang and M. Reiser, Phys. Fluids B **5** (7), 2286 (1993).
[3] A. Hofmann, CERN Report No. 77-13, 139 (1976).
[4] C. Birdsall and J. R. Whinnery, J. Appl. Phys. **24**, 314 (1953).
[5] A. G. Ruggiero and M. Blaskiewicz, BNL/NSNS Technical Note No. 027 (1997).
[6] A. Faltens and L. J. Laslett, Proc. of the 1975 ISABELLE Summer Study, vol. II, p. 486.
[7] T. S. Wang, CERN Report No. PS 94-08 (DI) (1994).
[8] V. K. Neil and A. Sessler, Rev. Sci. Instr. **36** (4), 429 (1965).
[9] A. G. Ruggiero, CERN Report, ISR-SRM/67-13 (1967).
[10] J. G. Wang, H. Suk, D.X. Wang, and M. Reiser, *Phys. Rev. Lett.* **72** (13), 2029 (1994).
[11] A. W. Trivelpiece and R. W. Gould, J. Appl. Phys. **30**, 1784 (1959).
[12] R. Baartman, TRI-DN-92-K206, TRIUMF, 1992.
[13] C. K. Allen, N. Brown, and M. Reiser, Part. Accel. **45**, 149 (1994).
[14] J. G. Wang and M. Reiser, Rev. Sci. Instrum. **65**(11), 3444 (1994).
[15] D. X. Wang, J. G. Wang, D. Kehne, and M. Reiser, Appl. Phys. Lett., **62**(25), 3232 (1993).

# Impedance Considerations for the Intense Pulsed Neutron Source (IPNS) Rapid Cycling Synchrotron (RCS)

J. C. Dooling, F. R. Brumwell, G. E. McMichael

*Argonne National Laboratory, Argonne, IL USA*

**Abstract.** The use of Second Harmonic (SH) rf is being investigated to increase the RCS current limit. Hofmann-Pedersen distributions are employed to provide analytical guidance. The SH phase θ, is optimized using a numerical analysis to maximize transmission and minimize instabilities. The effect of the RCS stainless steel liner on the impedance of the machine is also discussed.

## I. Introduction

The Intense Pulsed Neutron Source (IPNS) provides neutrons for over 300 science experiments per year and is oversubscribed by almost a factor of two. One method for increasing the number of experiments performed or increasing signal to noise in the present array of instruments is to raise the proton current from the accelerator. Presently, the IPNS accelerator produces 14-15 μA of 450 MeV protons which are directed to a depleted uranium target. By adding a third rf cavity operating at the second harmonic, it is hoped to increase the current in the Rapid Cycling Synchrotron (RCS) by 20-40 percent. Before the RCS current can be raised, it is important to understand the present limitations on the device.

The RCS is a small compact synchrotron. The beam pipe in the bending magnets has an elliptical cross section, 20.3 cm (8 in.) wide by 7.6 cm (3 in) high, but in the extraction section this narrows to approximately 10 cm (4 in) wide by 5 cm (2 in) high. Presently, two rf cavities, diametrically opposite in the ring, provide up to 22 kV per turn for bunching and acceleration. Plans are underway to add a third rf cavity, operable at either the fundamental or second harmonic (SH) frequency capable of providing an additional 10-12 kV per turn. Other parameters for the RCS are given in Table 1.

## II. Analysis—Longitudinal Equations of Motion

A. Numerical Recipes

Solution to the coupled set of the synchrotron equations of motion is sought where damping and modulation are ignored but space charge and SH excitation are included. $V_o$ is the amplitude of the fundamental rf voltage, δ is ratio of second harmonic to fundamental voltage amplitudes, and θ is the phase between the two waveforms.

CP496, *Workshop on Instabilities of High Intensity Hadron Beams in Rings,*
edited by T. Roser and S. Y. Zhang
1999 American Institute of Physics 1-56396-910-6

| Parameter | Value |
|---|---|
| Injection/extraction energy ($W_{inj}$, $W_{extr}$) | 50 MeV, 450 MeV |
| Injection/extraction frequency ($f_{inj}$, $f_{extr}$) | 2.21 MHz, 5.14 MHz |
| Average machine radius, R | 6.83 m |
| Bending radius, $\rho$ | 3.68 m |
| Initial Tunes $v_x$, $v_y$ | 2.25, 2.35 |
| Lattice | DOFDFO (combined fn. dipoles $|n|$=11.7) |
| Injected pulsewidth | 70-80 µs |
| Injected charge | $3.6 \times 10^{12}$ H$^-$ (0.58 µC) |
| Extracted pulsewidth (FWHM) | 38 ns |
| Extracted charge | $3.2 \times 10^{12}$ H$^+$ (0.51 µC) |
| Rep rate | 30 Hz |
| Peak current | 14 A |
| Bunching factor at extraction | 0.19 |

Table 1: IPNS RCS parameters

$$\frac{d}{dt}\left[\frac{\Delta E}{\omega_o}\right] = \frac{q}{2\pi}[V(\phi,\theta) - V(\phi_s,\theta) + V_{sc}(\phi)] \quad (1a)$$

$$\frac{d}{dt}\Delta\phi = \frac{h\omega_o\eta}{\beta^2 E_s}\Delta E \quad (1b)$$

where $V(\phi,\theta) = V_o f(\phi,\theta)$, with

$$f(\phi,\theta) = \sin(\phi) - \delta\sin(2\phi + \theta). \quad (2)$$

The space-charge voltage term is expressed as,

$$V_{sc}(\phi) = -q\frac{d\lambda(\phi)}{d\phi}\left[\frac{Rg_o}{2\varepsilon_o\gamma^2} - L\beta^2 c^2\right]\frac{1}{R^2} \quad (3)$$

To quantitatively evaluate the effects of space-charge and second harmonic (SH) rf on the bunch, the synchrotron numerical analysis program, CAPTURE_SPC[1,2] is employed. CAPTURE_SPC solves the preceding equations with a second-order "leap-frog" differencing algorithm:

$$W_{n+1/2} = W_{n-1/2} + \frac{qV_n\tau}{2\pi}(f(\phi_n,\theta_n) - f(\phi_{s,n},\theta_n)) + \frac{q^2 g_o}{4\pi\varepsilon_o}\frac{h^2\tau}{R\gamma_{s,n}^2}\left[\frac{d\lambda(\phi)}{d\phi}\right]_n \quad (4a)$$

$$\phi_{n+1} = \phi_n + h\tau \left[ \frac{\eta_s \omega_s^2 W}{\beta_s^2 E_s} \right]_{n+1/2} + \phi_{s,n+1/2} - \phi_{s,n-1/2} \tag{4b}$$

A separate predictor-corrector program uses the algorithm[3]:

$$y_n = 2y'_{n-1}(s_{n-1})\Delta s + y_{n-2} \tag{5}$$

where,

$$y'(s) = \frac{dy}{ds}, \tag{6}$$

and solves the system of equations:

$$\begin{bmatrix} E_j \\ \phi_j \end{bmatrix} = \begin{bmatrix} 2qV_o\big(f(\phi_{j-1},\theta) - f(\phi_s,\theta)\big) + E_{j-2} \\ \dfrac{4\pi\eta}{\beta^2 E_s} E_{j-1} + \phi_{j-2} \end{bmatrix} \tag{7}$$

Presently, the predictor-corrector program generates phase-space contours ignoring space charge. The bucket area predicted by the predictor-corrector program is in excellent agreement with the CAPTURE_SPC program with space-charge turned off. A comparison of two analyses with and without SH is given in Figure 1; also shown in the figure are the bucket sizes including the effects of space charge ($3.6\times 10^{12}$ protons=0.58 µC, injected) determined by CAPTURE_SPC. As indicated in Fig. 1, the minimum stable phase-space area in the fundamental rf only case is 0.29 eV-sec approximately 7 ms after injection. The maximum proton phase-space density occurs at this time reaching a value of $11\times 10^{12}$ protons/eV-sec.

B. Hofmann-Pedersen Distributions

Hofmann-Pedersen[4] elliptical distributions have been used to examine current profiles and limits in the RCS. Although the particle distributions are "stationary" and therefore strictly correct only for time-invariant situations, they provide a useful approximation. In the presence of space-charge, the total voltage becomes,

$$V_t(\phi,\theta) = V_o\left( f(\phi,\theta) - 2\pi h^2 I_b \, \text{Im}\left\{\frac{Z_e}{n}\right\} \frac{f(\phi,\theta) - f(\phi_s,\theta)}{u(\phi_1,\phi_2,\theta)} \right) \tag{8}$$

where,

$$u(\phi_1,\phi_2,\theta) = \int_{\phi_1}^{\phi_2}\left[ U(\xi,\theta) - U(\phi_2,\theta) \right] d\xi, \tag{9}$$

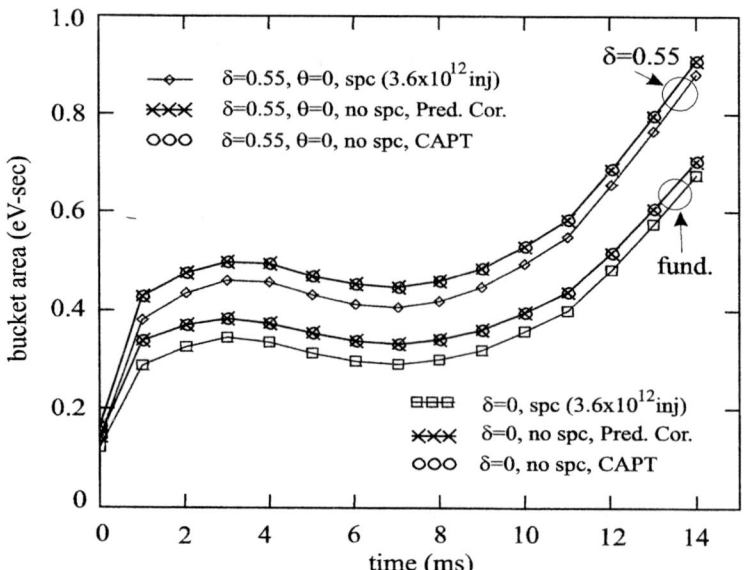

Figure 1: Evolution of RCS stable phase-space comparing CAPTURE_SPC and predictor-corrector analyses for fundamental rf only and SH with $\delta=0.55$.

$$\text{Im}\left\{\frac{Z_e}{n}\right\} = \left[\omega L - \frac{g_o Z_o}{2\beta\gamma^2}\right] \tag{10}$$

with $g_o = 1 + 2\ln(b/a)$, $Z_o = 120\pi\ \Omega$ and

$$U(\xi,\theta) = V_o\left(f_2(\xi,\theta) - f_2(\phi_s,\theta) - (\xi - \phi_s)f(\phi_s,\theta)\right) \tag{11}$$

with $f_2(\phi,\theta) = \int f(\phi,\theta)d\phi$. The limiting current is expressed as,

$$I_{b,\max} = \frac{u(\phi_1,\phi_2,\theta)}{2\pi h^2\ \text{Im}\{Z_e/n\}} \tag{12}$$

The Keil-Schnell (KS) [5] microwave instability limit is given by:

$$\frac{|Z_e|}{n} \leq F \frac{E|\eta|}{q\beta^2} \frac{[\Delta E(\phi,\theta)/E]^2}{I(\phi,\theta)} \tag{13}$$

This limit reduces to,

$$I_b \le \frac{V_o |f(\phi_1,\phi_2,\theta)|}{5\pi h^2 |Z_e|/n} = 0.4 I_{b,max} \qquad (14)$$

The current limit is examined 8 ms after injection, just prior to $B_{max}$, the time of maximum rate of change in the guiding magnetic field strength, when the stable phase-space is near a minimum. Up to and including this time, the bucket is completely filled by the proton bunch; however, beyond this point, the stable phase-space of the bucket grows more rapidly than the bunch.

## C. SH Phasing

One might expect that up to the time of minimum phase-space, SH would have the greatest positive effect maximizing the stable bucket size; whereas, later in the cycle, SH should be phased to maximize the bunching factor, $B_f$ to reduce the growth of instabilities. Bunching factor, which is the ratio of the mean to peak current in the bunch, is defined as,

$$B_f = \frac{1}{I_{max}} \oint I(\phi)d\phi \qquad (15)$$

A study has been made using CAPTURE_SPC to investigate the proper phasing for the SH. When SH ($\delta=0.55$) is used at zero phase with respect to the fundamental ($\theta=0$), acceleration losses during the first half of the cycle are reduced compared with both fundamental rf only and phasing to maximize the $B_f$, $\theta_{max}(t)$. The reduced loss is probably due to the increased bucket size for this case. Also, no additional loss is observed if the phase is ramped from zero to $\theta_{max}$ between 8 and 10 ms and then maintained at $\theta_{max}(t)$ until the end of the cycle. This phasing scenario is presented in Figure 2.

KS microwave instability limits, 8 ms after injection, are presented in Table 2 for three cases: 1) fundamental rf only and SH with $\delta=0.55$ for 2) $\theta=0$ and 3) $\theta=-1.05$ rad. Also given in the Table are the stable phase-space area and bunching factor ignoring the effects of space-charge. Though the current limit is lower when phase is set to maximize $B_f$, the peak current for the same charge will be smaller because $B_f$ is greater. The maximum charge that can be held in a bucket is proportional to $B_f I_b/f$ and is given in the last line of Table 2. Also shown in Table 2 is a comparison of the same SH parameters at 10 ms. The revolution frequencies at 8 ms and 10 ms are 4.04 MHz and 4.55 MHz, respectively. Analytical current and rf voltage profiles for the 8 ms cases given in Table 2 are presented in Figure 3; the space-charge reduced voltage waveform is also shown in the figure.

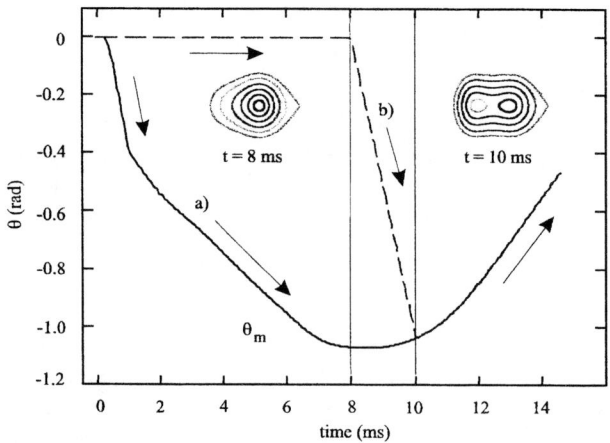

Figure 2: SH phasing to minimize acceleration losses during the first half of the acceleration cycle and maximize $B_f$ during the remainder of the cycle.

|  | Fund. rf only | 8 ms | | 10 ms | |
|---|---|---|---|---|---|
|  |  | $\delta=.55$, $\theta=0$ | $\delta=.55$, $\theta=\theta_m$ | $\delta=.55$, $\theta=0$ | $\delta=.55$, $\theta=\theta_m$ |
| $A_b$ (eV-s) | 0.388 | 0.521 | 0.469 | 0.618 | 0.564 |
| $B_f$ | 0.306 | 0.354 | 0.487 | 0.363 | 0.492 |
| $I_b$ (A) | 9.23 | 14.03 | 10.36 | 14.48 | 11.11 |
| $B_f \cdot I_b/f$ (μC) | 0.698 | 1.229 | 1.249 | 1.156 | 1.202 |

Table 2: Comparison of bucket parameters at 8 ms and 10 ms after injection in the RCS ($A_b$ and $B_f$ are calculated for zero space charge).

## III. Impedance

A. Inductance and Compensation

Impedance estimates can be determined for the RCS using Eq. 9. The inductance, L is assumed to be the inductance of the ring, ignoring the compensating effect of the liner; i.e., $L=\mu_o \pi R/2=13.5$ μH. At 8 ms, $\omega=2\pi f=2.538 \times 10^7$ rad/s; therefore the magnitude of the inductive impedance is 342 Ω. For the capacitive space charge term, let the average beam radius, a=20.9 mm, and the average wall radius, b=49.4 mm. The beam energy at 8 ms is 212 MeV, so $\beta=0.578$ and $\gamma=1.226$; the magnitude of the capacitive impedance is 590 Ω. With these values, the maximum peak current according to Eq. 13 in the case of fundamental rf is 8.0 A. In addition, two blocks of ferrite making up the kicker magnets are also exposed to the time-varying fields of the beam. The kicker inductance is 1.3 μH. Adding kicker inductance to that for the ring

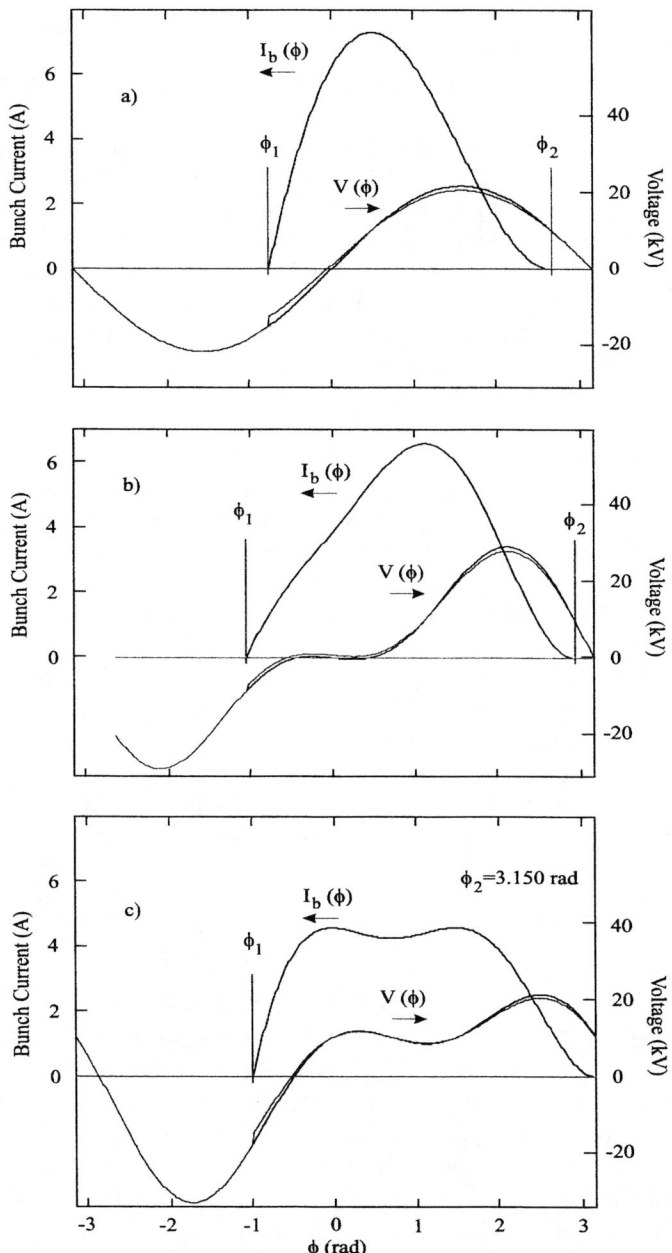

Figure 3: RCS longitudinal current and rf voltage profiles using Hofmann-Pedersen distributions for a) fundamental rf only, b) $\delta=0.55$, $\theta=0$, and c) $\delta=0.55$, $\theta=-1.07$ rad (maximizing $B_f$) 8 ms after injection.

results in a peak current limit of 9.2 A. This suggests that adding still more unshielded ferrite around the RCS ring might produce a larger current limit. According to the above analysis, an additional 8 µH could be added to the ring to cancel the calculated capacitive impedance.

Attempts to provide inductive compensation have met with mixed results to date[6,7,8, & 9]. Insert studies so far have looked at either the synchrotron frequency or the required rf voltage for stability. Though the addition of inductive inserts have had positive effects on both frequency and voltage (raising and lowering, respectively) some deleterious results were noted at the Proton Storage Ring at Los Alamos in the form of cavity-cavity oscillations on the bunch. Nevertheless, inductive inserts potentially offer a relatively inexpensive path to higher currents and the installation of such a device in the RCS will be investigated further.

## B. RCS Liner

To hide magnet laminations, the RCS uses a stainless steel liner within the combined function dipoles[10]. Since the bending radius of the dipoles is 3.68 m, the total length of the liner is 23.1 m, or just over half of the total path around the RCS (42.9 m). The liner is fabricated from 16 mil (0.41 mm) thick sheet approximately 50 cm (20 in.) in width. The sheet is sheared (no metal removed) across the width to within 2.5 cm (1 in.) of either edge making 2.5 cm bands in the sheet in the direction of the beam. The liner is formed by tack welding both edges to one another creating a 15.2 cm (6 in.) diameter tube. The welded edge forms the inner radius of the liner, while the bands on the outer edge separate along the shear as the liner is curved to match the bending radius of the magnets. Final liner dimensions result from inserting two pipes of smaller diameter through the length of the tube and pulling the pipes in opposite directions until an elliptical cross section is created approximately 20.3 cm (8 in.) wide by 7.6 cm (3 in.) high. The gaps between the bands vary from 0.7 mm at the center of the liner 1.4 mm at the outer radius. A view of the liner looking into one of the combined function triplet magnets is given in Figure 4.

The measured liner resistivity, $\rho$ is about 40 times that of copper or $6.5 \times 10^{-7}$ $\Omega$-m. The resistance of the liner, a function of the skin depth, $\delta = \sqrt{2\rho/\mu\omega}$, increases during the acceleration cycle. The effective resistance is a function of both the revolution frequency as well as the bunch pulsewidth. The revolution frequency, as indicated in Table 1, varies from 2.21 MHz at injection to 5.14 MHz at extraction. The pulsewidth of the bunch falls from 190 ns near injection to 38 ns at extraction. Temporal evolution of the FWHM bunch pulsewidth is presented in Figure 5. For comparison, the FWHM pulsewidth determined from CAPTURE_SPC is also shown in the figure. Because of the discrete nature of the simulated charge—$3.6 \times 10^{12}$ injected protons are modeled as 2464 macrocharges—longitudinal CAPTURE_SPC profiles are often not as smooth as the measured distributions, causing numerically-induced noise in the pulsewidth results; this is evident in Figure 5. However overall, the numerical analysis is in good agreement with the measured data. A full description of the roll played by

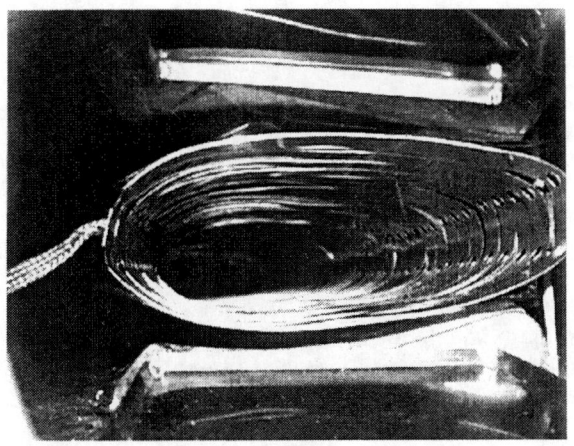

Figure 4: The RCS stainless steel liner. The approximate dimensions are 20.3 cm in width by 7.6 cm in height.

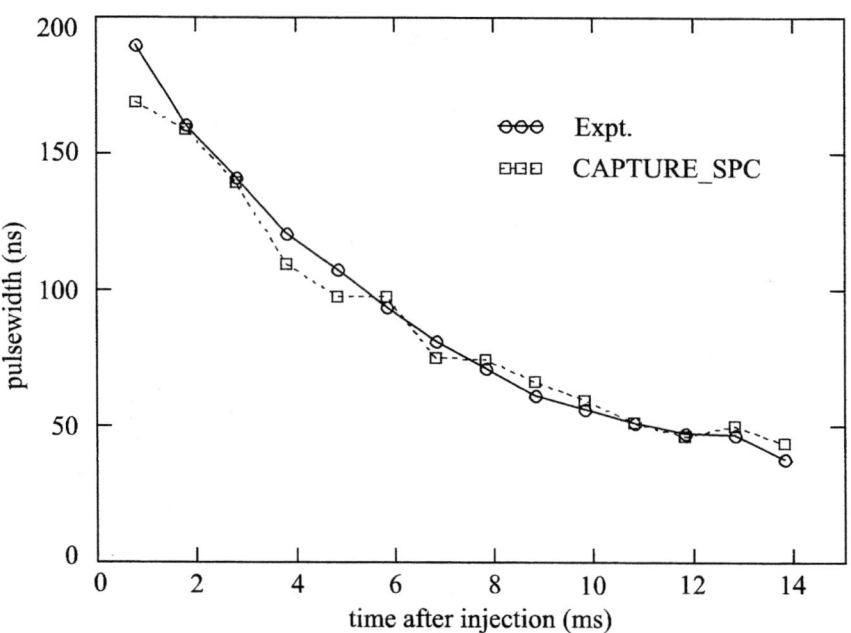

Figure 5: Comparison of the RCS bunch pulsewidth (FWHM) predicted by CAPTURE_SPC with Fast-Q data. The simulation follows the motion of 2464 macroparticles used to represent an injected charge of 0.58 µC.

the liner in the RCS, including contributions to impedance and bunch stability, is yet to be determined and is currently under study.

## IV. Discussion

Numerical and analytical tools are being used to determine whether a second harmonic cavity can be employed to increase the current stability limit in the RCS and hence provide a higher flux for neutron scattering instruments. In order to utilize these modeling tools, good impedance measurements are necessary. From modeling done so far with observed machine current limits, the RCS appears to exhibit a significant amount of inductive loading. Calculations also indicate that adding more inductance might increase the present current limit. It is hoped to explore the inductive loading question in the near future.

## Acknowledgement

This work is supported by the United States Department of Energy under contract No.: W-31-109-ENG-38.

## References

1. Y. Cho, E. Lessner, K. Symon, Proc. European Particle Accelerator Conf., p. 1228, (1994).
2. E. Lessner and K. Symon, Computational Accelerators Physics, Williamsburg, VA, Sept., 1996, AIP Conf. Proc. 391, p.185 (1997).
3. J. C. Dooling, et al., Proc. of the 1999 Part. Accel. Conf., March 28-April 2, 1999, New York City
4. A. Hofmann and F. Pedersen, IEEE Trans. Nuc. Sci., **26**(3), 3526 (1979).
5. E. Keil and W. Schnell, CERN-ISR-TH-RF/69-48 (1969).
6. K. Y. Ng and Z. B. Qian, Proc. 1999 Part. Accel. Conf., March 28-April 2, 1999, New York City.
7. K. Koba, et al., Rev. Sci. Instrum., **70**(7), 2988(1999)
8. M. A. Plum, et al., Phys. Rev. Spec. Topics-AB, **2**, 064201(1999)
9. R. J. Macek, these proceedings
10. R. L. Kustom, private communication

# 3D Multispecies Nonlinear Perturbative Particle Simulations of Collective Instabilities in Intense Particle Beams

Hong Qin, Ronald C. Davidson, and W. Wei-li Lee

*Plasma Physics Laboratory*
*Princeton University, Princeton, NJ 08543*

**Abstract.** Collective instabilities in intense charged particle beams described self-consistently by the Vlasov-Maxwell equations are studied using a 3D multispecies nonlinear perturbative particle simulation method. The electron-proton (e-p) two-stream instability is observed in the simulations carried out with the newly-developed Beam Equilibrium Stability and Transport (BEST) code. This code provides an effective numerical tool to investigate collective instabilities, periodically-focused beam propagation in alternating-gradient focusing fields, halo formation, and other important nonlinear processes in intense beam propagation.

## I INTRODUCTION AND THEORETICAL MODEL

For accelerator applications to spallation neutron sources, hadron colliders, waste transmutation, and heavy ion fusion, space-charge effects on beam equilibrium, stability, and transport properties become increasingly important. To understand these collective processes at high beam intensities, it is necessary to treat the nonlinear beam dynamics self-consistently using the nonlinear Vlasov-Maxwell equations [1–3]. Recently, the $\delta f$ formalism, a low-noise, nonlinear perturbative particle simulation technique, has been developed for intense beam applications, and applied to matched-beam propagation in a periodic focusing field [4–7] and other related studies. The present paper reports recent advances in applying the $\delta f$ formalism to simulate collective instabilities in an intense beam. The BEST code described here is a newly-developed 3D multispecies nonlinear perturbative particle simulation code [8], which can be applied to a wide range of important collective processes in intense beams, such as the electron-ion two-stream instability [9,10], periodically-focused beam propagation [11–13], and halo formation.

In the theoretical model [9,10,14], we consider a thin, continuous, high-intensity ion beam ($j = b$), with characteristic radius $r_b$ propagating in the $z$-direction through background electron and ion components ($j = e, i$), each of which is described by a distribution function $f_j(\boldsymbol{x}, \boldsymbol{p}, t)$. The charge components ($j = b, e, i$)

propagate in the z-direction with characteristic axial momentum $\gamma_j m_j \beta_j c$, where $V_j = \beta_j c$ is the directed axial velocity, $\gamma_j = (1 - \beta_j^2)^{-1/2}$ is the relativistic mass factor, $e_j$ and $m_j$ are the charge and rest mass, respectively, of a j'th species particle, and $c$ is the speed of light in *vacuo*. For each component ($j = b, e, i$), the transverse and axial particle velocities in a frame of reference moving with axial velocity $\beta_j c \hat{e}_z$ are assumed to be *nonrelativistic*. While the nonlinear $\delta f$ formalism outlined here is readily adapted to the case of a *periodic* applied focusing force, for present purpose we make use of a *smooth-focusing* model in which the applied focusing force is described by

$$\boldsymbol{F}_j^{foc} = -\gamma_j m_j \omega_{\beta j}^2 \boldsymbol{x}_\perp, \tag{1}$$

where $\boldsymbol{x}_\perp = x\hat{e}_x + y\hat{e}_y$ is the transverse displacement, and $\omega_{\beta j} = const$ is the effective applied betatron frequency for transverse oscillations. For example, in the absence of background ions ($f_i = 0$), to describe the two-stream interaction between the beam ions ($j = b$) and background electrons ($j = e$), we normally assume stationary electrons with $V_e = \beta_e c \simeq 0$. The space-charge intensity is allowed to be arbitrarily large, subject only to transverse confinement of the beam ions by the applied focusing force, and the background electrons are confined in the transverse plane by the space-charge potential $\phi(\boldsymbol{x}, t)$ due to the excess ion charge. In the electrostatic approximation, we represent the self-electric and self-magnetic fields by $\boldsymbol{E}^s = -\nabla \phi(\boldsymbol{x}, t)$ and $\boldsymbol{B}^s = \nabla \times A_z(\boldsymbol{x}, t)\hat{e}_z$, respectively. The nonlinear Vlasov-Maxwell equations in the six-dimensional phase space $(\boldsymbol{x}, \boldsymbol{p})$ can be approximated by [9,10,14]

$$\left\{ \frac{\partial}{\partial t} + \boldsymbol{v} \cdot \frac{\partial}{\partial \boldsymbol{x}} - [\gamma_j m_j \omega_{\beta j}^2 \boldsymbol{x}_\perp + e_j(\nabla \phi - \beta_j \nabla_\perp A_z)] \cdot \frac{\partial}{\partial \boldsymbol{p}} \right\} f_j(\boldsymbol{x}, \boldsymbol{p}, t) = 0, \tag{2}$$

and

$$\begin{aligned} \nabla^2 \phi &= -4\pi \sum_j e_j \int d^3 \boldsymbol{p} f_j(\boldsymbol{x}, \boldsymbol{p}, t), \\ \nabla^2 A_z &= -4\pi \sum_j e_j \beta_j \int d^3 \boldsymbol{p} f_j(\boldsymbol{x}, \boldsymbol{p}, t). \end{aligned} \tag{3}$$

Here, $\nabla_\perp = \hat{e}_x \partial/\partial x + \hat{e}_y \partial/\partial y$.

## II  NONLINEAR δF SIMULATION METHOD AND THE BEST CODE

In the nonlinear δf formalism, we express the total distribution function as $f_j = f_{j0} + \delta f_j$, where $f_{j0}$ is a *known* solution to the nonlinear Vlasov-Maxwell equations (2) and (3), and determine the detailed evolution of the perturbed distribution function $\delta f_j \equiv f_j - f_{j0}$. This is accomplished by advancing the weight function defined by $w_j \equiv \delta f_j/f_j$, together with the particles' positions and momenta. The equations of motion for the particles, obtained from the characteristics of the nonlinear Vlasov equation (2), are given by

$$\frac{d\boldsymbol{x}_{ji}}{dt} = (\gamma_j m_j)^{-1} \boldsymbol{p}_{ji},$$

$$\frac{d\boldsymbol{p}_{ji}}{dt} = -\gamma_j m_j \omega_{\beta j}^2 \boldsymbol{x}_{\perp ji} - e_j(\nabla \phi - \beta_j \nabla_\perp A_z). \tag{4}$$

Here the subscript "$ji$" labels the i'th simulation particle of the j'th species. The weight functions $w_j$, as functions of phase space variables, are carried by the simulation particles, and the dynamical equations for $w_j$ are easily derived from the definition of $w_j$ and the Vlasov equation (2). Following the algebra in Refs. [4–7], we obtain

$$\frac{dw_{ji}}{dt} = -(1-w_{ji})\frac{1}{f_{j0}}\frac{\partial f_{j0}}{\partial \boldsymbol{p}} \cdot \delta\left(\frac{d\boldsymbol{p}_{ji}}{dt}\right),$$

$$\delta\left(\frac{d\boldsymbol{p}_{ji}}{dt}\right) \equiv \left.\frac{d\boldsymbol{p}_{ji}}{dt}\right|_{(\phi,A_z)\to(\delta\phi,\delta A_z)}, \tag{5}$$

where $\delta\phi = \phi - \phi_0$ and $\delta A_z = A_z - A_{z0}$. Here, the equilibrium solutions ($\phi_0$, $A_{z0}$, $f_{j0}$) solve the steady-state ($\partial/\partial t = 0$) Vlasov-Maxwell equations (2) and (3) with $\partial/\partial z = 0$ and $\partial/\partial \theta = 0$. A wide variety of axisymmetric equilibrium solutions to Eqs. (2) and (3) have been investigated in the literature. The perturbed distribution $\delta f_j$ is obtained through the weighted Klimontovich representation [1]

$$\delta f_j = \frac{N_j}{N_{sj}} \sum_{i=1}^{N_{sj}} w_{ji}\delta(\boldsymbol{x}-\boldsymbol{x}_{ji})\delta(\boldsymbol{p}-\boldsymbol{p}_{ji}), \tag{6}$$

where $N_j$ is the total number of actual j'th species particles, and $N_{sj}$ is the total number of *simulation* particles for the j'th species. Maxwell's equations are also expressed in terms of the perturbed fields and perturbed density according to

$$\nabla^2 \delta\phi = -4\pi \sum_j e_j \delta n_j,$$

$$\nabla^2 \delta A_z = -4\pi \sum_j e_j \beta_j \delta n_j, \tag{7}$$

where

$$\delta n_j = \int d^3p\, \delta f_j(\boldsymbol{x},\boldsymbol{p},t) = \frac{N_j}{N_{sj}} \sum_{i=1}^{N_{sj}} w_{ji} U(\boldsymbol{x},\boldsymbol{x}_{ij}). \tag{8}$$

Here, $U(\boldsymbol{x},\boldsymbol{x}_{ij})$ represents the method of distributing particles on the grids in configuration space. The nonlinear particle simulations are carried out by iteratively advancing the particle motions, including the weights they carry, according to Eqs. (4) and (5), and updating the fields by solving the perturbed Maxwell's equations (7) with appropriate boundary conditions at the cylindrical conducting wall.

Even though it is a perturbative approach, the δf method is *fully nonlinear* and simulates the original nonlinear Vlasov-Maxwell equations. Compared with conventional particle-in-cell simulations, the noise level in δf simulations is significantly reduced. In addition, the δf method can be used to study *linear* stability properties provided the factor $(1 - w_{ji})$ in Eq. (5) is approximated by unity, and the forcing term in Eq. (4) is replaced by the unperturbed force (i.e., advancing particles along the unperturbed orbits). Implementation of the 3D multispecies nonlinear δf simulation method described above is embodied in the BEST code at the Princeton Plasma Physics Laboratory. For the simulations presented here, we assume perturbations with long axial wavelength ($k_z^2 r_b^2 \ll 1$) and neglect the perturbed axial force on the charge components. The distribution function is reduced to a function on the five-dimensional phase space $(\boldsymbol{x}, \boldsymbol{p}_\perp)$. The code advances the particle motions using a 4th-order Runge-Kutte method, and solves Maxwell's equations by a fast Fourier transform and finite-difference method in cylindrical geometry. Written in Fortran 90/95, the code utilizes extensively the object-oriented features provided by the computer language. The NetCDF scientific data format is implemented for large-scale diagnostics and visualization. The code has achieved an average speed of 40μs/(particle×step) on a DEC alpha personal workstation 500au computer.

## III  TWO EXAMPLES OF COLLECTIVE BEHAVIOR

In this section, we present some illustrative simulation results for a single-species thermal equilibrium ion beam in a constant focusing field. In this case, equilibrium properties depend on the radial coordinate $r = (x^2 + y^2)^{1/2}$. The thermal equilibrium distribution function in the transverse phase space is given by

$$f_{b0}(r, \boldsymbol{p}_\perp) = \frac{\hat{n}_b}{2\pi\gamma_b m_b T_b} \exp\left\{-\frac{p_\perp^2/2\gamma_b m_b + \gamma_b m_b \omega_{\beta b}^2 r^2/2 + e_b(\phi_0 - \beta_b A_{z0})}{T_b}\right\}, \quad (9)$$

where $\hat{n}_b$ is the density of beam particles at $r = 0$, and $T_b$ is the transverse temperature of the beam ions in energy units. It is also assumed that the beam is centered inside a cylindrical chamber with perfectly conducting wall located at $r = r_w$. The equilibrium self-field potentials $\phi_0$ and $A_{z0}$ can be determined numerically from the nonlinear Maxwell's equations in Eq. (3).

First, we examine the nonlinear propagation properties of the beam. A random initial perturbation is introduced into the system, and the beam is propagated from $t = 0$ to $t = 500\tau_\beta$, where $\tau_\beta \equiv \omega_{\beta b}^{-1}$. The simulation results show that the perturbations do not grow and the beam propagates quiescently, which agrees with the nonlinear stability theorem [15,16] for the choice of equilibrium distribution function in Eq. (9). Shown in Fig. 1 are plots of the change in transverse emittance-squared (normalized by $V_b^2/\omega_{\beta b}^2$), $\delta\epsilon^2 = \epsilon^2(t) - \epsilon_0^2$, and the change in mean-square radius,

$\delta r_b^2 = r^2(t) - r_0^2$, versus normalized time $t/\tau_\beta$, for perturbations about the thermal equilibrium distribution in Eq. (9). The system parameters in Fig. 1 correspond to protons with $\gamma_b = 1.85$, and normalized beam intensity $K\beta_b c\tau_\beta/\epsilon_0 = 0.025$, where $K = 2N_b e_b^2/\gamma_b^3 m_b \beta_b^2 c^2$ is the self-field perveance, and $N_b$ is the number of beam ions per unit axial length. The amplitudes of the initial random perturbation in weights in Fig. 1 is $10^{-4}$, which leads to the very small offset in beam emittance and mean-square radius. It is evident from Fig. 1 that the variations in beam emittance $\delta\epsilon^2$ and the variations in mean-square radius $\delta r_b^2$, remain extremely small for perturbations about a thermal equilibrium beam, and the beam propagates quiescently over large distances.

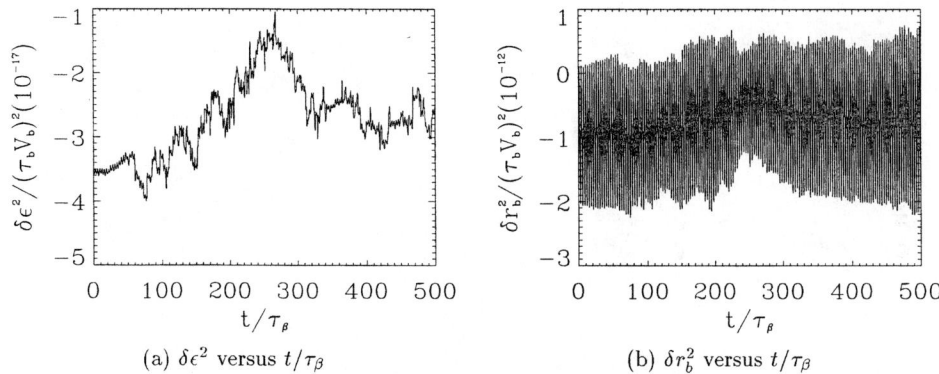

**FIGURE 1.** Time History of $\delta\epsilon^2$ and $\delta r_b^2$

As a second example, we study the linear surface mode for perturbations about a thermal equilibrium ion beam in the space-charge-dominated regime, with flat-top density profile and $K\beta_b c\tau_\beta/\epsilon_0 \gg 1$. These modes are of practical interest because they can be destabilized by a two-stream electron-ion interaction when background electrons are present [9,10]. The BEST code, operating in its linear stability mode, has recovered very well-defined eigenmodes with mode structures and eigenfrequencies which agree well with theoretical predications. For $K\beta_b c\tau_\beta/\epsilon_0 \gg 1$, and azimuthal mode number $l = 1$, the dispersion relation for these modes is given by [1,9,10]

$$\omega = k_z V_b \pm \frac{\hat{\omega}_{pb}}{\sqrt{2}\gamma_b}\sqrt{1 - \frac{r_b^2}{r_w^2}}, \qquad (10)$$

where $r_b$ is the radius of the beam edge, and $r_w$ is location of the conducting wall. In Eq. (10), $\hat{\omega}_{pb}^2 = 4\pi\hat{n}_b e_b^2/\gamma_b m_b$ is the ion plasma frequency-squared, and $\hat{\omega}_{pb}/\sqrt{2}\gamma_b \simeq \omega_{\beta b}$ in the space-charge-dominated limit. Shown in Fig. 2(a) is a comparison between plots of the eigenfrequency versus $r_w/r_b$ obtained from the simulations (diamonds and triangles) and that predicted by Eq. (10) (solid curves). The parameters for this case are chosen close to the space-charge limit with $K\beta_b c\tau_\beta/\epsilon_0 = 6.59$, and the perturbation has normalized axial wavenumber $k_z V_b/\omega_{\beta b} = 2\pi$. It is clear from Fig. 2 that the simulation results agree well with theory.

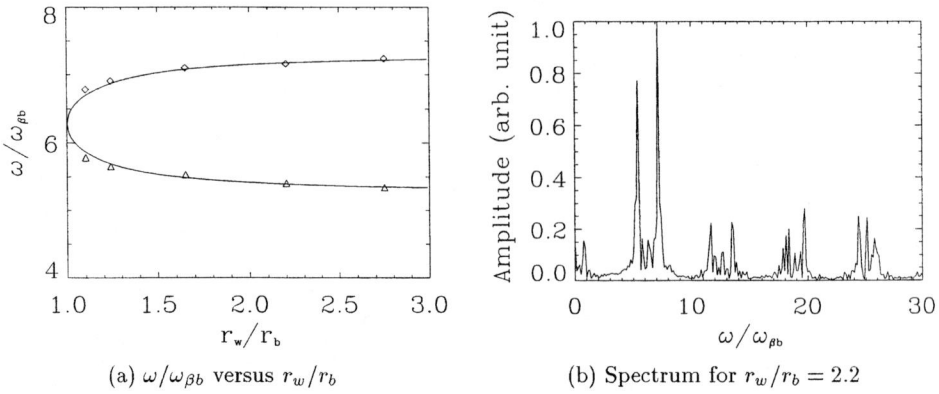

(a) $\omega/\omega_{\beta b}$ versus $r_w/r_b$

(b) Spectrum for $r_w/r_b = 2.2$

**FIGURE 2.** $l = 1$ Surface-Mode Excitation in a Uniform-Density Ion Beam

## IV  ELECTRON-PROTON INSTABILITY

In a high intensity ion beam, the collective mode described in the previous section can be destabilized by the presence of a background electron population [9,10]. This instability is basically of the two-stream type. The directed velocity difference, $V_b - V_e$, between the beam ions and the background electrons provides the free energy for the collective modes to grow. The instability observed in the Proton Storage Ring [17–19] is believed to have this two-stream characteristic. We present here initial simulation results for the electron-proton (e-p) two-stream instability obtained with the BEST code, choosing $f_{j0}$ to correspond to thermal equilibrium distributions for both species. As the first step, the simulations presented are performed for the *linear* phase of the instability.

First, shown in Fig. 3 is the eigenmode excited in a single-species ion beam with a near-Gaussian density profile. Generally, these is no analytical description of the eigenmodes in beams with nonuniform density profiles. From Fig. 3, however, we note that the mode is localized in the region where the density gradient is large, which is consistent with the surface mode in beams with uniform density profiles [9,10]. This simulation is performed for a proton beam with $\gamma_b = 1.85$. The normalized space-charge parameter is $\hat{\omega}_{pb}^2/\gamma_b^2\omega_{\beta b}^2 = 0.1$, where $\hat{\omega}_{pb}^2 = 4\pi\hat{n}_b e_b^2/\gamma_b m_b$ is beam plasma frequency-squared, and $\hat{n}_b$ is the beam density on axis. The normalized perpendicular beam temperature is taken to be $T_b/\gamma_b m_b V_b^2 = 2.25 \times 10^{-6}$. For the dominant mode excited in Fig. 3, the azimuthal mode number is $l = 1$, and the normalized axial wavenumber is $k_z V_b/\omega_{\beta b} = 22.86$. An FFT analysis shows that there are two dominant eigenfrequencies with $\omega/\omega_{\beta b} = 25.83$ and $\omega/\omega_{\beta b} = 26.60$.

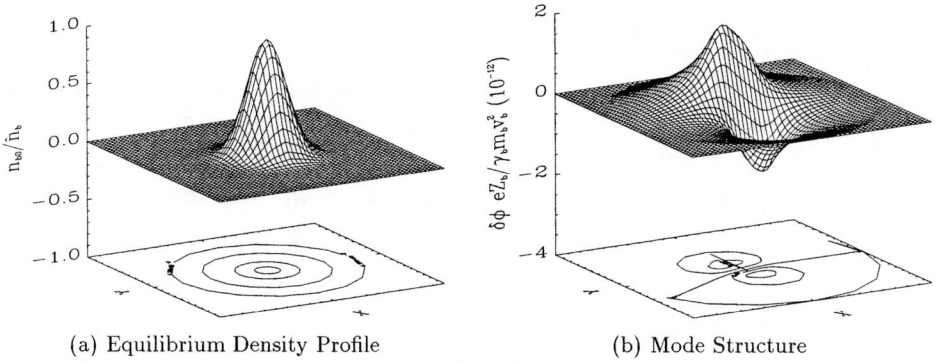

(a) Equilibrium Density Profile    (b) Mode Structure

**FIGURE 3.** $l = 1$ Eigenmode in a Single-Species Ion Beam with Near-Gaussian Density Profile

When a background electron component is introduced with $\beta_e = V_e/c \simeq 0$, the $l = 1$ "surface mode" can be destabilized for a certain range of axial wavenumber and a certain range of electron temperature $T_e$. Illustrated in Fig. 4 is a typical unstable case, where the initial perturbation grows exponentially during the linear phase. The system parameters in Fig. 4 are the same as those in Fig. 3, except that background electrons with $\hat{n}_e/\hat{n}_b = 0.1$ and $T_e/T_b = 0.183$ are introduced into the system.

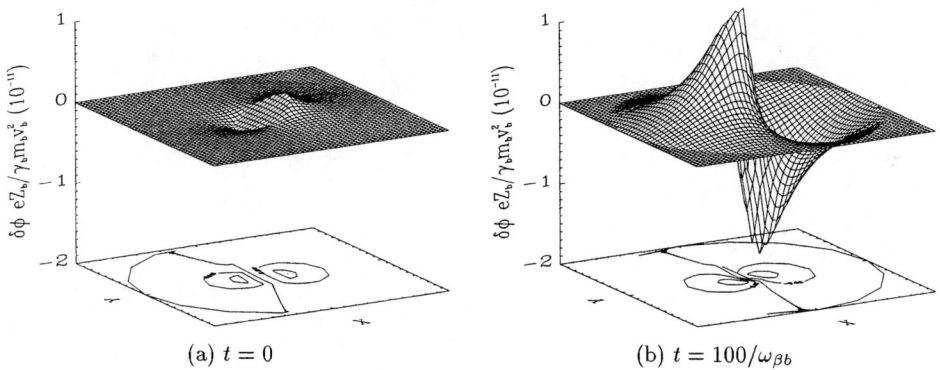

(a) $t = 0$    (b) $t = 100/\omega_{\beta b}$

**FIGURE 4.** Electron-Proton Two-Stream Instability Growing from Initial Noise

Plots of the instability growth rate $\gamma$ versus $T_e/T_b$ and $k_z V_b/\omega_{\beta b}$, with other parameters kept constant, are shown in Fig. 5. The $k_z V_b/\omega_{\beta b}$ dependence of the growth rate is qualitatively consistent with the analytical results obtained for uniform-density beams [9,10]. The important physics here is that only for a certain range of $k_z V_b/\omega_{\beta b}$ can the collective mode of the beam ions effectively resonate with the electrons and produce instability. From Fig. 5(b), $T_e/T_b$ obviously has an important effect on the growth rate. In order to maximize their energy exchange with the beam ions, the electrons must physically overlap the region where the eigenmode of beam ions is localized (approximately the region with the largest ion density gradient), which requires sufficiently large $T_e/T_b$. The electrons are radially confined by the space-charge potential of the beam ions, and the electron temperature determines the radial extent of the electron density profile. The growth rate is therefore strongly dependent on $T_e/T_b$.

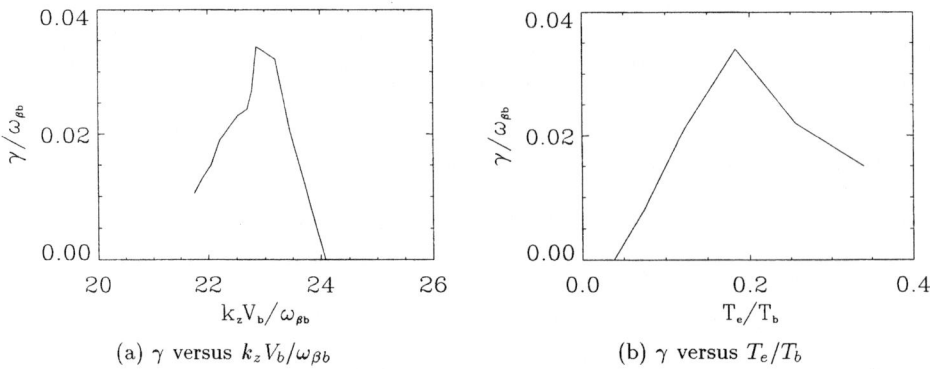

**FIGURE 5.** Plots of Instability Growth Rate

Shown in Fig. 6 and Fig. 7 is a comparison between two cases with identical parameters except for the values of $T_e/T_b$. When $T_e/T_b = 0.014$ (Fig. 6), the electrons are relatively cold and localized in the beam center, and no instability developed over $100\omega_{\beta b}^{-1}$. When $T_e/T_b = 0.183$ (Fig. 7), however, the electrons are sufficiently hot that the electron density profile overlaps that of the beam ions, and the onset of a strong e-p instability is observed.

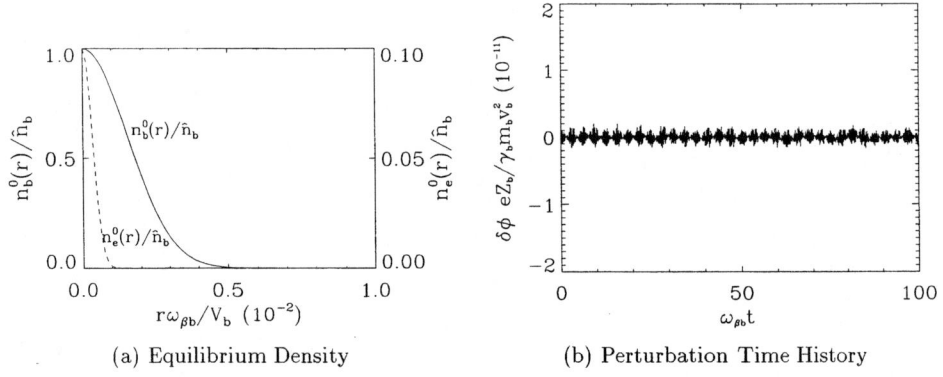

**FIGURE 6.** Cold Electrons with $T_e/T_b = 0.014$

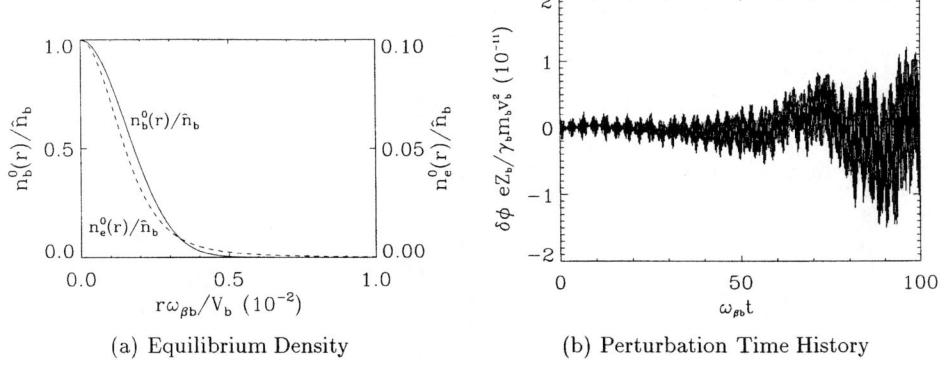

**FIGURE 7.** Warm Electrons with $T_e/T_b = 0.183$

# V  CONCLUSION AND FUTURE WORK

A 3D multispecies nonlinear perturbative particle simulation method has been developed to study collective instabilities in intense charged particle beams described self-consistently by the Vlasov-Maxwell equations. Simulation results show that a thermal equilibrium ion beam in a constant focusing field is nonlinearly stable and can propagate quiescently over hundreds of lattice periods. For surface eigenmodes excited in a uniform-density beam, the simulation results agree well with the analytical results [9,10]. Introducing a background component of electrons, the electron-proton (e-p) two-stream instability is observed in the simulations. Several properties of this instability are investigated numerically, and are found to be in qualitative agreement with theoretical predictions [9,10]. In the future, detailed simulations of the nonlinear phase of the e-p instability will be carried out. The newly-developed BEST code has been tested and applied in different parameter regimes. As a 3D multispecies perturbative particle simulation code, it provides several unique capabilities. Since the simulation particles are used to simulate only the perturbed distribution functions and the perturbed self-fields, the simulation noise is reduced significantly. The perturbative approach also enables the code to investigate different physics effects separately, as well as simultaneously. The code can be easily switched between linear and nonlinear operation, and used to study both linear stability properties and nonlinear beam dynamics. These features, combined with 3D and multispecies capabilities, provide an effective tool to investigate the electron-ion two-stream instability, periodically focused solutions in alternating gradient fields, halo formation, and many other important problems in nonlinear beam dynamics and accelerator physics. Finally, the BEST code is readily adapted to the case where the applied focusing force, $\boldsymbol{F}_j^{foc}$, corresponds to a periodic focusing quadrupole field or solenoidal field. Results of these studies will be reported in future publications.

## ACKNOWLEDGMENT

This research was supported by the Department of Energy and the Short Pulse Spallation Source Project and LANSCE Division of Los Alamos National Laboratory.

## REFERENCES

1. R. C. Davidson, *Physics of Nonneutral Plasmas* (Addison-Wesley Publishing Co., Reading, MA, 1990), and references therein.
2. T. P. Wangler, *Principles of RF Linear Accelerators* (John Wiley & Sons, Inc., New York, 1998).
3. M. Reiser, *Theory and Design of Charged Particle Beams* (John Wiley & Sons, Inc., New York, 1994).
4. W. W. Lee, Q. Qian, and R. C. Davidson, Physics Letters A **230**, 347 (1997).
5. Q. Qian, W. W. Lee, and R. C. Davidson, Physics of Plasmas **4**, 1915 (1997).
6. P. H. Stoltz, W. W. Lee, and R. C. Davidson, Nuclear Instruments and Methods in Physics Research **415**, 433 (1998).
7. P. H. Stoltz, R. C. Davidson, and W. W. Lee, Physics of Plasmas **6**, 298 (1999).
8. H. Qin, R. C. Davidson, and W. W. Lee, Proceedings of the 1999 Particle Accelerator Conference, in press (1999).
9. R. C. Davidson, H. Qin, and T. -S. Wang, Physics Letters A **252**, 213 (1999).
10. R. C. Davidson, H. Qin, P. H. Stoltz, and T. -S. Wang, Physical Review Special Topics on Accelerators and Beams **2**, 054401 (1999).
11. P. J. Channell, Physics of Plasmas **6**, 982 (1999).
12. R. C. Davidson, H. Qin, and P. J. Channell, Physics Letters A, in press (1999).
13. R. C. Davidson, H. Qin, and P. J. Channell, Physical Review Special Topics on Accelerators and Beams, in press (1999).
14. R. C. Davidson and C. Chen, Particle Accelerators **59**,175 (1998).
15. R. C. Davidson, Physical Review Letters **81**, 991 (1998).
16. R. C. Davidson, Physics of Plasmas **5**, 3459 (1998).
17. D. Neuffer, E. Colton, D. Fitzgerald, T. Hardek, R. Hutson, R. Macek, M. Plum, H. Thiessen, and T. -S. Wang, Nuclear Instruments and Methods in Physics Research **A321**, 1 (1992).
18. D. Neuffer and C. Ohmori, Nuclear Instruments and Methods in Physics Research **A343**, 390 (1994).
19. M. A. Plum, D. H. Fitzgerald, D. Johnson, J. Langenbrunner, R. J. Macek, F. Merrill, P. Morton, B. Prichard, O. Sander, M. Shulze, H. A. Thiessen, T. -S. Wang, and C. A. Wilkinson, Proceedings of the 1997 Particle Accelerator Conference, 1611 (1997).

# A SIMPLE SIMULATION OF ELECTRON-PROTON INSTABILITY*

Tai-Sen F. Wang

*Los Alamos National Laboratory, Los Alamos, NM 87545, USA*

**Abstract.** A computer program has been implemented for simulating the electron-proton instability of a partially neutralized long proton bunch. The simulation is based on a theoretical model that considers the motions of the proton-beam centroid and macro-particles representing electrons as well as the contribution of secondary emission of electrons due to impingement of particles on the beam pipe. Preliminary results from simulating the $e$-$p$ instability in the PSR are consistent with earlier studies using the centroid model and qualitatively agree well with experimental observations. It is found that enhancement of the instability due to the electron multiplication occurs only when oscillation of the proton beam has grown to large amplitude. This simulation program can be also used to find the beam-intensity threshold for multipactoring.

## INTRODUCTION

In recent years, sufficient evidence has been collected to confirm that the transverse instability observed in the Proton Storage Ring at Los Alamos (PSR) is an electron-proton ($e$-$p$) instability caused by the trapped electrons in the proton bunches.(1-4) The same kind of instability has been previously observed in the Bevatron at LBNL(5,6) and at CERN(7-9). Although the basic mechanism of the $e$-$p$ instability has been well understood (5-12), recent projects, including the PSR upgrade, the proposed Spallation Neutron Source (SNS), and the European Spallation Source (ESS) call for more detailed understanding of the process and the characteristics of the instability.

This paper is a progress report on studying the $e$-$p$ instability in a long proton bunch like the one in the PSR by using computer simulations. A previous simulation program(11,12) that considers the motions of the proton-beam centroid and the centroid of electrons has recently been modified to replace the centroid of electrons by a group of macro-electrons and to include the process of the possible electron multipactoring on the beam pipe. A further emphasis of the present work is to learn about the effects of electron multipactoring on the $e$-$p$ instability. In the following, the theoretical model and the numerical approach will be delineated first. Then an example of $e$-$p$ instability simulation in PSR and an example of using the program to find the threshold beam intensity for electron multipactoring will be given.

---

*Work supported by Los Alamos National Laboratory, under the auspices of the US Department of Energy.

# MODEL AND NUMERICAL APPROACH

We consider a bunched proton beam of total length $L$ with a round cross-section of radius $a$, propagating at constant speed $v$ inside a perfect conducting pipe of radius $b$. Protons are focused in the transverse direction by an external force that depends linearly on the radial distance. The proton bunch is partially neutralized by electrons possibly produced by secondary emission, gas ionization, or the charge-change injection process. To describe the system further, a Cartesian coordinate system is adopted such that the $z$ axis is in the direction of proton propagation and the $y$ axis is perpendicular to the ring plane. The origin of the coordinate system coincides with the center of the beam cross section. We then assume that in the equilibrium state, particle densities are uniform in the transverse direction but variable in the $z$ direction. We also assume the proton beam motion is unstable in only one transverse direction, say the $y$ direction, so our study will be focused on the motion in that direction. The axial motion of electrons and the synchrotron motion of protons are neglected for simplicity.

Defining the proton beam centroid $Y_p(z,t)$ as the average displacement of protons at the axial location $z$ and the time $t$, we can infer the following equation of motion:

$$\left(\frac{\partial}{\partial t} + v\frac{\partial}{\partial z}\right)^2 Y_p + \omega_\beta^2 Y_p = \sum_{j=1}^{n_e} \frac{F_{ej}}{m_p} - C_{dp}\left(\frac{\partial Y_P}{\partial t} + v\frac{\partial Y_P}{\partial z}\right), \quad (1)$$

where $\omega_\beta$ is the angular betatron frequency, $F_{ej}$ is the force due to the $j$th electron, $m_p$ is the relativistic mass of a proton, $n_e = n_e(z)$ is the number of electrons per unit distance at the axial location $z$, and we have made the simplification that protons only interact with the local electrons. The second term on the right hand side of Eq. (1) represents the damping force on the coherent proton oscillation due to the proton tune spread. Thus, $C_{dp}$ is the damping constant for the motion of the proton centroid. Neglecting the interaction among electrons, the equation of transverse motion for the $j$th electron located at $(y_{ej}, z)$ is

$$\frac{d^2 y_{ej}}{dt^2} = a_{ej}(z,t), \quad (2)$$

where $a_{ej}(z,t)$, to be detailed in the following, is the acceleration of the $j$th electron due to the proton beam.

The numerical computation is actually carried out in the coordinate frame moving with the proton bunch. We divide the proton bunch and the electron cloud each into $N$ slices (grid) along the axial direction. The time step $\Delta t$ is chosen such that $vN\Delta t = L$. Each proton slice contains one macro-proton with charge and mass assigned according to the proton line density $\lambda_p(z)$. An electron slice contains $n_e$ macro-electrons which have two components that differ by the locations of their creation as well as their charges and masses. The core macro-electrons, referred herein as core-electrons, each initially with charge $c_c$ and mass

$m_c$, are created in the region $|y| < b$. The wall macro-electrons, referred herein as wall-electrons, each initially with charge $c_w$ and mass $m_w$, are created on the wall where $|y| = b$. An electron slice has $n_{ec}$ core-electrons and $n_{ew}$ wall-electrons, so that $n_e = n_{ec} + n_{ew}$.

For a perfect conducting beam pipe, the acceleration of the $j$th macro-electron due to the field of the proton beam is approximated by

$$a_{ej}(z,t) \approx \begin{cases} -\dfrac{e^2 q_j \lambda_p}{2\pi \epsilon_o m_{ej}} \left( \dfrac{y_{ej} - Y_p}{a^2} + \dfrac{Y_p}{b^2 - y_{ej} Y_p} \right), & \text{for } |y_{ej} - Y_p| \leq a, \\ -\dfrac{e^2 q_j \lambda_p}{2\pi \epsilon_o m_{ej}} \left( \dfrac{1}{y_{ej} - Y_p} + \dfrac{Y_p}{b^2 - y_{ej} Y_p} \right), & \text{for } |y_{ej} - Y_p| \geq a, \end{cases} \quad (3)$$

where $e$ is the unit charge, $\epsilon_o$ is the permittivity of the free space; $q_j$ and $m_{ej}$ are, respectively, the "effective charge" and the "effective mass" of the $j$th macro-electron to be further discussed later. On the right hand side of Eq. (3), the first and the second terms inside the parentheses represent the direct force on electrons from the protons and the force due to the induced charge on the beam pipe, respectively. Similarly, the acceleration of a proton caused by the $j$th electron is approximated by

$$\dfrac{F_{ej}}{m_p} \approx \begin{cases} -\dfrac{e^2 q_j \lambda_p}{2\pi \epsilon_o m_p} \left( \dfrac{Y_p - y_{ej}}{R_e^2} + \dfrac{y_{ej}}{b^2 - y_{ej} Y_p} \right), & \text{for } |Y_p - y_{ej}| \leq R_e, \\ -\dfrac{e^2 q_j \lambda_p}{2\pi \epsilon_o m_p} \left( \dfrac{1}{Y_p - y_{ej}} + \dfrac{y_{ej}}{b^2 - y_{ej} Y_p} \right), & \text{for } |Y_p - y_{ej}| \geq R_e, \end{cases} \quad (4)$$

where $R_e$ is the hypothetical "radius" of a macro-electron, introduced to avoid the singularity. Here and hereafter, it should be understood that the index $j$ runs within an electron slice only.

Equations (1) and (2) are solved by using the Runge-Kutta-Gill method. In every time step, the axial positions of electron-slices are advanced by one grid space toward the tail of the proton bunch to simulate the drifting of electrons through the proton beam. As time proceeds, a slice of macro-electrons starting at the head of the proton bunch will travel through the bunch and exit at the tail. A random number generator is called at every time step to select a slice in which a pair of macro-electrons are created on the beam pipe with zero velocity to simulate the electron generation on the beam pipe due to the lost protons. The electrons created by background gas scattering can be treated similarly in principle. However, in order to expedite the computation, the number of macro-particles has to be limited. Most of the electrons created by gas scattering are trapped inside the proton bunch until they reach the later part of the bunch. Therefore, to model the electron generation by gas scattering and the accumulation of these electrons in the proton bunch, a weight function for the charge and the mass of macro-electrons, $W_e(z)$, is introduced according to

$$q_j = q_j(z,t) = c_j(z,t)W_e(z) \,, \tag{5}$$

where $q_j$ is the "effective charge" mentioned before and $c_j$ is the "variable charge". Depending on where the $j$-th electron is created, the initial value of $c_j$ is either $c_c$ or $c_w$ the values of which are determined by the weight function $W_e(z)$, the initial neutralization factor $\chi(z)$, the ratio $c_c/c_w$, as well as the values of $n_{ec}$ and $n_{ew}$. If the number of electrons generated per proton by gas scattering is a constant and the electron production rate is proportional to $\lambda_p(z)$, a guess for $W_e$ could be

$$W_e(z) \propto k + \int_0^z \lambda_p(z')dz' \,, \tag{6}$$

where $k$ is a constant representing the fraction of electrons created at the head of the proton bunch and/or electrons from other sources like the particles in the gap that are overtaken by the beam.

At every time step, the possibility of impacting on the beam pipe is checked for all macro-electrons. If an impact is detected, the charge $c_j$ and the mass of the impinging macro-electron is adjusted according to the secondary emission yield (SEY), and the transverse velocity of the macro-particle is set to zero. In doing so, the charge and mass ratio of macro-electrons, $q_j/m_{ej}$ is kept at a constant value of $e/m_{eo}$, where $m_{eo}$ is the rest mass of an electron. The SEY is calculated according to the universal fit to the curves of yields(13):

$$\delta_{ts}(E_0/\theta_0) = \hat{\delta}(\theta_0)D(E_0/\hat{E}(\theta_0)) \,, \tag{7}$$

where
$$D(x) = sx/(s-1+x^s) \,, \tag{8}$$

$\delta_{ts}(E_0/\theta_0)$ is the SEY, $E_0$ and $\theta_0$ are the energy and the incident angle of the impinging electron, respectively, $\hat{E}$ is the energy when $D$ reaches the maximum, and $\hat{\delta}$ is the maximal SEY for the incident angle $\theta_0$. In the present work, we assume normal incidence.

## EXAMPLES OF SIMULATIONS

### Instability Study

As an example, we considered a proton bunch in PSR with a clean gap. The proton line density was modeled by the parabolic distribution

$$\lambda_p(z) = 6N_p s(1-s)/L \,, \tag{9}$$

where $s = z/L$, and $z$ is the axial distance measured from the head of the bunch. The weight function was chosen as

$$W_e(z) = 0.1 + 1.8s^2(3-2s) \,, \tag{10}$$

so that $W_e(L/2) = 1$. The plots of $\lambda_p$ and $W_e$ versus $\tau = z/v$ are shown in Fig. 1. The initial neutralization factor $\chi(z)$ was assumed to be 4% at $z = L/2$. In the computation, $N$ and $\Delta t$ were chosen to be 520 and 0.5 ns, respectively. The number of core-electrons was fixed at 23 per slice ($n_{ec} = 23$). When created, these particles were evenly distributed in the region between the walls. The number of wall-electrons was initially chosen to be 2 per slice ($n_{ew} = 2$ at $t = 0$). The maxi -mal value of $n_e$ was limited to 29. The initial electron charges were assigned accor -ding to $c_c = [e\lambda_p\chi/(n_{ec} + 0.5 n_{ew})]|_{z=L/2}$ and $c_w = c_c/2$. The values of $R_e = 0.7$ cm and $C_{dp} = 5 \times 10^4$/s were used. For the SEY, we used the secondary emission parameter values of $\hat{\delta} = 2.0$ and $\hat{E} = 295$ eV for stainless steel.(14) The following PSR parameter values were considered: $\gamma = 1.85$, $a = 1.5$ cm, $b = 5$ cm, the circumference $C = 90$ m, $N_p = 2.6 \times 10^{13}$, $L = 65$ m, and betatron tune of 2.3. The maximal electron bouncing frequency at these parameter values is about 185 MHz. We assumed that at $t = 0$, all electrons were at rest, $Y_e(z) = 0$, and $Y_p(z) = 0.076 \sin(\pi z/\hat{z})$ cm, where $\hat{z} = 1$ m. This initial $Y_p$, when carried by the traveling proton beam, corresponds to a wave about 126 MHz in the laboratory frame.

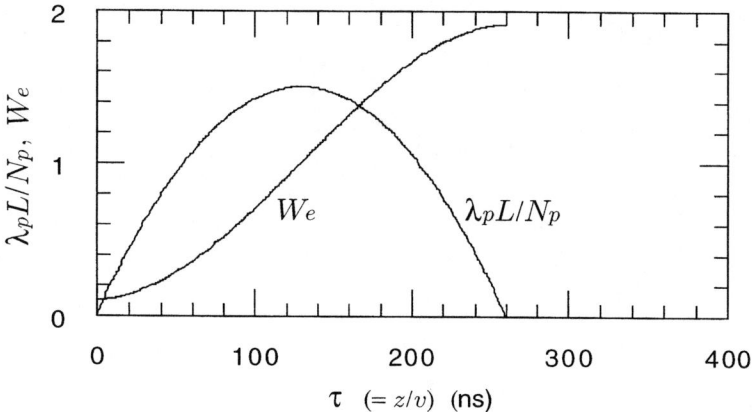

**FIGURE 1.** The proton line density $\lambda_p$ (normalized by $N_p/L$) and the electron weight function $W_e$ considered in the examples. The gap is between $\tau = 260$ ns and $\tau = 360$ ns.

Shown in Fig. 2(a) is a snapshot of $Y_p$ after tracking the motions for 30 revolutions of protons in the PSR ($\approx 10.8 \mu s$). At this stage, the wiggles corresponding to the electron bouncing frequencies have been developed over the proton bunch. The increase of the oscillation amplitude in the tail of the bunch indicates that the instability grows both in time and space. The dipole moment density $D_p(\tau)$ shown in Fig. 2(b) was computed according to the definition $D_p(\tau) = \lambda_p(v\tau, \tau) Y_p(v\tau, \tau)$. Note that $D_p(\tau)$ given here includes the time of flight effect, therefore corresponds to the quantity actually sensed by the beam position monitor (BPM). The plots in Figs. 2(a) and (b) indicate that the wavelength of the *e-p* oscillation is roughly proportional to the square root of the proton density. Figure 2(c) shows the ratio $\sum_j c_j/c_c$ that signifies the variation of electron numbers caused by the absorption or the emission process when interacting with the beam pipe. The result in Fig. 2(c) shows some electron multiplication at the tail of the proton bunch.

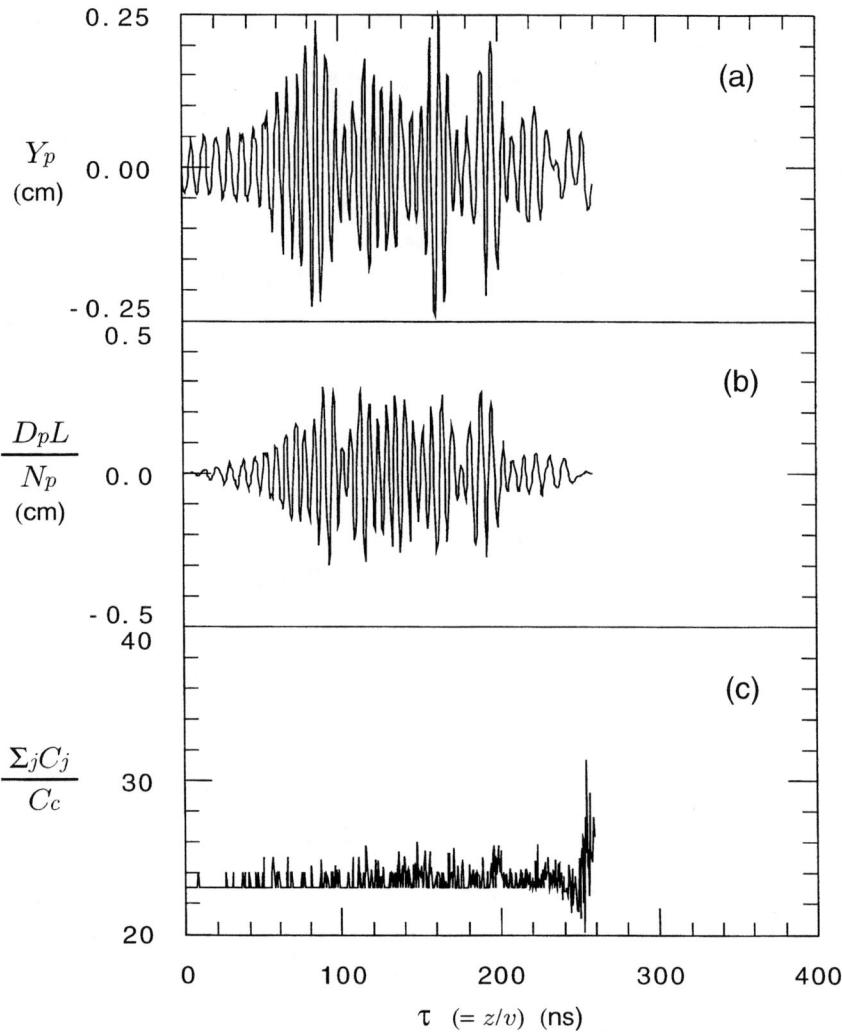

**FIGURE 2.** (a) A snapshot of $Y_p$ taken after tracking motion for 30 proton revolutions in the PSR ($\approx 10.8\mu s$), (b) the dipole moment density $D_p(\tau) = \lambda_p (v\tau,\tau) Y_p( v\tau,\tau)$ at the 30th revolution shown after normalization by $N_p/L$, and (c) the snapshot of the quantity $\sum_j c_j/c_c$ along the proton bunch taken at the same time as $Y_p$. The parameter values are described in the text. In (c), electron multiplications occur at the locations where $\sum_j c_j/c_c > 26$.

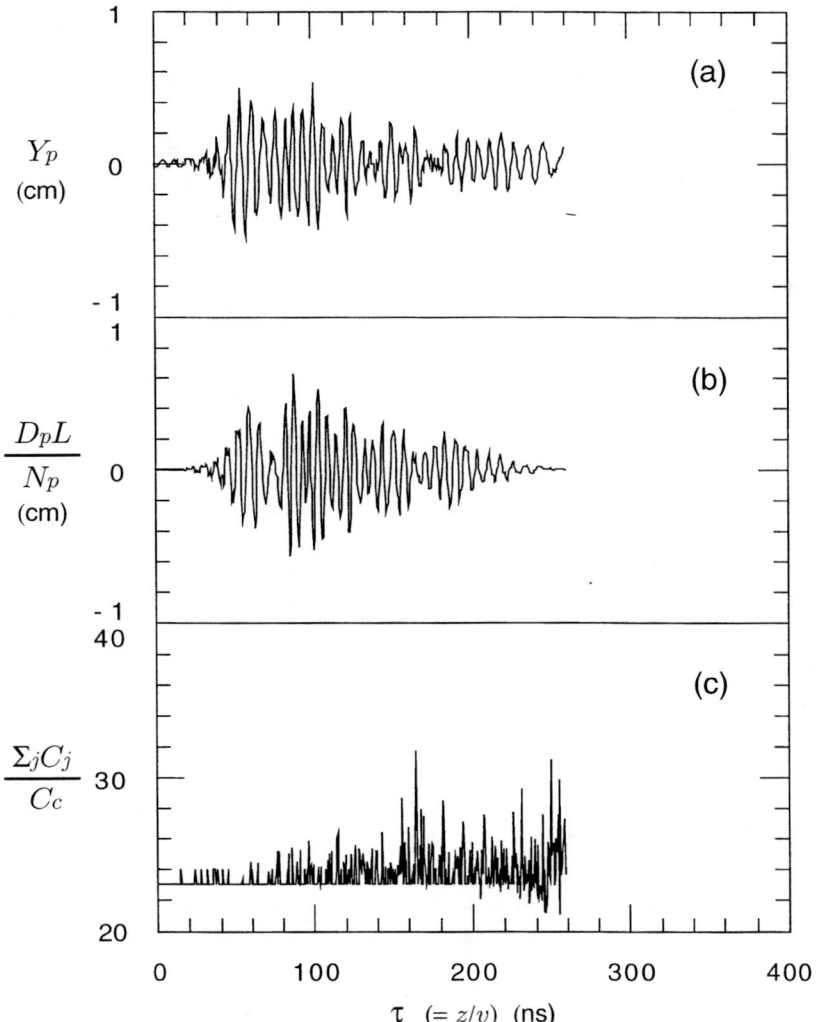

**FIGURE 3.** (a) A snapshot of $Y_p$ taken after tracking 120 turns in the PSR ($\approx 43.2\mu s$), (b) the dipole moment density $D_p(\tau) = \lambda_p(v\tau,\tau) Y_p(v\tau,\tau)$ at the 120th revolution shown after normalization by $N_p/L$, and (c) the snapshot of the quantity $\sum_j C_j/C_c$ along the proton bunch taken at the same time as $Y_p$. The parameter values are described in the text. In (c), electron multiplications occur at the locations where $\sum_j C_j/C_c > 26$.

Figure 3 shows the same kind of snapshots taken after tracking the motions for 120 turns in PSR ($\approx 13.3\mu s$). The instability is fully developed at this time. The electron multiplication now occurs in most of the proton bunch and the amount of secondary electrons is higher than that shown in Fig. 2. Here, the amplitude growth in the tail appears to be the same as or smaller than the earlier part of the bunch. This is due to the frequency spread and hence the decoherence of motions among electrons with large oscillation amplitude caused by the nonlinear beam field.

Some general results observed in the simulations are summarized here:

**(a)** In the parameter range of PSR, it takes only a few percent neutralization in the proton beam for the *e-p* instability to develop. The growth time can range from a few to a few tens of micro-seconds.

**(b)** In the PSR environment, multi-turn trapping of electrons is not a necessity for the *e-p* instability to occur. If the proton bunch is sufficiently neutralized (a few percent in the PSR case), the instability can still develop even when the gap is empty. With a small amount of protons in the gap, the growth rate is slightly higher and the threshold is slightly lower than those in the empty gap case.

**(c)** The wavelength of the *e-p* oscillation is roughly proportional to the square root of the local proton line density. The frequency spectra for proton bunches with nonuniform line density have large bandwidths.

**(d)** The *e-p* instability grows both in time and in space. Therefore, oscillations grow more in the tail than in the head of a proton bunch.

**(e)** Comparison made between even and uneven electron distributions around the ring shows that for the same amount of electrons, the *e-p* oscillation is slightly more unstable when electrons are concentrated in a segment of the ring.

## Multipactoring Threshold

The results in the last example show that electron multiplication can occur in the tail of the proton bunch at very small beam oscillation. This is because even if electrons were created with no initial kinetic energy, some of them can gain energy from the time-dependent beam field due to the nonuniform density profile so that when impinging on the wall, they become energetic enough to produce more electrons. Thus, for a given geometry of the beam environment and the material of the beam pipe, the electron multipactoring must depend on the number of protons in the bunch, and, presumably, there is a beam-intensity threshold for the multipactoring to occur. The following example is an exploration of this kind of threshold in PSR by using the instability-simulation program.

Since there is no proton-beam dynamics involved in this study, it is not necessary to use a simulation program to find the electron multiplication threshold. However, the features of multi-electron tracking and time-dependent beam field make the simulation program suitable for searching the threshold. If electron multiplication happens, it can be seen by tracking for less than one turn. The study was carried out with no perturbation of the proton beam ($Y_p = 0$). The multiplication threshold was searched by varying the total number of protons in runs with the same setup of electron initial conditions and SEY as in the instability study. It was found that for the beam geometry in PSR and a typical stainless-steel SEY, the threshold is around $1.3 \times 10^{13}$ protons per bunch. The quantity $\sum_j c_j/c_c$ for the proton densities below and above the threshold are shown in Fig. 4 as functions of distance/time behind the head of the proton bunch.

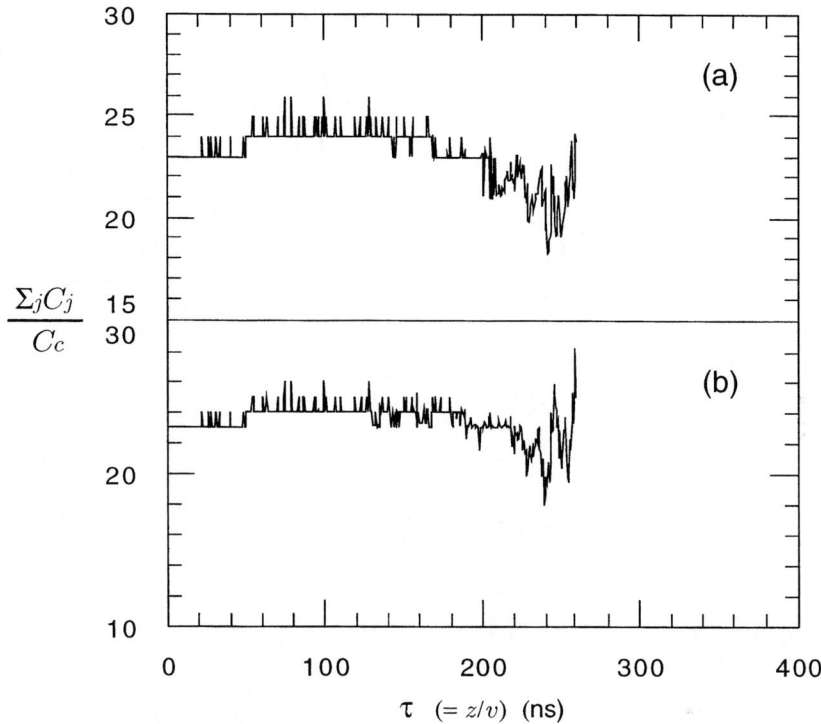

**FIGURE 4.** The quantity $\sum_j c_j/c_c$ as function of time behind the head of the proton bunch for (a) below and (b) above the multipactoring threshold. The system parameter values are the same as in the last example. The total numbers of protons in the bunch are $1.04 \times 10^{13}$ in case (a) and $1.69 \times 10^{13}$ in case (b). The threshold is about $1.3 \times 10^{13}$ protons per bunch. Here, electron multiplications ($\sum_j c_j/c_c > 26$) occur in the tail of the proton bunch when the beam intensity is above the threshold.

## CONCLUSIONS

The implementation of a multi-electron, one-dimensional computer program for simulating the $e$-$p$ instability in a long proton bunch has been initiated. The simulation is based on solving the equations of motion for the centroid of the proton beam using macro-electrons. The theoretical model considered includes the secondary electron emission and the electron absorption due to the interaction between particles and the beam pipe.

Preliminary application of this program for studying the e-p instability in PSR has produced results largely consistent with earlier simulations using the centroid model. These preliminary results also agree qualitatively very well with experimental observations. It was found that a few percent of neutralization in the PSR beam can lead to the e-p instability. It was also found that an empty gap does not always ensure the beam stability, and multi-turn trapping of electrons is not necessary for instability. Our simulation results confirm that the wavelength of the e-p oscillation is proportional to the square root of the proton line density and the wide-banded frequency spectrum of the e-p oscillation in a proton bunch with nonuniform line density. One of the results in the multi-electron tracking not found in simulations using a centroid model is the decoherence of electron motions due to the nonlinear beam field seen by electrons having large oscillation amplitude. We have found that for a typical stainless steel SEY, electron multipactoring initially occurs only in the tail of the proton bunch. Appreciable multipactoring occurs in the middle and the later part of the bunch after the proton oscillation has grown to large amplitude ($> 0.5$cm). We have also used the program to study the threshold for electron multipactoring and found that for PSR parameter values, multipactoring occur when the beam intensity is at $1.3 \times 10^{13}$ protons per bunch and higher.

## REFERENCES

1. R. Macek, these Proceedings.
2. M. Plum et al., Proc. of 1997 Particle Accelerator Conf., Vol 2, p. 1611.
3. T. Wang et al., Proc. of 1995 Particle Accelerator Conf., IEEE Catalog No. 95CH35843, Vol. 5, p. 3146.
4. D. Neuffer et al., Nucl. Instrum. and Meth. **A321**, 1 (1992).
5. H. Grunder and G. Lambertson, Proc. 8th Intl. Conf. on High Energy Accelerators, CERN, 1971, p. 308.
6. L. J. Laslett, A. M. Sessler and D. Möhl, Nucl. Instrum. Meth., **121**, 517 (1974).
7. H. G. Hereward, CERN Internal Report 71-15 (1971).
8. E. Keil and B. Zotter, CERN Internal Note CERN-ISR-TH/71-58, December 1971.
9. H. Schönauer and B. Zotter, CERN Internal Note, May 1972.
10. See, for example, the paper by R. C. Davidson et al., Phys. Rev. ST-AB, Vol. 2, No. 5, May 1999, and the references cited therein.
11. T. Wang, Proc. of 1995 Particle Accelerator Conf., IEEE Catalog No. 95CH35843, Vol. 5, p. 3143.
12. T. Wang, Los Alamos National Laboratory PSR Technical Notes 96-004, 96-005, Feb. 1996.
13. M. A. Furman and G. R. Lambertson, Lawrence Berkeley National Laboratory Report LBNL-41123, March 1998.
14. O. Gröbner, Proc. of the Santa Fe Workshop on Electron Effects, 1997, Los Alamos National Laboratory Report, LA-UR-98-1601, p. 169.

# MULTIPACTING ON THE TRAILING EDGE OF PROTON BEAM BUNCHES IN THE PSR AND SNS*

V.Danilov, A.Aleksandrov, J.Galambos, D.Jeon, J.Holmes, D.Olsen,

*ORNL SNS Project, Oak Ridge, TN 37831*

## Abstract

The Proton Storage Ring (PSR) in Los Alamos has a fast intensity-limiting instability, which may result from an electron cloud interaction with the circulating proton beam leading to a transverse mode coupling instability. The most probable mechanism of the electron creation is multipacting. Though the effect depends on many parameters, a model is presented which predicts a large electron creation in the vacuum chamber. A comparison of this effect between the PSR in Los Alamos and the Spallation Neutron Source (SNS) in Oak Ridge is given. In addition, several possibilities to reduce multipactor are discussed.

## 1 INTRODUCTION

The LANL PSR has a fast instability that limits the proton beam intensity per pulse. A probable explanation of this instability is that there exists a large electron density in the vacuum chamber resulting in an electron interaction with the proton beam leading to a transverse mode coupling instability between the circulating protons and oscillating electrons trapped in the proton potential well. Multipacting can drastically increase the electron density, increasing the instability. This physical phenomenon could be explained by two mechanisms:

(1) The first mechanism is clearest for a coasting beam. Electrons could accumulate during beam injection in the proton potential well, and after reaching some threshold density, could generate unstable coupled oscillations between themselves and the proton beam. In this case the lighter electrons gain large amplitudes and strike the vacuum chamber wall, producing an avalanche of secondary emission (SEM) electrons, resulting in the instability [1].

(2) The second mechanism is applicable to bunched beams. For the case of a constant longitudinal density, electrons with zero initial kinetic energy at the vacuum chamber wall oscillate across the vacuum chamber gap through the circulating beam with zero energy gain. If the longitudinal bunch density is increasing the electrons loss energy. If the longitudinal bunch density is decreasing the electrons gain energy.

CP496, *Workshop on Instabilities of High Intensity Hadron Beams in Rings,*
edited by T. Roser and S. Y. Zhang
© 1999 American Institute of Physics 1-56396-910-6/99/$15.00

It is speculated that a multipacting can significantly increase the number of electrons on the trailing edge of the proton bunch if the energy gain of the electrons is above 50 eV for an aluminium vacuum chamber. Instability measurements [2] show large proton beam oscillations on the bunch tail, though it may be related to the maximum of the electron density at this point.

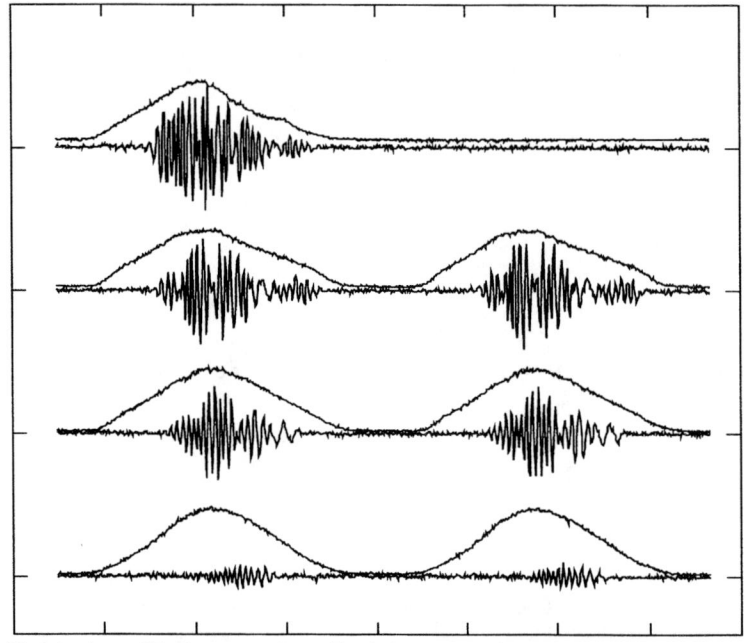

Fig. 1 Vertical oscillations on the tail of the proton bunch at PSR. (Courtesy Mike Plum, January, 1999.)

The lower left signals, presented at Figure 1, shows that initial vertical unstable motion (the lowest oscillating line) occur near the maximum slope of the longitudinal distribution (upper line). This is evidence in favour of the multipactor effect. On the other hand, well developed unstable motion (upper left signals) shows little dependence of the oscillations on the sign of the derivative of the longitudinal distribution. This fact could be related to other mechanisms of electron creation and their trapping.

## 2 MULTIPACTING ON THE TRAILING EDGE

To investigate the effect of secondary emission from multipacting, a computer code was created that calculates 1D electron trajectories, starting from the vacuum chamber wall. After striking the wall secondary emission electrons are produced depending on the primary initial energy. The secondary electrons start to oscillate with zero momentum in the proton potential since their initial energies are small in comparison with the average single-pass energy gain in the proton potential, about 100 eV.

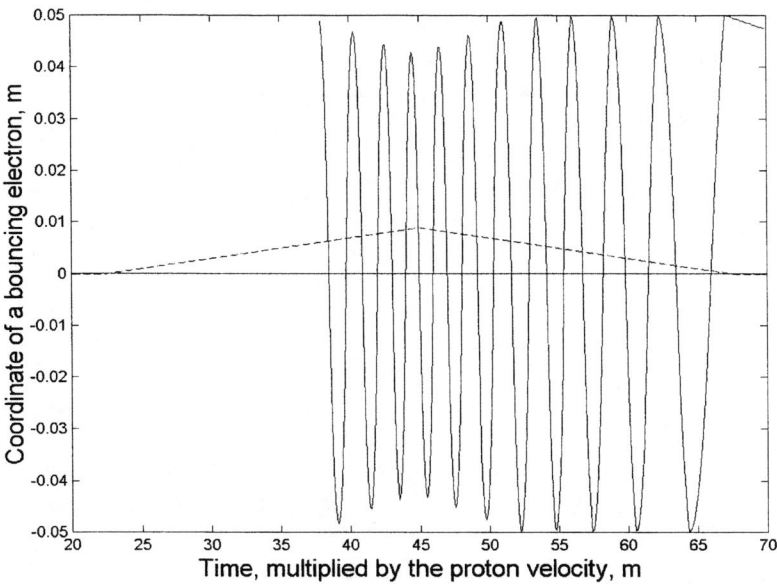

Fig. 2 Coordinate of an electron oscillation in the electric field of proton beam.

Figure 2 shows one example of the electron motion (solid line) with respect to the proton bunch distribution (dashed line). Zero longitudinal coordinate corresponds to the center of the beam gap, the initial electron vertical coordinate is about the vacuum chamber radius of 5 cm, the initial distance between the gap and the electron is 38 meters. One can see that initially the electron oscillation amplitude decreases due to the proton beam density increase. Once the proton beam center has passed the electron, the transverse amplitude increases and electron finally hits the wall several times, losing all its energy for each wall hit. Total number of secondary emission electrons is summed over all the collisions with the vacuum chamber using the formula for secondary emission yield (y) from Ref. 2, assuming the primary electrons are normal to the surface. That is:

$$y(E) = y_{max} 1.11 (E/E_m)^{-.35} (1 - e^{-2.3(E/E_m)^{1.35}}) \qquad (1)$$

where $E_m = 400\,eV$, $y_{max}$ depends on the vacuum chamber material. The proton beam transverse distribution is taken to be constant within the beam radius, and equal to zero otherwise.

The final result is presented as the secondary emission (SEM) coefficient, which is the natural logarithm of the average number of electrons, produced by one initial electron. The initial time of the test electron oscillation corresponds to the center of the proton bunch passing. The final time corresponds to the moment, when this number of produced secondary emission electrons is maximal. It is difficult to predict the threshold SEM value, for which the instability begins to occur. Nonetheless, some trends are examined. A likely scenario is that the electrons initially appear at the vacuum chamber wall due to the proton beam losses, and their interaction with the stripper foil, etc. Then, the electron density grows during the passage of the proton bunch. In addition, we assume that the electron cloud disappears in the proton beam gap due to their own space charge and momentum, so the multipacting process initiates and repeats itself each turn.

It was estimated in Ref. 3 that total number of initial electrons created per proton for one thousand turns due to losses could be from 0.01 up to 1. If we took the upper number, it would give us sufficient amount of electrons to explain the instability without additional processes of electron creation and trapping. In this case for each proton bunch revolution we have a newly created electron density of about 1/1000 of the proton density, namely, 0.1% of the proton density on average in the vacuum chamber. For the PSR we assume that 10% of the vacuum chamber consists of the elements with high secondary emission yield with a maximum of about 3.5. These elements are bellows, holes, the aluminium box near the stripper foil, etc. Also we assume the initial electron population grows 1000-fold, due to multipacting. Additionally, the fraction of the vacuum chamber, where these electrons build up is occurring, is already mentioned 10 % of the vacuum chamber. Thus, the final average compensation of the proton beam by the electron cloud is 0.1*.001*1000*.1=.01 or about 1% and, according to estimations in Ref. 4 that is enough for instability to occur. This means that for the PSR the threshold SEM coefficient should be about ln(1000) = 7. In fact, this number was obtained for the PSR with $2 \times 10^{13}$ protons. Consequently, the 1-D results are in reasonable agreement with the simple estimate. One can surely find additional facts and numbers for the initial number of electrons near the center of the proton bunch, since this value is uncertain. For the electron multipacting, the SEM coefficient is more predictable in a simplified geometry of the vacuum chamber. Therefore, in the next section we compare only SEM coefficients for different machines regarding prediction of the e-p instability threshold in the Spallation Neutron Source.

## 3 PSR AND SNS COMPARISON

The SEM coefficient calculations for the SNS parameters predict large electron production, even for relatively low secondary emission yield of a smooth stainless steel vacuum chamber. Table 1 shows materials, intensities and SEM coefficients, used for the calculations. The bunch length was taken to be about half of the rings. The SNS design is for 10 cm vacuum chamber radius (the last column), but we also show the 5 cm case (the third column) to highlight the effect of this parameter. The SNS has a larger aperture than PSR because of the larger acceptance and the need to reduce activation levels from halo. However, this larger aperture is bad for multipacting. It takes the electron a larger time to traverse a bigger gap: consequently, the electron sees a larger net acceleration as the proton bunch passes.

Table 1 Comparison of the SEM coefficients for the PSR and SNS. Negative coefficients mean that the electron density decreases.

| RING | Max SEM yield/ material | Vacuum chamber radius equal to 5 cm | Vacuum chamber radius equal to 10 cm |
|---|---|---|---|
| PSR $3 \times 10^{13}$ protons | 3.5/Aluminum 0.3% 3.5/Stainless Steel with holes and slots 10% | 10 | N/A |
| SNS $1 \times 10^{14}$ protons | 2/Stainless Steel | -17 | 3.8 |
| SNS $2 \times 10^{14}$ protons | 2/Stainless Steel | -15 | 11 |

One can see that for the SNS with $2 \times 10^{14}$ protons the SEM coefficient is higher than the case of the PSR at the e-p threshold intensity. There could be two other factors which strongly contribute in the e-p instability threshold in the SNS. First, the initial number of electrons at the vacuum chamber wall could be less due to the anticipated decreased proton losses. Secondly, in the SNS storage ring all parts of vacuum chamber are assumed to give equal contributions to multipactor, thus a longer relative part of the vacuum chamber emits electrons. The influence of these two factors is hard to predict since one gives an increase in threshold, while another gives a decrease. The most solid way to estimate the threshold SEM coefficient for the SNS is to use the PSR number for this coefficient as the reference. According to this assumption, the SNS ring with an ideal stainless steel smooth vacuum chamber with shielded bellows, ports, etc. is above the e-p instability threshold for protons equal to $2 \times 10^{14}$. This means that special efforts should be made to avoid multipacting in the SNS storage ring.

## 4 POSSIBLE WAYS TO ELIMINATE MULTIPACTING

### *4.1 TiN coating*

The conventional approach to multipacting in RF cavities is to coat the surface with TiN. In addition, because for corrugated surfaces the secondary emission coefficient could be higher than unity even with this TiN coating, additional shielding should be made for all vacuum chamber irregularities. This method is now under study for the PSR in Los Alamos.

### 4.2 Magnetic field near the vacuum chamber surface

Another approach includes a creation of the different multipole or longitudinal magnetic fields to prevent electrons from being accelerated towards the circulating beam. To keep the secondary emission electrons within 1 cm from the surface, a magnetic field of about 100 Gauss is needed for the SNS storage ring. In a variant with an alternating magnitude longitudinal magnetic field, the magnetic field action on the beam is negligible. Though it needs an additional aperture for this vacuum chamber, this method looks reliable and simple.

### 4.3 Longitudinal density variation

Additional possibility to reduce multipactor is to reduce the acceleration of the secondary electrons in the field of the proton bunch. Adding a small ripple to the main longitudinal distribution of the proton beam could reduce the electron acceleration on the trailing edge of a bunched beam [5].

## 5 CONCLUSION

We find that the multipacting effect is as important for the Spallation Neutron Source as for the Proton Storage Ring with intensities that correspond to the e-p instability. Consequently, special investigations and experiments should be made to determine appropriate methods to rise the e-p instability threshold in the PSR and avoid the e-p instability in the SNS.

*Research sponsored by the DOE, under contract no. DE-AC05-96OR22464 with LMER Corp. for ORNL.

## REFERENCES

[1] M. Blaskiewicz, "Instabilities in the SNS", PAC99, New York, March, 1999
[2] M. Plum, PSR Buncher Workshop, LANL, Jan 1999.
[3] T.-S. Wang, Talk on e-p instability, Los Alamos, January 26, 1998
[4] D. Neufer, et al., Particle Accelerators **23**, 133-148 (1988).
[5] V. Danilov, et al., "On the possibility to increase PSR instability threshold", PAC99, New York, March, 1999

# The Fast Loss Electron Proton Instability

Michael Blaskiewicz*
AGS Department Brookhaven National Laboratory

### Abstract

The fast loss electron proton instability is studied both experimentally and theoretically. It is shown that electron multi-pactoring is required for fast beam loss.

## 1 Introduction

Very fast, high frequency, transverse instabilities have been observed in the Los Alamos PSR[1, 2, 3] and the AGS Booster[4]. The e-folding times ( $\sim$ 10 turns) require a transverse resistance or order 5M$\Omega$/m which is significantly larger than can be accounted for by the impedance budget. Additionally, the frequency of the instability depends strongly on beam parameters like beam intensity and betatron tune which implies a broad band impedance. Other observations, detailed below, also argue against an impedance driven phenomena.

An alternate instability mechanism is the interaction between the proton beam and electrostatically trapped electrons[1, 2, 3, 4, 5, 6, 7, 8, 9, 10]. In the PSR and AGS Booster this e-p instability can result in losing more than half the beam in a few tens of microseconds. A less violent form of the ep instability was used to explain emittance growth in the ISR[10] and other machines have reported possible ep candidates. However, the PSR and AGS Booster are unique with the instability causing large losses in a short time.

## 2 Data from the AGS Booster

Instability studies in the AGS Booster were performed in 1998 and 1999. Data were taken with bunched beam using the normal magnet cycle and with coasting

---

*Work supported by the United States Department of Energy

Table 1: machine parameters during study

| parameter | value |
|---|---|
| circumference | $2\pi R = 202$m |
| kinetic energy | 200MeV |
| frequency spread | ±730 Hz |
| nominal betatron tunes | $Q_x = 4.8, Q_y = 4.95$ |
| beam pipe radius | $b = 5$cm |
| injected beam radius | ≈ 3cm |
| nominal chromaticity | $Q'_x = -3, Q'_y = -1$ |
| sextupoles off | $Q'_x = -7.5, Q'_y = -2.6$ |
| rf voltage | 0V |
| linac RF frequency | 200MHz |
| injected pulse length | 200 to 450$\mu$s |
| revolution period | 1207ns |

beam and a magnetic field fixed at the injection value. The coasting beam data will be discussed.

Table 1 summarizes the machine parameters. The data consisted of current transformer traces which measured the total beam current, and wall current monitor traces which measure the AC longitudinal component. Also, horizontal and vertical split can capacitive beam position monitors (BPMs) were used to measure high frequency signals. The data from the BPMs were sampled at 1GHz. The relative path lengths were measured to within 1ns and sum and difference signals were obtained. Various processing techniques were applied to the BPM signals. Evidence of a violent instability with nominal values of tune and chromaticity and with intensity just above the threshold value is shown in Figure 1. The beam current shows a rapid loss beginning after the rise of the narrow band signal. The narrow band signal was obtained by applying a numerical filter to the vertical difference BPM signal. Recursive equations were used to reduce memory requirements.

$$F_{n+1} = (\cos(\tilde{\omega}\tau)F_n - \sin(\tilde{\omega}\tau)G_n)\,e^{-\alpha\tau} + S_n \qquad (1)$$

$$G_{n+1} = (\sin(\tilde{\omega}\tau)F_n + \cos(\tilde{\omega}\tau)G_n)\,e^{-\alpha\tau} \qquad (2)$$

$$P_{n+1} = e^{-\tau/T_0}P_n + G_n^2 \qquad (3)$$

In equations (1), (2), and (3) $S_n$ are the input data, $\tau = 1$ns is the sampling interval, $\omega_r = \sqrt{\tilde{\omega}^2 + \alpha^2}$ is the center frequency, $\alpha$ is the half width at half power bandwidth, and $T_0$ is an additional smoothing time on the output signal $P_n$.

The center frequency and bandwidth were obtained by examining Figure 2, which shows the evolution of the spectral amplitude of the vertical difference signal.

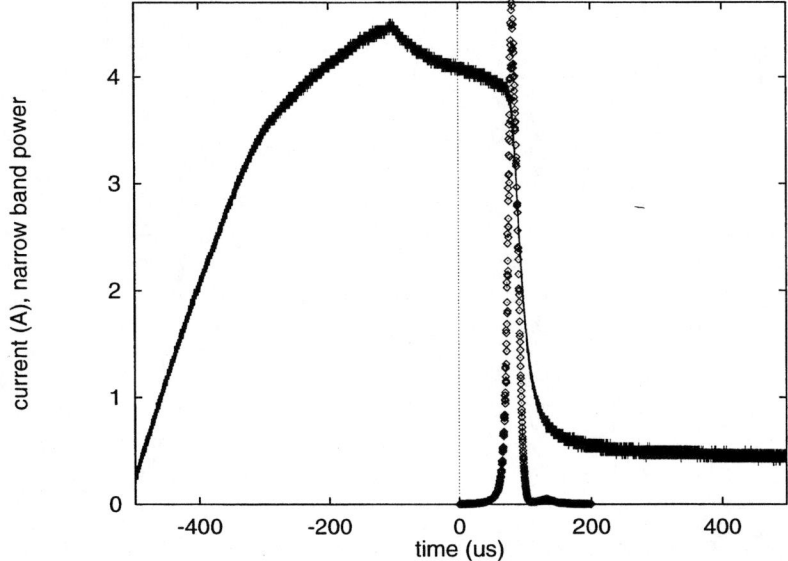

Figure 1: Beam current in amps and transverse power density measured using a capacitive BPM with bandpass between 75.6 and 76MHz and smoothed over 1 turn.

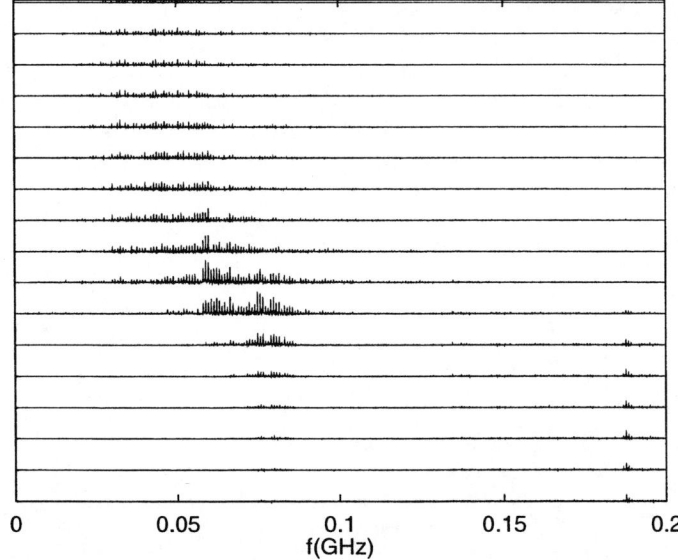

Figure 2: Spectral amplitude of vertical difference signal. The individual FFTs used ten turns of data ($12\mu s$ between traces).

During the rise time of the instability the narrow band data are reasonably well fitted by an exponential with a 5.7µs e-folding time. Using the formula for transverse growth rates of a cold coasting beam,

$$Im(\Omega) = \frac{qcI_{peak}Re(Z_\perp)}{4\pi E_0 Q_\beta}, \tag{4}$$

an e-folding time of $2 \times 5.7 = 11.4\mu s$ implies $Re(Z_\perp) = 5.4M\Omega/m$. Note that including frequency spread will increase the necessary impedance. Also, since many unstable lines are apparent in Figure 2 any driving impedance must be broad band as well. For comparison purposes the coherent transverse space charge impedance is

$$-i\frac{RZ_0}{\beta^2\gamma^2 b^2} = -i8.4M\Omega/m.$$

With $\beta\gamma = 0.69$ one expects transverse space charge to greatly dominate any other broad band impedance.

For other machine conditions the required impedance is larger still. Figure 3. shows the beam current and two narrow band signals. The vertical difference signal centered at 73MHz rises first, followed by the vertical sum signal. During the rise the vertical difference grows faster than exponentially. The instantaneous growth rate of the vertical signal, defined via $d\log P/dt$, peaks at 350/ms. Using equation (4) a transverse resistance of $8.8M\Omega/m$ is required.

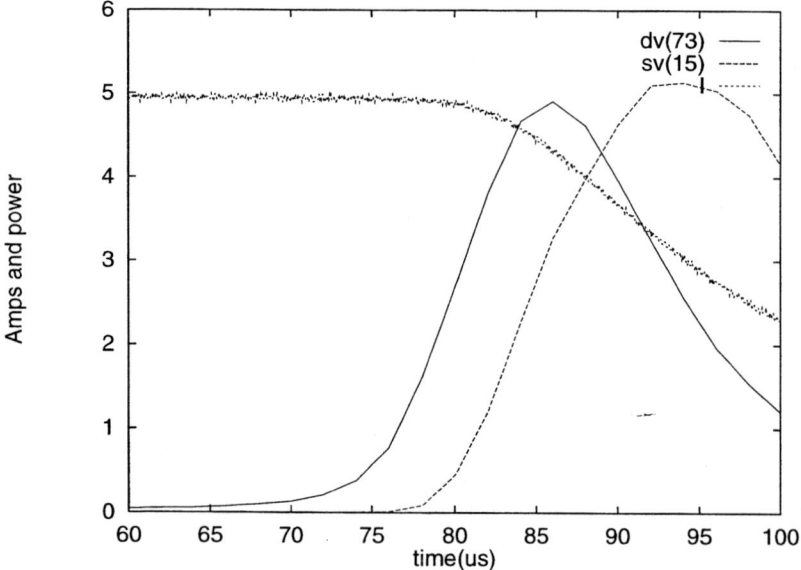

Figure 3: Beam current in amps, $P$ using vertical difference BPM centered at 73MHz, and $P$ using vertical sum BPM centered at 15MHz. Both are smoothed

with $\tau_0 = 1.2\mu s$.

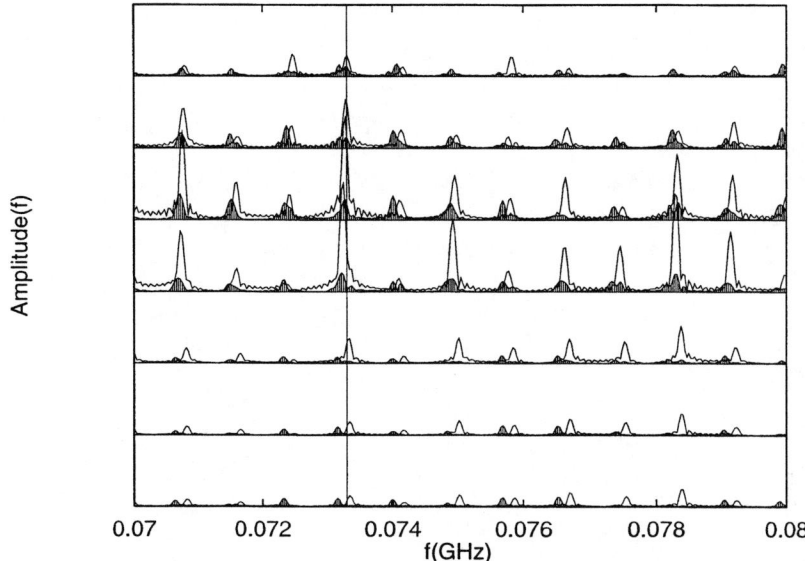

Figure 4: Spectral amplitude of vertical sum (cross hatched) and difference (smooth line) signals. The individual FFTs used ten turns of data ($12\mu s$ between traces). The vertical line is at 73.3MHz. The nearest vertical peak shifts down by $0.11 f_{rev} =$ 90kHz during the instability.

The setpoint tunes were $Q_x = 4.75$, $Q_y = 4.95$ and the sextupoles had zero current. The dependence of the instability threshold on tune was studied and it was found that these conditions had the largest threshold current. In particular, a smaller vertical tune did not increase the threshold. Figure 4 shows the evolution of the vertical sum and difference spectral amplitudes. There are several unstable lines implying that any driving impedance is broad band. From the figure it is also clear that the frequency of the peaks in the vertical difference spectrum shift down as the instability progresses. The shift is by 90kHz corresponding to a betatron tune shift of 0.1 which is very large given that the sextupoles were off. Additionally, it is difficult to imagine how such a strong impedance could be benign during normal operations. In the next section the coasting beam theory of the e-p instability is outlined and applied to these data.

## 3 Coasting Beam Model

The coasting beam theory in the linear limit with no electron multipactor has been carefully explored[5, 6, 7, 8]. The model involves a coasting beam which traps particles of opposite sign in its electrostatic potential well. Some early work focused on electron beams trapping protons but the equations have good symmetry and are easily modified to the relevant case. There are several variables.

Let $\theta$ denote machine azimuth, $x$ be the horizontal coordinate, and $y$ the vertical. Assume the proton and electron beams have a uniform charge density and are round with radius $a$. Only transverse dipole motion is considered, since these modes are the most dangerous [7, 8]. Let the proton beam have current $I$ and Lorentz factor $\gamma$. The line charge density of the protons is then $\lambda_p = I/v$. Take the fractional neutralization to be $f$, so that the line charge density of the electrons is $-f\lambda_p$. Assume the instability is vertical and take $\bar{y}_p(\theta,t)$ and $\bar{y}_e(\theta,t)$ to be the centroids of the proton and electron beams, respectively. Assume a round, perfectly conducting vacuum chamber of radius $b$. The $y$ component of the electric field due to the protons is

$$E_p(x,y) = \frac{\lambda_p}{2\pi\epsilon_0}\left(\frac{(y-\bar{y}_p)}{\max((y-\bar{y}_p)^2+x^2, a^2)} - \frac{(y-b^2/\bar{y}_p)}{(y-b^2/\bar{y}_p)^2+x^2}\right). \quad (5)$$

The proton beam also creates a magnetic field $B_x = -\beta E_p(x,y)/c$. The electrons create a $y$ component of electric field,

$$E_e(x,y) = \frac{-f\lambda_p}{2\pi\epsilon_0}\left(\frac{(y-\bar{y}_e)}{\max((y-\bar{y}_e)^2+x^2, a^2)} - \frac{(y-b^2/\bar{y}_e)}{(y-b^2/\bar{y}_e)^2+x^2}\right). \quad (6)$$

The electrons are non-relativistic and create negligible magnetic field. The vertical force on a proton is given by $F_p = e[E_p(x,y)/\gamma^2 + E_e(x,y)]$ while the force on an electron is $F_e = -e[E_p(x,y) + E_e(x,y)]$

Analytically solving the Vlasov equation with the non-linear force terms in equations (5) and (6) is intractable. Linearizing the equations one finds $E_p(x,y) = (\lambda_p/2\pi a^2\epsilon_0)[y - \bar{y}_p(1-a^2/b^2)]$ and $E_e(x,y) = (-f\lambda_p/2\pi a^2\epsilon_0)[y - \bar{y}_e(1-a^2/b^2)]$. For a non-round pipe there are additional geometrical factors associated with the quadrupole fields generated by a centered beam[7] but these are small and will be ignored.

Before considering the the general case assume that there is no frequency spread. The moment equations close [5] so only the average values of the coherent fields are relevant. The average force on a proton is given by

$$\bar{F}_p(\theta,t) = \frac{e\lambda_p}{2\pi a^2\epsilon_0}\left(\frac{a^2}{b^2}\frac{\bar{y}_p(\theta,t)}{\gamma^2} - f[\bar{y}_p(\theta,t) - \bar{y}_e(\theta,t)] - f\frac{a^2}{b^2}\bar{y}_e(\theta,t)\right).$$

The average force on an electron is

$$\bar{F}_e(\theta,t) = -\frac{e\lambda_p}{2\pi a^2\epsilon_0}\left(\frac{a^2}{b^2}\bar{y}_p(\theta,t) - [\bar{y}_p(\theta,t) - \bar{y}_e(\theta,t)] - f\frac{a^2}{b^2}\bar{y}_e(\theta,t)\right).$$

Seting $\bar{y} = Y(\theta, t)$ and applying the force laws give:

$$\ddot{Y}_p = \left(\frac{\partial}{\partial t} + \omega_0 \frac{\partial}{\partial \theta}\right)^2 Y_p = -\omega_\beta^2 Y_p + \bar{F}_p(\theta, t)/\gamma m_p \tag{7}$$

$$\ddot{Y}_e = \left(\frac{\partial}{\partial t}\right)^2 Y_e = \bar{F}_e(\theta, t)/m_e. \tag{8}$$

Setting $Y_e \equiv 0$ in (7) and $Y_p \equiv 0$ in (8) yields electron and proton frequencies of

$$\omega_e^2 = \frac{e\lambda_p}{2\pi a^2 \epsilon_0 m_e}\left(1 - f\frac{a^2}{b^2}\right). \tag{9}$$

and

$$\omega_p^2 = \omega_\beta^2 - \frac{e\lambda_p}{2\pi a^2 \epsilon_0 \gamma m_p}\left(\frac{a^2}{\gamma^2 b^2} - f\right). \tag{10}$$

Define the coupling frequencies

$$\Omega_e^2 = \frac{e\lambda_p}{2\pi a^2 \epsilon_0 m_e}\left(1 - \frac{a^2}{b^2}\right). \tag{11}$$

and

$$\Omega_p^2 = \frac{fm_e}{\gamma m_p}\Omega_e^2. \tag{12}$$

With these definitions $\ddot{Y}_p = -\omega_p^2 Y_p + \Omega_p^2 Y_e$ and $\ddot{Y}_e = -\omega_e^2 Y_e + \Omega_e^2 Y_p$. The equations are linear with constant coefficients. Barring double roots, any unstable solution is a linear combination of [7]

$$Y_p = \hat{Y}_p \exp(in(\theta - \omega_0 t) + i(\omega_p + \delta\omega_p)t),$$

and

$$Y_e = \hat{Y}_e \exp(in\theta - i(\omega_e + \delta\omega_e)t).$$

Of course the frequencies are the same with $n\omega_0 - (\omega_p + \delta\omega_p) = \omega_e + \delta\omega_e$. Assume $\delta\omega_p \ll \omega_p$ and $\delta\omega_e \ll \omega_e$ then

$$-2\omega_p \delta\omega_p Y_p = \Omega_p^2 Y_e, \quad -2\omega_e \delta\omega_e Y_e = \Omega_e^2 Y_p \tag{13}$$

Solving for $\delta\omega_p$ yields

$$\delta\omega_p = \frac{\Delta\omega}{2} \pm \sqrt{\left(\frac{\Delta\omega}{2}\right)^2 - \frac{\Omega_p^2 \Omega_e^2}{4\omega_p \omega_e}} \tag{14}$$

where $\Delta\omega = n\omega_0 - \omega_p - \omega_e$. For unstable conditions $|\delta\omega_e| = |\delta\omega_p|$ and the ratio of proton to electron amplitudes is given by

$$\left|\frac{\hat{Y}_p}{\hat{Y}_e}\right| = \sqrt{\frac{\Omega_p^2 \omega_e}{\Omega_e^2 \omega_p}} = \sqrt{\frac{fm_e\omega_e}{\gamma m_p \omega_p}}. \tag{15}$$

Since $|\Delta\omega| \leq \omega_0/2$ for some $n$ the beam will be unstable if the second term under the square root in the equation (14) is greater than $\omega_0^2/16$. Substituting this condition in (15) gives $\hat{Y}_p \sim \omega_0 \hat{Y}_e/\omega_e \ll \hat{Y}_e$. Hence, the electrons will reach the beam pipe radius before the proton beam gains a significant amplitude. If the electrons are lost at the walls the instability will stop and a slight increase in proton emittance will be observed, as in the ISR[10]. The instability seen in the AGS Booster and PSR is much more violent so there must be some mechanism to replace electrons which hit the walls.

An electron striking the beam pipe leads to secondary emission which can cause an electron cascade and beam loss. A key parameter is the kinetic energy the electron has when it strikes the wall. This is easily estimated by assuming an electron grazes the wall on one oscillation and hits it on the next. If only one side of the vacuum chamber is involved the electron velocity on impact is given by

$$v_e = b\sqrt{4\pi\omega_e\omega_I}\left(1 + O(\sqrt{\omega_I/\omega_e})\right) \tag{16}$$

where $\omega_I = Im(\delta\omega_e)$. Assuming $\Delta\omega = 0$, the electron kinetic energy when striking the wall is

$$KE_e = 2\pi m_e b^2 \omega_e \omega_I \approx \pi m_e (\omega_e b)^2 \sqrt{\frac{fm_e}{\gamma m_p}\frac{\omega_e}{\omega_p}}, \tag{17}$$

where the second equality neglects image terms. The largest growth rate seen in the AGS Booster gives $\omega_I = 175/\text{ms}$. With $\omega_e/2\pi = 80\text{MHz}$ and $b = 5\text{cm}$ the first expression for the electron kinetic energy gives $KE_e = 7.8\text{eV}$. This is a small kinetic energy and does not lead to a secondary emission yield greater than 1. However, the validity of the coasting beam dispersion relations is an open issue. Suppose for the moment that the electron cloud uniformly fills the pipe. The proton beam will suffer a frequency shift of

$$\delta\omega_p = \frac{feIZ_0}{4\pi\beta\gamma m_p c^2 \omega_\beta}\frac{c^2}{b^2}. \tag{18}$$

Setting this to the observed shift of $2\pi \times 90\text{kHz}$ gives $f \approx 4$. Setting $f = 4$ in the second expression of equation (17) results in $KE_e \approx 4\text{keV}$. Clearly, the theory is of limited value. Including frequency spreads will not help beyond the Landau damping threshold, so understanding multi-pactoring requires a non-linear theory.

A theory with multi-pactoring may be obtained by letting $i\delta\omega_p \to d/dt$ in equations (13).

$$\dot{X}_p = -i\frac{\Omega_p^2}{2\omega_p}X_e, \tag{19}$$

$$\dot{X}_e = i\Delta\omega X_e + i\frac{\Omega_e^2}{2\omega_e}X_p, \tag{20}$$

where $X_p$ and $X_e$ are the slowly varying parts of $Y_p$ and $Y_e$, respectively. These equations can be obtained by setting $Y = X(t)\exp(in[\theta-\omega_0 t]+i\omega_p t)$ in the coupled oscillator equations and neglecting the $\ddot{X}$ terms. When multipactoring is included, equations (19) and (20) must be supplemented by the conditions that $|X_e| \le b$ and the frequencies become time dependent due to variation in the fractional neutralization $f$. In particular, $\Delta\omega$ will change with $f$ due to the presence of the image term in (9). Suppose equations (19) and (20) are evolved until $|X_e| = b - \epsilon$, where $\epsilon$ is a small numerical factor. Let $X_e \to X_e b/|X_e|$ to limit the electron amplitude. To get the electron velocity when striking the wall use equation (20) to obtain $\omega_I = Real(\dot{X}_e/X_e)$. If $\omega_I > 0$, which is not always the case for $|X_e| > b - \epsilon$, use equation (16) to get the electron kinetic energy, $E_k$. The change in $f$ with respect to time depends on the secondary emission yield, $SEY(E_k)$ via,

$$\dot{f} = f\frac{\omega_e}{2\pi}\left(SEY(E_k) - 1\right) - D(f), \tag{21}$$

where the factor $\omega_e/2\pi$ implies that electrons hit the wall on every oscillation after multipactoring occurs, and the term $D(f)$ is present since when $f \gtrsim 1$ the mutual electrostatic repulsion of the electrons expels them from the vacuum chamber[11]. In the numerical work I simply set $f \to 1$ when the updated value was greater than 1. The $SEY$ depends on the chamber material and on the incidence angle of the electrons [11]. Bellows and sharp transitions in pipe radius yield enhanced values of $SEY$. Additionally, if the electrons hit both vacuum chamber walls equation (16) overestimates $E_k$ by a factor of 2 and $\omega_e \to 2\omega_e$ in equation (21). Therefore, any specific numerical results are of limited value.

Some general trends in the simulations of the AGS Booster have been observed.

- The system is usually unstable if $SEY > 1$ the first time it is called. In some cases, a first value of $SEY = 0.8$ results in unstable behavior after a few microseconds.

- The shift in $\Delta\omega$ with $f$ does not cause the system to stabilize after multipactoring begins.

- The image terms can reduce $E_k$ by a factor of 2 below the value obtained ignoring images.

- During instability, while $f$ is increasing, the growth in $|X_p|$ is roughly linear with time.
- For peak $SEY \sim 2$ and $f \lesssim 0.1$ initially, the characteristic time from the onset of multipactoring to $|X_p| = b$ is of order ten microseconds. (For numerical simulations I set $X_p(0) = 0$ and $X_e(0) \sim b/10$.)
- Reducing the peak value of $SEY$ increased the threshold value of $f$.
- Increasing the energy at which $SEY$ peaks increases the threshold value of $f$.

# 4 Conclusions

Instabilities observed in coasting beam data exhibit transverse growth rates $\sim 100$/ms. Assuming normal instabilities, broad band resistive components of order $5 \to 10 \text{M}\Omega/\text{m}$ are implied. When viewed in terms of the linear coasting beam electron proton instability the observed growth rates are too small. Adding multipactoring and limiting the electron amplitude to remain within the beam pipe yield a theory in reasonable agreement with the data.

This work has benefited from discussions with Thomas Roser and Y.Y. Lee.

# References

[1] D. Neuffer *et. al. NIM* **A321** p1 (1992).

[2] M. A. Plum *et. al* PAC97 p 1611.

[3] V. Danilov *et.al* PAC99 TUA 52.

[4] M. Blaskiewicz PAC 99 TUP 60.

[5] B.V. Chirikov, *Sov. Atomic Energy*, **19** p1149 (1965).

[6] E. Keil, B. Zotter, CERN/ISR-TH/71-58, (1971).

[7] L. J. Laslett, A. M. Sessler, D. Möhl, NIMA, **121**, p517, 1974.

[8] R. C. Davidson, H. Qin, P. H. Stoltz, Phys Rev ST Accel Beams, 054401, 1999.

[9] A. Ruggiero, M. Blaskiewicz, PAC 97 p1581 (1998).

[10] H. G. Hereward, CERN 71-15 (1971).

[11] G.V. Stupakov, LHC Project Report 141 (1997).

# Longitudinal Relaxation Oscillations Induced by HOMs[1]

### J. Sebek, C. Limborg
*SSRL/SLAC, Stanford, CA 94309, USA*

**Abstract.** Taking advantage of the vastly different time scales of the problem, a simple analytical model of HOM induced relaxation oscillations has been developed. First a continuous approximation of the impulsive discrete forces is made. Then only the components of the force with the synchrotron frequency are retained to describe the slowly varying amplitude and frequency of the relaxation oscillation. A two particle version of this model reproduces the main characteristics of the system.

## INTRODUCTION

The voltage induced by the beam on the cavity impedance, at the upper synchrotron sideband of the revolution harmonics has a destabilizing effect on the beam. The instability commonly known as the coupled bunch instability occurs when this force exceeds the net damping force. On the SPEAR electron storage ring, such an instability can be produced from a multi-traversal effect acting on a single bunch. Linear theory predicts that above the instability threshold, the amplitude of the bunch synchrotron oscillation would grow to infinity. However, it has been long observed [1,2] that the amplitude of oscillation can also saturate. In particular, at the onset of instability, there exists a regime in which the bunch performs relaxation oscillations. Extensive measurements were performed on the fundamental impedance. To explain these data a theoretical model explaining the physics of the mechanism was developed. A more complete paper is published elsewhere [3].

## EXPERIMENTAL RESULTS

### Motivations

While characterizing the RF cavities in order to improve the stability of SPEAR and focusing our attention on the growth of the HOM induced instabilities, a regular

---

[1] Work supported by the Department of Energy, contract DE-AC03-76SF00515.

**TABLE 1.** Machine parameters

| Energy | $U_0$ | $V_{RF}$ | $\tau_{damp}$ | $R_S$ |
|---|---|---|---|---|
| 2.3 GeV | 193 keV | 1.68 MV | 10 ms | 10 MΩ |

modulation about the saturation level was observed (Figure 1a). Its period was of the order of a radiation damping time. This modulation is often small, but certain machine parameters can make it very large, regular, and quite striking (Figure 1b). The possibility of adjusting the HOM frequency, by positioning a moveable RF cavity tuner in the passive RF cavity, made such observations very repeatable and convenient to study on the SPEAR ring.

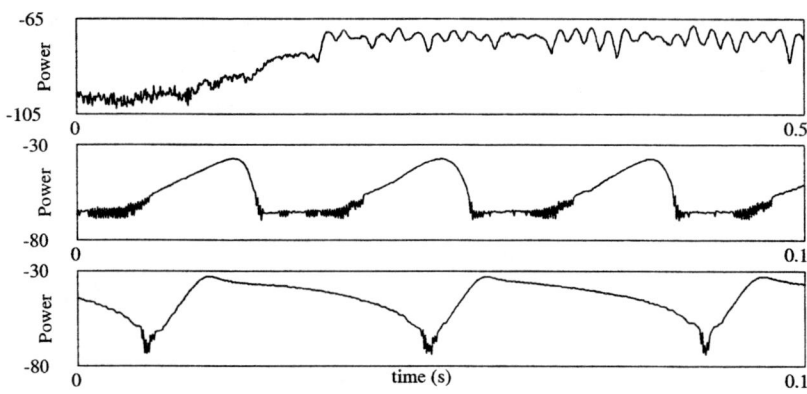

**FIGURE 1.** Output from the cavity signal at $hf_0 + f_s$. Note the different growth and damping rates in (b) and (c).

## Experimental parameters

The experimental parameters are presented in Table 1. The resonance studied is the fundamental resonance, $f_{HOM} = f_{RF}$, corresponding to the Robinson instability. The large variety of time scales involved in the relaxation mechanism is presented in Table 2.

## Description of measurements

Since the effect of interest depends on the HOM strength, the largest available impedance was chosen for the study. The fundamental mode of the idle SPEAR RF cavity was chosen for our study. The strength of this impedance means that the HOM produces a strong long-range wakefield at currents for which the short-range effects of the wakefield are not very important.

**TABLE 2.** System time scales

|  | Frequency | Period | $N_{turns}$ |
|---|---|---|---|
| $f_{sawtooth}$ | < 100 Hz | > 10 ms | > 12800 |
| $f_{so}$ | 28.4 kHz | 35 $\mu$s | 45 |
| $\alpha_R$ | 56 kHz | 17.8 $\mu$s | 23 |
| $f_o$ | 1.28 MHz | 0.78 $\mu$s | 1 |
| $f_{RF} = f_{HOM}$ | 358.5 MHz | 2.8 ns | 1/280 |
| $(1/\sigma_\tau)$ | 10 GHz | 100 ps | 1/7840 |

Data were first taken on an RF spectrum analyzer, tuned as a narrowband receiver, from a signal coming from an RF cavity probe (Figure 1). The evolution

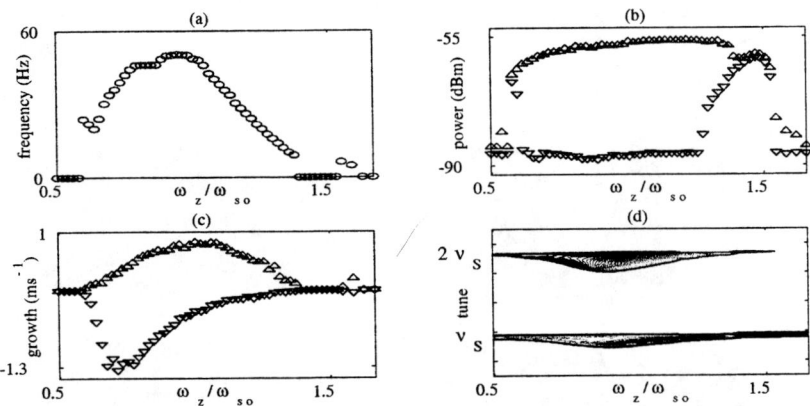

**FIGURE 2.** Relaxation oscillation parameters vs $f_{HOM}$: (a) relaxation oscillation frequency, (b) maximum ($\triangle$) and minimum ($\triangledown$) power of oscillation, (c) growth times ($\triangle$) and damping times ($\triangledown$), (d) $\nu_s$, showing $\sim$ 15% deviation over the range of $f_{HOM}$.

of crucial parameters of the relaxation oscillation as a function of the resonator frequency is summarized in Figure 2. From the amplitude information, (2b), one sees that the amplitude of oscillation is large along most of the resonance. The growth rate, as a function of frequency, is symmetric with respect to the center frequency; it tracks the resistive part of the resonator impedance (2c). The damping rate is not symmetric. It is very small over the second half of the resonance curve (2c). Because of this asymmetry, the frequency of this relaxation oscillation (2a) as a function of the HOM frequency is also asymmetric. The variation and/or spread of the synchrotron tune during these oscillations was also measured. The frequency deviation showed a decrease of 15% from the nominal synchrotron frequency, corresponding to the shift for large amplitude pendulum oscillations (2d).

A streak camera was used to observe the time distribution of the bunch. The bunch envelope visible in a slow scan shows the bunch to be concentrated near the extremes of the oscillation during its growth phase, while its charge density decreases with time (Figure 3a). The maximum amplitude of oscillation reached is about $\pi/2$ radians. In the damping phase, this macroparticle still exists and damps, but it has a much reduced intensity compared to its initial value. At the end of this phase, particles have accumulated around the center. In the particular case of $f_{HOM}$ slightly above $pf_0 + f_{so}$, when the damping is very slow, a second accumulation point clearly forms near the origin (3b). The charge at this point grows in both amplitude and intensity as the original macroparticle continues its decay. This second accumulation point, when seen, is phase-locked to the initial macroparticle, but approximately $\pi$ out of phase with it (3c,3d).

In summary, the oscillation starts when one center grows exponentially, saturates and loses particles, and finally begins to damp. These lost particles accumulate at a second center which grows in amplitude, accumulating more particles, as the first center migrates to the origin while losing its charge. The two centers have now exchanged roles in this oscillation, giving the system its bistable character.

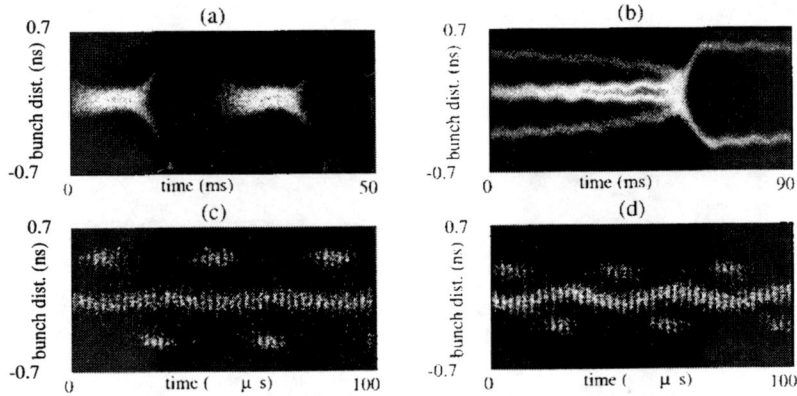

**FIGURE 3.** Relaxation cycles for two different values of $f_{HOM}$: (a) $f_z \sim f_s$. (b) $f_z > f_s$, showing appearance of second center. Note that (b) has much slower damping than (a). (c) Individual streaks are now visible. The second center has more charge than the original main body, yet its oscillation amplitude is still small. (d) The second center now has most of the charge. Its amplitude continues to grow while that of the original main body continues to damp. Oscillations are about $\pi$ out of phase.

# SIMULATIONS

The equation of motion of individual particles obeys a second order difference equation. The turn by turn difference equation of the code includes the synchrotron radiation emission through losses, radiation damping, and quantum fluctuations.

The long memory of the high Q cavity is retained by the use of propagators, which enable the accurate retention of the phase information of the rapidly oscillating wake over the comparatively long time scale of one revolution period.

The wake from the earlier particles is computed by propagating the fields from the previous bin over the time $\Delta t$. With $\Phi_t = \omega'_R \Delta t$, the propagator can be represented

$$\begin{pmatrix} W_{t+\Delta t} \\ W'_{t+\Delta t} \end{pmatrix} = e^{-\alpha \Delta t} \begin{pmatrix} \cos \Phi_t & \frac{1}{\omega'_R} \sin \Phi_t \\ -\omega'_R \sin \Phi_t & \cos \Phi_t \end{pmatrix} \begin{pmatrix} W_t \\ W'_t \end{pmatrix} + \begin{pmatrix} 2\alpha R_s n_{t+\Delta t} \\ -4\alpha^2 R_s n_{t+\Delta t} \end{pmatrix}$$

## Simulation results

The code results are in good agreement with the data. Simulations reproduce the very low frequency of the relaxation oscillation (always below 100 Hz in our case). They confirm qualitatively the evolution of frequency and amplitude as a function of the induced voltage. The program reproduces the $\pi/2$ limit cycle amplitude observed with the streak camera. Finally, the simulation corroborates the streak camera data, discussed above, that show the bunch grow as a macroparticle that loses charge density to an attractor at the center (Figure 4).

Based on these agreements, the predictions of the code can be viewed with confidence. They were used to gain further insight into the details of the oscillation too sensitive to be seen with our experimental setup. The tracking code phase space distribution shows that the filamentation starts from the head of the bunch. Particles spiral from the head of the bunch towards the center of phase space (Figure 4b). One can observe that the escaping particles perform synchrotron oscillations at a higher frequency than those still attached to the main body. These results gave important clues for the theoretical model.

## ANALYTICAL MODEL

### Continuous Approximation

The wakefield generated when a bunch passes through a cavity can be represented as the impulse response of a narrow-band resonator of resistance, $R_S$, frequency, $\omega_R$, and damping factor $\alpha_R$

$$W(t-\tau) \cong U(\tau) 2\alpha_R R_S e^{-\alpha_R(t-u)} \cos \omega_R (t-\tau)$$

where $U(\tau)$ is the step function. The wake seen by any particle is the sum of all wakes generated by all particles in all previous turns. This infinite sum can be approximated by an integral. When the bunch has $N$ particles of charge, $e$, giving a machine current, $I$, the electrical potential $V(t)$ generated by the wake is

$$V(t) = 2\alpha_R R_S I \int_{-\infty}^{t} e^{-\alpha_R(t-u)} \cos\left(\omega_z(t-u) + \omega_R(\tau(t) - \tau(u))\right) du$$

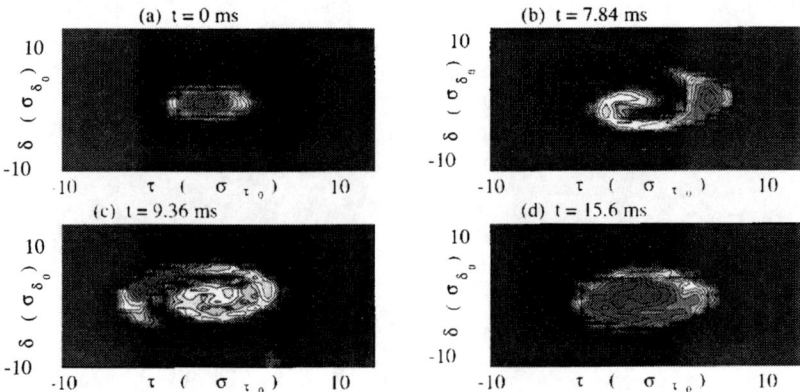

**FIGURE 4.** Simulations of bunch in four stages of relaxation oscillation. Intensity scale is logarithmic.

The continuous approximation of the synchrotron motion is that of common pendulum motion, for which oscillations as large as $\pi/2$ are still very sinusoidal. Therefore $\tau(t)$ and $\tau(u)$ can be represented as

$$\tau_t = \hat{\tau}_t \cos(\omega_{st}t + \phi_t); \quad \tau_u = \hat{\tau}_u \cos(\omega_{su}t + \phi_u)$$

with $\hat{\tau}_t$, $\hat{\tau}_u$, $\omega_{st}$, $\omega_{su}$, $\phi_t$, and $\phi_u$ all slowly varying with respect to a synchrotron period. The exponential damping in the integral means that only important contributions come from times no further back than a few damping times, during which time these quantities can be treated as constant. With the notation $r_t = \hat{\tau}_t \omega_R$, $r_u = \hat{\tau}_u \omega_R$, the integral in $V(t)$ can be expressed as the real part of

$$\sum_{p,k=-\infty}^{\infty} \frac{j^{p-k} J_p(r_t) J_k(r_u)}{\alpha_r + j(k\omega_{su} - \omega_z)} e^{j(k(\omega_{su}+\phi_u)+p(\omega_{st}+\phi_t))t}$$

## Application of KBM Method

The averaging method of Krylov, Bogoliubov, and Mitropolskii [4,5] is well suited to such an oscillatory problem with slowly varying parameters [6]. To solve a driven harmonic oscillator, $\ddot{x} + \omega_{so}^2 x = f_x(x, \dot{x})$, new variables, $(r(t), \phi(t))$, are defined in terms of $(x(t), \dot{x}(t))$ by

$$x = r\cos(\omega_{so}t + \phi); \quad \dot{x} = -\omega_{so} r \sin(\omega_{so}t + \phi)$$

The averaged evolution equations of the oscillation amplitude and phase become

$$\dot{\bar{r}} = -\frac{1}{2\omega_{so}} F_{S1}(\bar{r}, \bar{\phi}); \quad \dot{\bar{\phi}} = -\frac{1}{2\omega_{so}\bar{r}} F_{C1}(\bar{r}, \bar{\phi})$$

$F_{S1}$ and $F_{C1}$ are the Fourier coefficients of the frequency $\omega_{su}$ in the wakefield generated by the particle at $(r_u, \phi_u)$.on the particle at $(r_t, \phi_t)$

$$F_{S1} = -2A\alpha_R R_S I \sum_{k=1}^{\infty} J_k(r_u) \left[ J_{k-1}(r_t) + J_{k+1}(r_t) \right] \times \qquad (1)$$
$$\left[ \left(a_k^- - a_k^+\right) \cos(k\Delta\phi) - \left(b_k^- - b_k^+\right) \sin(k\Delta\phi) \right]$$

$$F_{C1} = 2A\alpha_R R_S I \left\{ 2b_0^+ J_0(r_u) J_1(r_t) + \sum_{k=1}^{\infty} J_k(r_u) \left[ J_{k-1}(r_t) - J_{k+1}(r_t) \right] \times \right. \qquad (2)$$
$$\left. \left[ \left(b_k^- - b_k^+\right) \cos(k\Delta\phi) + \left(a_k^- - a_k^+\right) \sin(k\Delta\phi) \right] \right\}$$

where $A = \frac{\omega_{s0}^2}{V_{RF}|\cos\phi_s|}$, $\Delta\phi = \phi_t - \phi_u$ and

$$a_k^{\pm} = \frac{\alpha_R}{\alpha_R^2 + (k\omega_{su} \pm \omega_z)^2}; \quad b_k^{\pm} = \frac{(k\omega_{su} \pm \omega_z)}{\alpha_R^2 + (k\omega_{su} \pm \omega_z)^2}$$

This paper discusses only the case when $\omega_R = \omega_{RF}$.

The wakefield is not the only perturbation to the harmonic equation. A radiation damping term, $-\alpha_{rad}.\bar{r}_t$, contributes to the $\dot{\bar{r}}_t$ equation. The amplitude dependent decrease in pendulum frequency can be approximated by a term quadratic in $\bar{r}_t$ [5]. The KBM method is applied by treating the three terms as independent contributions to the equations of motion of $\bar{r}$ and $\bar{\phi}$, giving the final, averaged equations of motion for a test particle at $(r_t, \phi_t)$ due to a macroparticle at $(r_u, \phi_u)$

$$\dot{\bar{r}} = -\frac{1}{2\omega_{st}} F_{S1}\left(\bar{r}_t, \bar{\phi}_t, \bar{r}_u, \bar{\phi}_u\right) - \alpha_{rad}.\bar{r}_t \qquad (3)$$

$$\dot{\bar{\phi}} = -\frac{1}{2\omega_{st}\bar{r}_t} F_{C1}\left(\bar{r}_t, \bar{\phi}_t, \bar{r}_u, \bar{\phi}_u\right) - \frac{1}{16}\bar{r}_t^2 \omega_{so} \qquad (4)$$

## ANALYSIS OF RELAXATION OSCILLATIONS

The convenient reference frame to use is one rotating in phase with the source particle with coordinates $(r_u, \phi_u = 0)$. In this frame, the test particle has coordinates $(r_t, \Delta\phi = \phi_t - \phi_u)$. As the test particle rotates in this frame, the forces from the wake of the source particle change character, from damping to anti-damping, and from frequency increase to frequency decrease. For narrow band resonators tuned with $\omega_z = \omega_{su}$, the line of maximal growth and zero frequency shift both lie close to $\Delta\phi = 0$.

For the case of a single macroparticle model, $r_t = r_u$, $\Delta\phi = 0$, and the source particle carries all of the charge. In most cases, the $m = 0$ and $m = 1$ terms of the series dominate the total force. Assuming a narrowband resonator, for small

$r$, after defining $Z_k^p = Z(p\omega_0 + k\omega_z)$, one recovers the standard formulae for growth and frequency shifts.

$$\dot{r} = \frac{\omega_{s0}}{2V_{RF}|\cos\phi_s|} I \operatorname{Re}\left\{Z_1^h - Z_{-1}^h\right\} \bar{r} - \alpha_{rad}\bar{r}$$

$$\dot{\phi} = \frac{\omega_{s0}}{2V_{RF}|\cos\phi_s|} I \operatorname{Im}\left\{2Z_0^h - Z_1^h - Z_{-1}^h\right\}$$

## Growth and Filamentation

Equations 1 and 2 give the driving force acting on a test particle $(r_t, \Delta\phi)$ produced by the source particle (main body) $(r_u, \Delta\phi = 0)$. $\Delta r$ is defined as $\Delta r \triangleq r_t - r_u$.

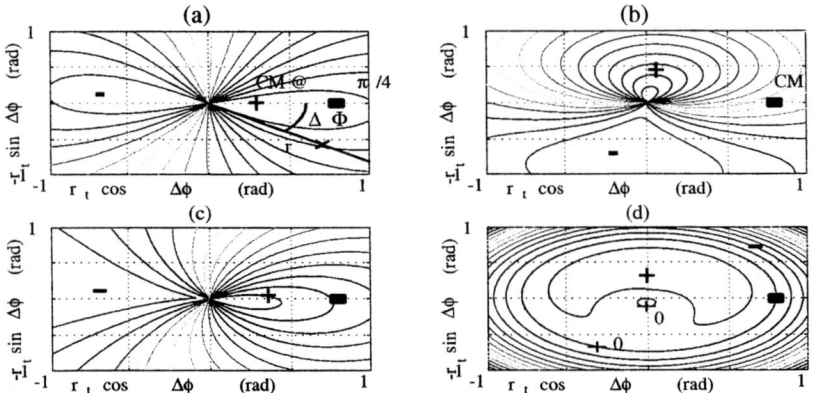

**FIGURE 5.** $\dot{r}$, $\dot{\phi}$ terms in $(r_t, \phi_t)$ plane. (a) $F_{S1}$; (b) $F_{C1}$; (c) $F_{S1}$ and $\alpha_{rad}\bar{r}$; (d) $\dot{F}_{C1}$ and pendulum shift

Figure 5 shows the effect of the growth and frequency shift due to the macroparticle wake in the rotating frame:

- particles ahead of (behind) the main body ($r_u = 0, \Delta\phi > (<) 0$) experience a lesser (greater) $\dot{\phi}$ than that body and will fall back (catch up) to it.(5b)

- particles at $(\Delta r < 0 \, (\Delta r > 0), \Delta\phi = 0)$ feel more (less) growth than the main particle and will grow (fall) toward it (5a,c)

This justifies the bunch cohesion during its growth; the main body is an attractor for all the particles of the bunch. As the oscillation amplitude increases, two nonlinear effects become important: the Bessel terms decrease the growth rate; the pendulum frequency shift starts to dominate the frequency term.

At moderate amplitudes, the pendulum frequency shift now causes a strong enough asymmetry in $\Delta r$ that the test particles start to escape from the front of the bunch.

- Particles with $\Delta r = 0$ experience the same growth as the main body, and will tend to group back towards it as during the growth.

- Particles with $(\Delta r > 0, \Delta\phi = 0)$ slow down more than the main body and acquire a $\Delta\phi < 0$. They go to an area of smaller radial growth, decreasing $\Delta r$, and hence increasing frequency. They return to the main body.

- But particles with $(\Delta r < 0, \Delta\phi = 0)$ speed up more than the main body and acquire a $\Delta\phi > 0$. Once they cross the angle of maximal growth, the particles at positive $\Delta\phi$ experience a smaller driving force and move to even more negative $\Delta r$. The particles escape from the front of the bunch.

The experimental data show the decrease in strength of the growth term, which comes partly from the weaker Bessel terms and partly from the loss of charge density of the main body. Consequently, the bunch saturates at lower amplitudes than predicted by the macroparticle model. The relaxation of the oscillation comes from the loss of growth due to the leakage of particles away from the main body and the formation of a second attractor close to the center of phase space.

## Damping of system

As the escaping particles spiral away from the main body towards the center, they alternately experience positive and negative forces from the main body. Over a rotation of $\Delta\phi = 2\pi$, the net growth due to the main body nearly vanishes, so the particles damp at about the radiation damping rate. The only growth they see is due to the wakefield generated by other particles synchronous with them.

The finite main body amplitude, $\tau_u \neq 0$, implies equations 1 and 2 are non-zero at the origin; therefore all values of $\bar{\phi}$ exist near there (Eqn. 4). On the locus of points in phase with the main body, the particles will again feel the main body's wake. On the locus exists a point at which the radial growth vanishes; very close by is an attractor for particles leaving the main body. To experience no growth from the wake, $\Delta\phi \sim \pi/2$ when $\omega_z = \omega_{so}$ (Figure 5). As charge accumulates at this point, its self-generated wake increases in strength and $\Delta\phi$ of the stable point increases to acquire damping from the main body. The second attractor is actually not fixed, but grows slowly in amplitude, attracting more and more particles until it becomes the new main body in the next relaxation cycle.

## Visualization of second attractor

An interesting case is observed when the lower edge of the resonance coincides with the synchrotron sideband ($\omega_z > \omega_{so}$). The second attractor forms away from the center nearly $\pi$ out of phase with the main initial body and the system damps much more slowly than when $\omega_z = \omega_{so}$. Defining $\phi_k^\pm \triangleq \arctan\left[(k\omega_{su} \pm \omega_z)/\alpha_R\right]$,

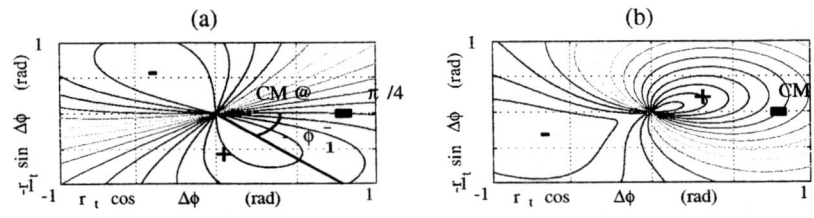

**FIGURE 6.** $\dot{r}$, $\dot{\phi}$ terms for $\omega_z > \omega_{su}$. (a) $F_{S1}$; (b) $F_{C1}$;

the line of maximal growth is now approximately $|\phi_1^-|$ ahead of the main body (figure 6). Test particles now trying to escape from the front have a more difficult time than for the case $\omega_z = \omega_{so}$ for two reasons. First, as $\Delta\phi$ of the particles increase, the particles move closer to the line of maximal growth, grow radially and slow down. Second, they need to precess $\Delta\phi = 2|\phi_1^-|$ before their radial growth is less than that of the macroparticle.

Longer escape times mean longer damping times of the relaxation oscillation for $\omega_z > \omega_{so}$. In this case the second attractor starts at $\phi \sim \pi/2 + |\phi_1^-|$ and increases in phase towards $\pi$ as it gains particles and grows in amplitude.

## ACKNOWLEDGEMENTS

We thank H. Wiedemann and M. Cornacchia for their support and encouragement, and we would like to thank our colleagues J. Hinkson and J. Byrd (LBNL), A. Fisher (SLAC) and A. Lumpkin (APS) for their assistance in obtaining the beautiful pictures with the streak camera.

## REFERENCES

1. G. Rakowsky, "Coherent synchrotron relaxation oscillation in an electron storage ring," in *Proc. of PAC85*, PAC, (Vancouver), pp. 2377 2379, IEEE-APS, 1985.
2. A. Wrulich et al., "Observation of multibunch instabilities in ELETTRA," in *Proc. of EPAC96*, 1996.
3. C. Limborg and J. Sebek, "Relaxation oscillations of the synchrotron motion due to narrow band impedances," *to be published in Physical Review E*, October 1999.
4. N. M. Krylov and N. N. Bogoliubov, *Introduction to nonlinear mechanics*. Annals of mathematical studies, Princeton: Princeton University Press, 1947.
5. E. A. Jackson, *Perspectives of nonlinear dynamics*, vol. 1. Cambridge: Cambridge University Press, 1989.
6. S. Krinsky, "Saturation of a longitudinal instability due to nonlinearity of the wake field," in *Proc. of PAC85*, PAC, (Vancouver), pp. 2320 2322, IEEE-APS, 1985.

# Surface Impedance and Synchronous Modes

## Gennady V. Stupakov

*Stanford Linear Accelerator Center, Stanford University, P.O. Box 4349, Stanford, CA 94309*

**Abstract.** The concept of the surface impedance is applicable to the case, when obstacles on the surface of a vacuum chamber are small compared to characteristic dimensions of the problem. We apply this concept to the calculation of a synchronous mode that can propagate in a tube with slightly corrugated walls. We also show that such a mode can propagate in a pipe with a rough surface, or a perforated pipe with a large number of holes.

## I  INTRODUCTION

There are cases in practice, when the impedance of a vacuum chamber of accelerator is determined by a small obstacle located on the surface of the wall. We call an obstacle small if its characteristic size $g$ is much smaller then both the transverse size of the chamber $a$ and the wavelength $\lambda/2\pi$ associated with the frequency $\omega$ for which the impedance is calculated. For the case of a single small axisymmetric obstacle, a general treatment of the problem was given in Ref. [1]. It was shown that in this case the obstacle is characterized by an electric and magnetic moments, which can be found from the solution of the respective electrostatic and magnetostatic problems, and both contribute to the interaction with the beam.

There are situations however, when many small obstacles densely populate the surface of the wall. Examples are: a rough perfectly conducting surface with a random profile of roughness [2,3]; a corrugated wall of vacuum bellows; and a wall perforated by a large number of small pumping holes [4]. To describe the interaction of such wall with the electromagnetic field, on can introduce an averaged characteristic of the surface in terms of the *surface impedance* $\zeta(\omega)$. If this surface impedance is known, it is easy to calculate the longitudinal and transverse impedances for the beam propagating in the chamber.

The concept of the surface impedance has been previously used by V. Balbekov [5,6] in the treatment of small obstacles on the pipe surface. More recently, M. Dohlus discussed the application of this concept to the calculation of the impedance of rough surfaces [7].

In this paper we will consider how the concept of surface impedance can be used to simplify calculations of the impedance for an ultrarelativistic beam. Following [7], we will show how the surface impedance can explain the appearance of a synchronous wave that propagates with the phase velocity equal to the speed of light, and can be excited by a relativistic beam. We will also apply this concept to a rough surface, and, directly from Maxwell's equations, will find $\zeta$ for a corrugated wall in the limit of shallow corrugations. For simplicity, in this paper, we will consider the longitudinal impedance only, although most of our results can be easily generalized for the transverse impedance as well.

## II  LONGITUDINAL IMPEDANCE OF ULTRARELATIVISTIC BEAM IN CIRCULAR TUBE

Consider an infinitely thin beam, traveling with the speed of light along the axis of a circular pipe of radius $a$. The only nonzero components of the fields generated by the beam are $E_r(r, z, t)$, $E_z(r, z, t)$, and $H_\theta(r, z, t)$, where $r$ and $z$ are cylindrical coordinates. We want to calculate the longitudinal impedance for such beam at frequency $\omega$.

In Fourier representation, the beam current is given by $I(z,t) = I_0 e^{-i\omega t + ikz}$, where $k = \omega/c$, corresponding to $v = c$. The Maxwell's equations *in vacuum*, that is for $0 < r < a$, are

$$\nabla \times \boldsymbol{H} = -\frac{i\omega}{c}\boldsymbol{E}, \quad \nabla \times \boldsymbol{E} = \frac{i\omega}{c}\boldsymbol{H}. \tag{1}$$

It follows from Eqs. (1) that

$$\frac{\partial}{\partial r}\frac{1}{r}\frac{\partial}{\partial r} r H_\theta = 0, \tag{2}$$

whose solution is

$$H_\theta = \frac{A}{r} + \frac{1}{2}Br, \tag{3}$$

where $A$ and $B$ are unknown constants. Using the relation

$$E_z = \frac{i}{k}(\nabla \times \boldsymbol{H})_z = \frac{i}{k}\frac{1}{r}\frac{\partial}{\partial r}r H_\theta, \tag{4}$$

one finds from Eq. (3) $E_z = iB/k$. It is important to emphasize here the the longitudinal electric field in the pipe does not depends on radius $r$ (this is only true in the limit $v = c$).

Using Ampere's law, one can relate the constant $A$ to the beam current, $A = 2I_0/c$, and express the magnetic field in the pipe in terms of the (yet unknown) longitudinal electric field

$$H_\theta = \frac{2I_0}{rc} + E_z \frac{kr}{2i}. \tag{5}$$

We see that the whole solution is now determined by a single constant – the electric field $E_z$. To find this constant, we need to specify a boundary condition on the surface of the pipe.

After the problem is solved, we can find the longitudinal impedance *per unit length of the pipe* with the help of the following relation

$$Z = -\frac{E_z}{I_0}. \tag{6}$$

## III  SURFACE IMPEDANCE

There are situations when the longitudinal electric field on the surface of the pipe can be directly related to the magnetic field $H_\theta$ at the same point. A well known example of such relation is the Leontovich boundary condition [8] on the surface of the metal when the skin depth associated with the frequency $\omega$ is much smaller than the tube thickness.

In general case, let us assume that a relation between the longitudinal electric field and the tangential magnetic field holds on the wall, $r = a$,

$$E_z|_{r=a} = -\zeta(\omega) H_\theta|_{r=a}, \tag{7}$$

where the complex variable $\zeta(\omega)$ is the *surface* impedance of the tube[1]. For the case of the Leontovich boundary condition (see [8], Eq. (87,6)) $\zeta = (1-i)\sqrt{\omega/8\pi\sigma}$, where $\sigma$ is the conductivity of the metal.

Using Eq. (7) we can now solve equation (5) to obtain

$$E_z = -\zeta \frac{2I_0}{ac} \frac{1}{1 + ka\zeta/2i}. \tag{8}$$

Let us consider first a situation when the second term in the denominator of Eq. (8) can be neglected, which is valid for small frequencies, $ka \ll |\zeta|^{-1}$. In this case, neglecting the second term in the denominator of Eq. (8) yields

$$E_z = -\zeta \frac{2I_0}{ac}, \tag{9}$$

and the longitudinal impedance becomes

$$Z = -\frac{E_z}{I_0} = \frac{2\zeta}{ac} = \frac{Z_0 \zeta}{2\pi a}. \tag{10}$$

---

[1] In the general case, $\zeta$ is a function of two variables, $\zeta = \zeta(k,\omega)$. In this paper, however, we consider only the application of $\zeta$ to ultrarelativistic beams, when $k = \omega/c$, and hence $\zeta$ can be considered as a function of frequency.

For the conducting wall, using the Leontovich boundary condition, one immediately obtains the well known result $Z = (1-i)\sqrt{\omega/2\pi\sigma}/ac$.

As a second example, consider a thin dialectric/magnetic layer of thickness $h \ll a$ covering the surface of a perfectly conducting metal [9,10]. As it follows from the Maxwell's equation, inside the layer the following equation holds

$$\frac{\partial E_z}{\partial r} = ik(\epsilon^{-1} - \mu)H_\theta, \qquad (11)$$

where $\epsilon$ is the dielectric constant, and $\mu$ is the magnetic permeability of the layer. Because the layer is thin, we can integrate this equation using the boundary condition on the surface of the metal, $E_z|_{r=a} = 0$, and assuming that the magnetic field is approximately constant within the layer,

$$E_z|_{r=a-h} = -ikh(\epsilon^{-1} - \mu)H_\theta. \qquad (12)$$

Since in vacuum $E_z$ does not depend on $r$, we can move the boundary condition (12) from $a-h$ to $a$ and impose it on the surface of the wall – this will produce a small error of the order of $h/a$. Eq. (12) is then equivalent to the surface impedance that is purely imaginary with the negative imaginary part (assuming $\epsilon$, $\mu > 1$),

$$\zeta = -ikh(\mu - \epsilon^{-1}). \qquad (13)$$

In the small-frequency approximation this gives the inductive impedance (per unit length of the pipe), $Z = -iL\omega$, where

$$L = \frac{2h}{bc^2}(\mu - \epsilon^{-1}). \qquad (14)$$

## IV  SYNCHRONOUS MODE

If we do not neglect the second term in the denominator of Eq. (5), we find for the impedance the following formula

$$Z(\omega) = \frac{2\zeta(\omega)}{ac} \frac{1}{1 + ka\zeta/2i}. \qquad (15)$$

An important feature of this impedance is that for a purely imaginary $\zeta$ with $\text{Im}\,\zeta < 0$, there is a singularity at the frequency $\omega = \omega_0$ that satisfies the following equation

$$\omega_0 \zeta(\omega_0) = -\frac{2ic}{a}. \qquad (16)$$

This singularity is due to the wave that can propagate in the tube with the phase velocity equal to the speed of light. Indeed, it is easy to check by going back to the derivation of Sec. 1, that if Eq. (16) is satisfied, an electromagnetic wave

$$E_r = H_\theta = H_0 \frac{r}{a} e^{-i\omega_0 t + ikz}, \qquad E_z = H_0 \frac{2i}{ka} e^{-i\omega_0 t + ikz}, \qquad (17)$$

satisfies both the Maxwell's equations and the boundary condition (7). For a thin dielectric layer, with $\zeta$ given by Eq. (13), $\omega_0 = c\sqrt{2/ah(\mu - \epsilon^{-1})}$.

This mode will be excited by the beam resulting in the wakefield that can be found by integration of the contribution from the pole of the impedance (15)

$$w(s) = \frac{2Z_0 c}{\pi a^2} \frac{\zeta(\omega_0)}{(\omega\zeta)'|_{\omega=\omega_0}} \cos(\omega_0 s/c). \qquad (18)$$

In the case of linear dependence, $\zeta \propto \omega$, Eq. (18) reduces to $w(s) = (Z_0 c/\pi a^2)\cos(\omega_0 s/c)$.

## V  ROUGH SURFACE

In a simple model [2], a rough surface can be considered as a collection of small densely packed bumps sitting on a plane perfectly conducting substance. As was pointed out in [1], a small obstacle on the surface is characterized by dipole and magnetic moments. A dense collection of such obstacles can be considered as an artificial dielectric layer, if the inverse wavenumber $k^{-1}$ is much larger then the size of obstacles (bumps). Not surprisingly, it was found that, similar to the dielectric layer, the rough surface is characterized by an inductive impedance, and can support propagation of a synchronous wave [11,12].

A quantitative theory of the rough wall impedance, applicable (within some limitations) for arbitrary random profile, was developed in Ref. [13]. The approach is based on the assumption that the angle between the normal to the rough surface and the radial direction is small compared to unity. If we assume that the rough surface is given by the equation $y = h(x, z)$, then the small-angle approximation means that $|\nabla h| \ll 1$. It is also required that the height of the bumps and their characteristic width $g$ be small compared to the radius of the tube $b$, $g, |h| \ll b$, and the frequency $\omega$ is small compared to $c/g$, $\omega \ll c/g$.

To describe a rough surface with a random profile, we assume that $h(x, y)$ is a random function with zero average, $\langle h(x,z)\rangle = 0$. Statistical properties of such a surface are characterized by the correlation function $K(x,y)$,

$$K(x - x', z - z') = \langle h(x', z')h(x, z)\rangle, \qquad (19)$$

where the angular brackets denote averaging over possible realizations of $h(x, z)$. Equation (19) implies that statistical properties of $h(x, z)$ do not depend on the position on the surface. An important statistical characteristic of the roughness is the *spectral density* (or *spectrum*) $R(\kappa_z, \kappa_x)$, defined as a Fourier transform of the correlation function,

$$R(\kappa_x, \kappa_z) = \frac{1}{(2\pi)^2} \int dx\, dz\, K(x,z) e^{-i\kappa_x x - i\kappa_z z}. \qquad (20)$$

The main result of Ref. [13] is that the longitudinal impedance (per unit length) of a circular tube of radius $a$ with a rough perfectly conducting surface characterized by the spectral function $R(\kappa_x, \kappa_z)$ in the frequency range limited by the condition $\omega \ll c/g$, is $Z(\omega) = -i\omega L_{\rm rs}$ with the inductance

$$L_{\rm rs} = \frac{Z_0}{2\pi ca} \int d\kappa_z \, d\kappa_x \, R(\kappa_x, \kappa_z) \frac{\kappa_z^2}{\kappa}. \tag{21}$$

The presence of the factor $\kappa_z^2$ in the integrand of Eq. (21) means that the contributions to $L_{\rm rs}$ of roughness in longitudinal ($z$) and azimuthal ($x$) directions are different. For example, bellow-type variations on the surface have spectral components with $\kappa_z \neq 0$ and $\kappa_x = 0$, and result in nonvanishing $L_{\rm rs}$. On the other hand, grooves of all sizes in the longitudinal direction, as described in the previous Section, generate a spectrum with $\kappa_x \neq 0$ and $\kappa_z = 0$, and according to Eq. (21) do not contribute to $L_{\rm rs}$.

Using Eq. (10), valid in the limit of small frequencies, we can infer the roughness surface impedance $\zeta_{\rm rs}$ from Eq. (21)

$$\zeta_{\rm rs} = -ik \int d\kappa_z \, d\kappa_x \, R(\kappa_x, \kappa_z) \frac{\kappa_z^2}{\kappa}. \tag{22}$$

As was pointed out in the previous section, this kind of impedance allows for the propagation of a synchronous mode at some frequency $\omega_0$. Indeed, such a mode was found in Ref. [11] (see also [12]), where its existence was explained in analogy with a mode in a pipe coated by a dielectric layer.

## VI  SURFACE IMPEDANCE FOR CORRUGATED PIPE

In a simple case of a corrugated pipe, the surface impedance can be found directly from Maxwell's equations, even when the wavelength is compared to the period of corrugation.

Let us assume that the pipe surface is given by

$$r = a - h \sin \kappa z, \tag{23}$$

where $2\pi/\kappa$ is the period of corrugation, and $h$ is its amplitude. We assume that both the wavelength and the amplitude are small, $h \ll a$ and $\kappa a \ll 1$. This allows us to neglect the curvature effects and to consider the surface locally as a plane one. We will also assume a shallow corrugation

$$h\kappa \ll 1, \tag{24}$$

that is the amplitude of the bumps is much smaller then their period.

Introducing a local Cartesian coordinate system $x$, $y$, $z$ with $y = a - r$ (directed from the wall toward the beam axis), and $x$ directed along $\theta$, the surface equation becomes $y = y_0(z) \equiv h \sin \kappa z$. The magnetic field near the surface $H_x(y, z)$ does not depend on $x$ (that is $\theta$) due to the axisymmetry of the problem. It satisfies the Helmholtz equation

$$\frac{\partial^2 H_x}{\partial y^2} + \frac{\partial^2 H_x}{\partial z^2} + k^2 H_x = 0 \tag{25}$$

with the boundary condition

$$(\mathbf{n}\nabla H)|_{y=y_0} = 0, \tag{26}$$

where $\mathbf{n}$ is the normal vector to the surface, $\mathbf{n} = (0, 1, -h\kappa \cos \kappa z)$.

Note that the longitudinal electric field $E_z$ can be expressed in terms of $H_x$,

$$E_z = -\frac{i}{k}\frac{\partial H_x}{\partial y}. \tag{27}$$

Using the small parameter $h/a$, we will develop a perturbation theory for calculation of $H_x$ near the surface and find how $E_z$ is related to $H_x$.

In the zeroth approximation, the $z$ dependence of $H_x$ is dictated by the beam current periodicity,

$$H_x(y, z) = \mathcal{H}(y) e^{ikz}. \tag{28}$$

Putting Eq. (28) into Eq. (25) we find that $d^2\mathcal{H}/dy^2 = 0$, hence $\mathcal{H}(y) = H_0 + Ay$, where the constant $A$ can be related, through Eq. (27), to the electric field on the surface, and hence $\zeta$, $A = ikE_z = -ik\zeta H_0$. We will see below that $A$ is of the second order in $h$.

For a flat surface, for which $\mathbf{n} = (0, 1, 0)$, from the boundary condition (26), we would conclude that $A = 0$, however, the corrugations result in a nonzero $A$, and hence $E_z$. Substituting the magnetic field (28) into the right hand side of Eq. (26) one finds

$$\mathbf{n}\nabla H = -\frac{1}{2}ihk\kappa H_0 \left[e^{i(k+\kappa)z} - e^{i(k-\kappa)z}\right] - ik\zeta H_0 e^{ikx}. \tag{29}$$

Clearly, the boundary condition is not satisfied in this approximation. To correct this, we have to add satellite modes to the fundamental solution (28)

$$H_x(y, z) = \mathcal{H}(y) e^{ikz} + \mathcal{H}_1(y, z), \tag{30}$$

where

$$\mathcal{H}_1(y, z) = B^+(y) e^{i(k+\kappa)z} + B^-(y) e^{i(k-\kappa)z}. \tag{31}$$

The dependence of $B^\pm$ versus $y$ can be found from the Helmholtz equation,

$$B = B_0^{\pm} e^{-y\sqrt{\kappa^2 \pm 2\kappa k}}, \tag{32}$$

where $B_0^{\pm}$ are constants. In order for $B^{\pm}$ to exponentially decay in $y$, we have to assume here that $k < \kappa/2$.

Substituting $\mathcal{H}_1$ terms into the boundary condition (26) generates first order terms that have $x$-dependence $\exp i(k \pm \kappa)x$, and second order terms proportional to $\exp(ikx)$. From the former one finds that

$$B_0^{\pm} = -\frac{ik\kappa H_0 h}{2\sqrt{\kappa^2 \pm 2k\kappa}}, \tag{33}$$

and the latter gives an expression for the surface impedance $\zeta$,

$$\zeta(k) = -\frac{1}{4} ikh^2 \kappa \frac{\sqrt{\kappa^2 + 2k\kappa} + \sqrt{\kappa^2 - 2k\kappa}}{\sqrt{\kappa^2 - 4k^2}}. \tag{34}$$

## VII  SYNCHRONOUS WAVE IN CORRUGATED PIPE

First, consider the case of small frequencies, $k \ll \kappa$. In this limit, Eq. (34) reduces to

$$\zeta = -\frac{1}{2} ikh^2 \kappa. \tag{35}$$

Being purely imaginary with the negative imaginary part, this $\zeta$ allows for a propagation of the synchronous mode. Its frequency is given by Eq. (16)

$$\omega_0 = \frac{2c}{h\sqrt{a\kappa}}. \tag{36}$$

For the general case, when $k$ can be comparable with $\kappa$, one has to solve the exact dispersion relation. The result is shown in Fig. 1, where we also plot the formfactor $f$ that gives the wakefield of the mode $w(s) = (2f Z_0 c/\pi a^2) \cos(\omega_0 s/c)$. We see that decreasing the height of the corrugation results in smaller wakes, and hence leads to the suppression of the interaction of the synchronous wave with the beam.

We have to mention here that our theory breaks down for very small values of $h$. Indeed, we implicitly assumed that the satellite harmonics in Eq. (32) are localized near the surface, otherwise our approximation of plain surface becomes invalid. Hence, we have to require that $\kappa - 2k \gg a^{-1}$, which gives the following condition of applicability

$$h \gtrsim a^{-1/4} \kappa^{-5/4}. \tag{37}$$

This condition explains why this mode has not been found in earlier papers [14,15] that treated the problem of a corrugated pipe: being purturbative in parameter $h$ the approach developed in those papers is applicable only when $h$ can be made arbitrarily small.

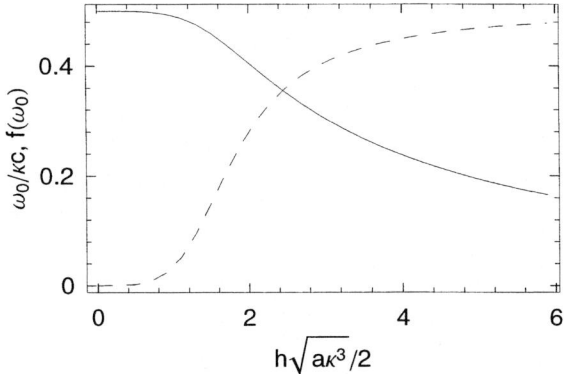

**FIGURE 1.** Frequency $\omega_0$ (solid curve) and the loss form-factor $f$ (dashed curve) for the synchronous mode as a function of height $h$.

## VIII  PERFORATED PIPE

The notion of the surface impedance can be also applied to the case of a pipe with a large number of small holes perforated in the pipe wall. Again, we have to assume that the wavelength $c/\omega$ is much larger then the hole size $b$ and the average distance between the holes $d$, $c/\omega \gg b, d$.

The value of $\zeta$ can be found from the correspondence between magnetic and electric properties of a thin dielectric layer considered in Section III and a large collection of small holes.

Indeed, the electric polarizability $\alpha_e$ of a single round hole of radius $b$ is $\alpha_e = -b^3/3\pi$, and its magnetic permeability $\alpha_m$ is $\alpha_m = 2b^3/3\pi$, see [16], Sec. 9.5. Assuming that the holes do not interact with each other, we conclude that a tangential magnetic field $H_\theta$ on the wall induces in holes the magnetic moment per unit area equal to $\alpha_m \rho H_\theta$ where $\rho = d^{-2}$ is the number of holes per unit area. For a magnetic layer of thickness $h$ the magnetic moment per unit area in the tangential field would be $H_\theta(\mu - 1)/4\pi$, hence we find the correspondence between the product $(\mu - 1)h$ and $\alpha_m$, $(\mu - 1)h \to 4\pi\rho\alpha_m$. Similarly, a radial electric field $E_r$ induces in the holes the radial dipole moment per unit area equal to $\alpha_e \rho E_r$, whereas in the dielectric layer it would induces the dipole moment $E_r(\epsilon - 1)/4\pi\epsilon$, hence $(1 - \epsilon^{-1})h \to 4\pi\rho\alpha_e$. These two relations allow us to convert the surface impedance of the dielectric layer (13) into the surface impedance of holes

$$\zeta_{\text{holes}} = -4\pi\rho ik(\alpha_e + \alpha_m) = -\frac{4}{3}\rho ikb^3. \tag{38}$$

This result has been also recently found in Ref. [17].

Again, we see that this kind of surface impedance allows propagation of synchronous modes in the perforated pipe with the frequency

$$\omega_0 = c\sqrt{\frac{3}{2ab^3\rho}}. \tag{39}$$

This formula is applicable in the limit of noninteracting holes when the wavelength of the mode is much larger than the distance between the holes.

## IX ACKNOWLEDGEMENTS

The author thanks K. Bane and V. Dolgashev for useful discussions.

This work was supported by the US Department of Energy, Office of Basic Energy Sciences, under contract # DE-AC03-76SF00515.

## REFERENCES

1. S. S. Kurennoy and G. V. Stupakov, Particle Accelerators **45**, 95 (1994).
2. K. L. F. Bane, C. K. Ng, and A. W. Chao, Report SLAC-PUB-7514, SLAC (1997).
3. K. L. F. Bane and G. V. Stupakov, Tech. Rep. SLAC-PUB-8023, SLAC (December 1998), presented at International Computational Accelerator Physics Conference (I-CAP 98), Monterey, CA, 14-18 Sep 1998.
4. R. L. Gluckstern, Phys. Rev. **A46**, 1110 (1992).
5. V. I. Balbekov, Tech. Rep. IFVE-93-55, IHEP, Protvino, Russia (April 1993).
6. V. I. Balbekov, Tech. Rep. IFVE-93-56, IHEP, Protvino, Russia (April 1993).
7. M. Dohlus, private communication, (1999).
8. L. D. Landau and E. M. Lifshitz, *Electrodynamics of Continuous Media*, (Pergamon, London, 1960), 2nd ed.
9. A. V. Burov and A. V. Novokhatskii, Tech. Rep. BUDKERINP-90-28, Budker Inst. of Nucl. Physics, Novosibirk, Russia (February 1990).
10. K.-Y. Ng, Phys. Rev. **D42**, 1819 (1990).
11. A. Novokhatski and A. Mosnier, in *Proceedings of the 1997 Particle Accelerator Conference* (IEEE, Piscataway, NJ, 1997), pp. 1661–1663.
12. K. L. F. Bane and A. Novokhatskii, Tech. Rep. SLAC-AP-117, SLAC (March 1999).
13. G. V. Stupakov, Phys. Rev. ST Accel. Beams **1**, 064401 (1998).
14. M. Chatard-Moulin and A. Papiernik, IEEE Trans. Nucl. Sci. **26**, 3523 (1979).
15. S. Krinsky, in W. S. Newman, ed., *Proc. International Conference on High-Energy Accelerators, Geneva, 1980*, CERN, European Lab. for Particle Physics (Birkhäuser Verlag, Basel, Switzerland, 1980), p. 576.
16. J. D. Jackson, *Classical Electrodynamics* (Wiley, New York, 1975), 2nd ed.
17. S. Petracca, Tech. Rep. CERN-SL-99-003, CERN (February 1999).

# Nonlinear Longitudinal Waves in High Energy Stored Beams *

Stephan I. Tzenov

*Stanford Linear Accelerator Center, Stanford University,
STANFORD, CA 94309-0210*

## Abstract

We solve the Vlasov equation for the longitudinal distribution function and find stationary wave patterns when the distribution in the energy error is Maxwellian. In the long wavelength limit a stability criterion for linear waves has been obtained and a Korteweg-de Vries- Burgers equation for the relevant hydrodynamic quantities has been derived.

## 1 Introduction.

Nonlinear wave interaction in high energy synchrotrons has recently received a great deal of attention (see e.g. [1], [2], [3]), since it has proven its importance for understanding a variety of phenomena in high intensity beams.

Perhaps, the simplest problem to study is the evolution in longitudinal direction only of a intense coasting beam influenced by a broad-band resonator type impedance. This model exhibits a surprisingly vast variety of interesting features, part of which have already been experimentally observed and theoretically investigated [1], [2], [3]. Different types of beam equilibria can be detected due to the collective (nonlinear) interaction between beam particles and resonator waves, the latter being induced by the beam itself. Solutions describing similar types of plasma equilibria [Bernstein-Greene-Kruskal (BGK) modes] are well-known in plasma physics [4]. Structures of arbitrary shape can be formed in the nonlinear stationary regime, which substantially depend on the type of the initial velocity distribution.

It is the purpose of the present paper to apply techniques borrowed from plasma physics to study nonlinear patterns in coasting beams that are in close analogy with BGK modes. In sections 3 and 4 we solve the Vlasov equation by expanding the distribution function in a power series of the resonator potential [5], and in the case

---

*Work supported by Department of Energy contract DE–AC03–76SF00515.

of initial Maxwellian energy error distribution we obtain an equation, describing the evolution of stationary waves on the resonator. In section 5 we find a stability criterion for linear waves in the long wavelength limit and derive a Korteweg-de Vries-Burgers equation for the beam density, current velocity and resonator voltage.

## 2  Model.

We consider the longitudinal dynamics of a high energy stored beam governed by the set of equations [2], [3]:

$$\frac{\partial f}{\partial T} + v\frac{\partial f}{\partial \theta} + \lambda V \frac{\partial f}{\partial v} = 0, \qquad (2.1)$$

$$\frac{\partial^2 V}{\partial T^2} + 2\gamma \frac{\partial V}{\partial T} + \omega^2 V = \frac{\partial I}{\partial T}, \qquad (2.2)$$

$$I(\theta; T) = \int dv\, v f(\theta, v; T). \qquad (2.3)$$

The first equation (2.1) is the Vlasov equation for the longitudinal distribution function $f(\theta, v; T)$ of an unbunched beam, while the second equation (2.2) governs the variation per turn of the voltage $V(\theta; T)$ on a resonator. All dependent and independent variables, as well as free parameters in equations (2.1-3) are dimensionless and have been rescaled according to the relations:

$$T = \omega_s t \quad ; \quad v = \frac{1}{\omega_s}\frac{d\theta}{dt} = 1 + \frac{k_0 \Delta E}{\omega_s} \quad ; \quad \omega = \frac{\omega_R}{\omega_s}, \qquad (2.4a)$$

$$\gamma = \frac{\omega}{2Q} \quad ; \quad \lambda = \frac{e^2 \mathcal{R}\gamma k_0 \rho_0}{\pi}. \qquad (2.4b)$$

Here $\omega_s$ is the angular revolution frequency of the synchronous particle, $\Delta E$ is the energy error, $\omega_R$ is the resonant frequency, $Q$ is the quality factor of the resonator, $\mathcal{R}$ is the resonator shunt impedance and $\rho_0$ is the uniform beam density distribution in the thermodynamic limit. Furthermore

$$k_0 = -\frac{\eta \omega_s}{\beta_s^2 E_s} \qquad (2.5)$$

is the proportionality constant between the frequency deviation of a non synchronous particle with respect to the synchronous one, while $\eta = \alpha_M - \gamma_s^{-2}$ ($\alpha_M$ - momentum compaction factor) is the phase slip coefficient. The voltage variation per turn $V(\theta; T)$, the beam current $I(\theta; T)$ and the longitudinal distribution function $f(\theta, v; T)$ entering equations (2.1-3) have been rescaled as well from their actual values $V_a(\theta; T)$, $I_a(\theta; T)$ and $f_a(\theta, v; T)$ as follows:

$$V_a = 2e\omega_s \rho_0 \gamma \mathcal{R} V \quad ; \quad I_a = e\omega_s \rho_0 I \quad ; \quad f_a = \rho_0 f. \qquad (2.6)$$

From the Vlasov equation (2.1) it is straightforward to obtain the continuity equation:

$$\frac{\partial}{\partial T}\int dv f + \frac{\partial}{\partial \theta}\int dv v f = 0, \qquad (2.7)$$

which will be needed for the exposition in the next section.

## 3 Solution of the Vlasov Equation.

Let us now try to solve the Vlasov equation by the simple separation of variables ansatz:

$$f(\theta, v; T) = g(v)\psi(\theta; T). \qquad (3.1)$$

Substitution of (3.1) into the continuity equation (2.7) yields:

$$\frac{\partial \psi}{\partial T} + \Omega \frac{\partial \psi}{\partial \theta} = 0, \qquad (3.2)$$

where

$$\Omega = \frac{\int dv v g(v)}{\int dv g(v)}. \qquad (3.3)$$

The Vlasov equation (2.1) with (3.1-3) in hand can be further transformed to

$$\frac{\partial \psi}{\partial \theta} = \frac{\lambda V \psi}{g(\Omega - v)}\frac{dg}{dv}.$$

The separation of variables ansatz (3.1) implies

$$\frac{dg}{dv} = \frac{\Omega - v}{\sigma_v^2} g, \qquad (3.4a)$$

which leads to the well-known equilibrium Maxwell-Boltzmann distribution:

$$g(v) = \frac{1}{\sigma_v \sqrt{2\pi}}\exp\left[-\frac{(v-\Omega)^2}{2\sigma_v^2}\right], \qquad (3.4)$$

$$\psi(\theta; T) = \mathcal{Z}\exp\left[\frac{\lambda\varphi(\theta; T)}{\sigma_v^2}\right] \quad ; \quad V(\theta; T) = \frac{\partial \varphi(\theta; T)}{\partial \theta}, \qquad (3.5)$$

where

$$\mathcal{Z}^{-1} = \int_0^{2\pi} d\theta \exp\left[\frac{\lambda\varphi(\theta; T)}{\sigma_v^2}\right]. \qquad (3.6)$$

The solution (3.4-6) suggests further generalization [5] of the separation of variables ansatz (3.1)

$$f(\theta, v; T) = \sum_{k=0}^{\infty} g_k(v)\varphi^k(\theta; T). \tag{3.7}$$

Instead of equations (3.2) and (3.3) we now have

$$\frac{\partial \varphi}{\partial T} + \Omega(\theta; T)\frac{\partial \varphi}{\partial \theta} = 0, \tag{3.2a}$$

where

$$\Omega(\theta; T) = \frac{\sum_{k=1}^{\infty} k\mathcal{A}_k \varphi^{k-1}(\theta; T)}{\sum_{k=1}^{\infty} k\mathcal{B}_k \varphi^{k-1}(\theta; T)}. \tag{3.3a}$$

$$\mathcal{A}_k = \int dv\, v g_k(v) \quad ; \quad \mathcal{B}_k = \int dv\, g_k(v). \tag{3.8}$$

In order to determine the yet unknown functions $g_k(v)$ we make the assumption:

$$\Omega(\theta; T) = const, \tag{3.9}$$

which will be proved *a posteriori* to hold and substitute (3.7) into the Vlasov equation (2.1). Taking into account (3.2a) we obtain:

$$(v - \Omega)\sum_{k=1}^{\infty} k g_k(v)\varphi^{k-1}(\theta; T) + \lambda \sum_{k=0}^{\infty} \frac{dg_k(v)}{dv}\varphi^k(\theta; T) = 0. \tag{3.10}$$

Equating coefficients in front of powers of $\varphi$ yields the following recurrence relation

$$(v - \Omega)(k+1)g_{k+1}(v) = -\lambda \frac{dg_k(v)}{dv},$$

or

$$g_{k+1}(v) = \frac{\lambda}{k+1}\widehat{\mathcal{D}}g_k(v), \tag{3.11}$$

where we have introduced the operator [5]

$$\widehat{\mathcal{D}} = \frac{1}{\Omega - v}\frac{d}{dv}. \tag{3.12}$$

Noting that the formal solution of the recurrence relation (3.11) has the form

$$g_k(v) = \frac{\lambda^k}{k!}\widehat{\mathcal{D}}^k g_0(v) \tag{3.13}$$

we finally arrive at the general solution of the Vlasov equation

$$f(\theta, v; T) = \sum_{k=0}^{\infty} \frac{\lambda^k \varphi^k(\theta; T)}{k!} \widehat{\mathcal{D}}^k g_0(v). \tag{3.14}$$

What remains now is to verify the condition (3.9). It suffices to note that [5]

$$\mathcal{A}_k = \frac{\lambda^k}{k!} \int dv\, v \widehat{\mathcal{D}}^k g_0(v) = \frac{\lambda^k}{k!} \int dv\, \frac{v}{\Omega - v} \frac{d}{dv} \left[ \widehat{\mathcal{D}}^{k-1} g_0(v) \right] =$$

$$= -\frac{\lambda^k \Omega}{k!} \int \frac{dv}{(\Omega - v)^2} \widehat{\mathcal{D}}^{k-1} g_0(v),$$

and similarly

$$\mathcal{B}_k = -\frac{\lambda^k}{k!} \int \frac{dv}{(\Omega - v)^2} \widehat{\mathcal{D}}^{k-1} g_0(v).$$

Thus

$$\mathcal{A}_k = \Omega \mathcal{B}_k, \tag{3.15}$$

which proves equation (3.9).

Clearly the solution (3.14) is uniquely determined by the generic function $g_0(v)$. The simplest choice is when $g_0(v)$ is the Maxwellian (3.4), that is $g_0(v)$ itself is an eigenfunction of the operator $\widehat{\mathcal{D}}$ with an eigenvalue $\sigma_v^{-2}$ [c.f. equation (3.4a)]. In this case we immediately recover the distribution (3.1) with (3.4-6).

## 4 Nonlinear Stationary Waves.

In order to derive an equation for the potential $\varphi(\theta; T)$ we insert (3.1) and (3.4-6) into (2.2) and obtain:

$$\frac{\partial^3 \varphi}{\partial \theta \partial T^2} + 2\gamma \frac{\partial^2 \varphi}{\partial \theta \partial T} + \omega^2 \frac{\partial \varphi}{\partial \theta} = \mathcal{Z}\Omega \frac{\partial}{\partial T} \left[ \exp\left(\frac{\lambda \varphi}{\sigma_v^2}\right) \right]. \tag{4.1}$$

Making use of relation (3.2a) we cast equation (4.1) into the form

$$\frac{\partial^3 \varphi}{\partial T^3} + 2\gamma \frac{\partial^2 \varphi}{\partial T^2} + \omega^2 \frac{\partial \varphi}{\partial T} = -\mathcal{Z}\Omega^2 \frac{\partial}{\partial T} \left[ \exp\left(\frac{\lambda \varphi}{\sigma_v^2}\right) \right]. \tag{4.2}$$

Integrating once equation (4.2) with due account of the initial condition

$$\varphi(\theta; T = 0) = \frac{\partial \varphi(\theta; T = 0)}{\partial T} = \frac{\partial^2 \varphi(\theta; T = 0)}{\partial T^2} = 0 \tag{4.3}$$

we obtain

$$\frac{\partial^2 \varphi}{\partial T^2} + 2\gamma \frac{\partial \varphi}{\partial T} + \omega^2 \varphi = \mathcal{Z}\Omega^2 \left[1 - \exp\left(\frac{\lambda \varphi}{\sigma_v^2}\right)\right]. \tag{4.4}$$

Expanding the factor in square brackets on the right-hand-side of equation (4.4) around the stationary solution $\varphi_s = 0$ yields

$$\frac{\partial^2 \varphi}{\partial T^2} + 2\gamma \frac{\partial \varphi}{\partial T} + \omega^2 \varphi = -\frac{\lambda \mathcal{Z}\Omega^2}{\sigma_v^2} \varphi \left(1 + \frac{\lambda \varphi}{2\sigma_v^2} + \frac{\lambda^2 \varphi^2}{6\sigma_v^4} + ...\right). \tag{4.5}$$

Above the transition energy $\gamma_s > \gamma_T$ $\left(\gamma_T = \alpha_M^{-1/2}\right)$ the parameter $\lambda$ is negative, so that two cases can be distinguished. Defining

$$\omega_0 = \omega^2 - \frac{|\lambda|\mathcal{Z}\Omega^2}{\sigma_v^2}, \tag{4.6}$$

we can state the two cases mentioned above in a more explicit way:

Case I: Provided $\omega_0 > 0$, equation (4.5) can be transformed to a damped Duffing equation with an additional quadratic nonlinearity

$$\frac{\partial^2 \varphi}{\partial T^2} + 2\gamma \frac{\partial \varphi}{\partial T} + |\omega_0| \varphi = -\frac{\lambda^2 \mathcal{Z}\Omega^2}{2\sigma_v^4} \left(\varphi^2 - \frac{|\lambda|}{3\sigma_v^2} \varphi^3\right). \tag{4.7}$$

Case II: For $\omega_0 < 0$ equation (4.5) takes the form

$$\frac{\partial^2 \varphi}{\partial T^2} + 2\gamma \frac{\partial \varphi}{\partial T} - |\omega_0| \varphi = -\frac{\lambda^2 \mathcal{Z}\Omega^2}{2\sigma_v^4} \left(\varphi^2 - \frac{|\lambda|}{3\sigma_v^2} \varphi^3\right). \tag{4.8}$$

In the limit $\gamma \to 0$ equation (4.8) can be solved when neglecting the cubic term. The result is:

$$\varphi(\theta; T) = \frac{3|\omega_0|\sigma_v^4}{\lambda^2 \mathcal{Z}\Omega^2 \cosh^2\left[\frac{\sqrt{|\omega_0|}}{2\Omega}(\theta - \Omega T)\right]}. \tag{4.9}$$

This is a drifting hump-like structure that is well-known as a solitary wave of the Korteweg-de Vries (KdV) type.

## 5 The Korteweg-de Vries-Burgers Equation.

The exact solution of the Vlasov equation obtained in the preceding sections was found based on the stationary wave condition given by the continuity equation (3.2). In order to provide a more general treatment of the problem we introduce the new coordinates and variables along with the moving beam particles

$$z = \theta - T \quad ; \quad u = v - 1. \tag{5.1}$$

Then the basic equations (2.1-3) can be written as:

$$\frac{\partial f}{\partial \theta} + u\frac{\partial f}{\partial z} + \lambda V \frac{\partial f}{\partial u} = 0, \qquad (5.2)$$

$$\frac{\partial^2 V}{\partial z^2} - 2\gamma \frac{\partial V}{\partial z} + \omega^2 V = -\frac{\partial}{\partial z} \int du (1+u) f(z, u; \theta). \qquad (5.3)$$

Let us now pass to the hydrodynamic description of the longitudinal beam motion. The gas dynamic equations read as

$$\frac{\partial F}{\partial \theta} + \frac{\partial}{\partial z}(FU) = 0, \qquad (5.4)$$

$$\frac{\partial U}{\partial \theta} + U\frac{\partial U}{\partial z} = \lambda V - \frac{\sigma_v^2}{F}\frac{\partial F}{\partial z}, \qquad (5.5)$$

$$\frac{\partial^2 V}{\partial z^2} - 2\gamma \frac{\partial V}{\partial z} + \omega^2 V = \frac{\partial F}{\partial \theta} - \frac{\partial F}{\partial z}, \qquad (5.6)$$

where

$$F(z;\theta) = \int du f(z,u;\theta) \quad ; \quad F(z;\theta)U(z;\theta) = \int du u f(z,u;\theta). \qquad (5.7)$$

Obviously the stationary solution of the gas dynamic equations (5.4-6) is given by

$$F_0 = 1 \quad ; \quad U_0 = 0 \quad ; \quad V_0 = 0.$$

The dispersion law of linear waves of the form

$$(F, U, V) = (F_L, U_L, V_L) \exp\left[i\left(\Omega \theta - kz\right)\right]$$

is governed by the following equation

$$1 - i\lambda Z(k) \frac{k+\Omega}{\Omega^2 - k^2 \sigma_v^2} = 0, \qquad (5.8)$$

where $Z(k)$ is the well-known impedance function

$$Z(k) = \frac{ik}{k^2 + 2i\gamma k - \omega^2}. \qquad (5.9)$$

In the long wavelength limit (small $k$) the dispersion equation (5.8) has two roots given by the expression

$$\Omega_{1,2} = \frac{k}{2\omega^2}\left(\lambda \pm \sqrt{\lambda^2 + 4\lambda\omega^2 + 4\omega^4 \sigma_v^2}\right), \qquad (5.10)$$

which are real below transition energy. However, the situation when the energy of the synchronous particle is above transition energy is different. The solutions (5.10) to the dispersion equation are real, provided

$$|\lambda| \leq \lambda_1 \quad ; \quad |\lambda| \geq \lambda_2 \quad ; \quad \lambda_{1,2} = 2\omega^2\left(1 \mp \sqrt{1-\sigma_v^2}\right). \tag{5.11}$$

An instability occurs when $\Omega_{1,2}$ are complex, that is when

$$\lambda_1 < |\lambda| < \lambda_2. \tag{5.12}$$

In what follows we will study the case when our system is linearly stable, that is either below transition energy or in the stability region (5.11).

The solution of the dispersion equation in the long wavelength limit suggests that new scaled coordinates should be introduced [6], [7]

$$\sigma = \sqrt{\epsilon}(z - \alpha\theta) \quad ; \quad \chi = \epsilon^{3/2}\theta, \tag{5.13}$$

where $\epsilon$ is a formal small parameter. Then the gas dynamic equations can be rewritten as

$$-\alpha\frac{\partial F}{\partial \sigma} + \frac{\partial}{\partial \sigma}(FU) + \epsilon\frac{\partial F}{\partial \chi} = 0, \tag{5.14}$$

$$-\alpha\frac{\partial U}{\partial \sigma} + U\frac{\partial U}{\partial \sigma} + \epsilon\frac{\partial U}{\partial \chi} = \lambda\widetilde{V} - \frac{\sigma_v^2}{F}\frac{\partial F}{\partial \sigma}, \tag{5.15}$$

$$\epsilon\frac{\partial^2 \widetilde{V}}{\partial \sigma^2} - 2\epsilon\gamma_0\frac{\partial \widetilde{V}}{\partial \sigma} + \omega^2\widetilde{V} = -(1+\alpha)\frac{\partial F}{\partial \sigma} + \epsilon\frac{\partial F}{\partial \chi}, \tag{5.16}$$

where

$$V = \sqrt{\epsilon}\widetilde{V} \quad ; \quad \gamma = \sqrt{\epsilon}\gamma_0, \tag{5.17}$$

$$\omega^2\alpha^2 - \lambda\alpha - \lambda - \omega^2\sigma_v^2 = 0 \quad ; \quad \left(\alpha = \frac{\Omega_{1,2}}{k}\right). \tag{5.18}$$

Assuming the perturbation expansions:

$$F = 1 + \sum_{m=1}^{\infty}\epsilon^m F_m \quad ; \quad U = \sum_{m=1}^{\infty}\epsilon^m U_m \quad ; \quad \widetilde{V} = \sum_{m=1}^{\infty}\epsilon^m V_m \tag{5.19}$$

for the first and second-order terms in $\epsilon$ we obtain respectively

$$\frac{\partial U_1}{\partial \sigma} = \alpha\frac{\partial F_1}{\partial \sigma} = \frac{\alpha\lambda V_1}{\sigma_v^2 - \alpha^2}, \tag{5.20}$$

or

$$U_1(\sigma, \chi) = \alpha F_1(\sigma, \chi) + G(\chi), \tag{5.21}$$

where $G(\chi)$ is a generic function of the variable $\chi$, and

$$-\alpha\frac{\partial F_2}{\partial \sigma} + \frac{\partial U_2}{\partial \sigma} + \frac{\partial}{\partial \sigma}(F_1 U_1) + \frac{\partial F_1}{\partial \chi} = 0, \qquad (5.22a)$$

$$-\alpha\frac{\partial U_2}{\partial \sigma} + U_1\frac{\partial U_1}{\partial \sigma} + \frac{\partial U_1}{\partial \chi} = \lambda V_2 - \sigma_v^2\frac{\partial F_2}{\partial \sigma} + \sigma_v^2 F_1\frac{\partial F_1}{\partial \sigma}, \qquad (5.22b)$$

$$\frac{\partial^2 V_1}{\partial \sigma^2} - 2\gamma_0\frac{\partial V_1}{\partial \sigma} + \omega^2 V_2 = -(1+\alpha)\frac{\partial F_2}{\partial \sigma} + \frac{\partial F_1}{\partial \chi}. \qquad (5.22c)$$

Eliminating $F_2$, $U_2$ and $V_2$ from equations (5.22) we finally arrive at the Korteweg-de Vries-Burgers equation

$$\frac{\partial F_1}{\partial \chi} + (c_1 F_1 + c_2 G)\frac{\partial F_1}{\partial \sigma} + D\frac{\partial^3 F_1}{\partial \sigma^3} - 2\gamma D\frac{\partial^2 F_1}{\partial \sigma^2} = h\frac{dG}{d\chi}, \qquad (5.23)$$

where

$$c_1 = \frac{\omega^2(3\alpha^2 - \sigma_v^2)}{2\alpha\omega^2 - \lambda} \quad ; \quad c_2 = \frac{2\alpha\omega^2}{2\alpha\omega^2 - \lambda}, \qquad (5.24a)$$

$$D = \frac{\sigma_v^2 - \alpha^2}{2\alpha\omega^2 - \lambda} \quad ; \quad h = \frac{\omega^2}{\lambda - 2\alpha\omega^2}. \qquad (5.24b)$$

It is important to note that $\alpha^{-1}U_1$ and $\lambda(\sigma_v^2 - \alpha^2)^{-1}\int d\sigma V_1$ satisfy exactly the same equation (5.23).

Similar Korteweg-de Vries-Burgers equation in the case below transition energy has been recently derived by A. Aceves employing the method of multiple scales [8].

# 6 Concluding Remarks.

We have studied the longitudinal dynamics of a high energy coasting beam moving in a resonator. The coupled Vlasov equation for the longitudinal distribution function and the equation for the resonator voltage have been solved by closely following the method of Karimov and Lewis [5]. The key point of this method consists in the representation of the distribution function as a power series in the resonator potential. Further self-consistent stationary wave patterns have been found in the simplest equilibrium case of Maxwellian distribution in the energy error.

In the long wavelength (small wavenumber) limit a stability criterion for linear waves has been obtained and a Korteweg-de Vries-Burgers equation for the relevant hydrodynamic quantities has been derived.

An important (and interesting) extension of the results obtained here involves the longitudinal dynamics of a bunched beam. These will be reported elsewhere.

# Acknowledgements.

I would like to thank A. Aceves and P. Colestock for many helpful discussions concerning the subject of the present paper.

This work was supported by the US Department of Energy, Office of Basic Energy Sciences, under contract DE-AC03-76SF00515.

# References

[1] P.L. Colestock, L.K. Spentzouris and S.I. Tzenov, *"Coherent Nonlinear Phenomena in High Energy Synchrotrons: Observations and Theoretical Models."*, In **International Symposium on Near Beam Physics**, R.A. Carrigan and N.V. Mokhov eds., Fermilab, June 1998, pp 94-104.

[2] S.I. Tzenov and P.L. Colestock, *"Solitary Waves on a Coasting High-Energy Stored Beam."*, FERMILAB-Pub-98/258, Fermilab, September 1998.

[3] S.I. Tzenov, *"Formation of Patterns and Coherent Structures in Charged Particle Beams."*, FERMILAB-Pub-98/275, Fermilab, October 1998.

[4] I.B. Bernstein, J.M. Greene and M.D. Kruskal, Phys. Rev. **108**, (1957) p. 546.

[5] A.R. Karimov and H.R. Lewis, Phys. Plasmas, **6**, (1999) p. 759.

[6] H. Washimi and T. Taniuti, Phys. Rev. Lett., **17**, (1966) p. 996

[7] Lokenath Debnath, *"Nonlinear Partial Differential Equations for Scientists and Engineers."*, Birkhauser, Boston, 1997.

[8] A. Aceves, To be published.

# Space-Charge Impedance Calculations in Long-Wavelength Approximation

Sergey S. Kurennoy

*Los Alamos National Laboratory, Los Alamos, NM 87545, USA*

**Abstract.** Space-charge impedance calculations for smooth vacuum chambers with an arbitrary cross-section and perfectly conducting walls are considered in the long-wavelength approximation, when $\omega b/(\beta\gamma c) \ll 1$, where $b$ is a typical transverse size. For the SNS beam energies $\beta\gamma \leq 1.8$, and the wavelengths are long when $\lambda \gg b$. Within the long-wavelength approximation, the fields can be found by solving a 2-D electrostatic problem. Two examples are presented: the space-charge impedance of screening wires (RF-cage) and of a ceramic chamber with inner metal stripes. In addition, we explore the transverse space-charge impedance of a circular pipe with account of betatron oscillations in a wide frequency range.

## INTRODUCTION

In relatively low-energy proton machines with high currents — e.g., spallation sources — the space-charge and resistive-wall impedances are typically dominant. They influence the beam stability in such machines much stronger than so-called geometrical impedances due to various discontinuities of the vacuum chamber such as RF cavities, insertion devices or cross-section variations. Since high-frequency resonances due to the chamber discontinuities can be dangerous for the long-term stability of the beam, their consideration is important for storage rings or colliders, but not for the synchrotrons and accumulator rings in spallation sources with very short beam cycles. The low-frequency (inductive) geometrical contributions are usually small compared to the space-charge impedances for the low-energy accelerators. The positive side of this situation is that the space-charge and resistive-wall contributions to the impedance budget are easier to calculate in general, since these coupling impedances are distributed over the machine circumference and their estimates can be obtained with a good accuracy using a homogeneous model of the vacuum chamber.

Calculations of the coupling impedances due to the finite wall conductivity are described elsewhere, see in reviews [1,2] or papers [3,4]. Here we concentrate mostly on calculating the space-charge impedance for homogeneous chambers with an arbitrary cross section. We will demonstrate that in the long-wavelength regime, which is important for the low-energy machines like the Spallation Neutron Source (SNS), this problem can be greatly simplified by reducing it to an easily solvable 2-D electrostatic problem.

For a frequency of interest $\omega$ and a typical transverse size $b$ of the vacuum chamber, we consider the case of $\omega b/(\beta\gamma c) \ll 1$, where $\beta c$ is the beam velocity, and $\gamma = (1-\beta^2)^{-1/2}$. This condition contains two different physical cases: (i) the ultra-relativistic limit, when $\beta \to 1$ and $\gamma \to \infty$, and (ii) the low-frequency (or long-wavelength) limit, when the condition holds because of $\omega b/(\beta c) \ll 1$. In the last case, which corresponds to the parameter range of low-energy proton rings, the wave number $k = \omega/(\beta c)$ satisfies $kb \ll 1$, and corresponding wavelengths are large compared to the chamber dimension, $\lambda \gg b$.

# I  SPACE-CHARGE IMPEDANCE OF HOMOGENEOUS CHAMBERS

Let us consider a sheet current perturbation of the form $\rho(\vec{r})\delta(z - \beta ct)$ that propagates in the longitudinal direction $z$ with the speed $\beta c$. Here $\vec{r}$ belongs to the plane transverse to the $z$-axis. The transverse distribution $\rho$ is normalized to contain a charge $q$ by the relation $\int d\vec{r} \rho(\vec{r}) = q$.

For an axisymmetric case $\rho(\vec{r}) = \rho(r)$, where $r = |\vec{r}|$, one obtains the following equation for the corresponding Fourier harmonics $E_{z,\omega}(r)$ of the longitudinal electric field, see e.g. [1],

$$\left[\nabla_\perp^2 - \left(\frac{\omega}{\beta\gamma c}\right)^2\right] E_{z,\omega} = \frac{ik}{\gamma^2} \frac{\rho(r)}{\varepsilon_0 \beta c} . \tag{1}$$

Above and further we implicitly assume the same dependence $\exp(ikz)$ for all Fourier harmonics under consideration. In the long-wavelength approximation, when $\omega b/(\beta\gamma c) = kb/\gamma \ll 1$, Eq. (1) can be reduced to the following one,

$$\nabla_\perp^2 E_{z,\omega} = \frac{ik}{\gamma^2} \frac{\rho(r)}{\varepsilon_0 \beta c} . \tag{2}$$

The equation (2) is, up to normalization, an equation for a 2-D electrostatic potential produced by the charge distribution $\rho(r)$, and therefore can be solved by electrostatic methods.

## A  Longitudinal Impedance

It is convenient to write the longitudinal space-charge impedance per unit length of the vacuum chamber in the form

$$\frac{Z(\omega)}{L} = \frac{i\omega Z_0}{2\pi c} \frac{1}{\beta^2\gamma^2} g_L , \tag{3}$$

where $g_L$ is the so-called $g$-factor. As defined here, it is equal to $\ln(b/a)$ for a smooth circular pipe of radius $b$ with perfectly conducting walls, with $a$ being the beam radius. One should note that the value of $g_L$ depends on the current density distribution in the beam, i.e.

$$g_L = \begin{cases} \ln(b/a) & \text{for a hollow thin beam of radius } a, \\ \ln(b/a) + 1/4 & \text{for a uniform beam of radius } a, \\ \ln(b/a) + 1/2 & \text{the same, using } E_z(r=0) \text{ instead of } E_z(r) \text{ averaged.} \end{cases} \quad (4)$$

For the case of a hollow thin circular beam of radius $a$, which corresponds to the first line in (4), the transverse charge density in Eq. (1) is $\rho(r) = \delta(r-a)/(2\pi a)$. We will use this particular distribution in the following calculations. As follows from Eq. (1), the space-charge factor $g_L$ in Eq. (3) for a homogeneous vacuum chamber with an arbitrary cross section $S$ can be found by solving the following 2-D Dirichlet boundary problem in $S$

$$\nabla^2 \psi = -\delta(r-a)/a ; \qquad \psi|_{\partial S} = 0 . \quad (5)$$

In other words, a solution $\psi$ to the Eq. (5) is an electrostatic potential created by the charge $\lambda = 2\pi\varepsilon_0$ per unit length of the beam, evenly distributed along the ring of radius $a$. The potential $\psi$ is normalized this way to be dimensionless, and its (average) value at $r = a$ gives us directly the $g$-factor, i.e. $g_L = \psi(a)$.

For some simple particular cases the boundary problem (5) has an analytical solution. However, it is rather simple to solve it numerically for an arbitrary geometry of the cross section, using one of numerous electrostatic codes, like POISSON [5] or the static solver in the MAFIA code package [6].

As a practical example relevant to low-energy machines like SNS or ISIS, we consider the space-charge impedance of an RF cage for screening the chamber walls from the intense beam. In the simplest case, the RF shield consists of metal wires placed symmetrically around the beam. Let $n$ wires be spaced azimuthally by the angle $\Delta\theta = 2\pi/n$ at some distance $d$ from the chamber axis, such that $a < d < b$. More precisely, our "wires" are rather the segments of the coaxial cylinder with inner radius $d$ and of the radial thickness $t$. For a circular chamber one needs to solve the problem (5) only in one sector $\Delta\theta$ with periodic boundary conditions. We apply the MAFIA static solver to this problem.

The results for the longitudinal screening factor $g_L$ are shown in Fig. 1, in the case of $b/a=10$ and $d/a=5$, for a few values of $n$. We used $a = 10$ mm in our computations. The $g_L$-factor depends on the subtended-angle fraction $\alpha$, which is defined as a fraction of the total azimuthal angle screened by all wires: $\alpha = nw/(2\pi d)$, where $w$ is the arc length covered by a single wire. The upper dot-dashed line corresponds to an unscreened chamber ($\ln(b/a)$, no wires), the lower dot-dashed line to completely screened one ($\ln(d/a)$, a "metal wall" at $r = d$). The star at $\alpha = 0$ is the numerical result without wires. In calculations we assumed the wire thickness $t = 2$ mm in the radial direction. The thin dotted line connects the results for wires with approximately square cross sections, $w \approx t$.

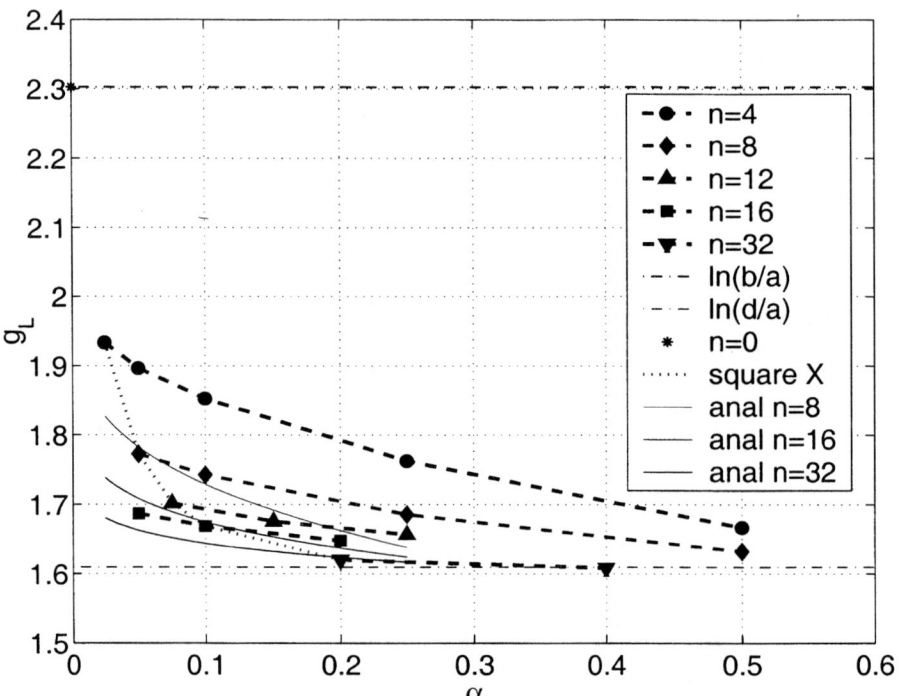

**FIGURE 1.** The longitudinal screening factor $g_L$ in circular chamber versus the subtended-angle fraction $\alpha$ for the varying number $n$ of axisymmetric wires (see the legend and text).

Analytical results have been obtained for an axisymmetric chamber with a large number of thin round wires, see [7] and references therein. The thin solid lines in Fig. 1 show the analytical results [7], which are valid for $\alpha \ll 1$ and $n \gg 1$. One can see a good agreement between the analytical approximation and our numerical results for wires with square cross sections, which means that for thin wires the shape of their cross section is not important.

To demonstrate some applications of our approach, we calculate the longitudinal screening factor $g_L$ for a few different layouts of the vacuum chamber. First, we insert an inner circular ceramic pipe in the circular chamber of Fig. 1. The ceramics inner radius is $d = 50$ mm, its thickness is $t_{cer} = 8$ mm, and the dielectric constant is $\varepsilon = 10$. In addition, we put $n = 8$ metal strips of thickness 1 mm on the inner surface of the ceramic pipe. The Figure 2 shows the results for $g_L$.

Also, for comparison, we consider a chamber with the square cross section $2b \times 2b$, with $b = 100$ mm, equal to the radius of the circular pipe. A square ceramic chamber with the inner cross section $2d \times 2d$ is inserted inside the metal one, again with $t_{cer} = 8$ mm and $\varepsilon = 10$. A few cases are studied, including a thin metal coating on the ceramics inner vertical sides, as well as $n = 8$ metal strips with

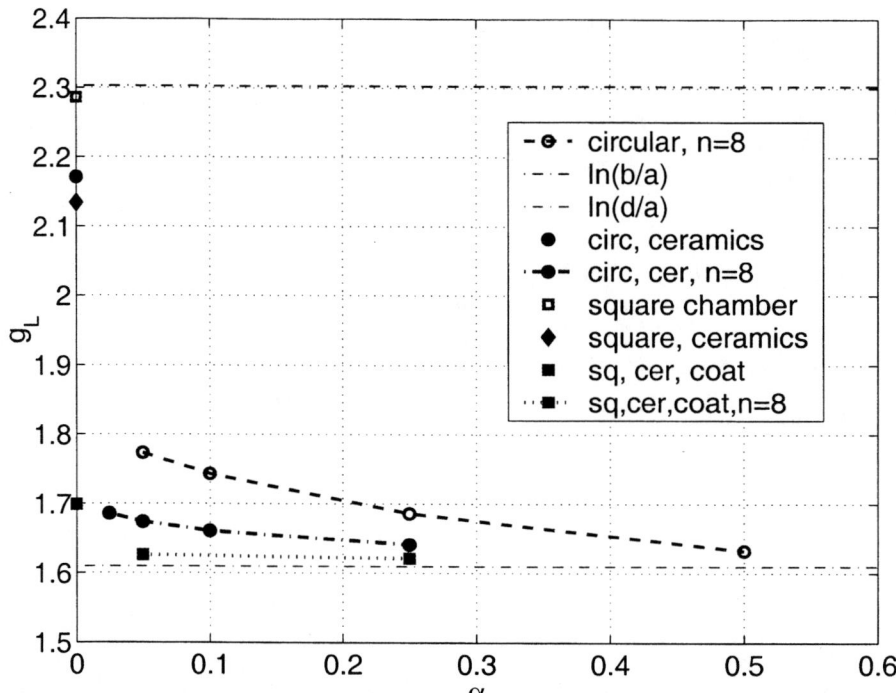

**FIGURE 2.** The longitudinal screening factor $g_L$ versus the subtended-angle fraction $\alpha$ for circular and square chambers with inserted ceramics and wires or strips (see the legend and text).

thickness 1 mm on the top and bottom inner surface of the ceramics. This last layout resembles the ISIS vacuum chamber [8]. Using an analogy to the subtended-angle fraction in the axisymmetric case, for the last layout we define $\alpha = nw/(2d)$, where $w$ is the strip width. One can see that inserting the ceramic chamber reduces $g_L$ by a few percent. The ceramics with metal stripes is certainly more effective. The strongest screening, of course, is provided by metal side coating plus stripes in the square case. However, the direct comparison with the circular pipe here is not quite fair, since the metal side coating by itself already covers 50% of the azimuth.

## B  Transverse Impedance

Similarly to the longitudinal case, the transverse space-charge impedance per unit length of a homogeneous chamber can be written as

$$\frac{Z_t(\omega)}{L} = \frac{iZ_0}{2\pi\beta\gamma^2}\frac{1}{a^2}g_T, \tag{6}$$

where $g_T$ is the transverse g-factor. It is equal to $1 - (a/b)^2$ for a smooth circular pipe of radius $b$ with perfectly conducting walls. Arguments similar to those in the beginning of this section show that for a hollow thin circular beam of radius $a$ the factor $g_T$ of a homogeneous vacuum chamber with an arbitrary cross section $S$ can be found by solving the following 2-D Dirichlet boundary problem in region $S$

$$\nabla^2 \phi = -\cos\theta \, \delta(r-a)/a \; ; \qquad \phi|_{\partial S} = 0 \; . \tag{7}$$

In a general case, the $g_T$ factor is calculated from the normalized dimensionless potential $\phi(r, \theta)$, which satisfies (7), as $g_T = 2 \int_0^{2\pi} d\theta \, \phi(a, \theta) \cos\theta$. For an axisymmetric case, $\phi(r, \theta) = \phi(r) \cos\theta$, so that $g_T$ is simply $\phi(a)$. A discrete approximation was used to simulate a cos-distribution of the charge in the rhs of Eq. (7): the ring region was split into a large number of segments ($m = 40$), each segment carrying an evenly distributed charge proportional to its $\cos\theta$.

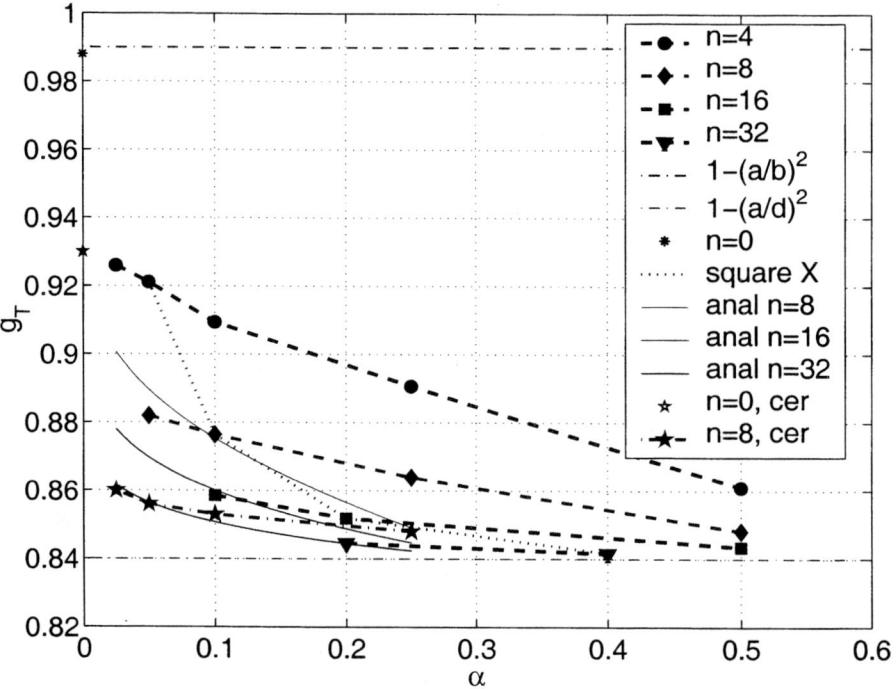

**FIGURE 3.** The transverse screening factor $g_T$ versus the subtended-angle fraction $\alpha$ for the varying number $n$ of axisymmetric wires (see the legend and text).

The results for the transverse screening factor $g_T$ are shown in Fig. 3, for the case when $b/a=10$ and $d/a=2.5$ ($a = 10$ mm was used in calculations). The radial thickness of the metal wires is $t = 2$ mm. The upper dot-dashed line, $1 - (a/b)^2$, corresponds to an unscreened chamber (no wires), the lower dot-dashed line, $1 -$

$(a/d)^2$, to the completely screened one (a metal wall at $r = d$). The star at $\alpha = 0$ is the numerical result without wires. The thin dotted line connects the results for wires with approximately square cross sections. The thin solid lines in Fig. 3 show the analytical results for $g_T$ [7], valid for $\alpha \ll 1$ and $n \gg 1$. Again, as well as in the longitudinal case, there is a good agreement between the analytical approximation and numerical results for wires with square cross sections. For comparison, we included some results for the ceramic pipe (inner radius $d = 25$ mm, ceramics thickness $t_{cer} = 8$ mm, $\varepsilon = 10$) with $n = 8$ evenly spaced metal strips of thickness 0.5 mm on its inner surface, see the legend.

## II  TRANSVERSE SPACE-CHARGE IMPEDANCE WITH ACCOUNT OF BETATRON OSCILLATIONS

Transverse beam oscillations with a small amplitude $d$, say, in the vertical direction (along $Oy$-axis, the transverse-azimuth angle is $\theta = \pi/2$) are simulated by introducing a surface perturbation of the beam charge density [9]:

$$\rho = \rho_n d \sin\theta \delta(a - r) \exp(ik_n z - i\omega_n t) ,$$

which propagates with the phase velocity $\omega_n/k_n = \beta c$, where $\omega_n = n\omega_0$ and $k_n = n/R$, with $2\pi R$ being the ring circumference. The corresponding current density is

$$\vec{j} = (j\sin\theta, j\cos\theta, \rho\beta_p c) ,$$

where

$$j = i\rho_n dk_n c(\beta_p - \beta)\Theta(a - r)\exp(ik_n z - i\omega_n t) .$$

Here $\beta_p c$ is the longitudinal velocity of the beam particles, which, in general, differs from the phase velocity $\beta$ of the perturbation wave. Taking into account betatron oscillations allows one to put the constraint $\beta = (1 \pm \nu/n)\beta_p$, where $\nu$ is the corresponding betatron tune.

For a smooth circular vacuum chamber of radius $b$ with perfectly conducting walls and a beam of radius $a$, one can derive (e.g., as a particular case from the equations (8)-(10) in [10]) the following expression of the transverse impedance

$$\frac{Z_t(\omega, \beta, \beta_p, r)}{L} = \frac{iZ_0 \gamma^2}{\pi a^2 \beta_p^2} \left\{ -\beta_1^2 + I_1(rx)K_1(rx)\left(\beta_1^2 + \beta_2^2\right) \right.$$
$$\left. + I_1^2(rx)\left[\beta_1^2 \frac{xK_0(x) + K_1(x)}{xI_0(x) - I_1(x)} - \beta_2^2 \frac{K_1(x)}{I_1(x)}\right] \right\} , \qquad (8)$$

where $\beta_1 = \beta_p - \beta$, $\beta_1 = 1 - \beta_p\beta$, $x = \omega b/(\beta\gamma c)$ is a dimensionless frequency parameter, $\gamma_i^2 = 1/(1 - \beta_i^2)$, and $r = a/b$; $I_n(z)$ and $K_n(z)$ are the modified Bessel functions of the first and second kind, respectively.

In the long-wavelength limit, where $x = \omega b/(\beta\gamma c) \ll 1$, the expansion of the Eq. (8) has the following form

$$\frac{Z_t(\omega, \beta, \beta_p, a/b)}{L} = \frac{iZ_0}{2\pi a^2} \frac{1}{\beta_p^2 \gamma_p^2} \left\{ 1 - \frac{a^2}{b^2} - \right.$$
$$\left. - \left(\frac{\omega b}{\beta \gamma c}\right)^2 \frac{a^2}{b^2} \left[\frac{1}{2}\left(\ln\frac{b}{a} - \frac{1}{2} + \frac{1}{2}\frac{a^2}{b^2}\right) + (\beta_p - \beta)^2 \gamma_p^2 \gamma^2 \left(\ln\frac{b}{a} + \frac{1}{2}\right)\right] + \right.$$
$$\left. + O\left(\left(\frac{\omega b}{\beta \gamma c}\right)^4\right) \right\}. \tag{9}$$

The first line in Eq. (9) is the well-known result for the transverse space-charge impedance of a circular pipe, and it turns out to be independent of the relation between $\beta_p$ and $\beta$. The lowest correction to this term in the long-wavelength limit is small, being of the second order in the small expansion parameter $\omega b/(\beta \gamma c)$. The correction is always negative, and the coefficient of $(\omega b/(\beta \gamma c))^2$ is minimal when $\beta_p = \beta$. One should note that the lowest frequency correction to the longitudinal $g$-factor is also negative and quadratic in frequency, see [11].

The expression (8) for the transverse space-charge impedance $Z_t$ is rather general and certainly valid far beyond the low-frequency range. To illustrate the frequency dependence of $Z_t$, we plot it in Fig. 4 after normalizing to its low-frequency limit $Z_{t0}$, given by the first line of Eq. (9). The results presented for a particular case of $\beta_p = 0.85$ and $a/b = 0.5$, in the SNS parameter range. For long wavelengths, which correspond to $kb \ll 1$ in Fig. 4, the corrections to the usual result $Z_{t0}$ are small and negative, in agreement with the expansion (9). However, at higher frequencies the value of $Z_t$ decreases significantly. Calculations show smaller frequency corrections for smaller values of the beam to pipe radius ratio $a/b$.

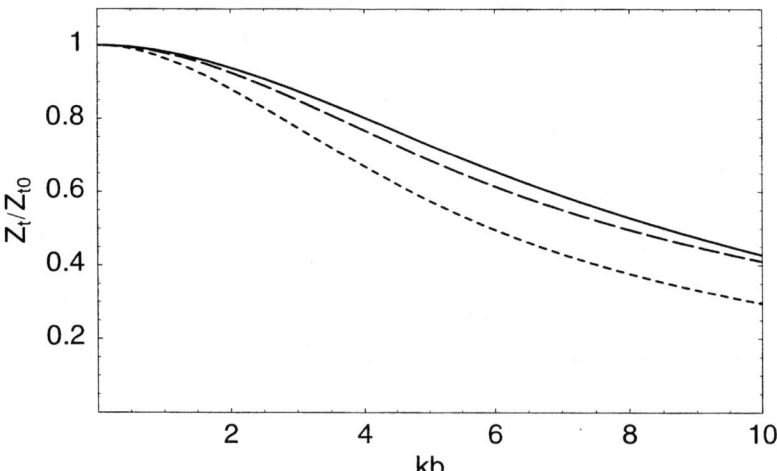

**FIGURE 4.** The normalized transverse impedance (8) versus the dimensionless frequency parameter $kb = \omega b/(\beta c)$ for fixed $\beta_p = 0.85$ and $a/b = 0.5$. The solid line is for $\beta = \beta_p$, long-dashed for $\beta = \beta_p + 0.05$, and short-dashed for $\beta = \beta_p - 0.05$.

To illustrate better how $Z_t$ depends on the relation between $\beta$ and $\beta_p$, we plot it in Fig. 5 for a few fixed values of the frequency parameter $kb$, again for $\beta_p = 0.85$ and $a/b = 0.5$. Obviously, in the low-frequency region the dependence is rather weak, except for very small values of $\beta$. One should note also that all three curves have maxima near $\beta = \beta_p$, that agrees with the lowest order correction in Eq. (9).

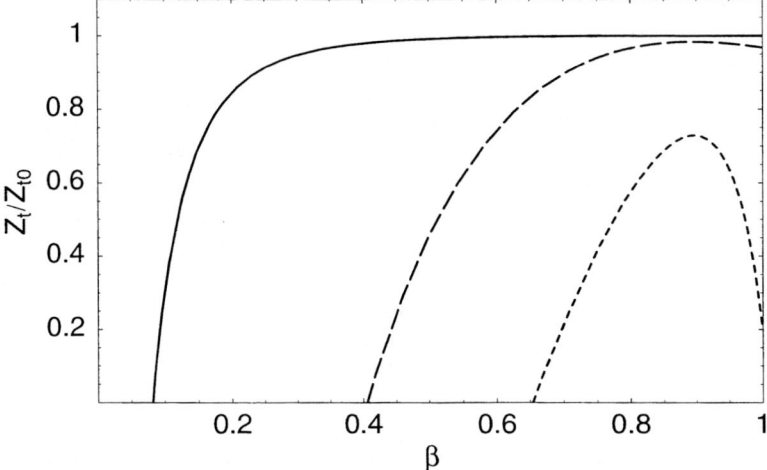

**FIGURE 5.** The normalized transverse impedance (8) versus $\beta$ for fixed values of $kb$=0.1 (solid), 1 (long-dashed), 5 (short-dashed). As in Fig. 4, $\beta_p = 0.85$ and $a/b = 0.5$.

## CONCLUSIONS

A simple method is developed to calculate the longitudinal and transverse space-charge impedances of homogeneous chambers with an arbitrary cross section in the low-frequency (long-wavelength) region. The condition of the method applicability is $\omega b/(\beta \gamma c) \ll 1$, where $b$ is a typical size of the chamber cross section, and $\omega$ is the frequency of interest. Obviously, it works as well in the ultra-relativistic limit, $\gamma \to \infty$. As examples of the method application, the space charge impedances of an RF-cage and of a ceramic chamber with inside metal stripes have been calculated in Sect. I. One should note that our approach assumes the perfectly conducting wires and stripes. We believe that contributions due to a finite conductivity can be added as perturbations.

A general expression of the transverse space-charge impedance of a circular pipe with perfectly conducting walls, that takes into account the beam betatron oscillations and is valid for any frequency, is presented in Sect. II. Its low-frequency limit is shown to produce the well-known result for the transverse space-charge impedance, independent of the relation between $\beta_p$ and $\beta$. The frequency corrections to that limit are derived.

The author would like to acknowledge useful discussions with Prof. R.L. Gluckstern and Dr. T.F. Wang.

## REFERENCES

1. B.W. Zotter and S.A. Kheifets, "Impedances and Wakes in High-Energy Particle Accelerators", World Scientific, 1998.
2. S.S. Kurennoy, *Phys. Part. Nuclei* **24**, 380 (1993); also CERN Report SL/91-31 (AP), 1991.
3. R.L. Gluckstern, J. van Zeijts, and B. Zotter, *Phys. Rev. E* **47**, 656 (1993).
4. S.S. Kurennoy, in *Proc. of EPAC* (Barcelona, 1996), p.1449.
5. K. Halbach, in *Proc. of the Second Intl. Conf. on Magnet Technology*, Oxford, UK (1967), p. 47; J.H. Billen and L.M. Young, "SUPERFISH/Poisson Group Of Codes". Report LA-UR-96-1834, Los Alamos, 1996 (revised 1999).
6. "MAFIA, release 4.00", CST (Darmstadt, 1997).
7. T.F. Wang and R.L. Gluckstern. Report LA-UR-99-1350, Los Alamos, 1999; presented at PAC99.
8. G.H. Rees, "Aspects of Beam Stability on ISIS", these proceedings.
9. B. Zotter, in CERN 77-13, p.175-218, Geneva, 1977.
10. S.S. Kurennoy and S.V. Purtov, *Part. Accel.* **34**, 155 (1990).
11. R.L. Gluckstern and A.V. Fedotov, "Analytic Methods for Impedance Calculations", these proceedings.

# Bunch Stabilization using rf Phase Modulation in the Intense Pulse Neutron Source (IPNS) Rapid Cycling Synchrotron (RCS)

J. C. Dooling, F. R. Brumwell, G. E. McMichael

*Argonne National Laboratory, Argonne, IL USA*

**Abstract.** Phase modulation (PM) is used to increase the current limit in the IPNS RCS. A device referred to as a scrambler introduces a small oscillating phase between the two RCS rf cavities at approximately twice the synchrotron frequency, $f_s$. The modulation introduced by the scrambler generates longitudinal oscillations in the bunch at $2f_s$. Modulations in the bunch are also observed transversely indicating a coupling between longitudinal and transverse motion. Comparing PM with amplitude modulation (AM), coupling to the beam is roughly equivalent at $2f_s$.

## I. Introduction.

The Intense Pulsed Neutron Source (IPNS) Rapid Cycling Synchrotron (RCS) delivers 450 MeV protons to a heavy-metal, spallation target at 30 Hz with an average current of 15 μA. Negative hydrogen ions are injected into the RCS at 50 MeV during a 70-80 μs period near the magnetic guide field minimum, $B_{min}$. A carbon stripper foil converts the injected beam into protons. The frequency of the rf voltage varies from 2.21 MHz to 5.14 MHz during the acceleration cycle[1]. Two rf cavities, located 180 degrees apart around the ring, bunch and accelerate the beam. During early operation, the achievable RCS current limit was substantially less than had been predicted. It was thought that an instability was causing the beam to grow in size transversely eventually leading it to strike the wall. Using a technique developed at KEK[2], phase modulation (PM) was added to the rf system to provide a small oscillation between the two RCS rf cavities at approximately twice the synchrotron frequency. The PM device, referred to as a "scrambler," was successful in increasing the RCS current limit[3]. Prior to installation of the scrambler, the current limit was 10-11 μA; with the scrambler, the current limit is now in excess of 15 μA. ($3.2 \times 10^{12}$ ppp). The PM induces a longitudinal disturbance in the bunch, which oscillates near the scrambler frequency. The bunch oscillation couples into the perpendicular plane making it visible on transverse monitors.

## II. Longitudinal Oscillations, Instability, and Stabilization

Fast Q signals from the beam in the RCS are detected with a toroid, then electronically assembled into a familiar "mountain range" display as shown in Figure 1. The mountain range in Figure 1a) occurs for a period just prior to the beginning of the phase modulation provided by the scrambler. The effect of the scrambler on longitudinal bunch shape is presented in Figure 1b). The PM signal begins approximately 10 ms after injection when the synchrotron frequency in the bunch is approximately half the modulation frequency. The predicted synchrotron frequency (excluding space-charge) is plotted in Figure 2; also shown in this figure are synchrotron frequency data obtained with a spectrum analyzer. The measured values fall below the calculation, indicating the defocusing effect of space-charge.

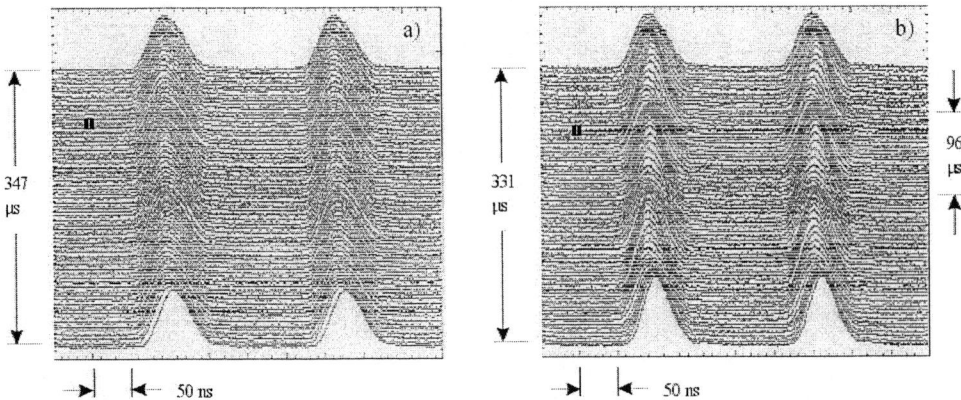

Figure 1: Mountain range display of the longitudinal bunch profiles a) before and b) during the scrambler.

The effect of the scrambler is significant with regard to both bunch dynamics and the current limit of the RCS. Without the scrambler, the maximum average current at 30 Hz is 10-11 μA; with the scrambler, in excess of 15.4 μA has been achieved. The reason for the current enhancement is believed to be the result of an increased energy spread in the bunch during quadrupole oscillations. The scrambler operating at approximately twice the synchrotron frequency drives the oscillations. The increased energy spread helps to slow the growth of destructive instabilities via Landau damping. Examples of bunch profiles just prior to extraction, with and without the scrambler, are given in Figure 3. The normal quadrupolar oscillations with scrambler are in evidence in Fig. 3a); however, with the scrambler turned off, a destructive, higher-order mode is seen in Figure 3b). Losses in the latter case are significantly higher and it is not unusual to spill half the beam.

## III. Observed Transverse Effects

As the proton beam is accelerated in the RCS, growing image currents perturb the bunch and initiate oscillations. It is known that without the scrambler, some of these oscillations grow into a destructive instability which quickly dumps 50-70 percent of the beam near the end of the cycle[3].

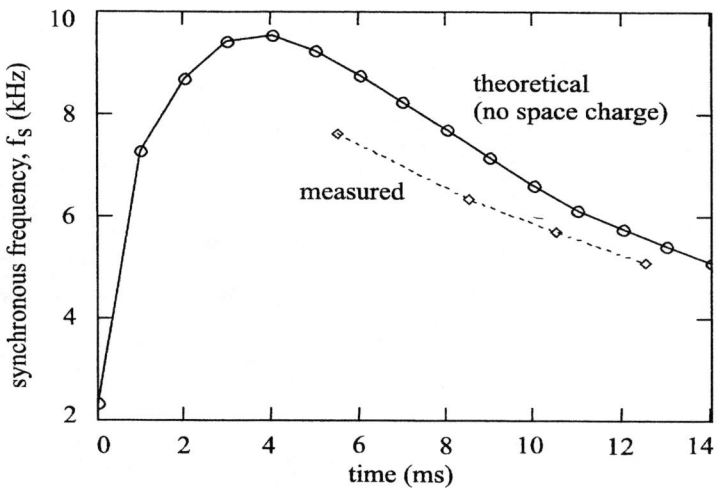

Figure 2: Synchrotron frequency in the RCS.

In addition to the scrambler, a second method is used to avoid the instability: early extraction. As indicated in Figure 4, extraction takes place approximately 2.7 ms prior to $B_{max}$ (t=16.67 ms). Though early extraction places greater demand on magnet supplies and rf voltage, it provides a more reliable extracted beam, by giving instabilities less time to grow. Transverse oscillations present in the RCS are most evident after initiation of the scrambler. The oscillations are observed in both loss monitors and split-ring electrode diagnostics described below.

A. Split-Ring ("Pie") Electrodes.

Split ring or "pie" electrodes are placed in a number of locations in the RCS vacuum chamber. Each pair provides horizontal or vertical beam position information. A spectral analysis of pie data has recently been performed to examine synchrotron frequencies in the RCS. A spectrum analyzer is gated to look at spectra from the pie electrodes during various periods in the RCS acceleration cycle. The gate period is held constant at 3 ms while the delay time is varied to cover the full range of the acceleration cycle as shown in Figure 4. Data from the spectrum analyzer is presented in Figure 5. Ostensibly, the scrambler supplies the perturbation, but in addition, discontinuities in the RCS beam pipe can act as a perturbing source via impedance mismatch and wakefield generation. Initially, the pie spectra are quiescent; however, as acceleration progresses, spectral features begin to appear close to the depressed fundamental and second harmonic synchrotron frequencies. Note that these occur prior to the scrambler. Because of the dynamic nature of the RCS, it would be desirable to examine the bunch spectra with a shorter gate. It is hoped that these tests can be conducted in the near future.

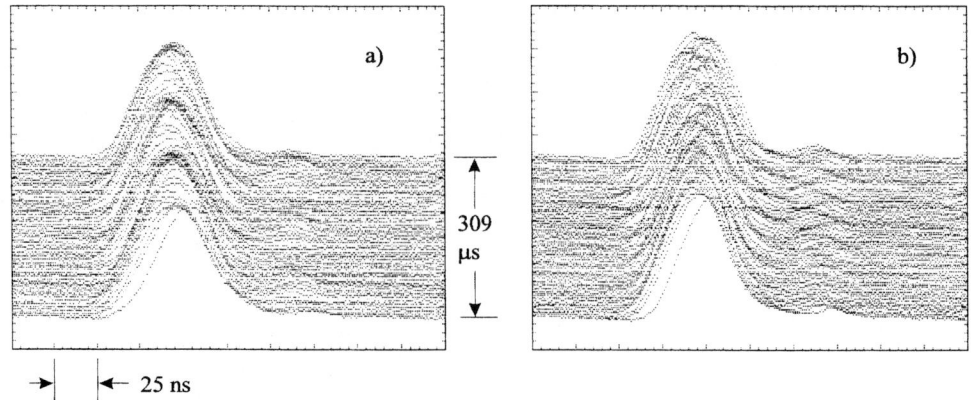

Figure 3: Mountain Range display beginning 13.55 ms after the end of injection, a) with the phase modulating scrambler and b) with the scrambler off.

B. Loss Monitors

The loss monitor detector front-end consists of a photomultiplier tube immersed in a bath of scintillating fluid. Loss monitors are placed at a number of locations in the RCS tunnel around the ring. After the scrambler is initiated, fluctuations are seen on the loss monitor signals. These fluctuations have a period very close to the longitudinal bunch modulations shown in Figure 1. An example of a loss monitor signal is presented in Figure 6; also indicated in the figure are the sample periods over which FFT analyses of the data are performed.

## IV. Analysis

### A. Longitudinal Equations of Motion

As mentioned above, the scrambler employs phase modulation (PM)[4,5] of the rf voltage to excite the RCS bunch longitudinally. The purpose of the scrambler is to broaden the energy spread and increase Landau damping of destructive oscillations that would otherwise grow resulting in large losses from the bunch. Longitudinal equations of motion, including PM and damping, may be expressed in terms of coupled difference equations (accurate to second order) as follows,

$$\phi_n = \phi_{n-1} + \frac{2\pi\eta}{\beta^2 E_s}\Delta E_{n-1} \qquad (1)$$

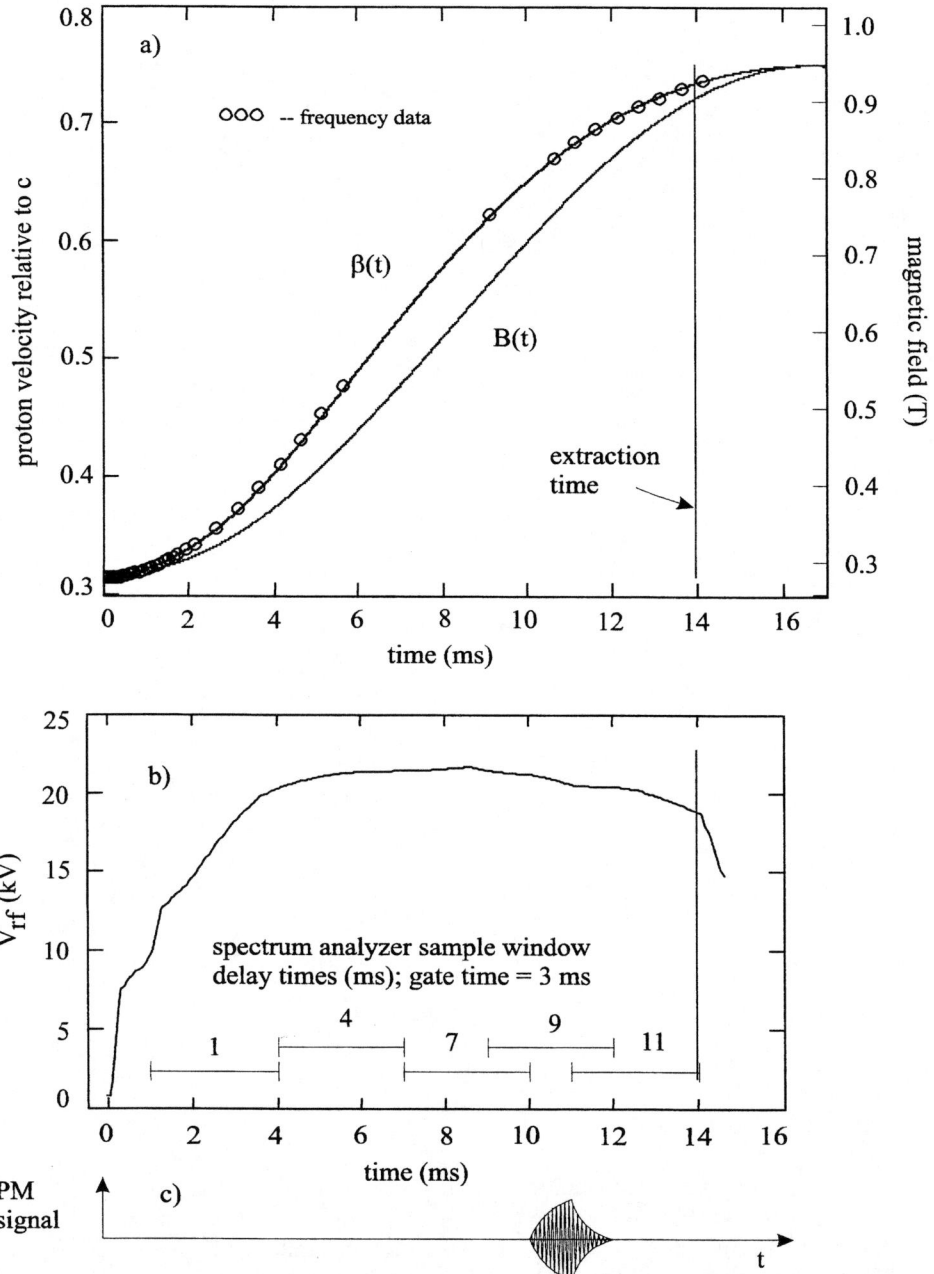

Figure 4: a) Magnetic field and proton velocity during the acceleration cycle, b) total applied RF voltage per turn with spectrum analyzer sample times on the S5 pie electrode indicated, and c) the scrambler pulse.

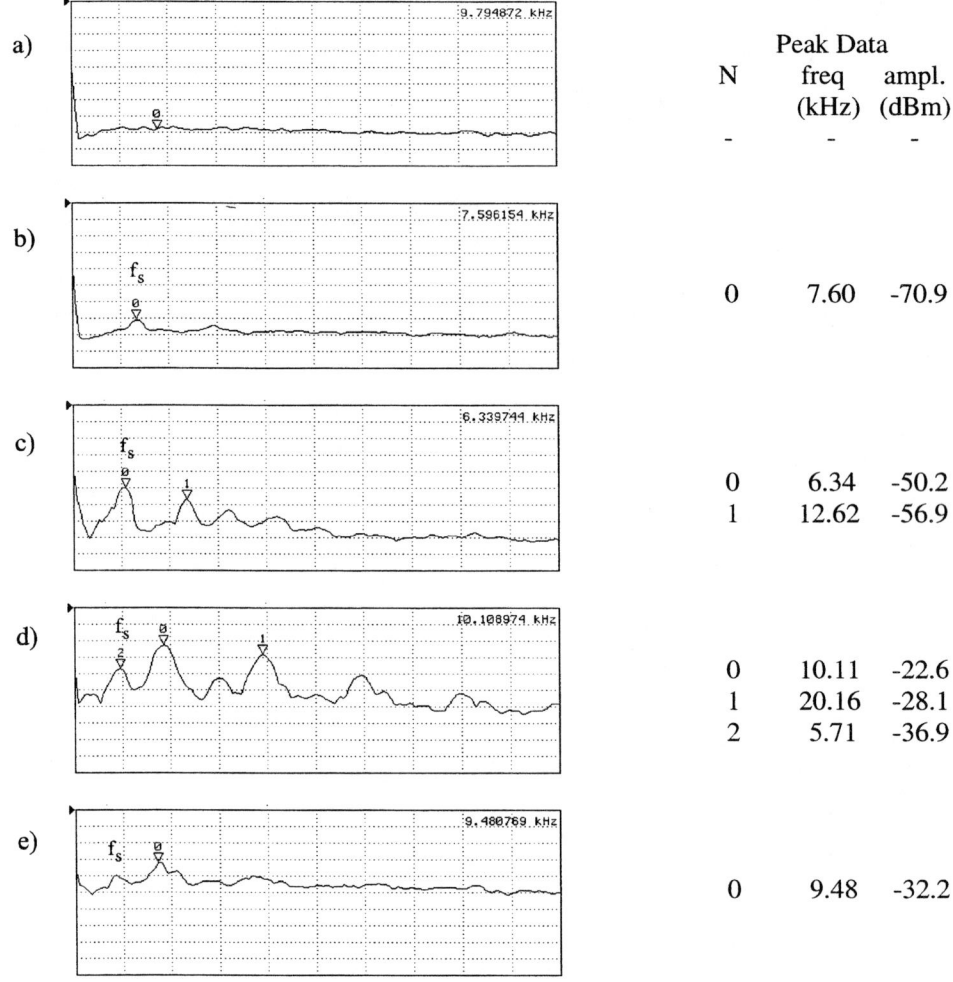

Figure 5: Spectrum Analyzer signals from horizontal split ring (or "pie") electrodes in the RCS, where a) through e) correspond to delay times of 1, 4, 7, 9, and 11 ms after injection, respectively; the gate width is 3 ms (see Fig. 4). Vertical axis: 10 dB/div.; Horizontal axis: start freq.: 1 kHz, stop freq.: 50 kHz. The approximate location of the synchrotron frequency is indicated.

$$\Delta E_n = \left(1 - \frac{4\pi\alpha}{\omega_{rf}}\right)\Delta E_{n-1} + qV(\sin[\phi_n + \varepsilon\sin(\nu_m f_s \tau_n + \theta)] - \sin\phi_s) \qquad (2)$$

where $\alpha$ is the damping rate, $\omega_{rf}$ is the rf radian frequency, $\nu_m$ is harmonic number of the modulation relative to the synchrotron tune, $\nu_s$ (i.e., for second harmonic PM, $\nu_m = 2$), the synchrotron frequency $f_s = \nu_s f_{rf}$, $\tau_n = n/f_{rf}$, and $\eta$ is the slip factor. The synchrotron tune is defined as,

$$\nu_s = \sqrt{\frac{h|\eta|qV}{2\pi\beta^2 E_s}} \qquad (3)$$

Equations 1 and 2 introduce PM through the voltage term rather than directly in the particle phase equation since bunch motion is affected through the applied field. The IPNS RCS accelerates a single bunch so the harmonic number, h=1. Acceleration is implied by a nonzero $\phi_s$; however, over the period of analysis it is assumed that $\beta$ and $E_s$ are constant.

Amplitude modulation (AM) of the rf voltage has also been studied for the purpose of exciting parametric oscillations in longitudinal bunch shape[6]. Experimental evidence of parametric oscillations in the AGS have recently been presented[7] Figure 7 shows the effects of comparable AM ($\varepsilon_{AM}$=0.018, see preceding references) and PM excitation ($\varepsilon_{PM}$=0.02 rad) near the second harmonic of the synchrotron frequency. In both cases, damping is neglected ($\alpha$=0). The modulation depth for AM is made equivalent to a 0.02 radian shift in PM, equivalent to the value presently used by the RCS scrambler at a synchronous phase angle of 0.472 radians; i.e., $\Delta V/V = \Delta\phi\cos(\phi_s) = 0.018$.

## B. Transverse Coupling

The appearance of both transverse and longitudinal oscillations at roughly the same frequency indicates a coupling between longitudinal and transverse motion in the bunch. Such a coupling can arise from dispersion into the horizontal plane or from intrabeam scattering caused by space-charge into either transverse directions. Scattering drives both longitudinal and transverse temperatures toward equilibrium. Assuming that the total energy in the bunch is conserved, then one can write[8],

$$\gamma m \overline{v_x^2} + \gamma m \overline{v_y^2} + \gamma^3 m \overline{(\Delta v_z)^2} = K \qquad (4)$$

where K is a constant. Assume, for simplicity, a KV distribution in four-dimensional transverse phase space[9],

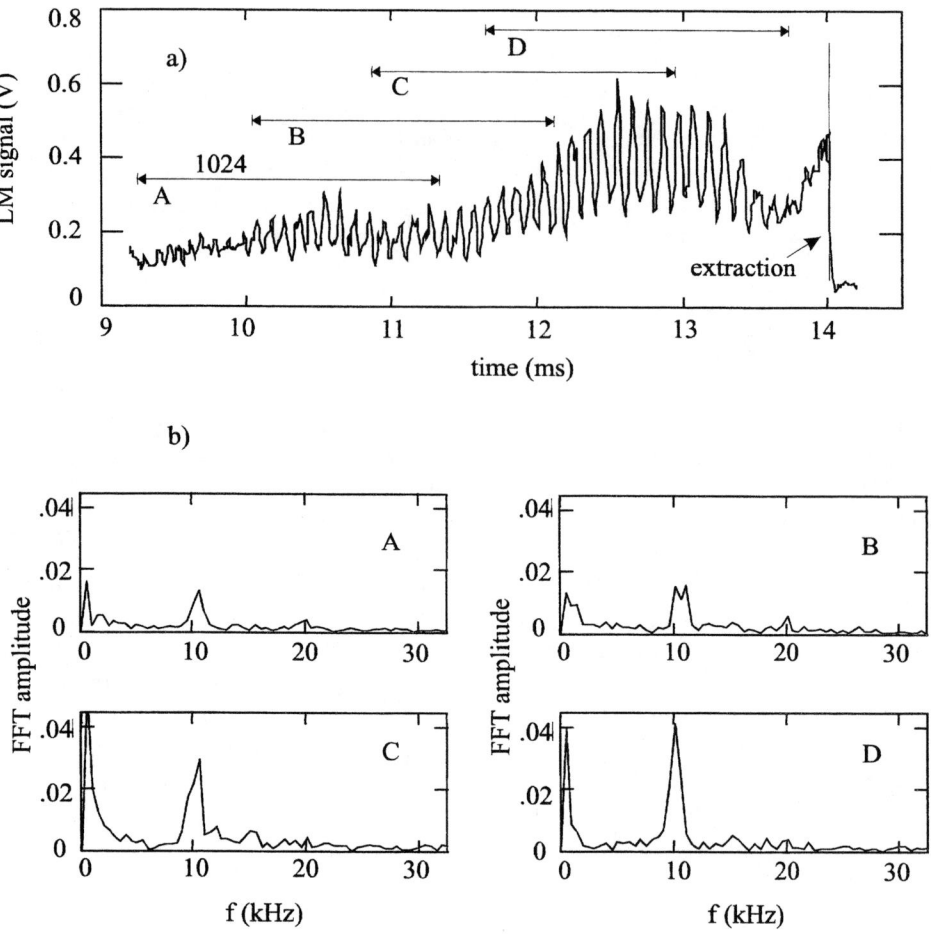

Figure 6: a) Loss monitor data indicating sample windows for 1024-word FFT and b) selected FFT spectra for times shown in a).

$$\frac{a^2}{x^2} + \left(\frac{ax'}{\varepsilon_x}\right)^2 + \frac{b^2}{y^2} + \left(\frac{by'}{\varepsilon_y}\right)^2 = 1 \tag{5}$$

where the x and y beam sizes and emittances are a, b, $\varepsilon_x$, and $\varepsilon_y$, respectively; and $x'=v_x/v_z$ and $y'=v_y/v_z$. As PM excites a quadrupolar oscillation in longitudinal phase space (as suggested from Figs. 1 and 2), $\Delta E$ and hence the average of $(\Delta v_z)^2$ oscillate as well. Since total energy is conserved, the average perpendicular energy must vary

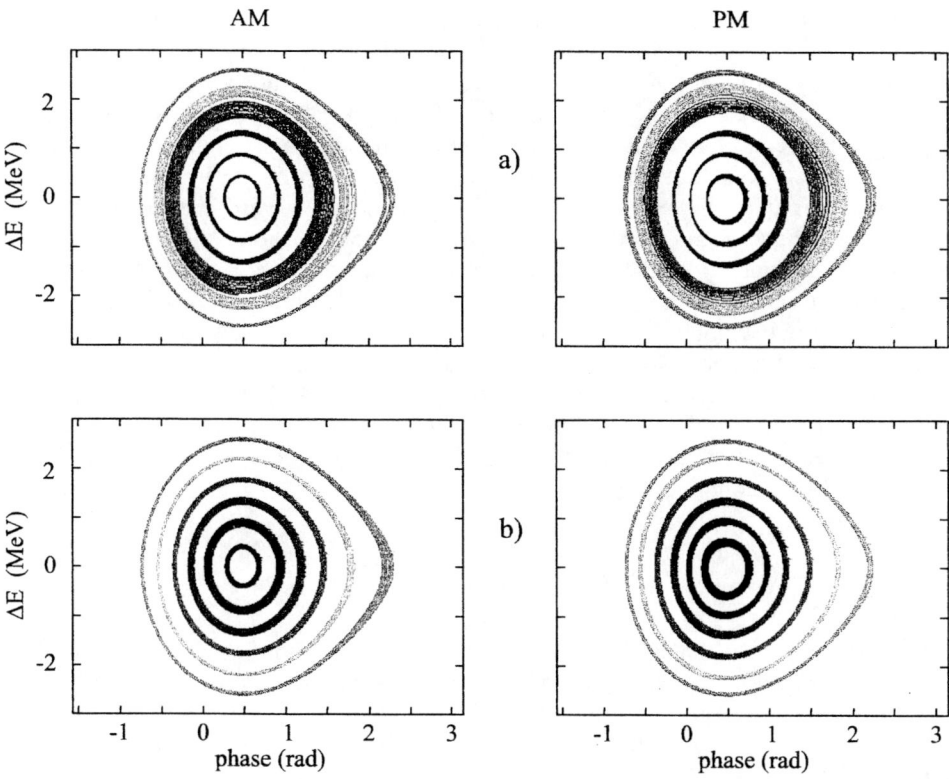

Figure 7: Comparison of six AM (left) and PM (right) phase-space trajectories for a) $\nu_m=1.8$ and b) $\nu_m=2.0$, with $\varepsilon_{AM}=0.018$, $\varepsilon_{PM}=0.02$ rad, $f_s=6.6$ kHz, and $\phi_s=0.472$ rad. The starting phase-space coordinates of the six trajectories are repeated in each of the four cases.

in accordance with Eq. 4. If transverse emittances or x' and y' vary, then to satisfy Eq. 5, it is reasonable to expect transverse beam sizes will also be affected. Measured transverse emittance in the beam extracted from the RCS does show an increase with injected charge or circulating current, suggesting that space-charge induced intrabeam scattering may be at work; however, other factors may also contribute to the rise in emittance.

## V. Discussion

A figure of merit for synchrotrons is the charge per phase-space that can be accelerated successfully. In the RCS, the peak phase-space-charge density occurs 7 ms after injection at a value of $11.0 \times 10^{12}$ protons/eV-sec ($3.2 \times 10^{12}$ p/0.29 eV-sec).

The high space-charge phase density leads to a substantial reduction of the synchrotron frequency in the bunch as indicated in Figure 2.

From the preliminary analysis presented here, the RCS bunch appears to have a synchrotron frequency just above half that of the scrambler when the PM signal begins. The second harmonic initially couples to the outermost edges of the bunch where the local synchrotron frequency is lowest. As time progresses and the synchrotron frequency decreases, the oscillation moves toward the inner phase-space of the bunch. This effect is illustrated in Figure 7 for both AM and PM. For both types of modulation, the effect on the outer trajectories is roughly equivalent. The modulation varies the energy amplitude of the trajectories within the stable phase-space of the bucket and would presumably extend to all charged particles within this band of phase-space in the bunch.

Presently, we are adding AM and PM to the CAPTURE_SPC analysis[10,11] to determine if quadrupolar oscillations in the bunch can be generated when space-charge and acceleration are included. In addition, these effects will be examined in the presence of second harmonic rf[12].

## Acknowledgements

The authors wish to thank Dr. S. Y. Lee for his helpful comments and the IPNS Operations Staff for their continued dedication and support.

This work is supported by the United States Department of Energy under contract no.: W-31-109-ENG-38.

## References

[1] T. W. Hardek, and F. E. Brandeberry, IEEE Trans. Nuc. Sci., **26**(3), 3021(1979).
[2] Y. Mizumachi and K. Muto, IEEE Trans. Nuc. Sci., **28**(3), 2563(1981).
[3] C. Potts, et al., IEEE Trans. Nuc. Sci., **32**(5), 3107(1985).
[4] M. Ellison, et al., Phys. Rev. Lett., **70**(5), 591(1993).
[5] H. Huang, et al., Phys. Rev. E, **48**(6), 4678(1993).
[6] Li et al., Nuc. Instrum. Meth., **364 A**, 205(1995).
[7] M. Bai, et al., Proc. of the 1999 Particle Accelerator Conf., New York, March 28-April 2, 1999
[8] M. Reiser, *Theory and Design of Charged Particle Beams*, Wiley, New York, 1994, p. 531.
[9] Ibid, p. 61.
[10] Y. Cho, E. Lessner, K. Symon, Proc. European Particle Accelerator Conf., p. 1228, (1994).
[11] E. Lessner and K. Symon, Computational Accelerators Physics, Williamsburg, Virginia, September 1996, AIP Conf. Proc. 391, p.185 (1997).
[12] J.C.Dooling, et al., Proc. of the 1999 Particle Accelerator Conf. New York, March 28-April 2, 1999

# Comment to the Kinematics of e-p Multipactoring

## S. Heifets, G. Stupakov

*Stanford Linear Accelerator Center, Stanford University, P.O. Box 4349, Stanford, CA 94309*

## V. Danilov

*Oakridge National Laboratory*

**Abstract.** Simple comments are given on the mechanism of e-p multipactoring. We emphasize that beam instability is a factor which may be important for this effect. The comments are results of the discussion which authors have during the Workshop.

It was emphasized during the Workshop that the beam induced multipactoring may be important in hadron machines. Indeed, the avalanche multiplication of electrons in the beam pipe is well known [1] and has been extensively studied recently in positron storage rings [2]. The electron cloud may lead to coherent transverse oscillations in the long train of bunches with the growth rate by an order of magnitude larger than that of conventional instabilities. The flux of electrons to the walls leads also to additional heating of the beam pipe and out-gasing of neutrals. The last process may deteriorate pressure but, at the same time, provides additional mechanism of surface cleaning.

Primary electrons are generated by synchrotron radiation or in the inelastic collisions with residual gas. They can be accelerated by the electric field of a bunch and produce secondary electrons provided their energy exceeds threshold energy $E_{th}$.

Secondary electrons have energy distribution centered at $E \simeq 5 - 10$ eV with width also of a few eV. A gap in the train provides cleaning if its length $L$ is large enough to let secondary electrons reach the wall at radius $b$, $L > b\sqrt{mc^2/E} \simeq 30$ m for $b \simeq 10$ cm.

If bunch spacing is smaller, then secondary electrons get a kick from the beam and can produce the next generation of electrons. The threshold energy $E_{th}$ may be defined as energy where yield of secondary electrons exceeds one. The energy dependence of the yield $\eta$ is more or less the same for all metals: it growth initially and then saturates at the level $\eta \simeq 1.2$ at $E > 150 - 200$ eV. Aluminum has exceptionally low $E_{th} \simeq 50$ eV and high maximum yield $\eta \simeq 1.5$ at $E \simeq 400$ eV. The process may be enhanced if time of flight to the wall is multiple of the bunch spacing.

Acceleration by the beam in the lepton machines with short bunches can be described as a kick. The kinematics in hadron machines is different because bunches are long. As a result, electrons trapped in the potential well of the beam oscillate with frequency $\omega/c = \sqrt{2\pi n r_0}$ where $n$ is the local charge density in the beam. This frequency is high, and electron can make $10-20$ oscillations within the length of the bunch.

In this case, longitudinal variation of the bunch density $n(s)$ is adiabatic compared to the fast oscillations. Hence, energy and amplitude of oscillations $A$ of a trapped electron changes along the bunch while $\omega A^2$ remains (approximately) constant. Hence, $A \simeq n^{-1/4}$, and electron energy $E \simeq \sqrt{n(s)}$. Therefore, electrons trapped in the head of the bunch with large amplitude and low energy increase their energy approaching the center of the bunch. The maximum energy can be quite high, much larger than $E_{th}$. However, due to adiabaticity, amplitude and the energy go to their initial values in the tail of the bunch. Such low energy electrons can not multiply. This is totally opposite to the situation in the positron machines.

Situation may be drastically different if adiabaticity is approximate or does not valid at all. We want to emphasize that the effect of the trapped electrons on the beam may provide a mechanism to break adiabaticity. Indeed, interaction of the electrons of a cloud and the beam may lead to the two-stream instability [3] and distort distribution of the bunch. Even small ripples, especially enhanced to the tail of a bunch, may lead to violation of adiabaticity.

It is important, therefore, to consider the problem of multipactoring as a self-consistent dynamic problem where distribution of the beam is affected by the electron cloud. Computer simulations should include this effect to produce a reliable results.

This work was supported by the US Department of Energy, Office of Basic Energy Sciences, under contract # DE-AC03-76SF00515.

## REFERENCES

1. see, for example, O. Grobner, Beam Induced Multipacting, LHC Project Report 127, May 1997
2. Proceeding of the International Workshop on Collective Effects and Impedance for B-factories, (CEIBA-95), Tsukuba, 1995
3. D.G. Koshkarev and P.B. Zenkevich, Resonance of Coupled Transverse Oscillations in Two Circular Beams, Particle Accelerators, (1972), V. 3 pp 1-9
   D. Sagan and A. Temnykh , Observation of the coherent beam-ion interaction in CESR storage ring, Nuck. Inst, and Methods A 344 (1994), pp. 459-469

# SUMMARIES OF WORKING GROUPS

# Impedance group summary*

M. Blaskiewicz (secretary), J. Dooling, M. Dyachkov, A. Fedotov,
R. Gluckstern, H. Hahn, H. Huang, S. Kurennoy, T. Linnecar,
E. Shaposhnikova, G. Stupakov, T. Toyama, J.G. Wang,
W.T. Weng, S. Y. Zhang, B. Zotter (chairman)

The impedance working group was charged to reply to the following 8 questions relevant to the design of high-intensity proton machines such as the SNS or the FNAL driver. These questions were first discussed one by one in the whole group, then each one of them assigned to one member to summarize. On the last morning these contributions were publicly read, re-discussed and re-written where required - hence they are not the opinion of a particular person, but rather the averaged opinion of all members of the working group.

1) **High intensity rings require large apertures. Are impedance calculations reliable for large vacuum chambers, large steps, and large aperture kicker magnets?**

- No problems for large steps or vacuum chambers (dimensional scaling).

- With advances in analytical and numerical methods these problems are well understood.

- Computer mesh codes such as MAFIA (3D), HFSS (1D and scattering matrix), and ABCI (cylindrical symmetry, easy, fast) yield very good agreement with measurements.

- For kickers different aspects are calculated with different codes: MAFIA or HFSS for impedance, EM properties using PSPICE.

- problems:

  Behavior less well understood for $\beta < 1$ in low $Q$ structures (High $Q$ are no problem);

  Multi-layer structures like coatings, stripes, and wire cages can be calculated using a 2-D analysis when no transitions between different types of

---
*Work supported by the United States Department of Energy

chambers are present. When transitions are present the problem becomes much more difficult as e.g. the case of a change of pipe radius forming a cavity or iris.

## 2) Is it beneficial to have the rf shielding or vacuum chamber follow the betatron envelope?

- Both $Z_\|$ and $Z_\perp$ are reduced.

- For $Z_\perp$, benefits, if any, are unclear.

- For $Z_\|$, required rf voltage is reduced for given energy spread, but with realistic elliptical beams in elliptical pipes naive (circular beam and pipe) estimates may be optimistic.

- For machines where the beam remains in the machine for several synchrotron periods, a reduced $Z_\|$ can lead to Landau damping of coupled bunch modes.

- Once the cage is in place significant changes in optics are no longer possible.

- Cost and reliability must be considered.

## 3) What is the impedance of ceramic chambers with or without metallic strips?

- For constant cross section low frequency impedance can be calculated with 2D electrostatic codes or analysis. With aperture changes see question 1.

- The dielectric properties of the ceramic are not important in impedance calculation.

- Metallic stripes should be sufficiently thick to shield the fields also at low frequency. This determines their surface resistivity.

- Gaps between stripes are necessary to reduce eddy currents, but should be narrow enough to avoid charge accumulation on the ceramic.

- The stripes should be made of high-conductivity metal like copper to reduce the resistive wall impedance.

- The design of the metallic stripes requires a compromise between penetration of the applied magnetic field and the need for shielding to reduce space charge impedance and static charge buildup.

**4) Is it beneficial to reduce the broad-band impedance to a few Ohm for high intensity proton machines?**

- The low frequency, inductive part of $Z_{\parallel}/n$ leads to potential well distortion, bunch lengthening etc. via interaction with the stable bunch spectrum. This inductance may be useful in compensating the "capacitive" space charge impedance, but it is produced by ill-defined sources such as discontinuities or bellows which may also introduce resistance and resonances and thus may lead to instability.

- Steps and discontinuities like pumping ports lead to accidental cavities. Frequencies are important to $\sim 3$ times the beam-pipe cutoff frequency and can lead to instabilities with many nodes on long bunches.

- Higher order modes in rf cavities must be carefully measured and damped.

- It is recommended to go for a smooth vacuum chamber leading to a minimum impedance machine. In any case an accurate impedance inventory up to and beyond pipe cutoff is required.

- It is difficult and costly to shield retroactively.

**5) How do ferrite, window frame, C frame, traveling wave, and strip-line kickers compare in terms of impedance and engineering requirements?**

- Dominant parameter is the rise time requirement

    Lumped ferrite kicker preferred, has simplest construction and power supply;

    Traveling wave ferrite kicker is faster but more complex;

    Compromise - hybrid solution: lumped kicker with traveling wave power supply (IPNS);

    Strip-line kicker is very fast but expensive in power requirement $\rightarrow$ use for transverse dampers;

- General recommendations:

    Keep aperture as small as possible to minimize current/voltage, but consistent with beam size requirements;

    Kicker impedance should not drive design choice. An inductive impedance contribution is OK for operation below transition;

    If possible avoid ceramic beam tube which reduces aperture and may cause high voltage arcing;

Contra-indication: ferrite has a resistive impedance which may lead to instabilities and over-heating. If tests/calculations indicate requirement for shielding then use metallic stripes on inside a ceramic pipe.

- Impedance estimates:

    Handbook formulae for ferrite kickers are suspect/incorrect;

    No obvious difference in $Z_\parallel$ for C-type and window frame designs;

    Transverse impedances differ in horizontal and vertical directions, window frame with metallic shielding strips seems preferable;

    Expect a small contribution from a strip-line damper ($\sim$ one BPM).

- Measurements are important

    Impedance measurements to assure absence of resonances;

    wire measurement at design level current to determine heating.

## 6) Is it practical and/or useful to compensate the longitudinal space charge impedance?

- Preliminary tests of inductive inserts at the PSR have shown that longitudinal instabilities can result from high frequency resonances ($\sim$ 70MHz). Reducing the quality factor of the resonant modes might help.

- The longitudinal coasting beam stability diagram shows a large stable region for negative inductance which can be $\sim$ 10 times greater than the simplified "Keil-Schnell" circle criterion. However, resistance may lead to instability and ferrite inserts will introduce resistance.

- For storage times shorter than the synchrotron period, as in SNS, compensation is less valuable than increased RF voltage.

- For storage times $\gg$ the synchrotron period some benefits could be realized due to increased Landau damping of coupled bunch modes.

## 7) What are the best methods to measure longitudinal and transverse impedance?

- Reliable "estimates" can be obtained if the right methods and correct interpretation are used.

- To measure narrow resonances (high Q), the best method is bead pulling, which is valid for both $v = c$ and $v < c$.

- Low and distributed longitudinal impedances (kickers, BPMs, bellows ...)

    forward transmission coefficient S21 should be interpreted with the "log-formula" and NOT treated as lumped element in a transmission line

    use smallest wire consistent with mechanical stability so $\delta Z/Z \ll 1$.

    match input and output port to $50\Omega$ of instrument, preferably by tapered cone adaptors or with a resistive match at low frequencies. De-embedding by TSD or TRL techniques to get the impedance of the device from several scattering measurements is not easy.

    Measure device with attached transitions to beam tube as a single unit

- Low and distributed transverse impedances (kickers, BPMs, bellows ...):

    horizontal and vertical measurements require using same precautions as for longitudinal measurements;

    Single wire measurement coupled with Panofsky-Wenzel theorem is prone to error;

    "Lecher-line" (two-wire) measurement: smaller errors but requires broadband 180° hybrid;

    Two-wire measurement easiest with $100\Omega$ transmission line structure, note that closely spaced, narrow wires allow $\delta Z/Z \ll 1$.

- Direct measurement using a high intensity electron beam (test facility) worth considering if not too expensive.

## 8) What are the key impedance issues for high power, short bunch, and low loss machines?

- Losses in parasitic resonances can lead to significant heating by the beams with high power or short bunches.

- Compensation of space charge impedance can help to stabilize coupled-bunch modes. Naturally occurring inductances such as bellows and BPMs could be helpful, but resonant structures should be avoided (see question 4).

- A realistic impedance model - valid also at higher frequencies, not just $Z/n$, is required.

- Turn-by-turn simulation using macro-particles or a Vlasov equation solver should be used to predict stability limits.

- Transverse damper designers should consider the variation of bunch length throughout the cycle.

This concludes the findings of the impedance working group which were reported in the final session. The discussion of some of the points was quite animated, in particular the recommendation not to use inductive inserts for space charge compensation when additional resistive or resonant impedances are unavoidable.

In addition to answering these questions, a number of short presentations were given in the working group on related subjects:

- G. Stupakov (SLAC): Wall roughness impedance.
- H. Hahn (BNL): Coupling impedance of RHIC injection kicker.
- S. Kurennoy (LANL): Space-charge impedance in long wavelength approximation.
- J. G. Wang (ORNL): Calculation of longitudinal space-charge impedance.
- J. Dooling (ANL): Impedance calculation.
- M. Dyachkov (TRIUMF): RF screening by thin resistive layer.

In spite of their interest, the discussion of these subjects had to be kept very short due to lack of time, in particular when they had only marginal bearing on high-intensity proton machines, the topic of this workshop.

# Summary of Instabilities and Damping Group

Shane Koscielniak
TRIUMF, 4004 Wesbrook Mall, Vancouver, B.C. V6T 2A3, Canada

## 1 Summary

There are a number of new high intensity proton synchrotrons proposed: the SNS[6] at Oakridge and ESS[7] in Europe, as neutron sources; the JHF/NSP[9] in Japan, a multi-disciplinary facility sponsored by KEK and JAERI; the New FNAL Booster[5]; and the Proton Driver[4, 8] as first stage of a $\mu - \mu$ collider. In addition there are operating machines such as ISIS[1] at RAL, the Brookhaven AGS[3] and its Booster, and the IPNS[35] at Argonne with high intensity and beam power. Moreover, at the energy frontier the beam quality requirements of CERN LHC[10] will pose new and challenging operation modes for its injector chain comprising Booster, PS[12], and SPS[11]. All will encounter beam current limitations arising from beam instability; and the lower energy machines are also limited by space-charge collective effects. Though many anticipated problems are shared in common between these machines, the impetus for the theme of this workshop, "Instabilities of High Intensity Hadron Beams in Rings", comes from the SNS and Proton Driver which were the inspiration for the charge to the "Instabilities Working Group" to answer the following questions.

1. Stability against longitudinal microwave instability at high intensity: Is a large momentum spread sufficient to stabilize the beam? Does it occur below transition?

2. Transverse stability at high intensity: What is the effect of space charge?

3. What is an appropriate description of fast transverse instabilities for long bunches and large space charge tune shifts?

4. What is an appropriate description of fast instabilities for short bunches and very large space charge tune shifts?

5. What is the effect of uneven longitudinal phase space distributions?

6. Do space-charge stabilized "hot spots" exist and do they affect overall beam stability?

7. E-P instability: How to identify it and how to cure it?

8. What is the status of synchro-betatron resonances?

Very briefly, we summarize the answers as follows:

1. Though $\Delta p/p_0$ is important, so is the distribution function. Instability may occur below $\gamma_t$ if large enough $R$ or $L$, but usually the $C$ from space-charge is enough to stabilize. Misuse of the circle-criterion can give overly conservative tolerances for allowable machine impedance.

2. Space charge induced tune shift and spread causes the beam foot-print to span imperfection resonances. For coasting beams, the picture (Sacherer[39], Hofmann[62], Baartman[63], Machida[64]) is quite clear[1], it is the tune for the various *coherent* modes that matters. For the bunched beam case, the problem is more complicated and there is no good conceptual framework.

3. For coasting beams there is the transverse microwave instability. There was no consensus as how to include the space-charge effect self-consistently: should space-charge impedance and tune spread both be included, or just one, or neither. For bunched beams, there is the fast head-tail instability. Experimental evidence at ISR and numerical eigenmode analysis both suggest space-charge stabilization, but neither result was considered definitive.

4. Short bunches are an unusual occurrence in hadron machines and not a great deal is known. The number of excitable modes is likely to be small[40], but complicated by 3D coupling. PIC tracking and envelope models[55] may help increased understanding.

5. The term 'uneven' it presumed to mean large scale inhomogeneity arising from imperfect painting schemes. Possible effects are:

    - large seeding of coherent instability (rather than growing from noise),
    - halo formation by space-charge shocks (core-halo model) if very mismatched.

    If these effects are not important, then there is no need for phase space painting schemes.

6. It is not entirely clear what is meant by "hot spot". No strong evidence for 'hot spots' stabilized by space-charge. However, some theoretical evidence[38] (speculative) for soliton waves in coasting beams interacting with a resonator impedance. If hot spots do exist, one can imagine locally large betatron incoherent tune spread and possibility for loss on resonances.

---

[1]Provided there are no image charge forces.

7. Although a two-stream instability accounts for some features of e-p, the observed fast growth at PSR and AGS Booster seems to require an additional mechanism of electron multipactoring initiated by electron liberation at the wall by proton halo. This is contentious.

8. Status of synchro-betatron understanding is captured in the ICFA workshop in Madeira[41], 1993.

## 2 Introduction

Stability analysis proceeds as follows. The beam distribution ($\Psi$) and wakefield ($W$) are decomposed into steady state parts $\Psi_0$ and $W_0$ and small (perturbative) time dependent parts $\Psi_1$ and $W_1$. It is a prerequisite of the analysis to show that $\Psi_0$ exists and is consistent with $W_0$. Introducing the perturbations, one finds a growth/damping rate and/or tune-shift that are correct in the limit of $\Psi_1 \to 0$. Stability versus steady state requirements can generate conflicts as in the case of inductive inserts. It is worth noting that a pure space charge impedance (below $\gamma_t$) is *astable* and (in the absence of Landau damping) oscillations will continue indefinitely.

A. Hofmann[14] gave a concise compendium of classical instability theory of particle beams in rings. In particular he gave physical/intuitive *aide memoire* for deciding which upper/lower sidebands are stabilizing/destabilizing depending as above/below $\gamma_t$.

## 3 Longitudinal Coasting Beam and Fast Bunched Beam Instability

Microwave instability is ubiquitous above transition energy; instability seen in the KEK-PS[13] jut above $\gamma_t$, at CERN PS[12] during debunching, and anticipated in the SPS[11] with LHC beams were reported at this workshop.

### 3.1 Theory

The Keil-Schnell Theory is appropriate for coasting beam; and for bunched beams when the growth rate is much larger than the synchrotron frequency. Let $Z$ be the driving impedance. When the $\Re[Z] \gg |\Im[Z]|$, the 'circle-criterion' may be adopted. However, when $|\Im[Z]| \gg \Re[Z]$, the theory must be applied with care using the 'stability diagram' which is a graphical solution of the dispersion relation and depicts a 'threshold contour'. Typically, the stable region has a tear-drop shape aligned with the $\Im[Z]$ axis. The theory applies correctly both above and below transition, except the stability region flips about the $\Re[Z]$ axis. Below transition energy, $\gamma_t$, a pure capacitive impedance (e.g. space-charge) is stable; whereas above $\gamma_t$ a pure inductive (e.g. wall) impedance is stable. Though these facts have been known (by many) since the time of Keil-Schnell, nevertheless some of the community had forgotten and I. Hofmann[17] attempted a reminder in 1992.

### 3.1.1 Comments

In general, though the microwave instability is possible below $\gamma_t$, it is unlikely to be observed because the space-charge impedance $Z_{s.c.}^{\|} \propto 1/(\beta\gamma^2)$ is typically large at lower energy. For example, there is no longitudinal microwave instability in the Rutherford ISIS. However, a large inductive impedance below $\gamma_t$ could prove fatal to the particle beam.

Although the momentum spread $\Delta p/p_0$ of the beam is important, nevertheless the shape of the distribution function $\Psi_0(\Delta p)$ is very important as tails (i.e. smooth higher order derivatives) contribute greatly to stability. However, if there is non-zero $\Re[Z] > 0$, then how well the $\Im[Z]$ stabilizes the beam (particularly near the cusp of the tear-drop) is quite sensitive to the exact form of $\Psi_0$. Fortunately, there is a self-stabilizing effect: as a result of instability, the beam will acquire a single-sided tail and a new, larger threshold contour. However, depending on the form of the impedance (how it rolls off with frequency[17]), the tails can grow indefinitely – leading to beam halo.

It is worth pointing out that there is no mode-coupling in the Keil-Schnell theory, and this extension[2] is left as an exercise for the student.

## 3.2 Simulations

I. Hofmann[15] presented elegant simulation of coasting beam longitudinal microwave instability that:

- confirm the small amplitude instability theory beautifully,

- extend our understanding to the non-linear regime,

- agree with controlled experiments in ESR machine at GSI using a tuneable resonator impedance.

Let $D$ be the convective derivative. Then $D\Psi_0 = 0$ and in the linear regime $D\Psi_1 \propto \Psi_0 W_1$. In the non-linear regime evolution is proportional to the product of the perturbation and its wake: $D\Psi_1 \propto \Psi_0 W_1 + \Psi_1 W_1$. In this regime, one may see either saturation or linear growth depending as $Z$; also wave-breaking and sub/super-harmonic generation effects (reminiscent of the work of Colestock[43, 44, 45, 46]) are possible. Further, as is obvious, when $\Re[Z] > 0$ one sees a slow

---

[2]Some work in this direction by Colestock[44].

decrease in the beam mean energy due to a.c. losses. Saturation effects have been discussed analytically by Ng[18].

K.Y. Ng[5] also presented simulations of the microwave instability (far from transition energy). Although these simulations showed the formation of RF buckets with beamlets captured within, as is expected of a 'self-bunching' instability, the simulations showed other features not in agreement with theoretical predictions.

## 3.3 Microwave instability near transition

Several times (usually in connection with electron machines) it has been suggested to make short bunches by operating the ring with beam energy equal to the transition energy. In particular, the Proton Driver[4] for a $\mu$-factory needs short bunches just before extraction to the target. It has been recognized for some time that the chromatic dependence of the momentum compaction factor $\alpha_p = \alpha_0 + \alpha_1(\Delta p/p_0) + \ldots$ with $\alpha_0 = 1/\gamma_t^2$ must be included in the slip factor $\eta = \alpha_p - 1/\gamma_s^2$. Based on simulations, Ng[5] drew attention to the fact that applying coasting beam microwave theory to a bunched beam at transition is on shaky ground because the assumed steady state distributions $\Psi_0$ are not consistent with the steady state wakefields[3] $W_0$. Although beam particles cannot move in RF phase, and so the bunch shape and wake do not change, nevertheless particles do move in energy and so $\Psi_0$ shears linearly with time – there is no steady state $\Psi_0$ unless the instability is very fast. [This effect is well known in isochronous cyclotrons, and 'flat-topping' RF waveforms are superimposed to cancel the wake-effect.] Ng used additional harmonic waveforms to cancel $W_0$ and thereby artificially reproduce the (incorrect) theoretical assumption of a constant $\Psi_0$. When this was done, the simulations were still not in agreement with the theory of Bogacz[42] and Colestock[47].

### 3.3.1 Comments

Perhaps I. Hofmann's group could run these 'near transition' cases with their computer simulation programs. Further, theorists should also take a look at this condition once more.

### 3.3.2 Historical note

This omission of the steady-state effect parallels work (not close to $\gamma_t$) done during TRIUMF KAON. Ignoring the steady-state non-linear wake, analytic instability theory was used to show that a hollow bunched beam is stable under space-charge even though the beam current was so high that (parts of) the RF bucket had collapsed to zero. When, in simulations, the steady state wake was artificially

---

[3]Of course, for a coasting beam there is no steady state wake.

compensated to reproduce the same un-physical situation (no steady state wake) the beam became stable.

# 4 Steady State Impedance Compensation

Just as we detune the accelerating cavity fundamental resonance away from the the beam RF, to compensate the reactive component of beam-loading at the fundamental, so we can imagine[20, 21] to include additional passive elements in the ring to compensate the effect of *reactive* steady state wakes upon the beam *incoherent* motion.

In many machines built (ISIS, CERN PSB, AGS Booster, etc.) and planned (SNS, Proton Driver, JHF, etc.), the space-charge impedance produces severe potential well distortion and substantial reduction of RF bucket area below $\gamma_t$. For example, the proposed new FNAL Booster would have to double $V_{\rm rf}$ and this is an expensive proposition as power $\propto V^2/R$. This problem has led to the suggestion of Griffin[22] for an "inductive compensator". Ideally, the compensator should be tuneable to follows changes in the space-charge impedance, $Z^{\|}_{s.c.} \propto \ln(b/a)/(\beta\gamma^2)$. The beam momentum $\beta\gamma$ changes during acceleration, and the ratio of pipe to beam size $b/a$ varies during multi-turn injection. Candidate devices for an inductive compensator are ferrite and finemet loaded structures, and specially shaped RF resonators. Ng[5] discussed methods to reduce the a.c. hysteresis losses of ferrite structures by careful attention to the spin dynamics, etc.

## 4.1 Experiments

Several experiments with inductive inserts have confirmed that at least partial compensation of the space-charge impedance is achievable. At this workshop, we heard of experiments at KEK with finemet[19] and ferrite[2] at the LANL PSR. In the KEK experiments, an inferred increase in the incoherent tune was attributed to an insert. Measuring the longitudinal incoherent tune based on bunch length oscillations is a difficult procedure, and it was recommended to consider measuring $\Omega_s$ either by the head-tail instability or via a beam transfer function measurement. In the PSR experiments, there has been a definite increase in the slope of the curve "captured beam current versus peak RF voltage" and evidence ?? for a cleaner gap between the two ends of the bunch; both effects suggest reduction of space-charge. Regarding the effect on beam stability, the PSR experience is "no better, perhaps slightly worse".

## 4.2 Additional advantages

One possibly very useful application of the inductive insert is to cancel the longitudinal space-charge mismatch that occurs at transition energy by arranging

$Z^\|_{s.c.} = Z^\|_L$ at $\gamma_t$. This cancellation happens by good fortune in the CERN PS and ease of transition crossing has been attributed to this fact.

A more speculative benefit of the inductive insert may be to restore Landau damping of the dipole mode by restoring the incoherent frequencies closer to the coherent tune.

### 4.3 Comments

One must no overlook the possibly substantial conflict between using an inductive insert to compensate the steady-state wake an the possibility for microwave instability below transition energy. Do not overdo the compensation!

## 5 Slow Head-Tail Instability

The head-tail instability is important because it is so ubiquitous. At this workshop, observation of head-tail was reported at RAL-ISIS[1] (vertical, $m = -1$), CERN-PS[12] (horizontal $m = 6$ through $m = 10$, depending on chromaticity $\xi$, driven by resistive wall) and at KEK-PS[13] ($m = 0, 1, 2$ modes close to extraction, driven by kicker magnet impedance and resistive wall). The instability is also anticipated at LHC, SPS with LHC beam, JHF/NSP and ESS.

### 5.1 Cures/avoidance

- Landau damping via octupole-induced tune spread (LHC, CERN-PS, KEK-PS) with octupoles of the correct sign. Easier if separation between coherent and incoherent tune shifts is small.

- Use natural chromaticity (close to zero).

- Betatron tune jumping – use for narrowband impedance when tune is just below the integer. Because fast quadrupoles are required, there are eddy currents to consider unless a ceramic chamber is used[1].

- Reduce wall impedance.

- Careful attention to kicker impedance – matched terminations, etc.[9, 7]

- Dedicated narrowband feedback[9] – may be reactive or negative resistive. Berg[50] has drawn attention to a potential problem with feedback dampers.

- Horizontal-vertical betatron coupling by Skew quadrupoles (CERN PS)[31]

## 5.2 Theoretical advances

- Landau damping in both planes (with octupoles) gives synergistic effect[49, 10].

- Damping by coupling[31] with skew quadrupoles provided $\alpha_H + \alpha_V < 0$ where $\alpha$ are the growth rates $e^{\alpha t}$. This is experimentally confirmed in the CERN PS.

- General binomial distribution and Landau damping.[61]

## 6 Fast Transverse Instability

There is the transverse analogue of the microwave instability in coasting beams; and for bunched beams there is the head-tail instability. Hubner-Vaccaro formulated[57][58] the transverse microwave dispersion relation; the model contains chromatic spread of the betatron tune through slip factor $\eta$ and chromaticity $\xi$. Because these terms appear as $[(n - \nu_\beta)\eta + \xi]$ so it follows that the threshold contour of the instability contour does not simply flip at transition energy. The model also contains tune spread through amplitude dependence of tune as might be caused by octupoles.

Later the model was extended to include the transverse space-charge impedance and the transverse incoherent tune spread (for non K.V. beams). At the workshop there was greatly divergent opinion as to which of these effects should be included and justified – particularly because of the issue of coherent versus incoherent tune shift and uncertainty amongst some individuals as to the meaning/interpretation of transverse space-charge impedance.

S. Ruggiero drew attention to Möhl's paper[59] which, in an attempt to explain experimental experience at CERN PS Booster, claims that neither space-charge tune spread does nor the space-charge impedance (proportional to the difference[4] of incoherent and coherent tune shifts) enters into the dispersion relation (as a stabilizing mechanism) unless octupoles are also present. Later, Möhl extended this work to head-tail modes[60] in bunched beams. The community was not entirely comfortable with this interpretation and recommended further controlled measurements/experiments at the PSR and AGS, etc.

### 6.1 Fast Head-Tail Instability

The fast head-tail (HT) is observed in LEP, SPS[51] and other *electron* machines with short bunches, but not so far in hadron machines with long bunches; because the beam power spectra and its interaction[65] with the impedance differ between these cases. Other stabilizing mechanisms have been suggested. Y. Chin treated analytically the H-T with betatron tune spread[52] and synchrotron frequency

---

[4]In the case of a round beam in a round pipe, $Z \propto [1/a^2 - 1/b^2]$ with radii $a, b$ respectively.

spread[53, 54] (due to RF non-linearity) and found the beam stable for resonator impedance. $\Delta\Omega_s$ spread is usually more effective that $\Delta\omega_\beta$ spread.

### 6.1.1 Fast head-tail with strong space-charge

A. Hofmann reported experiments at the ISR where conditions were deliberately arranged to promote a fast head-tail: reducing $\Omega_s < 10\omega_\beta$ and set chromaticity positive. No HT was observed, and the stability was attributed to transverse incoherent tune spread arising from space charge.

Blaskiewicz reported posing the head-tail instability as an eigen-value problem with resonator and space-charge impedances included. He drew attention to problems with lack of convergence[32]; it is not clear if this is due to the choice of basis or impedance (inadequate "roll-off"), or both, but it does point to a need for careful use of basis functions. However, other work[33] suggests that longitudinal space-charge tune spread suppresses the instability. Ng[5] also reported numerical eigenmode analysis by Alexei Burov which indicate stabilization of the mode coupling up to some critical beam current at which point modes couple differently and there is instability.

Regarding the lack of convergence as the dimension of the eigenvalue matrix is expanded, it was pointed out that strong off-diagonal coupling implies the basis is too far from the true eigen-functions, and that the Legendre basis for parabolic beams (known to be exact solutions for space-charge) be used. Shaposhnikova noted that if coupling of azimuthal modes up to $m$ is important, then one must retain at least $m/2$ radial modes.

## 7 Transverse Instability in Short bunches

Comments of the 'Instability Working Group' are as follows.

In the absence of synchrotron radiation, short bunches are not a 'natural' state encountered for hadron beams. Even in the Proton Driver where longitudinal bunch rotation is used to shorten the bunches, these bunches are only short for 1/4 synchrotron oscillation and are then immediately extracted to a target in which case final beam quality is not terribly important.

One of the conclusions of the ICFA workshop[40] was that short bunches should suffer fewer problems because high order within bunch modes are harder to excite – assuming that the impedances are conventional and roll-off. This property can be seen either from the Sacherer/Laclare form factors or by appealing to the intuitive argument that a short bunch is more like a point charge. However, growth rates scale as the peak current (not average) and so viable modes will grow very quickly.

One can still anticipate transverse dipole and quadrupole modes. Because the bunches are short and non-relativistic there will be 3D coupling of the normal modes. The quadrupole (or envelope) modes may be treated using the model of Pabst[55, 56] developed for linacs.

Because space-charge impedances are normally derived for the limit of long bunches, and because of the anticipated 3D coupling, PIC models with finite-element solutions of Maxwell's field equations could be advantageous for understanding phenomena.

Although longitudinal potential well distortion is severe, and $\Psi_0^{\|}$ must be found self-consistently before instability analysis, there is little requirement to use a $\Psi_0^{\perp}$ much different from the single-particle case because transverse focusing is typically very strong: tune spread/shift of $\delta\nu \simeq 0.4$ must be compared with the integer part of the tune which is normally large.

## 8 E-P Instability

The electron-proton (or e-p) instability was first diagnosed in the CERN ISR[24] and LBL Bevatron[23]; and has been seen in the LANL PSR for a decade or more. Recently, the e-p instability has been observed[26] with coasting beams ?? in the AGS Booster at BNL. The e-p instability is considered a possibility at the proposed SNS, ESS and the Proton Driver. However, there are other high current machines notably RAL ISIS where where e-p has not been encountered.

### 8.1 Experimental results

The PSR experience is of a vertical[5] transverse instability resulting in fast beam loss, typically within about 600 $\mu$s. The momentum spread $\Delta p/p_0$ of the beam, which contributes Landau damping, is an important parameter as is also the vertical betatron tune, $\nu_V$. The instability proceeds from tail to head of the bunch. The instability displays thresholding. Longitudinal modulation of the beam current (i.e. gap) is important but mot essential to the instability. Recent experiments[2] indicate an anticipatory rise in the electron current before the instability really takes off. The AGS Booster experience[26] confirms the anticipatory electron current and an extremely fast growth rate of approx 20 $\mu$s. The fast growth rates indicate that although longitudinal current modulation may be important, synchrotron motion is irrelevant.

### 8.2 Electron sources

Multipactoring has recently been recognized as an important source of electrons. Two possible models of secondary emission yield (SEY) are being used in simulations and one should compare between them. Zhang[6] reported experimental data for electron production due to ion impacts versus impact angle and suggested to scale to protons. Certainly there is a case for large production rates.

---

[5]Typically, the e-p instability is vertical with the electrons describing helical motion around the magnetic field lines of dipole magnets.

Danilov[30] claims a 'natural' neutralization (arising from proton halo) of about $10^{-3}$ and that this is enhanced to 1% by SEY – though there is a strong dependence on pipe/beam radius.

## 8.3 Analytical models

There is continuing progress with theoretical models. The early model of Keil-Zotter[25] included the Landau damping effect. The later model of Neufer with rigid beams showed some of the signatures seen at the PSR. At this workshop, Davison[27] described a two-stream instability of fluid beams, but with no longitudinal momentum spread. The coupled Vlasov-Maxwell equations for two coasting beams with K.V. distributions is solved for arbitrary longitudinal and transverse mode numbers. Growth rates in the "ball park" of PSR are obtained, but the electrons are trapped inside the potential well of the proton beam and take 'longish' times to reach the wall.

### 8.3.1 Proposed directions for theorists

Davidson proposed the following directions for two-beam studies.

1. To use non-linear $\delta f$ simulations in combination with analytical estimates to determine the dependence of instability threshold, growth rate and nonlinear dynamics on:

   - Input ion and electron distributions
   - Density profile shapes
   - Transverse mode structure
   - Axial momentum spread
   - Fractional charge neutralization

2. To identify *control knobs* to eliminate or minimize deleterious effects of e-p instability at beam intensities of interest for PSR and SNS.

3. To understand key features of e-p instability both for continuous coasting beams, and for finite length bunches.

## 8.4 Numerical models

### 8.4.1 $\delta f$ method

H. Qin[28] reported results from the "BEST" computer program which solves the Vlasov-Maxwell equations of two (or more) coasting beams for arbitrary $\Psi_0$ using the $\delta f$ method which has the advantage that no computation time is wasted with $\Psi_0$. For the instability to develop, it is found that 'hot' electrons are required with the electron distribution function overlapping the (real space) proton eigen-mode which is concentrated where the radial derivative $d\Psi/dr$ is largest.

### 8.4.2 Macroparticle tracking

Wang[29] reported a model in which the proton beam is represented by transverse slices and macroparticles are substituted for electrons. The model includes electron liberation by halo protons and SEY arising from electrons striking the wall. The model predicts many qualitative phenomena in agreement with PSR observations; e.g. the instability proceeds from tail to head of the bunch.

Danilov[30] reported a slightly simpler model but with a different model for SEY. He finds multipactoring to be an important source of free electron production with electrons bouncing repeatedly across the beam pipe and not trapped in the potential well of the protons. A triangular modulation of the longitudinal beam current is assumed.

### 8.4.3 Comments on simulations

Suggest a detailed comparison of the predictions of Danilov's and Wang's codes using identical (or equivalent) parameters wherever possible.

Suggest to check fidelity of instability predictions against machines such as ISIS where there is known to be no e-p instability.

## 8.5 Comments

There is no doubt that a two-stream instability can account for some features of an e-p instability. However, there was quite strong disagreement as to what is the mechanism for the *observed fast growth*. The disagreement is partially about "where are the electrons liberated?" In fact, free electrons are produced everywhere: inside the beam by collision of protons with residual gas atoms/molecules and at the wall by impacts of the beam halo; so its more a question of what is the dominant process.

In the two-beam model[27, 28], electrons are initially trapped inside the potential well of the proton beam, the two beams start to oscillate, and when the electron oscillations grow to large amplitude they hit the wall and initiate multipactoring depending on the SEY.

In the halo-model[30, 29], electrons liberated at the wall by proton halo are not trapped and can bounce from side-to-side across the beam pipe; these electrons need a mechanism to increase their kinetic energy high enough for secondary emission to occur. One possible mechanism is shaking of the potential well (through e-p instability or initial coherent betatron motion). This mechanism might not be large enough unless there is some non-adiabatic effect to pump the electrons to kinetic energy > 50 eV. S. Heifets suggested such an effect arising from a longitudinal modulation of the beam current. What is needed is a ripple wavelength comparable with the electron bounce wavelength.

There was also some discussion of the possibility that the PSR 'e-p' instability could be a transverse microwave instability; this was linked to the debate of whether transverse space-charge impedance and incoherent tune spread should be included in the transverse microwave criterion.

## 8.6 Cures

- Surface conditioning to reduce the SEY of vacuum chamber wall, proposed for LHC[10].

- Clearing electrodes; possible only in straight sections and somewhat ineffective at LANL PSR.

- Solenoids – untried.

- Hard pumping or baking – helps clean up wall surface.

- Profiling of the vacuum chamber and the RF cage leads to less trapping of electrons because of the reduced potential well[1, 7].

- transverse feedback – as coherent lines 100-200 MHz appears on bpm signal spectra – untried.

# 9 Longitudinal Bunched Beam Instability

## 9.1 Coupled bunch

Longitudinal coupled bunch instability seen in the SPS[11] and new variations anticipated with new filling pattern for LHC. Problem is loss of Landau damping toward end of acceleration cycle. Have used 4th harmonic cavity in bunch shortening mode[34] to increase synchrotron frequency spread and promote damping; bunch lengthening mode is problematic. In recent years SPS staff have introduced a new technique to identify offending impedances: by observing mode growth while beam is allowed to debunch. Sources identified so far include septa and pumping ports.

A common problem in identifying driving impedance is that mode number only identified to within harmonic number $h$. For example, a (transverse?) instability in CERN-PS[12] could be $n = 2$ or $n = 6$ as $h = 8$. Shaposhnikova[34] reminded the community that the Sacherer/Laclare form factors contain a dependence on impedance frequency and within bunch mode number that can help settle this question; based on Taylor and asymptotic expansions, she gave two rules of thumb for identification.

## 9.2 Within bunch

Both at KEK PS[13] and IPNS RCS[35], bunches have been stabilized by increasing the longitudinal emittance; at KEK RF noise is used, whereas at IPNS phase modulation is employed. Similar methods are widespread.

Unless otherwise indicated, the references are to presentations reproduced in these proceedings.

# References

[1] G.H. Rees: *Aspects of Beam Stability on ISIS*

[2] R. Macek: *Status of the PSR e-p Instability*

[3] T. Roser: *High Intensity Beam Operation of the Brookhaven AGS*

[4] W. Chou: *Proton Driver*

[5] K.Y. Ng: *Some Stability Issues of Intense Beams*

[6] S.Y. Zhang: *Instability Issues at SNS Accumulator Ring*

[7] G.H. Rees: *Instability Issues for ESS Linac and Rings*

[8] K.Y. Ng: *Short Bunch Production Near Transition*

[9] M. Yoshii: *Beam Instabilities in the JHF Synchrotrons*

[10] F. Ruggiero: *Collective Instabilities in the LHC*

[11] T. Linnecar: *CERN SPS Impedance Issues*

[12] R. Cappi: *Collective Effects in the CERN-PS Machine*

[13] T. Toyama: *Head-Tail Instability and Microwave Instability in KEK-PS*

[14] A. Hofmann: *Overview of Beam Instabilities*

[15] I. Hofmann: *Nonlinear Features of the Longitudinal Instability*

[16] I. Hofmann: *Suppression of Microwave Instabilities*, Laser and Particle Beams, Vol.3, Part 1, pp. 1-8, 1985.

[17] I. Hofmann et al: *Impedances and Instability Studies at the ESR*, Proc. 3rd EPAC, Berlin, pp. 123-125, 1992.

[18] S. Bogacs & K.Y. Ng: *Nonlinear saturation of the longitudinal modes of the coasting beam in a storage ring*, Phys. Rev. D, **36**, No. 5, p. 1538-1542, 1987.

[19] K.Koba: *Longitudinal Impedance Tuner*

[20] A. Sessler & V. Vaccaro: *Passive Compensation of Longitudinal Space Charge Effects*, CERN 68-1, ISR Div.

[21] R. Cooper: *Passive Compensation of Beam Space Charge by Insertion of Ferrite Inductors*, PSR Tech Note 100, LANL, 1982.

[22] J. Griffin & W. Chou: *Impedance Scaling and Impedance Control*, Proc. IEEE PAC, Vancouver, BC, p. 1724, 1997.

[23] H. Grunder & G. Lambertson: Proc. 8th Int. Conf. HEACC, p.308, 1971.

[24] H.G. Hereward: CERN-71-15 (1971)

[25] E. Keil & B. Zotter: CERN-ISR-TH/71-58 (1971).

[26] M. Blaskiewicz: *The Fast Loss Electron Proton Instability.*

[27] R. Davidson: *E-P Instability in High Intensity Linacs and Storage Rings*

[28] H. Qin: *3D multispecies nonlinear perturbative particle simulation of collective instabilities in intense particle beams.*

[29] T.S. Wang: *A Simple Simulation of Electron-Proton Instability*

[30] S. Danilov: *A study on the Possibility to Increase the PSR E-P Instability Threshold.*

[31] E. Metral: *Stabilization of CPS Head-Tail Instabilities by Linear Coupling*

[32] M. Blaskiewicz: *Convergence of Basis Expansions*

[33] M. Blaskiewicz: *Fast head-tail instability with space charge*, Phys. Rev. ST Accel. Beams 1, 044201 (1998).

[34] E. Shaposhnikova: *Study of Coupled Bunch Instabilities*

[35] J. Dooling: *IPNS, Longitudinal oscillations*

[36] S. Heifets: Relaxation Oscillations in accelerators

[37] J. Sebek: *Relaxation Oscillations of the Synchrotron Motion Caused by Narrow-Band Impedances.*

[38] S. Tzenov: *Pattern Formation in High-Energy Accelerators and Storage Rings.*

[39] F.J. Sacherer: *Transverse Space-Charge Effects in Circular Accelerators,* Ph.D. Thesis, UCRL-18454, 1968.

[40] K. Hirata and T. Suzuki, Eds.: *Collective Effects in Short Bunches*, KEK Report 90-21 February 1991.

[41] A. Hofmann editor: *Synchro-Betatron Resonances*, Madeira, 1993, to be published from Madeira University.

[42] S.A. Bogacz: *Microwave Instability at Transition*, Proc. IEEE PAC, San Francisco, p. 1815, 1991.

[43] P. Colestock et al: *Observation and Analysis of Nonlinear Parametric Coupling of Longitudinal Modes in Synchrotrons*, Proc. 3rd EPAC, p. 126, 1992.

[44] P. Colestock et al: *A Generalized Model for Parametric Coupling of Longitudinal Modes in Synchrotrons*, Proc. IEEE PAC, Washington DC, p. 3384, 1993.

[45] L. Spentzouris & P. Colestock: *Coherent Nonlinear Longitudinal Phenomena in Unbunched Beams*, Proc. IEEE PAC, Vancouver BC, p. 16, 1997.

[46] P. Colestock et al: *Nonlinear Wave Phenomena in Coasting Beams*, Proc. IEEE PAC, Dallas, Texas, p. 2757, 1995.

[47] P. Colestock & J.A. Holt: *Microwave Stability at Transition*, Proc. IEEE PAC, Dallas, Texas, p. 3067, 1995.

[48] P. Colestock: *Experimental Observation of Nonlinear Coupling of Longitudinal Modes in Unbunched Beams*, ibid, p. 3070.

[49] J.S. Berg & F. Ruggiero: *Stability Diagrams for Landau Damping*, Proc. IEEE PAC, Vancouver BC, p. 1712, 1997.

[50] J.S. Berg: *Head-Tail Mode Instability Caused by Feedback*, Proc. 6th EPAC, Stockholm, p. 942, 1998.

[51] Y. Chin: *Transverse Mode Coupling Instabilities in the SPS*, CERN/SPS/85-2 (DI-MST), 1985.

[52] Y. Chin: *Formulation for Transverse Bunched Beam Instabilities in the Presence of Betatron Tune Spread*, CERN SPS/85-9 (DI-MST), 1985.

[53] Y. Chin et al : *Instability of a Bunched Beam with Synchrotron Frequency Spread*, Particle Accelerators, Vol.13 pp. 45-66, 1983.

[54] Y.H. Chin: CERN/SL/93-03 (AP), *Transverse mode coupling instability in a double rf system*, 1993.

[55] M. Pabst et al: *Progress on Intense Proton Beam Dynamics and Halo Formation*, Proc. 6th EPAC, Stockholm, p. 146, 1998.

[56] M. Pabst et al: *Halo Formation of Unbunched Beams In Periodic Focusing Systems*, IEEE PAC, New York City, TUP 121, 1999.

[57] Laslett, Keil & Sessler: *Transverse Resistive Instabilities*, Rev. Sci. Instr. **36**, No. 4, p. 436-448, 1965.

[58] K. Hubner & V. Vaccaro: *Dispersion Relation and Stability of Coasting Beam*, CERN-ISR-TH/70-44.

[59] D. Möhl & H. Schonauer: *Landau Damping by Nonlinear Space-Charge Forces and Octupoles*, Proc. 9th Int. Conf. HEACC, Stanford, 1974, p. 380,

[60] D. Möhl: *On Landau Damping of Dipole Modes by Nonlinear Space-Charge and Octupoles*, Proc. Workshop on Collective Effects in LHC, Montruex, 1994; CERN/PS 95-08 (DI).

[61] H. Tran: *Single-Particle and Collective Effects of Cubic Nonlinearity...*, Ph.D. Thesis, U.B.C., 1998.

[62] I. Hofmann: *Stability of Anisotropic Beams with Space Charge*, Phys. Rev. E **57**, p. 4713, 1998.

[63] R. Baartman: *Betatron Resonances with Space Charge*, AIP Conf Proc. 448, Workshop on Space-Charge Physics, Shelter Island N.Y., p. 56, 1998.

[64] S. Machida: *Simulation of Space-Charge Effects in a Synchrotron*, ibid, p. 73.

[65] J. Gareyte: *Beam Observation and the Nature of Instabilities*, CERN SPS/87-18 (AMS), pgs. 77-82.

# Short High-Intensity Bunches Group Summary*

M. Brennan, K. Brown, R. Cappi, W. Chou, T. Linnecar, C. Prior, G. Rees,
T. Roser, E. Shaposhnikova, M. Yoshi, W. VanAsselt.

Making very high intensity short bunches is one of the major challenges facing proton synchrotrons these days. The most demanding requirement comes from the muon collider where the phase space of the produced muons must be minimized by making as short as possible pulses from the proton driver. Bunches of $10^{14}$ protons with an rms pulse length of 1-2 ns will be needed.

This group discussed the current thinking on possible techniques for shortening bunches and tried to evaluate their relative merits and shortcomings. Special effort was made to identify any particular method as having potential to far exceed what heretofore has been achieved or which may have unique advantages for scaling to the required very high intensities.

Five techniques were described.

1. **High rf voltage:**
   The bunch will get shorter as the rf voltage is increased but this is not practical because the bunch length goes down only as the fourth root of the voltage. The required voltage quickly becomes astronomical and cannot be achieved.

2. **High rf frequency:**
   High rf frequency will naturally lead to shorter bunches but the longitudinal emittance will be small. The significance of the charge per unit of longitudinal phase space area was discussed at length. From the experience of high intensity machines one could infer that 1-2 $\times 10^{12}$ charges/ eV-s is somewhat of a limit for machines that pass through transition. This is an empirical observation and no one was aware of a theoretical basis for the value of this limit. But for high frequency rf it would be very hard to get enough beam is a single bunch.

3. **Bunch rotation**
   Bunch rotation is essentially a "conventional" technique that has many years of accumulated experience. In this context bunch rotation means a slow adiabatic decrease of the rf voltage to reduce the momentum spread of the beam without emittance blow-up. When the momentum spread is low and the bunch is long the rf voltage is rapidly (non-adiabatic) raised as high as possible. The bunch in the miss-matched phase space bucket rotates, and in one quarter of a synchrotron period

---

* Work performed under the auspices of the U.S. Department of Energy.

reaches a minimum length. The technique can reliably compress the bunch by a factor of two and under good conditions by up to a factor of four. Since the process is inherently non-adiabatic it will cause large emittance growth if the bunch is left in the miss-matched bucket. However, if the bunch is used (extracted from the ring) just when it becomes short then the emittance growth can be negligible. This makes bunch rotation applicable when transferring beam from one ring to the next or when sending the beam to the production target. It cannot, however, be used multiple times in one ring to cascade the bunch compression.

The difficulties of this technique for high intensity come from instabilities of two types. One is beam loading. When the rf voltage is adiabatically reduced beam loading gets very strong because the ratio of beam current to amplifier current gets high. At the same time voltage control must be good to keep the process adiabatic. High feedback gain can help the beam loading at the fundamental cavity frequency when the bunch is long but broadband beam loading becomes strong when the bunch compresses and it is difficult to correct for this at high frequency by feedback. The other instability comes within the beam itself. When the momentum spread is small there is a loss of Landau damping because the frequency spread in the beam is small. The threshold for instability (eg: microwave) generally goes up with the square of the frequency spread and down linearly with the local beam current. So as the bunch gets longer while preserving emittance the threshold goes down. High frequency instabilities within the bunch will increase the longitudinal emittance and defeat the bunch compression. Beam experiments in the AGS in 1999 showed signs of these effects at intensities near $4 \times 10^{12}$.

At the highest intensities new phenomena are likely to be important. For example, even at high energy (>10 GeV) transverse space charge tune spread will be large when the bunch gets short. One must examine how the variation within the bunch of transverse tune will affect the longitudinal dynamics through orbit distortions and path length variations.

4. **Unstable fixed point:**
The bunch can also be made long before rotation by switching the phase of the rf voltage to the unstable fixed point for some time. The advantage is that the rf voltage does not have to be lowered to the point where beam loading is aggravated. The rf phase is then switched back to the stable fixed point and the bunch rotates in phase space until it reaches minimum length. Since the beam is not adiabatically de-bunched the total frequency spread is maintained, which helps stability. The local frequency spread may, nevertheless, become small and so microwave instability is still a possibility. Microwave instability has not been seen at CERN where the technique is routinely used to shorten bunches in PS. The compression factor is limited by distortion in phase space as the bunch spreads out from the unstable fixed point. This effect can be reduced by adding a second harmonic component to linearize the net rf voltage with some extra cavities.

At high intensity another type of beam loading problem becomes important. The rf phase switching must be fast and so it is unlikely that the cavity compensation for reactive beam current will be fast enough change the sign the impedance offset. This demands twice the reactive beam current to be delivered by the power amplifier if the voltage and phase control are to be maintained.

5. **Flexible Momentum Compaction lattice gymnastics:**

On the more exotic side are gymnastics that involve FMC lattices. The basic notion is that explicit control of the frequency slip parameter, $\eta$, can be used to compress the bunch by manipulating the phase space trajectories without changing the rf voltage. For example, Jim Norem of ANL (not present) has proposed using the FMC to switch $\eta$ to zero with the rf voltage on to impose a large momentum spread on the bunch while preserving the bunch length. The bunch length does not change because $\eta = 0$ implies particles are frozen in phase. When the momentum spread is large enough (limited by the momentum aperture of the machine) the $\eta$ is switched back on and the bunch rotates into a short length. What is gained by switching $\eta$ is that the beam momentum spread is only increased, never reduced, and the rf beam loading is not increased by lowering the voltage. Furthermore, the high frequency beam loading and the machine longitudinal coupling impedance can be used to advantage by choosing the correct sign of $\eta$. For example, if the impedance is inductive and $\eta$ is negative then the beam-induced voltage will tend to further bunch the beam.

Existing machines do not have the full flexibility that would be required to test this scheme but an approximation to it has been studied in the AGS using the transition jumping quadrupoles. The beam was taken very close to the natural transition energy while holding the modified transition above the beam energy by about one unit with the jump quads. Then the jump quads were switched off fast, making $\eta$ very small but not zero. The beam then found itself in a miss-matched bucket just as in bunch rotation. Since the beam momentum spread was much smaller than the new bucket height the bunch rotated into a shorter length. A compression ratio on the order of three was found, giving an rms bunch length of ~2.5 ns for $5 \times 10^{12}$ protons. The test highlighted the key challenge of the technique. The second order terms in the frequency slip parameter, $\alpha_1$ and $\alpha_2$, have to be controlled. Since they dominate when $\eta$ is near zero their effect was the limiting factor in the AGS experiment. In the full-blown scheme $\eta$ is non-zero at the crucial time so their effect is reduced. Nevertheless, $\alpha_1$ and $\alpha_2$ are likely to be the determinants of the ultimate compression ratio.

## Conclusions

Techniques 3, 4, and 5 can be expected to yield compression ratios up to a factor of four but no one can envision a method for reaching qualitatively better performance in the order of a factor of ten or beyond. Some key questions pertain as to the behavior of $10^{14}$ charges which should be addressed by further study and machine experiments. For example does the beam become unstable at low momentum spread below transition? And, does the beam self bunch above some threshold with the appropriate sign of $\eta$? Representatives of four labs (PS at KEK, PS at CERN, Main Injector at FNAL, and AGS at BNL) expressed willingness to attempt to explore the questions at their respective laboratories. It was agreed that big shot bragging rights will belong to the lab that achieves top results in; 1. Maximum peak current in the bunch (the muon collider spec implies 4000 Amps), 2. Maximum charges per eV-s (> 10 GeV), 3. Highest bunch compression ratio.

# Author Index

## A
Aleksandrov, A., 315
Arakawa, D., 182

## B
Blaskiewicz, M., 231, 321, 385
Boine-Frankenheim, O., 86
Brennan, M., 407
Brown, K., 407
Brumwell, F. R., 285, 371
Burov, A. V., 49

## C
Cappi, R., 116, 407
Chou, W., 30, 407

## D
Danilov, V., 315, 381
Davidson, R. C., 295
Dooling, J. C., 285, 371, 385
Dyachkov, M., 385

## F
Fedotov, A. V., 77, 385

## G
Galambos, J., 315
Garoby, R., 116
Gluckstern, R. L., 77, 385

## H
Hahn, H., 266, 385
Heifets, S., 240, 381
Hofmann, A., 3

Hofmann, I., 86
Holmes, J., 315
Huang, H., 385

## I
Igarashi, S., 182

## J
Jeon, D., 315

## K
Kishiro, J., 182
Koba, K., 169, 182
Koscielniak, S., 391
Kurennoy, S. S., 361, 385

## L
Lee, W. W., 295
Limborg, C., 331
Linnecar, T., 64, 385, 407

## M
McMichael, G. E., 285, 371
Métral, E., 116
Mori, Y., 131

## N
Nakamura, E., 182
Ng, K. Y., 49, 213
Norem, J., 213

## O
Olsen, D., 315

## P

Prior, C., 407

## Q

Qin, H., 295

## R

Rees, G. H., 17, 151, 407
Roser, T., 22, 407
Ruggiero, F., 40

## S

Sebek, J., 331
Shaposhnikova, E., 256, 385, 407
Stupakov, G. V., 341, 381, 385

## T

Takayama, K., 182
Toyama, T., 182, 385
Tzenov, S. I., 351

## V

Van Asselt, W., 407

## W

Wang, J. G., 276, 385
Wang, T. F., 305
Wei, J., 197
Weng, W. T., 385

## Y

Yoshi, M., 407
Yoshii, M., 131, 182

## Z

Zhang, S. Y., 136, 385
Zhang, X., 40
Zotter, B., 99, 385